Graduate Texts in Mathematics

Graduate Texts in Mathematics bridge the gap between passive study and creative understanding, offering graduate-level introductions to advanced topics in mathematics. The volumes are carefully written as teaching aids and highlight characteristic features of the theory. Although these books are frequently used as textbooks in graduate courses, they are also suitable for individual study.

More information about this series at http://www.springer.com/series/136

Jet Nestruev

Smooth Manifolds
and Observables

Second Edition

 Springer

Jet Nestruev
Lizzano in Belvedere, Italy
Moscow, Russia

ISSN 0072-5285   ISSN 2197-5612   (electronic)
Graduate Texts in Mathematics
ISBN 978-3-030-45649-8   ISBN 978-3-030-45650-4   (eBook)
https://doi.org/10.1007/978-3-030-45650-4

Mathematics Subject Classification: 58-01, 58Jxx, 55R05, 13Nxx, 08-01, 14F40, 58Axx, 17A70, 13B30, 37J05

This Springer imprint is published by the registered company Springer Nature Switzerland AG
The registered company address is: Gewerbestrasse 11, 6330 Cham, Switzerland

Foreword

The author and his team undertook the revision of the book "Smooth Manifolds and Observables", published in 2002 by Springer Verlag, in 2016 at the suggestion of the Editors of Springer. It immediately became clear that the 2002 book should not only be revised, but must also be significantly expanded, since the main underlying idea of the book—passing from the differential geometry of smooth manifolds to the differential calculus in commutative algebras—has been successfully developed in the last 20 years in many new areas of mathematics and physics.

As a result, the 11 original chapters have been expanded to 21, the number of pages has almost doubled (from 222 to 431), and the Preface has become much longer. Nevertheless, the main principles of the first edition, in particular the "observability principle", which determines the relationship of mathematics to physics, have been preserved. We have tried to make the book as self-contained as possible, the only prerequisites being a general mathematical culture on the level of two years of graduate school. In this edition, we have also included the Afterword, as well as the Appendix to the first edition, written by A. M. Vinogradov, without any changes.

Unlike the famous French general, Jet Nestruev is a civilian, and the names of the members of his team are not veiled in secrecy, as are those of the Bourbaki team. For the first edition, they were: Alexander Astashov, Alexey Bocharov, Sergei Duzhin, Alexandre Vinogradov, Mikhail Vinogradov, and Alexey Sossinsky. For the present second edition, they are Alexander Astashov, Alexandre Vinogradov, Mikhail Vinogradov, and Alexey Sossinsky. The figures for both editions were expertly prepared by Alexander Astashov.

The author ackowledges the invaluable help of Joseph Krasil'shchik, who read the entire first draft of this edition and noticed, besides misprints and other minor snafus, numerous parts of the text that required essential changes; these changes have considerably improved the exposition. A. Sossinsky acknowledges the partial support of a J. Simons Foundation grant.

Jet Nestruev
Lizzano in Belvedere, Italy and Moscow, Russia
December 2019

Alexandre Vinogradov
Александр Михайлович Виноградов
1938–2019

Alexandre Vinogradov, an extraordinary human being and outstanding mathematician, deserves a more thorough and detailed life story than can fit into this page[2], so here we will limit ourselves to a quote from a letter from his daughter Katya, who was at his bedside to the very end.

"A few days ago Father was asleep in his bed and I was sitting next to him. He was no longer able to swallow, even water. He slept all the time. Suddenly he opened his eyes and tried to say something. Gently I said:

'Daddy, do you want me to wet your lips?'

Unexpectedly he spoke, clearly and angrily:

'Katya, damn, are you out of your mind? Lips, what lips? Don't you understand anything? I have finally figured out **quantization!** and here you are wanting to wet my lips, stupid girl!'

Then I fully understood, once and for all, what my father is."

---

[2]A necrologue with a short biography of A. M. Vinogradov and a sketch of his contribution to mathematics appears in the second issue for 2020 of *Russian Mathematical Surveys*.

Preface

> *The limits of my language*
> *are the limits of my world.*
> —L. Wittgenstein

This book is a self-contained introduction to smooth manifolds, fiber spaces, and differential operators on them. It is not only accessible to graduate students specializing in mathematics and physics, but also intended for readers who are already familiar with the subject. Since there are many excellent textbooks in manifold theory, the first question that should be answered is: Why another book on manifolds?

The main reason is that the good old differential calculus is actually a particular case of a much more general construction, which may be described as the *differential calculus over commutative algebras.* And this calculus, in its entirety, is just the consequence of properties of arithmetical operations. This fact, remarkable in itself, has numerous applications, ranging from delicate questions of algebraic geometry to the theory of elementary particles. Our book explains in detail why the differential calculus on manifolds is simply an aspect of commutative algebra.

In the standard approach to smooth manifold theory, the subject is developed along the following lines. First one defines the notion of smooth manifold, say M. Then one defines the algebra $\mathcal{F}_M$ of smooth functions on M, and so on. In this book this sequence is reversed: we begin with a certain commutative $\mathbb{R}$-algebra[1] $\mathcal{F}$, and then define the manifold $M = M_{\mathcal{F}}$

---

[1] Here and below $\mathbb{R}$ stands for the real number field. Nevertheless, and this is very important, nothing prevents us from replacing it by an arbitrary field (or even a ring) if this is appropriate for the problem under consideration.

as the $\mathbb{R}$-spectrum of this algebra. (Of course, in order that $M_{\mathcal{F}}$ deserve the title of a smooth manifold, the algebra $\mathcal{F}$ must satisfy certain conditions; these conditions appear in Chapter 3, where the main definitions mentioned here are presented in detail.)

This approach is by no means new: it is used, say, in algebraic geometry. One of its advantages is that from the outset it is not related to the choice of a specific coordinate system, so that (in contrast to the standard analytical approach) there is no need to constantly check that various notions or properties are independent of this choice. This explains the popularity of the coordinate-free algebraic viewpoint among mathematicians attracted by sophisticated algebra, but its level of abstraction discourages the more pragmatically inclined applied mathematicians and physicists.

But what is really new in the book is the physical motivation of the algebraic approach. It is based, in particular, on the (physical) notion of *observable* and the related *observability principle*, which asserts

what exists is only that which can be observed

As we all know, in developing scientific theories, *homo sapiens* gather the necessary data from observation and experiment. At the beginning, this data comes from our sensory organs, eventually replaced by accurate *measuring devices*. An appropriate collection of these devices constitutes a *physical laboratory* that studies some physical system. Let us formalize this notion, having in mind that we are not interested in how the devices function, only in what data from the physical system under study they register.

Measuring devices provide us with numerical data, i.e., their readings are real numbers. The obtained numbers are then processed, mainly by using the four arithmetical operations. Now if two devices Π_1 and Π_2 give two readings, instead of adding these numbers by hand, we can construct (or imagine) a device, $\Pi_1 + \Pi_2$ that gives us the sum of the two readings. Since the choice of the scale of the readings is subjective, for the sake of objectivity, instead of a measuring device Π, we must use (virtual) devices $\lambda\Pi$, $\lambda \in \mathbb{R}$. Further, in order to fix the unit in our scales, we need a special device, $\mathbb{I}$, whose only possible reading is $\mathbf{1}_{\mathbb{R}}$. Indeed, as compared to Π, the scale of the device $\Pi + \lambda\mathbf{1}_{\mathbb{R}}$ is shifted by λ.

If $\Pi_1, \ldots, \Pi_n$ are real measuring devices in the laboratory, then, continuing this process, we can associate to any polynomial $p(x_1, \ldots, x_n)$ the device $p(\Pi_1, \ldots, \Pi_n)$, whose reading is $p(a_1, \ldots, a_n)$ when $a_1, \ldots, a_n$ are the readings of the devices $\Pi_1, \ldots, \Pi_n$, respectively. Thus the operations of addition and multiplication by real numbers defined above transform the set of all virtual measuring devices into an algebra isomorphic to the polynomial algebra $\mathbb{R}[x_1, \ldots, x_n]$.

But we can go even further and introduce virtual devices of the form $f(\Pi_1, \ldots, \Pi_n)$, where $f(x_1, \ldots, x_n)$ is a smooth (i.e., infinitely differentiable) function, as it was done above for polynomial functions. Then the laboratory consisting of all such virtual devices will be, from the mathematical point of view, an algebra isomorphic to the algebra $C^\infty(\mathbb{R}^n)$ of smooth functions in Euclidean space $\mathbb{R}^n$. The device $\mathbb{I}$ will be the unit of this algebra.

To summarize the above for future reference, let us agree that

the mathematical formalization of a physical laboratory is an appropriate commutative algebra with unit over $\mathbb{R}$.

In classical physics, we assess the state of a given physical system S only from the readings of all the measuring devices. In that sense "state" and "observation" are identical notions. Mathematically, an *observation* of the state of the system S may be understood as a map $h : \mathcal{A} \to \mathbb{R}$ of the algebra $\mathcal{A}$ of all measuring devices, which to any device Π assigns its reading $h(\Pi)$ in that state. On the other hand, from the definition of arithmetical operations on measuring devices, it obviously follows that h is a homomorphism of unitary algebras, in particular, we have $h(\mathbb{I}) = \mathbf{1}_\mathbb{R}$. It is important to understand that not every such homomorphism is an observation of the system S. To clarify this, let us consider the following construction.

Let the ideal $\mathcal{I} \subset \mathcal{A}$ be the intersection of the ideals $\mathbf{Ker}(h)$ for all observations from S, and $\mathcal{A}_S \overset{\text{def}}{=} \mathcal{A}/\mathcal{I}_S$. Then any such a homomorphism h may be represented in the form of the composition

$$\mathcal{A} \overset{pr}{\longrightarrow} \mathcal{A}_S \overset{\tilde{h}}{\longrightarrow} \mathbb{R},$$

where pr is the natural projection and $\tilde{h}\,(a \bmod \mathcal{I}_S) = h(a)$. Conversely, under certain restrictions on the algebra $\mathcal{A}$ (which we omit at this point), any homomorphism $\mathcal{A}_S \to \mathbb{R}$ composed with the projection pr is an observation of the system S. Thus the family of all states of the system S can be identified with the set of unitary algebra homomorphisms from $\mathcal{A}_S$ to $\mathcal{R}$. This set is denoted by $\mathbf{Spec}_\mathbb{R}\,\mathcal{A}_S$ and is called the $\mathbb{R}$-*spectrum of the algebra $\mathcal{A}_S$.*

Summarizing the above, let us agree that

the algebra $\mathcal{A}_S$ is the mathematical model of the physical system S, while $\mathbf{Spec}_\mathbb{R}\,\mathcal{A}_S$ is the set of its states

The algebra $\mathcal{A}_S$ is sometimes called the *algebra of observables*, while its elements are *observable quantities* or simply *observables*.

If we agree, as we did above, that $\mathcal{A} = C^\infty(\mathbb{R}^n)$, then $\mathbf{Spec}_\mathbb{R}\,\mathcal{A}/\mathcal{I}$, where $\mathcal{I} \subset C^\infty(\mathbb{R}^n)$ is the ideal naturally identified with some closed subset of $\mathbb{R}^n$. On the other hand, we know that the space of states of mechanical and other physical systems with a finite number of degrees of freedom is a smooth manifold. From this we can conclude that the algebra of observables of such a system is none other than $\mathcal{A}_S = C^\infty(M)$, the algebra of smooth functions on a manifold M.

Since Hilbert and Bourbaki, it is generally considered that set theory is the foundation of all mathematics. This popular point of view, however, has many drawbacks. In particular, working with sets supplied with a certain structure, mathematicians are unable to distinguish concrete elements of those sets within the framework of set theory, and are forced to label them in some external way. A good example is the introduction of Cartesian coordinates in geometry, which considerably widened the possibilities of the axiomatic geometry of Euclid. Note that Descartes' coordinate method is precisely equivalent to the interpretation of n-dimensional Euclidean space as the spectrum of the polynomial algebra $\mathrm{Spec}_{\mathbb{R}}\, \mathbb{R}[x_1, \ldots, x_n]$.

$$* \quad * \quad *$$

A few words about how the book is organized.

It has 21 chapters. Chapter 1 consists of a series of examples of smooth manifolds taken from various branches of mathematics, mechanics, and physics. Analyzing these examples, we give a preliminary idea of the notion of smooth manifold from the classical differential-geometrical viewpoint (the corresponding formal definition appears only in Chapter 5) as well as from our (algebraic) point of view (which is presented in Chapters 3 and 4). The equivalence of the two approaches is proved in Chapter 7.

In the framework of our algebraic approach, smooth (i.e., differentiable) manifolds appear as $\mathbb{R}$-spectra of a certain class of $\mathbb{R}$-algebras (the latter are therefore called *smooth*), and their elements turn out to be the smooth functions defined on the corresponding spectra. Here the $\mathbb{R}$-spectrum of some $\mathbb{R}$-algebra A is the set of all its unital homomorphisms into the $\mathbb{R}$-algebra $\mathbb{R}$, i.e., the set that is "visible" by means of this algebra. Thus smooth manifolds are "worlds" whose observation can be carried out by means of smooth algebras. Because of the algebraic universality of the approach described above, "nonsmooth" algebras will allow us to observe "nonsmooth worlds" and study their singularities by using the differential calculus. But this differential calculus is not the naive calculus studied in introductory (or even "advanced") university courses; it is a much more sophisticated construction.

It is to the foundations of this calculus that the part of this book beginning with Chapter 9 is devoted. In it we "discover" the notion of differential operator *over a commutative algebra* and carefully analyze the main notion of the classical differential calculus, that of the derivative (or more precisely, that of the tangent vector). Moreover, in that chapter, we deal with the other simplest constructions of the differential calculus from the new point of view, e.g., with tangent and cotangent bundles, as well as jet bundles. The latter are used to prove the equivalence of the algebraic and the standard analytic definitions of differential operators for the case in which the basic algebra is the algebra of smooth functions on a smooth manifold.

Chapter 10 is devoted to the classical notion of symbol of a differential operator. In it we show how this notion can be carried over to the general situation of commutative algebras. Moreover, as an illustration of the "algebraic differential calculus," we present the construction of the Hamiltonian formalism over an arbitrary commutative algebra.

In Chapters 11 and 12, we study fiber bundles and vector bundles from the algebraic point of view. In particular, we establish the equivalence of the category of vector bundles over a manifold M, and the category of finitely generated projective modules over the algebra $C^\infty(M)$. Chapter 11 is concluded by a study of jet modules of an arbitrary vector bundle and an explanation of the universal role played by these modules in the theory of differential operators.

The present edition of this book contains ten new chapters that further develop the differential calculus over commutative algebras, show that this calculus is a generalization of the differential calculus on smooth manifolds, and indicate some of the various branches of the mathematical sciences to which it can be (or is being) applied. In the process, several additional examples of the basic functors (that we call *diffunctors*) of the differential calculus are presented and discussed.

Chapter 13 shows how the localization of various operators (which is practically automatic in the classical theory) can be carried out in the general context of commutative algebras, thereby showing that the algebraic theory generalizes the classical one in that respect as well.

In Chapter 14, we study differential 1-forms and jet spaces, first in the classical context, and then in their general algebraic guise. It is stressed in the chapter that jet spaces possess natural smooth manifold structures and are the usual domain of definition of differential operators.

Chapter 15 is a general introduction to the functors of the differential calculus (both in classical and in general algebraic form), to their representing objects and to the universal maps that arise in that context.

In Chapters 16 and 17, the study of examples of diffunctors is continued along the lines of Chapter 15. Specifically, these chapters deal with cosymbols, tensors, Spencer complexes, and differential forms.

Chapter 18 is a systematic study of the fundamental notion of differential form—the basis of such concepts as de Rham cohomology, symplectic and contact manifolds, infinitesimal symmetries, Jordan algebras, Killing forms, etc. The chapter indicates numerous directions in which the classical and general algebraic theory can be applied to different specific branches of mathematics, mechanics, and physics.

The last three chapters, which deal with topics of recent (or even ongoing) research, are written in a style closer to that of a research monograph than to that of an expository one, they are not as self-contained as the previous ones, and involve several references to recent publications. The extremely important Chapter 19 (written by Alexandre Vinogardov) deals with differential n-forms and their applications, including symplectic, Poisson, contact manifolds, Monge–Ampère equations, Lie derivatives, infinitesimal symmetries,

Killing fields, Jacobi brackets, while Chapter 20 (written jointly by Alexandre and Mikhail Vinogardov) and Chapter 21 (written by Mikhail Vinogardov) are devoted to such topics as de Rham cohomology, cohomological integration theory, Poisson brackets, graded commutative algebras, Batalin–Vilkovisky operators.

Looking ahead beyond the framework of this book, let us note that the mechanism of "quantum observability" is in principle of cohomological nature and is an appropriate specification of those natural observation methods of solutions to (nonlinear) partial differential equations that have appeared in the *secondary differential calculus* (see [17]) and in the fairly new branch of mathematical physics known as *cohomological physics.*

$$* \quad * \quad *$$

In 1969 Alexandre Vinogradov, one of the authors of this book, started a seminar aimed at understanding the mathematics underlying quantum field theory. Its participants were his mathematics students, and several young physicists, the most assiduous of whom were Dmitry Popov, Vladimir Kholopov, and Vladimir Andreev. In a couple of years, it became apparent that the difficulties of quantum field theory come from the fact that physicists express their ideas in an inadequate language, and that an adequate language simply does not exist (see the quotation preceding the Preface). If we analyze, for example, what physicists call the covariance principle, it becomes clear that its elaboration requires a correct definition of differential operators, differential equations, and, say, second-order differential forms.

For this reason in 1971 a mathematical seminar split out from the physical one, and began studying the structure of the differential calculus and searching for an analog of algebraic geometry for systems of (nonlinear) partial differential equations. At the same time, the abovementioned author began systematically lecturing on the subject.

At first, the participants of the seminar and the listeners of the lectures had to manage with some very schematic summaries of the lectures and their own lecture notes. But after ten years or so, it became obvious that all these materials should be systematically written down and edited. Thus Jet Nestruev was born, and he began writing a series of books entitled *Elements of the Differential Calculus.* Detailed contents of the first installments of the series appeared, and the first one was written. It contained, basically, the first eight chapters of the first edition of the present book.

Then, after an interruption of nearly fifteen years, due to a series of objective and subjective circumstances, work on the project was resumed, and the second installment was written. Amalgamated with the first one, it constituted the first edition of "Smooth Manifolds and Observables", first published in Russian by MCCME Publishers, then in English by Springer Verlag. The circumstances of the writing (directly in English) of the second edition are explained in the Foreword.

The present second edition of the book is self-contained, but open-ended. The reader who wishes to have a look ahead without delay can consult the references appearing on page 423 or the more complete bibliography in [1] and [21]. But we hope for much more: that he/she will go fearlessly into *terra incognita*, choosing one of the numerous branches of mathematics or physics to which our approach can be applied, and armed with the methods developed in the book, will discover the appropriate (and hopefully beautiful!) mathematical explanations of real-life phenomena that seem mysterious at present.

Jet Nestruev
Lizzano in Belvedere, Italy and Moscow, Russia
December 2019

Contents

1

Introduction

1.0. This chapter is a preliminary discussion of *finite-dimensional smooth* (*infinitely differentiable*) *real manifolds*, the main protagonists of this book.

Why are smooth manifolds important?

Well, we live *in* a manifold (a four-dimensional one, according to Einstein) and *on* a manifold (the Earth's surface, whose model is the sphere S^2). We are surrounded by manifolds: The surface of a coffee cup is a manifold (namely, the torus $S^1 \times S^1$, more often described as the surface of a doughnut or an anchor ring, or as the tube of an automobile tire); a shirt is a two-dimensional manifold with boundary.

Processes taking place in nature are often adequately modeled by points moving on a manifold, especially if they involve no discontinuities or catastrophes. (Incidentally, catastrophes—in nature or on the stock market—as studied in "catastrophe theory" may not be manifolds, but then they are smooth maps of manifolds.)

What is more important from the point of view of this book is that *manifolds arise quite naturally in various branches of mathematics* (in *algebra* and *analysis* as well as in *geometry*) *and its applications* (especially *mechanics*). Before trying to explain what smooth manifolds are, we give some examples.

1.1. The configuration space Rot(3) **of a rotating solid in space.** Consider a solid body in space fixed by a hinge O that allows it to rotate in any direction (Figure 1.1). We want to describe the set of positions of the body, or, as it is called in classical mechanics, its *configuration space*.

© Springer Nature Switzerland AG 2020
J. Nestruev, *Smooth Manifolds and Observables*, Graduate Texts
in Mathematics 220, https://doi.org/10.1007/978-3-030-45650-4_1

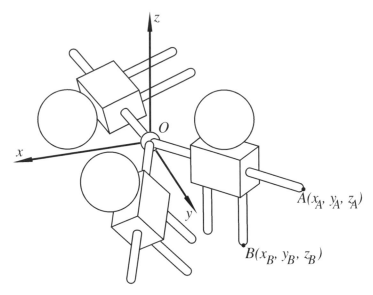

Figure 1.1. Rotating solid.

One way of going about it is to choose a coordinate system $Oxyz$ and determine the body's position by the coordinates (x_A, y_A, z_A), (x_B, y_B, z_B) of two of its points A, B. But this is obviously not an economical choice of parameters: It is intuitively clear that only *three* real parameters are required, at least when the solid is not displaced too greatly from its initial position OA_0B_0. Indeed, two parameters determine the direction of OA (e.g., x_A, y_A; see Figure 1.1), and one more is needed to show how the solid is turned about the OA axis (e.g., the angle $\varphi_B = B_0'OB$, where AB_0' is parallel to A_0B_0).

It should be noted that these are not ordinary Euclidean coordinates; the positions of the solid do not correspond bijectively in any natural way to ordinary three-dimensional space $\mathbb{R}^3$. Indeed, if we rotate AB through the angle $\varphi = 2\pi$, the solid does not acquire a new position; it returns to the position OAB; besides, *two* positions of OA correspond to the coordinates (x_A, y_A): For the second one, A is *below* the Oxy plane. However, *locally*, say near the initial position OA_0B_0, there is a bijective correspondence between the position of the solid and a neighborhood of the origin in 3-space $\mathbb{R}^3$, given by the map $OAB \mapsto (x_A, y_A, \varphi_B)$. Thus the configuration space $\mathrm{Rot}(3)$ of a rotating solid is an object that can be described locally by three Euclidean coordinates, but *globally* has a more complicated structure.

1.2. An algebraic surface V. In nine-dimensional Euclidean space $\mathbb{R}^9$ consider the set of points satisfying the following system of six algebraic equations:

$$\begin{cases} x_1^2 + x_2^2 + x_3^2 = 1; & x_1x_4 + x_2x_5 + x_3x_6 = 0; \\ x_4^2 + x_5^2 + x_6^2 = 1; & x_1x_7 + x_2x_8 + x_3x_9 = 0; \\ x_7^2 + x_8^2 + x_9^2 = 1; & x_4x_7 + x_5x_8 + x_6x_9 = 0. \end{cases}$$

This happens to be a nice three-dimensional surface in $\mathbb{R}^9$ ($3 = 9 - 6$). It is not difficult (try!) to describe a bijective map of a neighborhood of any point (say $(1, 0, 0, 0, 1, 0, 0, 0, 1)$) of the surface onto a neighborhood of the origin of Euclidean 3-space. But this map cannot be extended to cover the entire surface, which is compact (why?). Thus again we have an example of an object V locally like 3-space, but with a different global structure.

It should perhaps be pointed out that the solution set of six algebraic equations with nine unknowns chosen at random will not always have such a simple local structure; it may have self-intersections and other *singularities*. (This is one of the reasons why algebraic geometry, which studies such *algebraic varieties,* as they are called, is not a part of smooth manifold theory.)

1.3. Three-dimensional projective space $\mathbb{R}P^3$. In four-dimensional Euclidean space $\mathbb{R}^4$ consider the set of all straight lines passing through the origin. We want to view this set as a "space" whose "points" are the lines. Each "point" of this space—called *projective space $\mathbb{R}P^3$* by nineteenth century geometers—is determined by the line's directing vector (a_1, a_2, a_3, a_4), $\sum a_i^2 \neq 0$, i.e., a quadruple of real numbers. Since proportional quadruples define the same line, each point of $\mathbb{R}P^3$ is an equivalence class of proportional quadruples of numbers, which are denoted by $P = (a_1 : a_2 : a_3 : a_4)$, where (a_1, a_2, a_3, a_4) is any representative of the class. In the vicinity of each point, $\mathbb{R}P^3$ is like $\mathbb{R}^3$. Indeed, if we are given a point $P_0 = (a_1^0 : a_2^0 : a_3^0 : a_4^0)$ for which $a_4^0 \neq 0$, it can be written in the form $P_0 = (a_1^0/a_4^0 : a_2^0/a_4^0 : a_3^0/a_4^0 : 1)$ and the three ratios viewed as its three coordinates. If we consider all the points P for which $a_4 \neq 0$ and take $x_1 = a_1/a_4$, $x_2 = a_2/a_4$, and $x_3 = a_3/a_4$ to be their coordinates, we obtain a bijection of a neighborhood of P_0 onto $\mathbb{R}^3$. This neighborhood, together with three similar neighborhoods (for $a_1 \neq 0$, $a_2 \neq 0$, $a_3 \neq 0$), covers all the points of $\mathbb{R}P^3$. But points belonging to more than one neighborhood are assigned to different triples of coordinates (e.g., the point $(6 : 12 : 2 : 3)$ will have the coordinates $(2, 4, \frac{2}{3})$ in one system of coordinates and $(3, 6, \frac{3}{2})$ in another). Thus the overall structure of $\mathbb{R}P^3$ is not that of $\mathbb{R}^3$.

1.4. The special orthogonal group $\mathrm{SO}(3)$. Consider the group $\mathrm{SO}(3)$ of orientation-preserving isometries of $\mathbb{R}^3$. In a fixed orthonormal basis, each element $A \in \mathrm{SO}(3)$ is defined by an orthogonal positive definite matrix, thus by nine real numbers ($9 = 3 \times 3$). But of course, fewer than 9 numbers are needed to determine A. In canonical form, the matrix of A will be

$$\begin{pmatrix} 1 & 0 & 0 \\ 0 & \cos\varphi & \sin\varphi \\ 0 & -\sin\varphi & \cos\varphi \end{pmatrix},$$

and A is defined if we know φ and are given the eigenvector corresponding to the eigenvalue $\lambda = 1$ (two real coordinates a, b are needed for that, since eigenvectors are defined up to a scalar multiplier). Thus again three coordinates (φ, a, b) determine elements of SO(3), and they are Euclidean coordinates only locally.

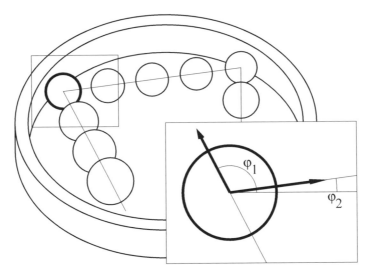

Figure 1.2. Billiards on a disk.

1.5. The phase space of billiards on a disk $\mathcal{B}(D^2)$. A tiny billiard ball P moves with unit velocity in a closed disk D^2, bouncing off its circular boundary C in the natural way (angle of incidence = angle of reflection). We want to describe the *phase space* $\mathcal{B}(D^2)$ of this mechanical system, whose "points" are all the possible *states of the system* (each state being defined by the position of P and the direction of its velocity vector). Since each state is determined by three coordinates $(x, y; \varphi)$ (Figure 1.2), it would seem that as a set, $\mathcal{B}(D^2)$ is $D^2 \times S^1$, where S^1 is the unit circle (i.e., $S^1 = \mathbb{R}$ mod 2π). But this is not the case, because at the moment of collision with the boundary, say at (x_0, y_0), the direction of the velocity vector jumps from φ_1 to φ_2 (see Figure 1.2), so that we must identify the states

$$(x_0, y_0, \varphi_1) \equiv (x_0, y_0, \varphi_2). \tag{1.1}$$

Thus $\mathcal{B}(D^2) = (D^2 \times S^1)/\sim$, where $/\sim$ denotes the factorization defined by the equivalence relation of all the identifications (1.1) due to all possible collisions with the boundary C.

Since the identifications take place only on C, all the points of

$$\mathcal{B}^0(D^2) = \text{Int } D^2 \times S^1 = (\text{Int } D^2 \times S^1)/\sim,$$

where $\text{Int } D^2 = D^2 \setminus C$ is the interior of D^2, have neighborhoods with a structure like that of open sets in $\mathbb{R}^3$ (with coordinates $(x, y; \varphi)$). It is a rather nice fact (not obvious to the beginner) that after identifications the "boundary states" $(x, y; \varphi)$, $(x, y) \in C$, also have such neighborhoods, so that again $\mathcal{B}(D^2)$ is locally like $\mathbb{R}^3$, but not like $\mathbb{R}^3$ globally (as we shall later show).

As a more sophisticated example, the advanced reader might try to describe the phase space of billiards in a right triangle with an acute angle of (a) $\pi/6$, (b) $\sqrt{2}\pi/4$.

1.6. The five examples of three-dimensional manifolds described above all come from different sources: classical mechanics 1.1, algebraic geometry 1.2, classical geometry 1.3, linear algebra 1.4, and mechanics 1.5. The advanced reader has not failed to notice that 1.1–1.4 are actually examples of *one and the same manifold* (appearing in different garb):

$$\text{Rot}(3) = V = \mathbb{R}P^3 = \text{SO}(3).$$

To be more precise, the first four manifolds are all "diffeomorphic," i.e., equivalent as smooth manifolds (the definition is given in Section 6.7). As for Example 1.5, $\mathcal{B}(D^2)$ differs from (i.e., is not diffeomorphic to) the other manifolds, because it happens to be diffeomorphic to the three-dimensional sphere S^3 (the beginner should not be discouraged if he/she fails to see this; it is not obvious).

What is the moral of the story? The history of mathematics teaches us that if the same object appears in different guises in various branches of mathematics and its applications, and plays an important role there, then it should be studied intrinsically, as a separate concept. That was what happened to such fundamental concepts as *group* and *linear space,* and is true of the no less important concept of *smooth manifold.*

1.7. The examples show us that a manifold M is a point set locally like Euclidean space $\mathbb{R}^n$ with global structure not necessarily that of $\mathbb{R}^n$. How does one go about studying such an object? Since there are Euclidean coordinates near each point, we can try to cover M with *coordinate neighborhoods* (or *charts,* or *local coordinate systems,* as they are also called). A family of charts covering M is called an *atlas.* The term is evocative; indeed, a geographical atlas is a set of charts or maps of the manifold S^2 (the Earth's surface) in that sense.

In order to use the separate charts to study the overall structure of M, we must know how to move from one chart to the next, thus "gluing

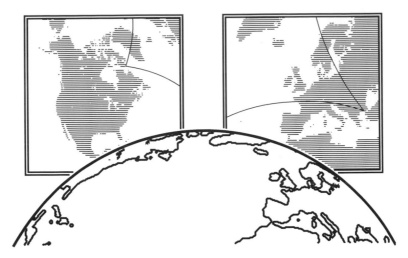

Figure 1.3. Charts for the surface of the Earth.

together" the charts along their common parts, so as to recover M (see Figure 1.3). In less intuitive language, we must be in possession of *coordinate transformations*, expressing the coordinates of points of any chart in terms of those of a neighboring chart. (The industrious reader might profit by actually writing out these transformations for the case of the four-chart atlas of $\mathbb{R}P^3$ described in 1.3.)

If we wish to obtain a *smooth* manifold in this way, we must require that the coordinate transformations be "nice" functions (in a certain sense). We then arrive at the *coordinate* or *classical approach* to smooth manifolds. It is developed in detail in Chapter 5.

1.8. Perhaps more important is the *algebraic approach* to the study of manifolds. In it we forget about charts and coordinate transformations and work only with the $\mathbb{R}$-*algebra* $\mathcal{F}_M$ of *smooth functions* $f\colon M \to \mathbb{R}$ on the manifold M. It turns out that $\mathcal{F}_M$ entirely determines M and is a convenient object to work with.

An attempt to give the reader an intuitive understanding of the natural philosophy underlying the algebraic approach is undertaken in the next sections.

1.9. In the description of a classical *physical system* or process, the key notion is the *state* of the system. Thus, in classical mechanics, the state of a moving point is described by its position and velocity at the given moment of time. The state of a given gas from the point of view of thermodynamics is described by its temperature, volume, pressure, etc. In order to actually assess the state of a given system, the experimentalist must use various *measuring devices* whose *readings* describe the state.

Suppose M is the set of all states of the classical physical system S. Then to each measuring device D there corresponds a function f_D on the set M, assigning to each state $s \in M$ the reading $f_D(s)$ (a real number) that the device D yields in that state. From the physical point of view, we are interested only in those characteristics of each state that can be measured in principle, so that the set M of all states is described by the collection Φ_S of all functions f_D, where the D's are measuring devices (possibly imaginary ones, since it is not necessary—nor indeed practically possible—to construct all possible measuring devices). Thus, theoretically, *a physical system S is nothing more that the collection Φ_S of all functions determined by adequate measuring devices (real or imagined) on S.*

1.10. Now, if the functions $f_1, \ldots, f_k$ correspond to the measuring devices $D_1, \ldots, D_k$ of the physical system S, and $\varphi(x_1, \ldots, x_k)$ is any "nice" real-valued function in k real variables, then in principle it is possible to construct a device D such that the corresponding function f_D is the composite function $\varphi(f_1, \ldots, f_k)$. Indeed, such a device may be obtained by constructing an auxiliary device, synthesizing the value $\varphi(x_1, \ldots, x_k)$ from input entries $x_1, \ldots, x_k$ (this can always be done if φ is nice enough), and then "plugging in" the outputs $(f_1, \ldots, f_k)$ of the devices $D_1, \ldots, D_k$ into the inputs $(x_1, \ldots, x_k)$ of the auxiliary device. Let us denote this device D by $\varphi(D_1, \ldots, D_k)$.

In particular, if we take $\varphi(x_1, x_2) = x_1 + x_2$ (or $\varphi(x) = \lambda x$, $\lambda \in \mathbb{R}$, or $\varphi(x_1, x_2) = x_1 x_2$), we can construct the devices $D_1 + D_2$ (or λD_i, or $D_1 D_2$) from any given devices D_1, D_2. In other words, if $f_i = f_{D_i} \in \Phi_S$, then the functions $f_1 + f_2$, λf_i, $f_1 f_2$ also belong to Φ_S.

Thus the set Φ_S of all functions $f = f_D$ describing the system S has the structure of an *algebra over* $\mathbb{R}$ (or $\mathbb{R}$-*algebra*).

1.11. Actually, the set Φ_S of *all* functions $f_D \colon M_S \to \mathbb{R}$ is much too large and cumbersome for most classical problems. Systems (and processes) described in classical physics are usually continuous or smooth in some sense. Discontinuous functions f_D are irrelevant to their description; only "smoothly working" measuring devices D are needed. Moreover, the problems of classical physics are usually set in terms of differential equations, so that we must be able to take derivatives of the relevant functions from Φ_S as many times as we wish. Thus we are led to consider, rather than Φ_S, the smaller set $\mathcal{F}_S$ of *smooth functions* $f_D \colon M_S \to \mathbb{R}$.

The set $\mathcal{F}_S$ inherits an $\mathbb{R}$-algebra structure from the inclusion $\mathcal{F}_S \subset \Phi_S$, but from now on we shall forget about Φ, since the *smooth $\mathbb{R}$-algebra $\mathcal{F}_S$* will be our main object of study.

1.12. Let us describe in more detail what the algebra $\mathcal{F}_S$ might be like in classical situations. For example, from the point of view of classical mechanics, a system S of N points in space is adequately described by the positions and velocities of the points, so that we need $6N$ measuring devices D_i to record them. Then the algebra $\mathcal{F}_S$ consists of all elements of the form $\varphi(f_1, \ldots, f_{6N})$, where the f_i are the "basic functions" determined by the devices D_i, while $\varphi \colon \mathbb{R}^{6N} \to \mathbb{R}$ is any nice (smooth) function.

In more complicated situations, certain *relations* among the basis functions f_i may arise. For example, if we are studying a system of two mass points joined by a rigid rod of negligible mass, we have the relation

$$\sum_{i=1}^{3}(f_i - f_{i+3})^2 = r^2,$$

where r is the length of the rod and the functions f_i (respectively f_{i+3}) measure the ith coordinate of the first (respectively second) mass point. (There is another relation for the velocity components, which the reader might want to write out explicitly.)

Generalizing, we can say that there usually exists a basis system of devices $D_1, \dots, D_k$ adequately describing the system S (from the chosen point of view). Then the $\mathbb{R}$-algebra $\mathcal{F}_S$ consists of all elements of the form $\varphi(f_1, \dots, f_k)$, where $\varphi \colon \mathbb{R}^k \to \mathbb{R}$ is a nice function and the $f_i = f_{D_i}$ are the relevant measurements (given by the devices D_i) that may be involved in relations of the form $F(f_1, \dots, f_k) \equiv 0$.

Then $\mathcal{F}_S$ may be described as follows. Let $\mathbb{R}^k$ be Euclidean space with coordinates $f_1, \dots, f_k$ and $U = \{(f_1, \dots, f_k) \mid a_i < f_i < b_i\}$, where the open intervals $]a_i, b_i[$ contain all the possible readings given by the devices D_i. The relations $F_j(f_1, \dots, f_k) = 0$ between the basis variables $f_1, \dots, f_k$ determine a surface M in U. Then $\mathcal{F}_S$ *is the $\mathbb{R}$-algebra of all smooth functions on the surface M.*

1.13. Example (thermodynamics of an ideal gas). Consider a certain volume of ideal gas. From the point of view of thermodynamics, we are interested in the following measurements: the volume V, the pressure p, and the absolute temperature T of the gas. These parameters, as is well known, satisfy the relation $pV = cT$, where c is a certain constant. Since $0 < p < \infty$, $0 < V < \infty$, and $0 < T < \infty$, the domain U is the first octant in the space $\mathbb{R}^3 \ni (V, p, T)$, and the hypersurface M in this domain is given by the equation $pV = cT$. The relevant $\mathbb{R}$-algebra $\mathcal{F}$ consists of all smooth functions on M.

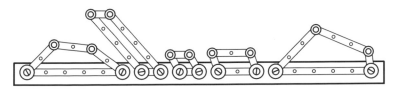

Figure 1.4. Hinge mechanisms $(5; 2, 2, 2)$, $(1; 4, 1, 4)$, $(1; 1, 1, 1)$, $(2; 1, 2, 1)$, $(5; 3, 3, 1)$.

1.14. Example: plane hinge mechanisms (also known as plane linkages). Such a mechanism consists of $n > 3$ ideal rods in the plane of lengths, say, $(l_1; l_2, \dots, l_n)$ (see Figure 1.4 for $n = 4$); the rods are joined in cyclic order to each other by ideal hinges at their end points; the hinges of the first rod (and hence the rod itself) are fixed to the plane; the other hinges and rods move freely (insofar as the configuration allows them to); the rods can sweep freely over ("through") each other. Obviously, the configuration space of a hinge mechanism is determined completely by the sequence of lengths of its rods. So, one can refer to a concrete mechanism just by indicating the corresponding sequence, for instance, $(5; 2, 3, 2)$. The reader is invited to solve the following problems in the process of reading the book. The first of them she/he can attack even now.

Exercise. Describe the configuration spaces of the following hinge mechanisms:

1. *Quadrilaterals*: $(5; 2, 2, 2)$; $(1; 4, 1, 4)$; $(1; 1, 1, 1)$; $(2; 1, 2, 1)$; $(5; 3, 3, 1)$.

2. *Pentagons*: $(3.9; 1, 1, 1, 1)$; $(1; 4, 1, 1, 4)$; $(6; 6, 2, 2, 6)$; $(1; 1, 1, 1, 1)$.

The reader will enjoy discovering that the configuration space of $(1; 1, 1, 1, 1)$ is the sphere with four handles.

Exercise. Show that the configuration space of a pentagon depends only on the set of lengths of the rods and not on the order in which the rods are joined to each other.

Exercise. Show that the configuration space of the hinge mechanism $(n - \alpha; 1, \ldots, 1)$ consisting of $n + 1$ rods is

1. The sphere S^{n-2} if $\alpha = \frac{1}{2}$.

2. The $(n - 2)$-dimensional torus $T^{n-2} = S^1 \times \cdots \times S^1$ if $\alpha = \frac{3}{2}$.

1.15. So far we have not said anything to explain what a state $s \in M_S$ of our physical system S really is, relying on the reader's physical intuition. But once the set of relevant functions $\mathcal{F}_S$ has been specified, this can easily be done in a mathematically rigorous and physically meaningful way.

The methodological basis of physical considerations is measurement. Therefore, two states of our system must be considered identical if and only if all the relevant measuring devices yield the same readings. Hence each state $s \in M_S$ is entirely determined by the readings in this state on all the relevant measuring devices, i.e., by the correspondence $\mathcal{F}_S \to \mathbb{R}$ assigning to each $f_D \in \mathcal{F}_S$ its reading (in the state s) $f_D(s) \in \mathbb{R}$. This assignment will clearly be an $\mathbb{R}$-algebra homomorphism. Thus we can say, by definition, that *any state s of our system is simply an $\mathbb{R}$-algebra homomorphism* $s \colon \mathcal{F}_S \to \mathbb{R}$. The set of all $\mathbb{R}$-algebra homomorphisms $\mathcal{F}_S \to \mathbb{R}$ will be denoted by $|\mathcal{F}_S|$; it should coincide with the set M_S of all states of the system.

1.16. Summarizing Sections 1.9–1.15, we can say that *any classical physical system is described by an appropriate collection of measuring devices, each* *state of the system being the collection of readings that this state determines on the measuring devices.*

The sentence in italics may be translated into mathematical language by means of the following dictionary:

- physical system = manifold, M;

- state of the system = point of the manifold, $x \in M$;

- measuring device = function on M, $f \in \mathcal{F}$;

- adequate collection of measuring devices = smooth $\mathbb{R}$-algebra, $\mathcal{F}$;

- reading on a device = value of the function, $f(x)$;

- collection of readings in the given state = $\mathbb{R}$-algebra homomorphism
$$x \colon \mathcal{F} \to \mathbb{R}, \quad f \mapsto f(x).$$

The resulting translation reads: *Any manifold M is determined by the smooth $\mathbb{R}$-algebra $\mathcal{F}$ of functions on it, each point x on M being the $\mathbb{R}$-algebra homomorphism $\mathcal{F} \to \mathbb{R}$ that assigns to every function $f \in \mathcal{F}$ its value $f(x)$ at the point x.*

1.17. Mathematically, the crucial idea in the previous sentence is the identification of points $x \in M$ of a manifold and $\mathbb{R}$-algebra homomorphisms $x \colon \mathcal{F} \to \mathbb{R}$ of its $\mathbb{R}$-algebra of functions $\mathcal{F}$, governed by the formula

$$x(f) = f(x). \tag{1.2}$$

This formula, read from left to right, defines the homomorphism $x \colon \mathcal{F} \to \mathbb{R}$ when the functions $f \in \mathcal{F}$ are given. Read from right to left, it defines the functions $f \colon M \to \mathbb{R}$, when the homomorphisms $x \in M$ are known.

Thus formula (1.2) is right in the middle of the important duality relationship existing between points of a manifold and functions on it, a duality similar to, but much more delicate than, the one between vectors and covectors in linear algebra.

1.18. In the general mathematical situation, the identification $M \leftrightarrow |\mathcal{F}|$ between the set M on which the functions $f \in \mathcal{F}$ are defined and the family of all $\mathbb{R}$-algebra homomorphisms $\mathcal{F} \to \mathbb{R}$ cannot be correctly carried out. This is because, first of all, $|\mathcal{F}|$ may turn out to be "much smaller" than M (an example is given in Section 3.6) or "bigger" than M, as we can see from the following example.

Example. Suppose M is the set $\mathbb{N}$ of natural numbers and $\mathcal{F}$ is the set of all functions on $\mathbb{N}$ (i.e., sequences $\{a(k)\}$) such that the limit $\lim_{k \to \infty} a(k)$ exists and is finite. Then the homomorphism

$$\alpha \colon \mathcal{F} \to \mathbb{R}, \quad \{a(k)\} \mapsto \lim_{k \to \infty} a(k),$$

does not correspond to any point of $M = \mathbb{N}$.

◀ Indeed, if α did correspond to some point $n \in \mathbb{N}$, we would have by (1.2)

$$n(a(\cdot)) = a(n),$$

so that

$$\lim_{k \to \infty} a(k) = \alpha(a(\cdot)) = n(a(\cdot)) = a(n)$$

for any sequence $\{a(k)\}$. But this is not the case, say, for the sequence $a_i = 0$, $i \leqslant n$, $a_i = 1$, $i > n$. Thus $|\mathcal{F}|$ is bigger than M, at least by the homomorphism α.   ▶

However, we can always add to $\mathbb{N}$ the "point at infinity" ∞ and extend the sequences (elements of $\mathcal{F}$) by putting $a(\infty) = \lim_{k \to \infty} a(k)$, thus viewing the sequences in $\mathcal{F}$ as functions on $\mathbb{N} \cup \{\infty\}$. Then obviously the homomorphism above corresponds to the "point" ∞.

This trick of adding *points at infinity* (or imaginary points, improper points, points of the absolute, etc.) is extremely useful and will be exploited to great advantage in Chapter 8.

1.19. In our mathematical development of the algebraic approach (Chapter 3), we shall start from an $\mathbb{R}$-algebra $\mathcal{F}$ of abstract elements called "functions." Of course, $\mathcal{F}$ will not be just any algebra; it must meet certain "smoothness" requirements. Roughly speaking, the algebra $\mathcal{F}$ must be smooth in the sense that locally (the meaning of that word must be defined in abstract algebraic terms!) it is like the $\mathbb{R}$-algebra $C^\infty(\mathbb{R}^n)$ of infinitely differentiable functions in $\mathbb{R}^n$. This will be the algebraic way of saying that the manifold M is locally like $\mathbb{R}^n$; it will be explained rigorously and in detail in Chapter 3. When the smoothness requirements are met, it will turn out that $\mathcal{F}$ entirely determines the manifold M as the set $|\mathcal{F}|$ of all $\mathbb{R}$-algebra homomorphisms of $\mathcal{F}$ into $\mathbb{R}$, and $\mathcal{F}$ can be identified with the $\mathbb{R}$-algebra of smooth functions on M. The algebraic definition of smooth manifold appears in the first section of Chapter 4.

1.20. Smoothness requirements are also needed in the classical coordinate approach, developed in detail below (see Chapter 5). In particular, coordinate transformations must be infinitely differentiable. The rigorous coordinate definition of a smooth manifold appears in Section 5.8.

1.21. The two definitions of smooth manifold (in which the algebraic approach and the coordinate approach result) are of course equivalent. This is proved in Chapter 7 below. Essentially, this book is a detailed exposition of these two approaches to the notion of smooth manifold and their equivalence, involving many examples, including a more rigorous treatment of the examples given in Sections 1.1–1.5 above.

2

Cutoff and Other Special Smooth Functions on $\mathbb{R}^n$

2.1. This chapter is an auxiliary one and can be omitted on first reading. In it we show how to construct certain specific infinitely differentiable functions on $\mathbb{R}^n$ (the $\mathbb{R}$-algebra of all such functions is denoted by $C^\infty(\mathbb{R}^n)$) that vanish (or do not vanish) on subsets of $\mathbb{R}^n$ of special form. These functions will be useful further on in the proof of many statements, especially in the very important Chapter 3.

2.2 Proposition. *There exists a function $f \in C^\infty(\mathbb{R})$ that vanishes for all negative values of the variable and is strictly positive for its positive values.*

◀ We claim that such is the function

$$f(x) = \begin{cases} 0 & \text{for } x \leqslant 0, \\ e^{-1/x} & \text{for } x > 0 \end{cases} \tag{2.1}$$

(see Figure 2.1 in the background). The only thing that must be checked is that f is smooth, i.e., $f \in C^\infty(\mathbb{R})$.

By induction over n, we shall show that the nth derivative of f is of the form

$$f^{(n)}(x) = \begin{cases} 0 & \text{for } x \leqslant 0, \\ e^{-1/x} P_n(x) x^{-2n} & \text{for } x > 0, \end{cases} \tag{2.2}$$

where $P_n(x)$ is a polynomial, and that $f^{(n)}$ is continuous.

For $n = 0$ this is obvious, since $\lim_{x \to +0} e^{-1/x} = 0$.

© Springer Nature Switzerland AG 2020
J. Nestruev, *Smooth Manifolds and Observables*, Graduate Texts in Mathematics 220, https://doi.org/10.1007/978-3-030-45650-4_2

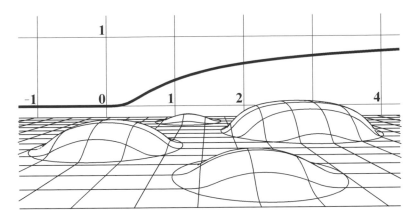

Figure 2.1. Special functions for Proposition 2.2 and Corollary 2.3.

If (2.2) is established for some $n \geqslant 0$, then obviously $f^{(n+1)}(x) = 0$ when $x < 0$, while if $x > 0$, we have

$$f^{(n+1)}(x) = e^{-1/x}\big(P_n(x) + x^2 P_n'(x) - 2nx P_n(x)\big)x^{-2n-2},$$

which shows that $f^{(n+1)}$ is of the form (2.2).

To show that it is continuous, note that $\lim_{x \to +0} e^{-1/x}x^{\alpha} = 0$ by L'Hospital's rule for any real α. Hence $\lim_{x \to +0} f^{(n+1)}(x) = 0$ and (again by L'Hospital's rule)

$$f^{(n+1)}(0) = \lim_{x \to 0} \frac{f^{(n)}(x) - f^{(n)}(0)}{x} = \lim_{x \to 0} \frac{f^{(n+1)}(x) - 0}{1},$$

so that $f^{(n+1)}$ equals 0 for $x \leqslant 0$ and is continuous for all x.   ▶

Exercise. Let f be the function defined in (2.1) and let $c_k = \max\left|f^{(k)}\right|$.

1. Prove that $c_k < \infty$ for all k.

2. Investigate the behavior of the sequence $\{c_k\}$ when $k \to \infty$.

2.3 Corollary. *For any $r > 0$ and $a \in \mathbb{R}^n$ there exists a function $g \in C^{\infty}(\mathbb{R}^n)$ that vanishes for all $x \in \mathbb{R}^n$ satisfying $\|x - a\| \geqslant r$ and is positive for all other $x \in \mathbb{R}^n$.*

◀ Such is, for example, the function

$$g(x) = f\left(r^2 - \|x - a\|^2\right),$$

where f is the function (2.1) from Section 2.2 (see Figure 2.1).   ▶

2.4 Proposition. *For any open set $U \subset \mathbb{R}^n$ there exists a function $f \in C^{\infty}(\mathbb{R}^n)$ such that*

$$\begin{cases} f(x) = 0, & \text{if } x \notin U, \\ f(x) > 0, & \text{if } x \in U. \end{cases}$$

◀ If $U=\mathbb{R}^n$, take $f \equiv 1$; if $U = \varnothing$, take $f \equiv 0$. Now suppose $U \neq \mathbb{R}^n$, $U \neq \varnothing$, and let $\{U_k\}$ be a covering of U by a countable collection of open balls (e.g., all the balls of rational radius centered at the points with rational coordinates and contained in U). By Corollary 2.3, there exist smooth functions $f_k \in C^\infty (\mathbb{R}^n)$ such that $f_k(x) > 0$ if $x \in U_k$ and $f_k(x) = 0$ if $x \notin U_k$. Put

$$M_k = \sup_{\substack{0 \leqslant p \leqslant k \\ p_1 + \cdots + p_n = p \\ x \in \mathbb{R}^n}} \left| \frac{\partial^p f_k}{\partial^{p_1} x_1 \cdots \partial^{p_n} x_n}(x) \right|.$$

Note that $M_k < \infty$, since outside the compact set $\overline{U}_k$ (the bar denotes closure) the function f_k and all its derivatives vanish.

Further, the series

$$\sum_{k=1}^{\infty} \frac{f_k}{2^k M_k}$$

converges to a smooth function f, since for all $p_1, \ldots, p_n$ the series

$$\sum_{k=1}^{\infty} \frac{f_k}{2^k M_k} \frac{\partial^{p_1 + \cdots + p_n} f_k}{\partial x_1^{p_1} \cdots \partial x_n^{p_n}}$$

converges uniformly (because whenever $k \geqslant p_1 + \cdots + p_n$, the absolute value of the kth term is no greater than 2^{-k}).

Clearly, the function f possesses the required properties. ▶

2.5 Corollary. *For any two nonintersecting closed sets A, $B \subset \mathbb{R}^n$ there exists a function $f \in C^\infty (\mathbb{R}^n)$ such that*

$$\begin{cases} f(x) = 0, & \text{when } x \in A; \\ f(x) = 1, & \text{when } x \in B; \\ 0 < f(x) < 1, & \text{for all other } x \in \mathbb{R}^n. \end{cases}$$

◀ Using Proposition 2.4, choose a function f_A that vanishes on A and is positive outside A and a similar function f_B for B. Then for f we can take the function

$$f = \frac{f_A}{f_A + f_B}$$

(see Figure 2.2). ▶

2.6 Corollary. *Suppose $U \subset \mathbb{R}^n$ is an open set and $f \in C^\infty(U)$. Then for any point $x \in U$ there exists a neighborhood $V \subset U$ and a function $g \in C^\infty (\mathbb{R}^n)$ such that $f\big|_V \equiv g\big|_V$.*

◀ Suppose W is an open ball centered at x whose closure is contained in U. Let V be a smaller concentric ball. The required function g can be defined

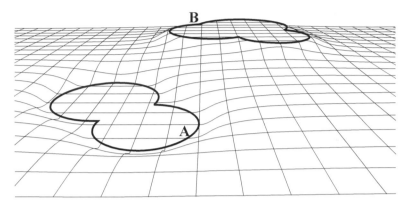

Figure 2.2. Smooth function separating two sets.

as

$$g(y) = \begin{cases} h(y) \cdot f(y), & \text{when } y \in U, \\ 0, & \text{when } y \in \mathbb{R}^n \setminus U, \end{cases}$$

where the function $h \in C^\infty(\mathbb{R}^n)$ is obtained from Corollary 2.3 and satisfies

$$h\big|_{\overline{V}} \equiv 1, \; h\big|_{\mathbb{R}^n \setminus W} \equiv 0. \blacktriangleright$$

2.7 Proposition. *On any nonempty open set $U \subset \mathbb{R}^n$ there exists a smooth function with compact level surfaces, i.e.,, a function $f \in C^\infty(U)$ such that for any $\lambda \in \mathbb{R}$ the set $f^{-1}(\lambda)$ is compact.*

◀ Denote by A_k the set of points $x \in U$ satisfying both of the following conditions:

(i) $\|x\| \leqslant k$,

(ii) the distance from x to the boundary of U is not less than $1/k$ (if $U = \mathbb{R}^n$, then condition (ii) can be omitted).

Obviously, all points of A_k are interior points of A_{k+1}. Hence A_k and the complement in $\mathbb{R}^n$ to the interior of the set A_{k+1} are two closed nonintersecting sets. By Corollary 2.5 there exists a function $f_k \in C^\infty(\mathbb{R}^n)$ such that

$$\begin{cases} f_k(x) = 0 & \text{if } x \in A_k, \\ f_k(x) = 1 & \text{if } x \notin A_{k+1}, \\ 0 < f_k(x) < 1 & \text{otherwise.} \end{cases}$$

Since any point $x \in U$ belongs to the interior of the set A_k for all sufficiently large k, the sum

$$f = \sum_{k=1}^{\infty} f_k$$

is well defined, and f is smooth on U (locally it is a finite sum of smooth functions).

Consider a point $x \in U \setminus A_k$. Since all the functions f_i are nonnegative, and for $i < k$ we have $f_i(x) = 1$, it follows that $f(x) \geqslant k - 1$. Hence, for any $\lambda \in \mathbb{R}$ the set $f^{-1}(\lambda)$ is a closed subset of the compact set A_k, where k is an integer such that $\lambda < k - 1$. A closed subset of a compact set is always compact, so that f is the required function. ▶

Let us fix a coordinate system $x_1, \ldots, x_n$ in a neighborhood U of a point z. Recall that a domain U is called *starlike* with respect to z if together with any point $y \in U$ it contains the whole interval (z, y).

2.8 Hadamard's lemma. *Any smooth function f in a starlike neighborhood of a point z is representable in the form*

$$f(x) = f(z) + \sum_{i=1}^{n}(x_i - z_i)g_i(x), \tag{2.3}$$

where g_i are smooth functions.

◀ In fact, consider the function

$$\varphi(t) = f(z + (x - z)t).$$

Then $\varphi(0) = f(z)$ and $\varphi(1) = f(x)$, and by the Newton–Leibniz formula,

$$\varphi(1) - \varphi(0) = \int_0^1 \frac{d\varphi}{dt}\,dt = \int_0^1 \sum_{i=1}^{n} \frac{\partial f}{\partial x_i}(z + (x_i - z_i)t)(x_i - z_i)\,dt$$

$$= \sum_{i=1}^{n}(x_i - z_i) \int_0^1 \frac{\partial f}{\partial x_i}(z + (x_i - z_i)t)\,dt.$$

Since the functions

$$g_i(x) = \int_0^1 \frac{\partial f}{\partial x_i}(z + (x_i - z_i)t)\,dt$$

are smooth, this concludes the proof of Hadamard's lemma. ▶

Let, as before, $x_1, \ldots, x_n$ be a fixed coordinate system of a point z in a neighborhood U and let $\tau = (i_1, \ldots, i_n)$ be a multiindex. Set

$$|\tau| = i_1 + \cdots + i_n, \qquad \frac{\partial^{|\tau|}}{\partial x^\tau} = \frac{\partial^{|\tau|}}{\partial x_1^{i_1} \cdots \partial x_n^{i_n}},$$

$$(x - z)^\tau = (x_1 - z_1)^{i_1} \cdots (x_n - z_n)^{i_n}, \qquad \tau! = i_1! \cdots i_n!.$$

2.9 Corollary. (Taylor expansion in Hadamard's form.) *Any smooth function f in a starlike neighborhood U of a point z is representable in the form*

$$f(x) = \sum_{|\tau|=0}^{n} \frac{1}{\tau!}(x - z)^\tau \frac{\partial^{|\tau|} f}{\partial x^\tau}(z) + \sum_{|\sigma|=n+1}(x - z)^\sigma g_\sigma(x), \tag{2.4}$$

where $g_\sigma \in C^\infty(U)$.

◀ In fact, using Hadamard's lemma for each function g_i in the decomposition (2.3), the function f can be represented in the form

$$f(x) = f(z) + \sum_{i=1}^{n}(x_i - z_i)g_i(z) + \sum_{i,j=1}^{n}(x_i - z_i)(x_j - z_j)g_{ij}(x).$$

Repeating this procedure for the functions g_{ij}, etc., we shall obtain the decomposition

$$f(x) = f(z) + \sum_{|\tau|=1}^{n}(x - z)^\tau \alpha_\tau + \sum_{|\sigma|=n+1}(x - z)^\sigma g_\sigma(x),$$

where α_τ are constants and $g_\sigma \in C^\infty(U)$. Applying to this equality all kinds of operators of the form $\partial^{|\tau|}/\partial x^\tau(z)$, $|\tau| \leqslant n$, we see that

$$\alpha_\tau = \frac{1}{\tau!}\frac{\partial^{|\tau|}f}{\partial x^\tau}(z). \qquad \blacktriangleright$$

2.10 Corollary. *Let* $f(x) \in C^\infty(\mathbb{R})$ *and* $f(0) = 0$. *Then the function* $f(x)/x$ *belongs to* $C^\infty(\mathbb{R})$.

◀ In fact, by Hadamard's lemma, any smooth function $f(x) \in C^\infty(\mathbb{R})$ is representable in the form $f(x) = f(0) + xg(x)$, where $g(x) \in C^\infty(\mathbb{R})$. If, in addition, $f(0) = 0$, then $f(x)/x = g(x)$. ▶

2.11 Lemma. *If* $f \in C^\infty(\mathbb{R}^n)$ *and* $f(z) = f(y) = 0$, *where* z *and* y *are two different points of the space* $\mathbb{R}^n$, *then the function* f *can be represented as a sum of products* $g_i h_i$, *where* $g_i(z) = 0$, $h_i(y) = 0$.

◀ By a linear coordinate change, the problem can be reduced to the case $z = (0, \ldots, 0, 0)$, $y = (0, \ldots, 0, 1)$. By Hadamard's lemma 2.8, any function f satisfying $f(z) = 0$ is representable in the form

$$f(x) = \sum_{i=1}^{n} x_i g_i(x).$$

Let us represent the function $g_i(x)$ in the form $g_i(x) = \big(g_i(x) - \alpha_i\big) + \alpha_i$, where $\alpha_i = g_i(y)$. Since in the product $x_i\big(g_i(x) - \alpha_i\big)$ the first factor vanishes at the point z, while the second one vanishes at the point y, the problem reduces to the case of a linear function

$$f(x) = \sum_{i=1}^{n} \alpha_i x_i.$$

The condition $f(y) = 0$ means that $\alpha_n = 0$. Representing now x_i, $i < n$, in the form $x_i = x_n x_i - x_i(x_n - 1)$, we conclude the proof of the lemma. ▶

2.12 Exercises.   1. Show that any function $f(x,y) \in C^\infty\left(\mathbb{R}^2\right)$ vanishing on the coordinate cross $\mathbf{K} = \{x = 0\} \cup \{y = 0\}$ is of the form

$$f = xy\, g(x,y), \quad g(x,y) \in C^\infty\left(\mathbb{R}^2\right).$$

2. Does a similar result hold if the cross is replaced by the union of the x-axis and the parabola $y = x^2$?

3
Algebras and Points

3.1. This chapter is a mathematical exposition of the algebraic approach to manifolds, which was sketched in intuitive terms in Chapter 1. Here we give a detailed answer to the following fundamental question: Given an *abstract* $\mathbb{R}$-algebra $\mathcal{F}$, find a set (smooth manifold) M whose $\mathbb{R}$-algebra of (smooth) functions can be identified with $\mathcal{F}$.

Further, $\mathcal{F}$ will always be a *commutative, associative algebra with unit over* $\mathbb{R}$, or briefly, an $\mathbb{R}$-*algebra*. All $\mathbb{R}$-*algebra homomorphisms* $\alpha \colon \mathcal{F}_1 \to \mathcal{F}_2$, i.e., maps of $\mathcal{F}_1$ into $\mathcal{F}_2$ preserving the operations

$$\alpha(f + g) = \alpha(f) + \alpha(g), \quad \alpha(f \cdot g) = \alpha(f) \cdot \alpha(g), \quad \alpha(\lambda f) = \lambda \alpha(f),$$

are assumed *unital* (i.e., α sends the unit in $\mathcal{F}_1$ into the one in $\mathcal{F}_2$).

We stress that the elements of $\mathcal{F}$, also called "functions," are not really functions at all; they are abstract objects of an unspecified nature. The point is to turn these objects into real functions on a manifold. In order to succeed in this undertaking, we shall successively impose certain conditions on $\mathcal{F}$. The key terms will be *geometrical* (Section 3.7), *complete* (Section 3.27), and *smooth* (Section 4.1) $\mathbb{R}$-algebras.

We begin with simple illustrations of how an abstractly defined $\mathbb{R}$-algebra can acquire substance and become a genuine algebra of nice functions on a certain set.

3.2. Example. Suppose $\mathcal{F}$ is the $\mathbb{R}$-algebra of all infinite sequences of real numbers $\{a_i\} = (a_0, a_1, a_2, \ldots)$ such that $a_i = 0$ for all i, except perhaps a finite number. The sum operation and multiplication by elements of $\mathbb{R}$ is defined term by term $(\lambda\{a_i\} = \{\lambda a_i\}$, etc.). The product $\{c_i\}$ of two

© Springer Nature Switzerland AG 2020
J. Nestruev, *Smooth Manifolds and Observables*, Graduate Texts
in Mathematics 220, https://doi.org/10.1007/978-3-030-45650-4_3

sequences $\{a_i\}$ and $\{b_i\}$ is defined by the formula

$$c_i = \sum_{k+l=i} a_k b_l.$$

Can this algebra $\mathcal{F}$ be realized as an algebra of nice functions on some set M?

We hope the reader has guessed the answer. By putting

$$\{a_i\} \mapsto \sum_{i \geqslant 0} a_i x^i$$

(this sum is always finite), we obtain an $\mathbb{R}$-algebra isomorphism $\mathcal{F} \to \mathbb{R}[x]$ of $\mathcal{F}$ onto the $\mathbb{R}$-algebra of polynomials in x, $\mathbb{R}[x]$. Thus any sequence $\{a_i\} \in \mathcal{F}$ may be viewed as the function on $M = \mathbb{R}$ given by $x \mapsto \sum_{i \geqslant 0} a_i x^i$.

3.3. Exercise. Suppose that the $\mathbb{R}$-algebras $\mathcal{F}_1$ and $\mathcal{F}_2$, as linear spaces, are isomorphic to the plane $\mathbb{R}^2 = \{(x,y)\}$. Let the multiplication in $\mathcal{F}_1$ and $\mathcal{F}_2$ be respectively given by

$$(x_1, y_1) \cdot (x_2, y_2) = (x_1 x_2, y_1 y_2),$$
$$(x_1, y_1) \cdot (x_2, y_2) = (x_1 x_2 + y_1 y_2, x_1 y_2 + x_2 y_1).$$

Find the set (manifold) M_i for which the algebra $\mathcal{F}_i$, $i = 1, 2$, is the algebra of smooth functions, explicitly indicating what function on M_i corresponds to the element $(x, y) \in \mathcal{F}_i$. Are the algebras $\mathcal{F}_1$ and $\mathcal{F}_2$ isomorphic?

3.4. We now return to our given abstract $\mathbb{R}$-algebra $\mathcal{F}$. Recalling the philosophy of Section 1.16 (a point of a manifold or state of a physical system is determined by all the relevant measurements), we introduce the following notations and definitions.

Denote by $M = |\mathcal{F}|$ the set of all $\mathbb{R}$-algebra homomorphisms of $\mathcal{F}$ onto $\mathbb{R}$:

$$M \ni x \colon \mathcal{F} \to \mathbb{R}, \quad f \mapsto x(f).$$

The elements of M will sometimes be called $\mathbb{R}$-*points* for the algebra $\mathcal{F}$ (they will indeed be the points of our future manifold), and $|\mathcal{F}|$ the *dual space of* $\mathbb{R}$-*points*. Further, set

$$\widetilde{\mathcal{F}} = \left\{ \tilde{f} \colon M \to \mathbb{R} \mid \tilde{f}(x) = x(f), \quad f \in \mathcal{F} \right\}. \tag{3.1}$$

The set $\widetilde{\mathcal{F}}$ has a natural $\mathbb{R}$-algebra structure given by

$$\left(\tilde{f} + \tilde{g} \right)(x) = \tilde{f}(x) + \tilde{g}(x) = x(f) + x(g),$$
$$\left(\tilde{f} \cdot \tilde{g} \right)(x) = \tilde{f}(x) \cdot \tilde{g}(x) = x(f) \cdot x(g), \tag{3.2}$$
$$\left(\lambda \tilde{f} \right)(x) = \lambda \tilde{f}(x) = \lambda x(f).$$

There is a natural map

$$\tau \colon \mathcal{F} \to \widetilde{\mathcal{F}}, \quad f \mapsto \tilde{f}.$$

We would like this map to be an isomorphism: Then we could view $\tilde{\mathcal{F}}$ as a realization of $\mathcal{F}$ in the form of an $\mathbb{R}$-algebra of functions on the dual space $M = |\mathcal{F}|$. But is this the case?

3.5. First we note that $\tau\colon \mathcal{F} \to \tilde{\mathcal{F}}$, $f \mapsto \tilde{f}$, *is a homomorphism.*
◀ Indeed, by definition of $\tilde{\mathcal{F}}$ (and because any $x \in M$ is a homomorphism),

$$\left(\widetilde{f+g}\right)(x) = x(f+g) = x(f) + x(g) = \tilde{f}(x) + \tilde{g}(x) = \left(\tilde{f} + \tilde{g}\right)(x).$$

The other two verifications are similar, and we leave them to the industrious reader.   ▶

It is also obvious that τ *is surjective.*

Thus it remains to show that τ is injective. Unfortunately, this is not so in the general case.

3.6. Example. Suppose $\mathcal{F}$ is the $\mathbb{R}$-algebra isomorphic (as a linear space) to the plane $\mathbb{R}^2 = \{(x,y) \mid x,y \in \mathbb{R}\}$ with the product

$$(x_1, y_1) \cdot (x_2, y_2) = (x_1 x_2, x_1 y_2 + x_2 y_1).$$

We shall show that the dual space $M = |\mathcal{F}|$ consists of a single point. This implies that τ is not injective, since $\tilde{\mathcal{F}}$ is then isomorphic to $\mathbb{R}$, while $\mathcal{F}$ is not ($\mathcal{F} \supset \{(x,0),\ x \in \mathbb{R}\} \cong \mathbb{R}$).

The element $(1,0)$ is obviously the unit of the algebra $\mathcal{F}$, and any element (x,y) has an inverse if $x \neq 0$, namely $(x,y)^{-1} = \left(x^{-1}, -yx^{-2}\right)$. Hence the only ideal of the algebra $\mathcal{F}$, other than the ideals $\{0\}$ and $\mathcal{F}$, is the *ideal* $\mathcal{I} = \{(0,y) \mid y \in \mathbb{R}\}$. The quotient algebra $\mathcal{F}/\mathcal{I}$ is naturally isomorphic to $\mathbb{R}$, the quotient map $q\colon \mathcal{F} \to \mathcal{F}/\mathcal{I} \cong \mathbb{R}$ being the projection $(x,y) \mapsto x$. This map is the only surjective $\mathbb{R}$-algebra homomorphism $\mathcal{F} \to \mathbb{R}$, so that $M = \{q\}$.

3.7. In order to be able to assert that $\tau\colon \mathcal{F} \to \tilde{\mathcal{F}}$ is injective (and hence an isomorphism), certain conditions must be imposed on $\mathcal{F}$. Note that τ *will be injective iff the ideal* $\mathcal{I}(\mathcal{F}) = \bigcap_{x \in M} \operatorname{Ker} x$ *is trivial.*
◀ Indeed,

$$f \in \operatorname{Ker}\tau \iff \tau(f) = \tilde{f} = 0$$

$$\iff \tilde{f}(x) = x(f) = 0 \quad \forall x \in M$$

$$\iff f \in \bigcap_{x \in M} \operatorname{Ker} x = \mathcal{I}(\mathcal{F}),$$

and therefore $\operatorname{Ker}\tau = 0 \iff \mathcal{I}(\mathcal{F}) = 0$.   ▶

This motivates (mathematically) the following definition.
Definition. An $\mathbb{R}$-algebra $\mathcal{F}$ is called *geometrical* if

$$\mathcal{I}(\mathcal{F}) = \bigcap_{x \in |\mathcal{F}|} \operatorname{Ker} x = 0.$$

(In the previous example, $\mathcal{I}(\mathcal{F}) = \mathcal{I} = \{(0,y) \mid y \in \mathbb{R}\} \neq 0$.)

It is worth noticing that an algebra with empty dual space is not geometrical.

Exercises.   1. Prove that the polynomial algebra $\mathbb{R}[x_1, \ldots, x_n]$ is geometrical.

2. Let V be a finite-dimensional vector space over $\mathbb{R}$ and let $G\colon V \to V$ be a linear operator. Consider the $\mathbb{R}$-algebra $\mathcal{F}_G$ generated by G^k, $k = 0, 1, \ldots$, as a vector space. Characterize the operators G for which $\mathcal{F}_G$ is geometrical.

3.8. In Sections 3.5, 3.7, we have in fact proved the following theorem.

Theorem. *Any geometrical $\mathbb{R}$-algebra $\mathcal{F}$ is canonically isomorphic to the $\mathbb{R}$-algebra $\widetilde{\mathcal{F}}$ of functions defined on the dual space $M = |\mathcal{F}|$ of $\mathbb{R}$-points $(M \ni x\colon \mathcal{F} \to \mathbb{R})$ by the rule $f(x) = x(f)$.*

Having this isomorphism in mind, we shall identify our abstract algebra $\mathcal{F}$ (which will usually be assumed geometrical) with the $\mathbb{R}$-algebra $\widetilde{\mathcal{F}}$ of functions on the dual space $M = |\mathcal{F}|$ once and for all. The notation $\widetilde{\mathcal{F}}$ will be abandoned; the elements $f \in \mathcal{F}$ will often be viewed as functions $M \to \mathbb{R}$.

3.9 Exercises. Check which of the following algebras are geometrical:

1. The formal series algebra $\mathbb{R}[[x_1, \ldots, x_n]]$.

2. The quotient algebra
$$\mathbb{R}[x_1, \ldots, x_n]/f^k \mathbb{R}[x_1, \ldots, x_n], \ f \in \mathbb{R}[x_1, \ldots, x_n].$$

3. The algebra of germs of smooth functions at $0 \in \mathbb{R}^n$.

4. The algebra of all smooth bounded functions.

5. The algebra of all smooth periodic functions (of period 1) on $\mathbb{R}$ (see Section 3.18).

6. The subalgebra of the previous algebra consisting of all even functions.

7. The algebras $\mathcal{F}_1$ and $\mathcal{F}_2$ described in Exercise 3.3.

8. The algebra of all differential operators in $\mathbb{R}^n$ with constant coefficients (multiplication in this algebra is the composition of operators).

3.10. The algebra $\mathcal{F}_S$ of functions corresponding to measuring devices of a classical physical system S (see Sections 1.9, 1.15) is always geometrical. This property is the mathematical formulation of the classical physical postulate asserting that if all the readings of two different devices for all the states of the system S are the same, then these two devices measure the same physical parameter (i.e., only one of the devices is needed).

3.11 Proposition. *For an arbitrary $\mathbb{R}$-algebra $\mathcal{F}$, the quotient $\mathbb{R}$-algebra*
$$\mathcal{F}/\mathcal{I}(\mathcal{F}), \quad \text{where} \quad \mathcal{I}(\mathcal{F}) = \bigcap_{p \in |\mathcal{F}|} \operatorname{Ker} p,$$

is geometrical and $|\mathcal{F}| = |\mathcal{F}/\mathcal{I}(\mathcal{F})|$.

◀ Define the map $\varphi \colon |\mathcal{F}/\mathcal{I}(\mathcal{F})| \to |\mathcal{F}|$ by assigning to each homomorphism $b \colon \mathcal{F}/\mathcal{I}(\mathcal{F}) \to \mathbb{R}$ the homomorphism $\varphi(b) = a = b \circ \mathrm{pr}$, where pr is the quotient map $\mathrm{pr} \colon \mathcal{F} \to \mathcal{F}/\mathcal{I}(\mathcal{F})$.

We claim that φ is bijective. Obviously, $b_1 \neq b_2$ implies $a_1 \neq a_2$, so φ is injective. Now suppose $a \in |\mathcal{F}|$. Then $\mathrm{Ker}\, a \supset \mathcal{I}(\mathcal{F})$. Hence the element $b([f]) = a(f)$, where $[f]$ is the coset of the element f modulo $\mathcal{I}(\mathcal{F})$, is well defined and determines a homomorphism $b \colon \mathcal{F}/\mathcal{I}(\mathcal{F}) \to \mathbb{R}$. Clearly, we have $a = \varphi(b)$, i.e., the map φ is surjective, so that φ identifies $|\mathcal{F}|$ with $|\mathcal{F}/\mathcal{I}(\mathcal{F})|$.

Suppose further that $b \in |\mathcal{F}/\mathcal{I}(\mathcal{F})|$ and $a = \varphi(b) = b \circ \mathrm{pr}$. In that case $\mathrm{Ker}\, b = \mathrm{Ker}\, a/\mathcal{I}(\mathcal{F})$. Hence

$$\mathcal{I}(\mathcal{F}/\mathcal{I}(\mathcal{F})) = \bigcap_{b \in |\mathcal{F}/\mathcal{I}(\mathcal{F})|} \mathrm{Ker}\, b = \bigcap_{a \in \mathcal{F}} \big(\mathrm{Ker}\, a/\mathcal{I}(\mathcal{F}) \big)$$

$$= \left(\bigcap_{a \in \mathcal{F}} \mathrm{Ker}\, a \right) / \mathcal{I}(\mathcal{F}) = \mathcal{I}(\mathcal{F})/\mathcal{I}(\mathcal{F}) = \{0\}. \quad ▶$$

3.12. Given a geometrical $\mathbb{R}$-algebra $\mathcal{F}$, we intend to introduce a topology in the dual set $M = |\mathcal{F}|$ of $\mathbb{R}$-points.

From the physical point of view, two states s_1, s_2 of a classical system S (two $\mathbb{R}$-points) are near each other if all the readings of the relevant measuring devices are close, i.e., for all measuring devices D we must have

$$f_D(s_2) \in \,]f_D(s_1) - \varepsilon, \; f_D(s_1) + \varepsilon[.$$

Mathematically, we express this by saying that the topology in M, called the *dual topology*, is given by the basis of open sets of the form $f^{-1}(V)$, where $V \subset \mathbb{R}$ is open and $f \in \mathcal{F}$. (The reader should recall at this point that the expression f^{-1} is meaningful only because we have identified $\mathcal{F}$ with an algebra of functions $f \colon M \to \mathbb{R}$.)

Another way of saying this is the following: *The topology in the dual space* $M = |\mathcal{F}|$ *is the weakest for which all the functions in* $\mathcal{F}$ *are continuous.*

3.13 Proposition. *The topology introduced in Section 3.12 in the dual space* $M = |\mathcal{F}|$ *is that of a Hausdorff space.*

◀ Suppose x and y are distinct points of $|\mathcal{F}|$, i.e., different homomorphisms of $\mathcal{F}$ into $\mathbb{R}$. This means there is an $f \in \mathcal{F}$ for which $f(x) \neq f(y)$, say $f(x) < f(y)$. Then the sets

$$f^{-1}\left(\, \left] -\infty, \frac{f(x) + f(y)}{2} \right[\right), \quad f^{-1}\left(\left] \frac{f(x) + f(y)}{2}, +\infty \right[\, \right)$$

are nonintersecting neighborhoods of the points x and y.   ▶

When speaking of the "space" $M = |\mathcal{F}|$, it is this topological (Hausdorff) structure that will always be understood.

3.14. In this section we assume that $\mathcal{F}_0$ is any $\mathbb{R}$-algebra of functions on a given set M_0. Then there is a natural map $\theta\colon M_0 \to |\mathcal{F}_0|$ assigning to each point $a \in M_0$ the homomorphism $f \mapsto f(a)$. In other words, $\theta(a)(f) = f(a)$, $a \in M_0$, and therefore if an element $f \in \mathcal{F}_0$, viewed as a function on the dual space $|\mathcal{F}_0|$, vanishes on $\theta(M_0)$, then f is the zero element of $\mathcal{F}_0$. In particular, the algebra $\mathcal{F}_0$ will be geometrical, and we have the following result.

Proposition. *If $\mathcal{F}_0$ is a subalgebra of the $\mathbb{R}$-algebra of continuous functions on the topological space M_0, then the map $\theta\colon M_0 \to |\mathcal{F}_0|$, $a \mapsto \big(f \mapsto f(a)\big)$, is continuous.*

◄ Suppose $U = f^{-1}(V)$ is a basis open set in $|\mathcal{F}_0|$. By definition U consists of all the homomorphisms $\mathcal{F}_0 \to \mathbb{R}$ that send the (fixed) function $f \in \mathcal{F}$ to some point of the open set $V \subset \mathbb{R}$. Then the inverse image $\theta^{-1}(U)$ consists of all points $a \in M_0$ such that $f(a) \in V$ and is therefore an open subset of M. ►

It should be noted that in our general situation (when $\mathcal{F}_0$ is any geometrical $\mathbb{R}$-algebra, $M_0 = |\mathcal{F}_0|$), M_0 is a topological space and the elements of $\mathcal{F}_0$ are continuous functions (see Section 3.12), so that the proposition proved above applies.

Exercise. Describe the dual space for each of the following algebras:

1. $\mathbb{R}[x,y]/xy\mathbb{R}[x,y]$;

2. $\mathbb{R}[x,y,z]/\left(x^2 + y^2 + z^2 - 1\right)\mathbb{R}[x,y,z]$.

3.15. Example. Suppose $\mathcal{F} = \mathbb{R}[x_1,\ldots,x_n]$ is the $\mathbb{R}$-algebra of polynomials in n variables. Every homomorphism $a\colon \mathcal{F} \to \mathbb{R}$ is determined by the "vector" $(\lambda_1,\ldots,\lambda_n)$, where $\lambda_i = a(x_i)$, since

$$a\left(\sum_{k_1,\ldots,k_n} c_{k_1\ldots k_n} x_1^{k_1} \cdots x_n^{k_n}\right) = \sum_{k_1,\ldots,k_n} c_{k_1\ldots k_n} \left(a(x_1)\right)^{k_1} \cdots \left(a(x_n)\right)^{k_n}$$

$$= \sum_{k_1,\ldots,k_n} c_{k_1\ldots k_n} \lambda_1^{k_1} \cdots \lambda_n^{k_n}.$$

Moreover, the map

$$\mathcal{F} \ni \sum_{k_1,\ldots,k_n} c_{k_1\ldots k_n} x_1^{k_1} \cdots x_n^{k_n} \longmapsto \sum_{k_1,\ldots,k_n} c_{k_1\ldots k_n} \lambda_1^{k_1} \cdots \lambda_n^{k_n} \in \mathbb{R}$$

is a homomorphism of the algebra $\mathcal{F}$ into $\mathbb{R}$ for all $\lambda_1,\ldots,\lambda_n \in \mathbb{R}$. Thus the dual space $|\mathcal{F}|$ in this case is naturally identified with $\mathbb{R}^n = \{(\lambda_1,\ldots,\lambda_n)\}$.

The topology defined in $|\mathcal{F}|$ (see Section 3.12) coincides with the usual topology of $\mathbb{R}^n$.

◄ Indeed, the sets $f^{-1}(V)$, where f is a polynomial and $V \subset \mathbb{R}$ is open, are open in $\mathbb{R}^n = |\mathcal{F}|$, since polynomials are continuous functions. Moreover, a ball of radius r with center $(b_1, \ldots, b_n)$ in $\mathbb{R}^n = |\mathcal{F}|$ is of the form $f^{-1}(\mathbb{R}_+)$, where $\mathbb{R}_+$ is the positive half of $\mathbb{R}$, if we take f to be

$$f(x_1, \ldots, x_n) = r^2 - \sum_{i=1}^{n}(b_i - x_i)^2.$$

Since such balls constitute a basis for the usual topology in $\mathbb{R}^n$, the two topologies coincide.   ▶

3.16. Example. Suppose $\mathcal{F} = C^\infty(U)$ is the $\mathbb{R}$-algebra of infinitely differentiable real-valued functions on an open subset U of $\mathbb{R}^n$. Consider the map

$$\theta\colon U \to |\mathcal{F}|, \quad x \mapsto (f \mapsto f(x)).$$

We claim that the *map θ is a homeomorphism*, so that the dual space $|C^\infty(U)|$ is homeomorphic to U.

◀ Since injectivity is obvious (elements of $\mathcal{F}$ being functions on U), we first prove the surjectivity of θ. Suppose $p \in |\mathcal{F}|$, i.e., $p\colon \mathcal{F} \to \mathbb{R}$ is an $\mathbb{R}$-algebra homomorphism onto $\mathbb{R}$. Choose a smooth function $f_c \in C^\infty(U)$ all of whose level surfaces are compact (such a function exists by Proposition 2.7). Then, in particular, the set $L = f_c^{-1}(\lambda)$, where $\lambda = p(f)$, is compact. Assume that $p \in |\mathcal{F}|$ does not correspond to any point of U. Then for any point $a \in U$ there exists a function $f_a \in \mathcal{F}$ for which $f_a(a) \neq p(f_a)$. The sets

$$U_a = \{x \in U \mid f_a(x) \neq p(f_a)\}, \quad a \in L,$$

constitute an open covering of L. Since L is compact, we can choose a finite subcovering $U_{a_1}, \ldots, U_{a_m}$. Consider the function

$$g = (f - p(f))^2 + \sum_{i=1}^{m}(f_{a_i} - p(f_{a_i}))^2.$$

This is a smooth nonvanishing function on U, so that $1/g \in \mathcal{F}$. Since p is a (unital!) $\mathbb{R}$-algebra homomorphism, we must have

$$p(1) = p(g \cdot (1/g)) = p(g) \cdot p(1/g) = 1. \tag{3.3}$$

But by the definition of g,

$$p(g) = (p(f) - p(f))^2 + \sum (p(f_{a_i}) - p(f_{a_i}))^2 = 0,$$

which contradicts (3.3), proving the surjectivity of θ.

The fact that θ is a homeomorphism is an immediate consequence of Proposition 3.14.   ▶

In particular, we have proved that $|C^\infty(\mathbb{R}^n)| = \mathbb{R}^n$.

3.17 Exercises. Describe the dual space for each of the following algebras:

1. $C^\infty(\mathbb{R}^3) / (x^2 + y^2 + z^2 - 1)\, C^\infty(\mathbb{R}^3)$.

2. $C^{\infty}\left(\mathbb{R}^{3}\right) / \left(x^{2} + y^{2} - z^{2}\right) C^{\infty}\left(\mathbb{R}^{3}\right)$.

3. Smooth even functions on the real line.

4. Smooth even functions of period 1 on the real line.

5. Smooth functions of rational period (not necessarily the same) on the real line.

6. Functions defined on the real line as the ratio of two polynomials $p(x)/q(x)$, where $q(x) \neq 0$ for all $x \in \mathbb{R}$.

7. The same functions as before, but with the additional requirement $\deg p(x) \leqslant \deg q(x)$.

8. Functions defined on the real line as the ratio of two polynomials $p(x)/q(x)$, where $q(x)$ is not identically zero (i.e., defined as rational functions).

9. The subalgebra $\{f \in C^{\infty}\left(\mathbb{R}^{2}\right) \mid f(x+1, y) = f(x, y)\}$ of $C^{\infty}\left(\mathbb{R}^{2}\right)$.

10. The subalgebra $\{f \in C^{\infty}\left(\mathbb{R}^{2}\right) \mid f(x+1, -y) = f(x, y)\}$ of $C^{\infty}\left(\mathbb{R}^{2}\right)$.

11. The subalgebra $\{f \in C^{\infty}\left(\mathbb{R}^{2}\right) \mid f(x, y+1) = f(x, y) = f(x+1, y)\}$ of $C^{\infty}\left(\mathbb{R}^{2}\right)$.

12. The subalgebra
$$\{f \in C^{\infty}(\mathbb{R}^{3} \setminus 0) \mid f(x, y, z) = f(\lambda x, \lambda y, \lambda z), \quad \forall \lambda \neq 0\}$$
of $C^{\infty}(\mathbb{R}^{3} \setminus 0)$.

13. The subalgebra $\{f \in C^{\infty}\left(\mathbb{R}^{2}\right) \mid f(x+1, -y) = f(x, y) = f(x, y+1)\}$ of $C^{\infty}\left(\mathbb{R}^{2}\right)$.

14. The subalgebra
$$\{f \in C^{\infty}(\mathbb{R}^{3} \setminus 0) \mid f(x, y, z) = f(\lambda x, \lambda y, \lambda z), \quad \forall \lambda \in \mathbb{R}_{+}\}$$
of $C^{\infty}(\mathbb{R}^{3} \setminus 0)$.

3.18. Example. Suppose $\mathcal{F}$ consists of all periodic smooth functions of period 1 on the line $\mathbb{R}$. Then, as usual, each point $a \in \mathbb{R}$ determines the homomorphism $\mathcal{F} \to \mathbb{R}$, $f \mapsto f(a)$. But different points can give rise to the same homomorphism; this happens iff the distance between the points is an integer.

We claim that there are no homomorphisms other than the ones determined by the points $a \in \mathbb{R}$.

◄ The proof is similar to the one in the previous section. Namely, if the homomorphism $p \colon \mathcal{F} \to \mathbb{R}$ is not determined by any point, then for any $a \in \mathbb{R}$ there exists a function $f_{a} \in \mathcal{F}$ such that $p(f_{a}) \neq f_{a}(a)$. From the open covering of the closed interval $[0, 1]$ by sets of the form
$$U_{a} = \{x \in \mathbb{R} \mid f_{a}(x) \neq p(f_{a})\}, \quad a \in \mathbb{R},$$

choose a finite subcovering $U_{a_1}, \ldots, U_{a_n}$. The function

$$g = \sum_{i=1}^{m} (f_{a_i} - p(f_{a_i}))^2$$

does not vanish anywhere on $[0, 1]$ and, by periodicity, anywhere on $\mathbb{R}$. Hence $1/g \in \mathcal{F}$, etc., just as in Section 3.16. ▶

Thus in our case $|\mathcal{F}|$ can be identified with the quotient space $\mathbb{R}/\mathbb{Z}$, where $\mathbb{Z}$ is the subgroup of integers in $\mathbb{R}$. Of course, $\mathbb{R}/\mathbb{Z} = S^1$ is the circle. Thus we have shown rigorously that *smooth periodic functions of period 1 on $\mathbb{R}$ are actually functions on the circle*, which is in accord with our intuitive understanding of such functions.

3.19. Now suppose $\mathcal{F}_1$ and $\mathcal{F}_2$ are two geometrical $\mathbb{R}$-algebras and $\varphi \colon \mathcal{F}_1 \to \mathcal{F}_2$ is an $\mathbb{R}$-algebra homomorphism. Then for the dual spaces of $\mathbb{R}$-points $|\mathcal{F}_1|$ and $|\mathcal{F}_2|$ the dual map $|\varphi|$ arises:

$$|\varphi| \colon |\mathcal{F}_2| \to |\mathcal{F}_1|, \quad x \mapsto x \circ \varphi.$$

We claim that *the map $|\varphi|$ is continuous*.

◀ Indeed, take a basis open set $U = f^{-1}(V) \subset |\mathcal{F}_1|$, where $f \in \mathcal{F}_1$ and $V \subset \mathbb{R}$ is open. Then U consists of all points $x \in |\mathcal{F}_1|$ such that $f(x) \in V$. The inverse image of U by $|\varphi|$ consists of all points $y \in |\mathcal{F}_2|$ such that $|\varphi|(y) \in U$, i.e., $f(|\varphi|(y)) \in V$. But

$$f(|\varphi|(y)) = f(y \circ \varphi) = (y \circ \varphi)(f) = y(\varphi(f)) = \varphi(f)(y)$$

(the reader should check each of these relations!). Therefore, the set $|\varphi|^{-1}(U)$ consists of all points $y \in |\mathcal{F}_2|$ for which $\varphi(f)(y) \in V$; thus the set $|\varphi|^{-1}(U)$ is open. ▶

3.20. If $\varphi_1 \colon \mathcal{F}_1 \to \mathcal{F}_2$, $\varphi_2 \colon \mathcal{F}_2 \to \mathcal{F}_3$ are $\mathbb{R}$-algebra homomorphisms of geometrical $\mathbb{R}$-algebras $\mathcal{F}_1, \mathcal{F}_2, \mathcal{F}_3$, then obviously $|\varphi_2 \circ \varphi_1| = |\varphi_1| \circ |\varphi_2|$ and $|\operatorname{id}_{\mathcal{F}_i}| = \operatorname{id}_{|\mathcal{F}_i|}$. Further, if $\varphi \colon \mathcal{F}_1 \to \mathcal{F}_2$ has an inverse homomorphism φ^{-1}, then

$$|\varphi^{-1}| = |\varphi|^{-1}.$$

In particular, if φ is an isomorphism, then $|\varphi|$ is a homeomorphism.

3.21. Having started from an abstract geometrical $\mathbb{R}$-algebra $\mathcal{F}$, we have constructed the (Hausdorff) topological space $M = |\mathcal{F}|$ (the dual space of $\mathbb{R}$-points), for which $\mathcal{F}$ is a subalgebra of the algebra of all continuous functions. It might now seem that we need only postulate that $\mathcal{F}$ be locally isomorphic to $C^\infty(\mathbb{R}^n)$ (i.e., cover M with a family of sets E_i, $M = \bigcup E_i$, such that the restriction of $\mathcal{F}$ to each E_i is isomorphic to $C^\infty(\mathbb{R}^n)$), and our program of defining a manifold in terms of its $\mathbb{R}$-algebra of functions will be carried out. Unfortunately, things are not as simple as they appear at first glance: certain technical difficulties, related to the notion of *restriction*, must be overcome before we succeed in implementing our program.

3.22. Example. Suppose $\mathcal{F} = C^{\infty}(\mathbb{R})$ and $\mathbb{R}_+ \subset \mathbb{R}$ is the set of positive real numbers. We would like to obtain the algebra of smooth functions on $\mathbb{R}_+$ as a "restriction" of the algebra $\mathcal{F}$. But consider the function $x \mapsto 1/x$ on R_+; it is certainly a smooth function on $\mathbb{R}_+$, but clearly is not the restriction of any function $f \in \mathcal{F} = C^{\infty}(\mathbb{R})$. How can such functions be obtained from $\mathcal{F}$?

3.23. Definition. Suppose $\mathcal{F}$ is a geometrical $\mathbb{R}$-algebra and $A \subset |\mathcal{F}|$ is any subset of its dual space $|\mathcal{F}|$; the *restriction* $\mathcal{F}\big|_A$ of $\mathcal{F}$ to A is the set of all functions $f \colon A \to \mathbb{R}$ such that for any point $a \in A$ there exists a neighborhood $U \subset A$ and an element $\bar{f} \in \mathcal{F}$ such that the (ordinary) restriction of f to U coincides with the restriction of $\bar{f}$ (understood as a function on $|\mathcal{F}|$) to U.

Obviously, $\mathcal{F}\big|_A$ is an $\mathbb{R}$-algebra.

Now we can return to Example 3.22. We claim that the function $x \mapsto 1/x$ belongs to $C^{\infty}(\mathbb{R})\big|_{\mathbb{R}_+}$.

◀ Indeed, for any point $a > 0$ there exists (see Section 2.5) a function $\alpha \in C^{\infty}(\mathbb{R})$ that vanishes when $x \leqslant a/3$ and equals 1 whenever $x \geqslant 2a/3$. For $\bar{f}$ take the function that vanishes when $x \leqslant 0$ and equals $\alpha(x)/x$ when $x > 0$. Obviously, $\bar{f}$ is smooth and coincides with the function $x \mapsto 1/x$ in the neighborhood $]2a/3, 4a/3[$ of the point a.   ▶

In a similar way we can show that any smooth function on $\mathbb{R}_+$ belongs to $C^{\infty}(\mathbb{R})\big|_{\mathbb{R}_+}$; i.e., we have

$$C^{\infty}(\mathbb{R})\big|_{\mathbb{R}_+} = C^{\infty}(\mathbb{R}_+).$$

This statement has the following generalization.

3.24 Proposition. *If $\mathcal{F} = C^{\infty}(U)$, where $U \subset \mathbb{R}^n$ is not empty and open, while V is open in $U = |\mathcal{F}|$, then $\mathcal{F}\big|_V = C^{\infty}(V)$.*

◀ The identification $U = |\mathcal{F}|$ in the statement of the proposition was established in Section 3.16. Suppose $f \in C^{\infty}(V)$ and $x \in V$. By 2.6 there exists a neighborhood W of the point x such that $W \subset V$ and a function $g \in C^{\infty}(\mathbb{R}^n)$ such that $g\big|_W = f\big|_W$. If $\bar{g} = g\big|_U$, then we will also have $\bar{g}\big|_W = f\big|_W$. Thus

$$f \in C^{\infty}(U)\big|_V; \text{ i.e., } C^{\infty}(V) \subset C^{\infty}(U)\big|_V.$$

The inverse inclusion immediately follows from the definition of the algebra $C^{\infty}(U)\big|_V$.   ▶

Exercise. For the subsets $A = \{1/n \mid n = 1, 2, 3, \dots\}$ and $B = A \cup \{0\}$ in $\mathbb{R}$ describe the restrictions $\mathbb{R}[x]\big|_A$ and $\mathbb{R}[x]\big|_B$.

3.25. In the general case, in which $\mathcal{F}$ is any geometrical $\mathbb{R}$-algebra and A is a subset of the dual space $|\mathcal{F}|$, we can assign to every function $f \in \mathcal{F}$ its restriction to $A \subset |\mathcal{F}|$, which obviously belongs to $\mathcal{F}\big|_A$. Thus we obtain the *restriction homomorphism*

$$\rho_A \colon \mathcal{F} \to \mathcal{F}\big|_A, \quad f \mapsto f\big|_A.$$

(Here as usual the element $f \in \mathcal{F}$ is viewed as a function on the dual space $|\mathcal{F}|$.)

Proposition. *Suppose $i \colon \mathcal{F}_1 \to \mathcal{F}_2$ is an isomorphism of two geometrical algebras, $A_2 \subset |\mathcal{F}_2|$, $A_1 = |i|(A_2)$. Then the map*

$$\mathcal{F}_1\big|_{A_1} \to \mathcal{F}_2\big|_{A_2}, \quad f \mapsto f\big(|i|\big|_{A_2}\big)$$

is an isomorphism.

◀ The proof is a straightforward verification of definitions.   ▶

3.26. Now, in the most important particular case $A = |\mathcal{F}|$, we can consider the restriction homomorphism $\rho \colon \mathcal{F} \to \mathcal{F}\big|_{|\mathcal{F}|}$. Since $\mathcal{F}$ is assumed geometrical (different elements $f \in \mathcal{F}$ are identified with different functions on $|\mathcal{F}|$), ρ *is injective.* Surjectivity, surprisingly enough, is *not* obvious: By the definition given in Section 3.23, $\mathcal{F}\big|_{|\mathcal{F}|}$ consists of all functions that are locally like those of $\mathcal{F}$, but it is not clear why all such functions belong to $\mathcal{F}$. Indeed, this is not always the case.

3.27. Example. Suppose $\mathcal{F}$ is the subalgebra of the algebra $C^\infty(\mathbb{R}^n)$ consisting of functions each of which is less in absolute value than some polynomial. Then *the dual space $|\mathcal{F}|$ is homeomorphic to $\mathbb{R}^n$.*

◀ The proof is similar to the one given in Section 3.16, except that the function f with compact level surfaces must be chosen so that it belongs to $\mathcal{F}$ but $f(x) \to \infty$ as $\|x\| \to \infty$; e.g., we can take $f \colon x \to \|x\|^2 + 1$. Then $1/g(x) \to 0$ as $\|x\| \to \infty$, so that $1/g$ also belongs to $\mathcal{F}$, and the proof proceeds as in Section 3.16.   ▶

Thus $\mathcal{F}\big|_{|\mathcal{F}|} = \mathcal{F}\big|_{\mathbb{R}^n}$ coincides with the algebra $C^\infty(\mathbb{R}^n)$ of *all* smooth functions on $\mathbb{R}^n$, since any function $f \in C^\infty(\mathbb{R}^n)$ in a neighborhood of some point $a \in \mathbb{R}^n$ coincides with the function $f\theta$, where θ is a smooth function that vanishes outside the ball of radius 2 and center a and equals 1 inside the concentric ball of radius 1 (see Section 2.5) and, obviously, $f\theta \in \mathcal{F}$. Hence $\rho \colon \mathcal{F} \to \mathcal{F}\big|_{|\mathcal{F}|} = C^\infty(\mathbb{R}^n)$ cannot be surjective.

3.28. Definition. A geometrical $\mathbb{R}$-algebra $\mathcal{F}$ is said to be *complete* if the restriction homomorphism $\rho \colon \mathcal{F} \to \mathcal{F}\big|_{|\mathcal{F}|}$ is surjective (and is therefore an isomorphism), i.e., if any function $|\mathcal{F}| \to \mathbb{R}$ locally coinciding with elements of $\mathcal{F}$ is itself an element of $\mathcal{F}$.

It is clear that the algebras $C^\infty(U)$, where $U \subset \mathbb{R}^n$ is open, are complete (see Section 3.24). The algebra in the previous example (Section 3.27) is not complete.

Exercise. Determine which of the following algebras are complete.

   1. $\mathcal{F} = \mathbb{R}[x]$.

2. The algebra of all smooth bounded functions.

3. The algebra of all smooth periodic functions (of period 1) on $\mathbb{R}$ (see Section 3.18).

3.29. We now return to the general situation in which A is a subset of the dual space $|\mathcal{F}|$ of a geometrical $\mathbb{R}$-algebra $\mathcal{F}$. It is natural to ask the following question: Can the set $\left|\mathcal{F}\big|_A\right|$ be identified with $A \subset \mathcal{F}$?

Proposition. *Suppose $\mathcal{F}$ is a geometrical $\mathbb{R}$-algebra and $A \subset |\mathcal{F}|$. Then the map*

$$\mu\colon A \to \left|\mathcal{F}\big|_A\right|, \quad (\mu(a))\,(f) = f(a),$$

is a homeomorphism onto a subset of the space $\left|\mathcal{F}\big|_A\right|$.

◀ Since all the elements of $\mathcal{F}$, understood as functions on the dual space $|\mathcal{F}|$, are continuous, $\mathcal{F}\big|_A$ is a subalgebra of the algebra of continuous functions on A, and therefore μ is continuous by Proposition 3.14. Further, μ is injective: If a_1 and a_2 are distinct points of A, there is a function $f_0 \in \mathcal{F}$ taking different values at these points; but then $f_0\big|_A$, which belongs to the algebra $\mathcal{F}\big|_A$, has the same values as f_0 at a_1 and a_2, and hence these points determine different homomorphisms $\mathcal{F}\big|_A \to \mathbb{R}$.

To prove that the inverse map $\mu^{-1}\colon \mu(A) \to A$ is continuous, consider a basis open set in A of the form $A \cap f^{-1}(V)$, where $f \in \mathcal{F}$ and $V \subset \mathbb{R}$ is open. It is mapped onto the set $\mu(A) \cap (f\big|_A)^{-1}(V)$, i.e., onto an open subset of $\mu(A)$. ▶

This proposition immediately implies that

$$A \subset B \subset |\mathcal{F}| \Rightarrow \left(\mathcal{F}\big|_B\right)\big|_A = \mathcal{F}\big|_A.$$

Should $\mu(A)$ coincide with $|\mathcal{F}\big|_A|$, we would have a positive answer to the question put at the beginning of this section. Proposition 3.24 implies that this is true for algebras $\mathcal{F} = C^\infty(U)$, where $U \subset \mathbb{R}^n$ is open, when $A \subset |\mathcal{F}| = U$ is also open. However, this is false in the general case.

3.30. Example. Suppose $\mathcal{F} = \mathbb{R}[x]$. Let $A = \mathbb{R}_+ \subset \mathbb{R}$ be the positive reals. Then the restriction homomorphism $\rho_A\colon \mathbb{R}[x] \to \mathbb{R}[x]\big|_{\mathbb{R}_+}$ is an isomorphism, since any nth degree polynomial is determined by its values at $(n+1)$ points, hence by its values on $\mathbb{R}_+$. On the other hand, the map $\mu\colon \mathbb{R}_+ \to \left|\mathbb{R}[x]\big|_{\mathbb{R}_+}\right|$ is the inclusion of $\mathbb{R}_+$ into $\mathbb{R} = \left|\mathbb{R}[x]\big|_{\mathbb{R}_+}\right| = |\mathbb{R}[x]|$, so that here $A = \mathbb{R}_+$ cannot be identified with $\left|\mathcal{F}\big|_A\right| = \mathbb{R}$.

3.31 Exercise. Describe an $\mathbb{R}$-algebra whose dual space is the configuration space of one of the given hinge mechanisms (see Section 1.14).

3.32. In order to avoid situations like the one in Example 3.30, we need a condition that would guarantee the bijectivity of the map

$$\mu\colon A \to \left|\mathcal{F}\right|_A\right|, \quad a \mapsto (f \mapsto f(a)).$$

Definition. A geometrical $\mathbb{R}$-algebra $\mathcal{F}$ is said to be *closed with respect to smooth composition, or C^∞-closed*, if for any finite collection of its elements $f_1, \ldots, f_k \in \mathcal{F}$ and any function $g \in C^\infty\left(\mathbb{R}^k\right)$ there exists an element $f \in \mathcal{F}$ such that

$$f(a) = g(f_1(a), \ldots, f_k(a)) \quad \text{for all } a \in |\mathcal{F}|. \tag{3.4}$$

Note that the function $f \in \mathcal{F}$ appearing in this definition is uniquely determined (since $\mathcal{F}$ is geometrical).

For the case in which $\mathcal{F}_S$ is the algebra of functions determined by measuring devices of a physical system S, the algebra $\mathcal{F}_S$ is always C^∞-closed. This is because the composite function (3.4) may be constructed by means of a device synthesizing the function $g(f_1, \ldots, f_k)$; see Section 1.10.

Exercise. Determine which of the algebras listed in Exercise 3.28 are closed.

3.33. We shall now show that *the map $\mu\colon A \to \left|\mathcal{F}\right|_A\right|$ (see Section 3.29) is surjective (and therefore a homeomorphism) for C^∞-closed algebras $\mathcal{F}$* in the case of any basis open set A:

$$A = \{a \in |\mathcal{F}| \mid \alpha < h(a) < \beta\}, \quad \alpha, \beta \in \mathbb{R}, \quad h \in \mathcal{F}.$$

(We shall not require this fact in more general form in what follows.)

◄ By Corollary 2.3, there exists a function $g \in C^\infty(\mathbb{R})$ such that $g \equiv 0$ on $\mathbb{R} \setminus \left]\alpha, \beta\right[$ and $g > 0$ on $\left]\alpha, \beta\right[$. Since $\mathcal{F}$ is C^∞-closed, there is a function $f \in \mathcal{F}$ such that $f(a) = g(h(a))$ for all points $a \in |\mathcal{F}|$. Then $f(a) > 0$ whenever $a \in A$, so that $f\big|_A$ is an invertible element of the algebra $\mathcal{F}\big|_A$.

Further, suppose $b' \in |\mathcal{F}|$ is the image of some point $b \in \left|\mathcal{F}\right|_A\right|$ under the natural map $\left|\mathcal{F}\right|_A\right| \to |\mathcal{F}|$. If $b' \notin A$, then

$$0 = f(b') = \left(f\big|_A\right)(b),$$

which contradicts the fact that $f\big|_A$ is invertible. Thus

$$\mu(A) = \left|\mathcal{F}\right|_A\right|. \quad ►$$

3.34. Having in mind the results of Section 3.33, we would like to modify a given geometrical $\mathbb{R}$-algebra $\mathcal{F}$ so as to obtain a C^∞-closed algebra $\overline{\mathcal{F}}$. The most direct way to do that is the following. Identifying $\mathcal{F}$ with the corresponding algebra of functions on $|\mathcal{F}|$, consider the set $\overline{\mathcal{F}}$ of functions on $|\mathcal{F}|$ that can be represented in the form

$$g(f_1, \ldots, f_l), \quad \text{where} \quad l \in \mathbb{N}, \quad f_1, \ldots, f_l \in \mathcal{F}, \quad g \in C^\infty\left(\mathbb{R}^l\right).$$

The set $\overline{\mathcal{F}}$ has an obvious $\mathbb{R}$-algebra structure, and $\mathcal{F}$ is a subalgebra of $\overline{\mathcal{F}}$. Denote the natural inclusion $\mathcal{F} \subset \overline{\mathcal{F}}$ by $i_{\mathcal{F}}$. Since the composition of smooth functions is smooth, the algebra $\overline{\mathcal{F}}$ is C^∞-closed. It is also geometrical, being the algebra of certain functions on a set (see Section 3.14). Thus we have constructed a natural inclusion map

$$i_{\mathcal{F}} \colon \mathcal{F} \hookrightarrow \overline{\mathcal{F}}$$

for any geometrical $\mathbb{R}$-algebra $\mathcal{F}$ into a C^∞-closed $\mathbb{R}$-algebra $\overline{\mathcal{F}}$, which we (temporarily) call the C^∞-closure of $\mathcal{F}$. This algebra possesses the following remarkable property.

3.35 Proposition. *Any homomorphism $\alpha \colon \mathcal{F} \to \mathcal{F}'$ of a geometrical $\mathbb{R}$-algebra $\mathcal{F}$ into a C^∞-closed $\mathbb{R}$-algebra $\mathcal{F}'$ can be uniquely extended to a homomorphism $\overline{\alpha} \colon \overline{\mathcal{F}} \to \mathcal{F}'$ of its C^∞-closure $\overline{\mathcal{F}}$.*

◀ Assume that the required extension $\overline{\alpha}$ exists, i.e., that $\alpha = \overline{\alpha} \circ i_{\mathcal{F}}$, where $i_{\mathcal{F}} \colon \mathcal{F} \hookrightarrow \overline{\mathcal{F}}$ is the natural inclusion. Then, by Section 3.20, we have $|\alpha| = |i_{\mathcal{F}}| \circ |\overline{\alpha}|$. Here $|\alpha|$ denotes the dual map (see 3.19). Further, for any point $a \in |\mathcal{F}'|$,

$$\begin{aligned}
\overline{\alpha}(g(f_1, \ldots, f_l))(a) &= g(f_1, \ldots, f_l)(|\overline{\alpha}|(a)) = g(i_{\mathcal{F}}(f_1), \ldots, i_{\mathcal{F}}(f_l))(|\overline{\alpha}|(a)) \\
&= g(f_1, \ldots, f_l)(|i_{\mathcal{F}}|(|\overline{\alpha}|(a))) = g(f_1, \ldots, f_l)(|\alpha|(a)) \\
&= g(f_1(|\alpha|(a)), \ldots, f_l(|\alpha|(a))) = g(\alpha(f_1), \ldots, \alpha(f_l))(a).
\end{aligned}$$

Since $\mathcal{F}'$ is geometrical, this implies

$$\overline{\alpha}(g(f_1, \ldots, f_l)) = g(\alpha(f_1), \ldots, \alpha(f_l)).$$

If $\overline{\alpha}$ exists, this last formula proves its uniqueness. To prove existence, we can use this formula as the definition of $\overline{\alpha}$, if we establish that the right-hand side is well defined, i.e., if we show that $g(f_1, \ldots, f_l) = g'(f_1', \ldots, f_{l'}')$ implies

$$g(\alpha(f_1), \ldots, \alpha(f_l)) = g'\left(\alpha\left(f_1'\right), \ldots, \alpha\left(f_{l'}'\right)\right).$$

Since $\mathcal{F}'$ is geometrical, it suffices to prove this at an arbitrary point $a' \in \mathcal{F}'$. But

$$\begin{aligned}
g(\alpha(f_1), \ldots, \alpha(f_l))(a') &= g\left(\alpha(f_1)\left(a'\right), \ldots, \alpha(f_l)\left(a'\right)\right) \\
&= g(f_1(a), \ldots, f_l(a)) = g(f_1, \ldots, f_l)(a),
\end{aligned}$$

where $a = |\alpha|(a')$. Similarly,

$$g'\left(\alpha\left(f_1'\right), \ldots, \alpha\left(f_{l'}'\right)\right)\left(a'\right) = g'\left(f_1', \ldots, f_{l'}'\right)(a).$$

Comparing the last two formulas, we see that $\overline{\alpha}$ is well defined, concluding our proof. ▶

3.36. It is remarkable that Proposition 3.35 entirely characterizes the C^∞-closure $\overline{\mathcal{F}}$ of a geometrical $\mathbb{R}$-algebra $\mathcal{F}$. To explain this in adequate terms, we need the following definition.

Definition. A C^∞-closed geometrical $\mathbb{R}$-algebra $\overline{\mathcal{F}}$ together with a homomorphism $i\colon \mathcal{F} \to \overline{\mathcal{F}}$ is called the *smooth envelope* of the $\mathbb{R}$-algebra $\mathcal{F}$ if for any homomorphism $\alpha\colon \mathcal{F} \to \mathcal{F}'$ of $\mathcal{F}$ into a geometrical C^∞-closed $\mathbb{R}$-algebra $\mathcal{F}'$ there exists a unique homomorphism $\overline{\alpha}\colon \overline{\mathcal{F}} \to \mathcal{F}'$ extending α (i.e., such that $\alpha = \overline{\alpha} \circ i$). In other words, under the above assumptions, the diagram

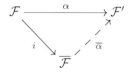

can always be uniquely completed (by the dotted arrow $\overline{\alpha}$) to a commutative one.

It now follows from Proposition 3.35 that the C^∞-closure (see 3.34) is a smooth envelope of $\mathcal{F}$.

3.37 Proposition. *The smooth envelope of any $\mathbb{R}$-algebra $\mathcal{F}$ is unique up to isomorphism. More precisely, if the pairs, $\big(i_k, \overline{\mathcal{F}}_k\big)$, $k = 1, 2$, are smooth envelopes of $\mathcal{F}$, there exists a unique isomorphism $j\colon \overline{\mathcal{F}}_1 \to \overline{\mathcal{F}}_2$ such that $i_2 = j \circ i_1$. In other words, the following diagram commutes:*

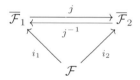

◀ First we note that for any given smooth envelope $(i, \overline{\mathcal{F}})$ of $\mathcal{F}$ any homomorphism $\alpha\colon \overline{\mathcal{F}} \to \overline{\mathcal{F}}$ satisfying $\alpha \circ i = i$ is the identity, $\alpha = \mathrm{id}_{\overline{\mathcal{F}}}$. (Indeed, by Definition 3.36, the "solution" of the "equation" $\alpha \circ i = i$ is unique, but this equation has the obvious solution $\mathrm{id}_{\overline{\mathcal{F}}}$.)

Further, according to the same definition for $(i_1, \mathcal{F}_1)$, the homomorphism $i_2\colon \mathcal{F} \to \overline{\mathcal{F}}_2$ can be uniquely represented in the form $i_2 = j_1 \circ i_1$, where $j_1\colon \overline{\mathcal{F}}_1 \to \overline{\mathcal{F}}_2$ is a homomorphism. Similarly, $i_1 = j_2 \circ i_2$, where $j_2\colon \overline{\mathcal{F}}_2 \to \overline{\mathcal{F}}_1$ is a homomorphism. Hence

$$i_2 = j_1 \circ i_1 = j_1 \circ (j_2 \circ i_2) = (j_1 \circ j_2) \circ i_2.$$

By the remark at the beginning of the proof, this implies $j_1 \circ j_2 = \mathrm{id}_{\overline{\mathcal{F}}_2}$. Similarly, $j_2 \circ j_1 = \mathrm{id}_{\overline{\mathcal{F}}_1}$. Thus j_1 and j_2 are isomorphisms inverse to each other, and we can put $j = j_1$ to establish the proposition. (The uniqueness of the isomorphism j follows from the definition of smooth envelopes.) ▶

A direct consequence of this proposition is that the temporary term "C^∞-closure" (see 3.34) is characterized by its universal property expressed in Proposition 3.35 and therefore coincides with the term "smooth envelope." It is the latter term that will be used from now on.

3.38. In accordance with its definition (see Section 3.36), the smooth envelope $\overline{\mathcal{F}}$ of a geometrical algebra $\mathcal{F}$ is an object that plays a universal role in its interactions (i.e., isomorphisms) with the "world" of C^∞-closed geometrical algebras. We can say that the smooth envelope is the "ambassador plenipotentiary" of the algebra $\mathcal{F}$ in this "world," and that $\mathcal{F}$ interacts with the latter exclusively via this ambassador.

The reader may have noticed that the arguments used in Section 3.37 are very general in nature. The art of finding and using such arguments is one of the main facets of *category theory*, familiarly known as *abstract nonsense*. We feel that one has to get used to it before learning it mathematically, so we shall not develop the theory, but use many of its standard arguments and tricks (e.g., see Sections 6.4, 6.6, 6.16, 6.17).

In the set-theoretic approach to mathematics, one studies the inner nature of mathematical objects, i.e., point sets supplied with certain structures. It is a biology of species. On the other hand, the categorical approach is a kind of sociology: one is no longer interested in the properties of individual objects, but in their relationships (called "morphisms" in the theory) with other objects of the same or similar type. One can also say that the categorical approach is similar to the experimental method in the natural sciences, when objects are not studied per se, but are analyzed in terms of their interaction with other objects.

3.39 Exercises. Find the smooth envelope of

(1) the algebra $\mathbb{R}[x_1, \dots, x_n]$;

(2) the algebra of functions on the line $\mathbb{R}$ of the form

$$f(x) = \sum_{k=0}^{n} a_k(x)|x|^k, \quad x \in \mathbb{R},$$

where $a_i(x) \in C^\infty(\mathbb{R})$.

4

Smooth Manifolds (Algebraic Definition)

4.1. A complete (Section 3.27) geometrical (Section 3.7) $\mathbb{R}$-algebra $\mathcal{F}$ is called *smooth* if there exists a finite or countable open covering $\{U_k\}$ of the dual space $|\mathcal{F}|$ such that all the algebras $\mathcal{F}|_{U_k}$ (Section 3.23) are isomorphic to the algebra $C^\infty\,(\mathbb{R}^n)$ of smooth functions in Euclidean space. The (fixed positive) integer n is said to be the *dimension* of the algebra $\mathcal{F}$.

Smooth n-dimensional algebras are our main object of study; they can be viewed as $\mathbb{R}$-algebras of smooth functions on n-dimensional smooth manifolds. From the viewpoint of formal mathematics, the $\mathbb{R}$-algebra $\mathcal{F}$ entirely determines the corresponding manifold M as the dual space $M = |\mathcal{F}|$ of its $\mathbb{R}$-points (Section 3.8) and is most convenient to work with: all of differential mathematics applies neatly to $\mathcal{F}$, so that the space M is not formally required. Nevertheless, in order to be able to visualize M as a geometrical object, we must learn to work simultaneously with the smooth algebra $\mathcal{F}$ and the space $M = |\mathcal{F}|$ of its $\mathbb{R}$-points.

Learning this will be the main goal of the present chapter.

Considering $\mathcal{F}$ and $M = |\mathcal{F}|$ simultaneously, we say that we are dealing with a *smooth manifold*. Although the second object in this pair is determined by the first, we make a concession to our geometrical intuition (and to traditional terminology) and say that $\mathcal{F}$ is the *algebra of smooth functions on the manifold* $M\,(=|\mathcal{F}|)$.

4.2. A somewhat more general concept is that of a *smooth algebra with boundary*. In this case, for each element U_k of the covering $\{U_k\}$ we require the algebra $\mathcal{F}|_{U_k}$ to be isomorphic either to $C^\infty\,(\mathbb{R}^n)$ or to $C^\infty(\mathbb{R}^n_H)$, where

$$\mathbb{R}^n_H = \{(r_1, \ldots, r_2) \in \mathbb{R}^n \mid r_1 \geqslant 0\},$$

© Springer Nature Switzerland AG 2020
J. Nestruev, *Smooth Manifolds and Observables*, Graduate Texts
in Mathematics 220, https://doi.org/10.1007/978-3-030-45650-4_4

and $C^\infty(\mathbb{R}^n_H)$ consists of the restrictions (in the usual sense) of all functions from $C^\infty(\mathbb{R}^n)$ to the set $\mathbb{R}^n_H$.

Exercise. Prove that the algebras $C^\infty(\mathbb{R}^n)$ and $C^\infty(\mathbb{R}^n_H)$ are not isomorphic.

The points of the space $|\mathcal{F}|$ that correspond to the boundary of the half-space $\mathbb{R}^n_H$ in the identification $U_k = \mathbb{R}^n_H$ are called *boundary points*.

Exercise. Prove that the set of boundary points $\partial|\mathcal{F}|$ has a natural structure of a smooth manifold (without boundary).

As above, emphasizing the geometrical viewpoint on this concept, we shall say that $\mathcal{F}$ is *the algebra of smooth functions on the smooth manifold* $M\ (= |\mathcal{F}|)$ *with boundary.*

4.3 Lemma. *If a geometrical $\mathbb{R}$-algebra $\mathcal{F}$ is isomorphic to $C^\infty(\mathbb{R}^n)$ or $C^\infty(\mathbb{R}^n_H)$, then it is C^∞-closed (see Section 3.32).*

◄ The lemma immediately follows from the following stronger statement. If $i\colon \mathcal{F}_1 \to \mathcal{F}_2$ is an isomorphism of geometrical $\mathbb{R}$-algebras, and $\mathcal{F}_1$ is C^∞-closed, then so is $\mathcal{F}_2$.

The verification of this statement is quite similar to the uniqueness proof of $\overline{\alpha}$ in Section 3.35, and the reader should have no difficulty in carrying it out.   ►

4.4 Proposition. *Smooth algebras are C^∞-closed. (The same is true for smooth algebras with boundary.)*

◄ Let $\mathcal{F}$ be a smooth $\mathbb{R}$-algebra (possibly with boundary), let $l \in \mathbb{N}$, $g \in C^\infty(\mathbb{R}^l)$, $f_1, \ldots, f_l \in \mathcal{F}$, and let $\{U_k\}$ be the covering that appears in Definition 4.1 (or 4.2).

Consider the function

$$h\colon |\mathcal{F}| \to \mathbb{R}, \quad h(a) = g(f_1(a), \ldots, f_l(a)).$$

By Lemma 4.3, for any k there exists an $h_k \in \mathcal{F}|_{U_k}$ such that

$$\forall a \in U_k, \quad h_k = g(f_1(a), \ldots, f_l(a)).$$

Thus in a neighborhood of each point the function coincides with a function from $\mathcal{F}$.

Since by 4.1 (or 4.2) $\mathcal{F}$ is complete (Section 3.28), it follows that $h \in \mathcal{F}$. ►

4.5. Example. Suppose $\mathcal{F}$ is the algebra of smooth periodic functions on the line $\mathbb{R}$ of period 1:

$$\mathcal{F} = \{f \in C^\infty(\mathbb{R}) \mid f(r+1) = f(r), \quad \forall r \in \mathbb{R}\}.$$

Being a subalgebra of the geometrical algebra $C^\infty(\mathbb{R})$, the algebra $\mathcal{F}$ is itself geometrical (see Section 3.19). It is not difficult to prove that $\mathcal{F}$ is complete. (In Section 4.22 we shall present a general argument implying the completeness of all the algebras considered in Examples 4.5–4.8.)

It was shown in 3.18 that the space $|\mathcal{F}|$ is the circle S^1. Now consider the functions g_1, $g_2 \in \mathcal{F}$,

$$g_1(r) = \sin^2 \pi r, \quad g_2(r) = \cos^2 \pi r,$$

and the open covering of the circle $|\mathcal{F}|$ by the sets

$$U_i = \{r \in |\mathcal{F}| \mid g_i(x) \neq 0\}, \quad i = 1, 2.$$

It is easy to establish bijections $U_i \leftrightarrow \,]0,1[$ that correspond to the isomorphisms

$$\mathcal{F}|_{U_i} \cong C^\infty(\,]0,1[\,) \quad (\cong C^\infty(\mathbb{R})).$$

Thus the algebra $\mathcal{F}$ is a smooth algebra of dimension 1, and the manifold it determines is the *circle* $S^1 = |\mathcal{F}|$.

In a similar way one establishes that the $\mathbb{R}$-algebra

$$\mathcal{F} = \left\{ f \in C^\infty(\mathbb{R})^2 \mid f(r_1 + 1, r_2) = f(r_1, r_2) \quad \forall (r_1, r_2) \in \mathbb{R}^2 \right\}$$

is a smooth algebra of dimension 2. In this case the space $M = |\mathcal{F}|$ is homeomorphic to the *cylinder*.

4.6 Exercise. Carefully review the previous example and find the mistake in the following argument: since a 1-periodic smooth function takes the same finite value at the end points of the closed interval $[0,1]$, while the algebra $C^\infty(\,]0,1[\,)$ contains unbounded functions as well as functions that have different limits at the points 0 and 1, the algebra $\mathcal{F}|_{]0,1[}$ (where $\mathcal{F}$ is the algebra of 1-periodic smooth functions on $\mathbb{R}$) cannot be isomorphic to $C^\infty(\,]0,1[\,)$.

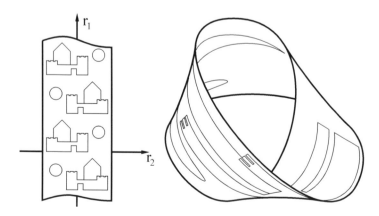

Figure 4.1. The Möbius band.

4.7. Examples. I. $\mathcal{F} = \{f \in C^\infty\left(\mathbb{R}^2\right) \mid f(r_1, r_2) = f(r_1 + 1, -r_2)$ for all $(r_1, r_2) \in \mathbb{R}^2\}$.

The space $|\mathcal{F}|$ is called the *open Möbius band*.

II. $\mathcal{F} = \{f \in C^\infty(\Pi) \mid f(r_1, r_2) = f(r_1 + 1, -r_2) \ \forall(r_1, r_2) \in \Pi\}$, where Π is the strip $\{(r_1, r_2) \in \mathbb{R}^r \mid -1 \leqslant r_2 \leqslant 1\}$. This is a smooth algebra with boundary, and the space $|\mathcal{F}|$ is known as the (*closed*) *Möbius band* (see Figure 4.1).

III. $\mathcal{F} = \{f \in C^\infty\left(\mathbb{R}^2\right) \mid f(r_1, r_2) = f(r_1 + 1, -r_2) = f(r_1, r_2 + 1)$ $\forall(r_1, r_2) \in \mathbb{R}^2\}$.

Using the functions

$$g_1(r_1, r_2) = \sin^2 \pi r_1, \qquad\qquad h_1(r_1, r_2) = \sin^2 \pi r_2,$$
$$g_2(r_1, r_2) = \cos^2 \pi r_1, \qquad\qquad h_2(r_1, r_2) = \cos^2 \pi r_2,$$

we can cover the space $|\mathcal{F}|$ by the four open sets

$$U_{ik} = \{(r_1, r_2) \in |\mathcal{F}| \mid g_i(r_1, r_2) \neq 0, \ h_k(r_1, r_2) \neq 0\}, \ i, k = 1, 2.$$

For each of these sets one can immediately construct a homeomorphism on the open square corresponding to the isomorphism of the $\mathbb{R}$-algebra $\mathcal{F}\big|_{U_{ik}}$ on the algebra of smooth functions on the open square. Therefore, $\mathcal{F}\big|_{U_{ik}} \cong C^\infty\left(\mathbb{R}^2\right)$, and $\mathcal{F}$ is a smooth $\mathbb{R}$-algebra of dimension 2. The space $|\mathcal{F}|$ is known as the *Klein bottle*.

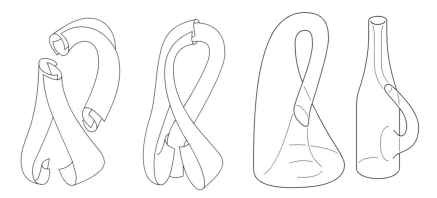

Figure 4.2. The Klein bottle.

It is useful to visualize how the squares U_{ik} are "glued together" when they are embedded into $|\mathcal{F}|$. The beginning of this process is pictured on the left-hand side of Figure 4.2. What happens if the process is continued in 3-space is shown on the right-hand side of the same figure (the little circular self-intersection does not really occur in the Klein bottle; it is due to the fact that the latter does not fit into 3-space).

4.8 Exercise. Prove that the $\mathbb{R}$-algebra

$$\mathcal{F} = \{f \in C^\infty\left(\mathbb{R}^2\right) \mid f(r_1, r_2) = f(r_1 + 1, -r_2) = f(-r_1, r_2 + 1),$$
$$(r_1, r_2) \in \mathbb{R}^2\}$$

is smooth. Find as many geometrical descriptions of the topological space $|\mathcal{F}|$ as you can. Prove, for example, that $|\mathcal{F}|$ is homeomorphic to the space whose "points" are the straight lines of $\mathbb{R}^3$ passing through the origin $(0, 0, 0)$; see Section 1.3.

4.9. Suppose $\mathcal{F}$ is the algebra of smooth functions on a manifold $M = |\mathcal{F}|$ with boundary. Recalling Definition 4.2, we say that a point $p \in M$ is a *boundary point* of M if it corresponds to a boundary point of $\mathbb{R}^n_H$ in the identification $U_k = \mathbb{R}^n_H$. The set of all boundary points is denoted by ∂M and called the *boundary* of M.

Exercise.   1. Prove that *if the boundary ∂M of the n-dimensional manifold $M = |\mathcal{F}|$ is nonempty, then $\mathcal{F}|_{\partial M}$ is a smooth algebra of dimension $(n-1)$ and ∂M has no boundary points* (see the second exercise in Section 4.2).

   2. Check algebraically that the manifold $\partial|\mathcal{F}|$ in Example 4.7, II, can be identified with the circle (cf. Sections 4.5 and 6.9).

4.10. Remark. It is far from obvious that the dimension of a smooth algebra is well defined, i.e., that it does not depend on the choice of the covering $\{U_k\}$ and of the isomorphisms $\mathcal{F}|_{U_k} \cong C^\infty\left(\mathbb{R}^n\right)$ (see 4.1). This almost immediately follows from the fact that *the algebras $C^\infty\left(\mathbb{R}^n\right)$ and $C^\infty\left(\mathbb{R}^m\right)$ are not isomorphic if $n \neq m$.*

The reader who has industriously worked his way through the previous examples undoubtedly feels that this is true. A more experienced reader will probably have no trouble in proving this fact by using "Sard's theorem on singular points of smooth maps" (advanced calculus). As for us, we shall prove this result in Chapter 9. Until then, our skeptical readers may consider dimension to be an invariant of the covering $\{U_k\}$ rather than that of the algebra $\mathcal{F}$ itself.

4.11. Definition. Suppose $\mathcal{F}$ is the algebra of smooth functions on the manifold M and $N \subset M = |\mathcal{F}|$ is a subset. If the algebra $\mathcal{F}_N = \mathcal{F}|_N$ is a smooth $\mathbb{R}$-algebra, then we say that N is a *smooth submanifold* of the smooth manifold M and that $\mathcal{F}_N$ is the *algebra of smooth functions on the submanifold N.*

If the restriction homomorphism $i \colon \mathcal{F} \to \mathcal{F}_N$ is surjective, the smooth submanifold $N \subset M = |\mathcal{F}|$ is called *closed.*

4.12. Let N be a closed submanifold of M. Were we being consistent in 4.11 when we spoke of $\mathcal{F}_N$ as "the algebra of smooth functions on the manifold N"? The answer to that question is given by the following result.

Proposition. *Suppose $\mathcal{F}$ is the algebra of smooth functions on the manifold M and $N \subset M = |\mathcal{F}|$ is a closed smooth submanifold. Then*

(i) *N is closed as a subset of the topological space M;*

(ii) *$N = |\mathcal{F}_N|$.*

◀ (i) Let $a \in M \smallsetminus N$ be a limit point of N, and $U \subset M$ a neighborhood of a such that $\mathcal{F}\big|_U \cong C^\infty(\mathbb{R}^n)$. This isomorphism may be chosen so that the elements of $\mathcal{F}\big|_U$ corresponding to the coordinate functions $r_1, \ldots, r_n \in C^\infty(\mathbb{R}^n)$ vanish at the point a. Consider the function on $\mathcal{F} \smallsetminus a$ corresponding to $1/\left(r_1^2 + \cdots + r_n^2\right)$. This function may be extended from the punctured neighborhood $U \smallsetminus a$ of a to a smooth function g on the submanifold $M \smallsetminus a$.

It is obvious that the restriction $g\big|_N$ belongs to the algebra $\mathcal{F}\big|_N = \mathcal{F}_N$ but it does not belong to the image of the restriction homomorphism $i \colon \mathcal{F} \to \mathcal{F}\big|_N$. This contradiction proves that $a \in N$.

(ii) Consider the $\mathbb{R}$-point $b \colon \mathcal{F}_N \to \mathbb{R}$ and take the composition $c = b \circ i \colon \mathcal{F} \to \mathbb{R}$. Assume that $c \notin N$. Generalizing Proposition 2.5, let us construct a function $f \in \mathcal{F}$ such that $f\big|_N \equiv 0$, $f(c) \neq 0$. But then $i(f) = 0$ and $f(c) = 0$: a contradiction. Therefore, the map $b \mapsto c = b \circ i$ is a surjection of $|\mathcal{F}_N|$ onto N. Together with Proposition 3.29 this gives the result. ▶

4.13. Examples. In $\mathbb{R}^2$ consider the set of points S^1 given by the equation

$$r_1^2 + r_2^2 - 1 = 0.$$

Let us check that the $\mathbb{R}$-algebra $\mathcal{F}_{S^1} = C^\infty(\mathbb{R}^2)\big|_{S^1}$ is isomorphic to the algebra of smooth periodic functions of period 1 on the line $\mathbb{R}$ (see Example 4.5).

◀ First, note that $\mathcal{F}_{S^1} = C^\infty(\mathbb{R}^2)\big|_{S^1} = C^\infty(\mathbb{R}^2 \setminus \{0\})|_{S^1}$ and consider the map

$$w \colon \mathbb{R} \to S^1 \subset \mathbb{R}^2, \quad r \mapsto (\cos 2\pi r, \sin 2\pi r).$$

Clearly, the corresponding $\mathbb{R}$-algebra homomorphism

$$|w| \colon \mathcal{F}_{S^1} \to C^\infty(\mathbb{R}),$$
$$|w|(f)(r) = f(w(r)) = f(\cos 2\pi r, \sin 2\pi r), \quad r \in \mathbb{R},$$

is injective and its image is contained in the subalgebra $C_{\mathrm{per}}^\infty(\mathbb{R})$ of smooth 1-periodic functions on $\mathbb{R}$. To prove that $|w|$ is surjective, consider the homomorphism $\iota \colon C_{\mathrm{per}}^\infty(\mathbb{R}) \to C^\infty(\mathbb{R}^2 \setminus \{0\})$ defined by

$$\iota(f)(r_1, r_2) = f\left(\frac{\arg z}{2\pi}\right), \quad z = r_1 + i r_2.$$

Obviously, $|w|(\iota(f)|_{S^1}) = f$. ▶

Exercise.   1. Show that any odd 2π-periodic smooth function is of the form $g(x) \sin x$, where $g(x)$ is an even 2π-periodic smooth function.

2. Is it true that any even 2π-periodic smooth function can be written as $f(\cos x)$ for $f \in C^\infty(\mathbb{R})$?

4.14. Recall that the restriction of elements of an $\mathbb{R}$-algebra $\mathcal{F}$ to a subset $N \subset |\mathcal{F}|$ was not defined in algebraic terms, but in geometrical ones (see Section 3.23). However, if $N \subset |\mathcal{F}|$ is a closed smooth submanifold, then there is a purely algebraic way to find the algebra $\mathcal{F}_N = \mathcal{F}\big|_N$.

Namely, suppose $A_N \subset \mathcal{F}$ is the set of elements of $\mathcal{F}$ that vanish on N, i.e.,

$$A_N = \{f \in \mathcal{F} \mid \forall a \in N, \ f(a) = 0\}.$$

This is obviously an ideal of the algebra $\mathcal{F}$, so we can consider the quotient algebra $\mathcal{F}/A_N$. There exists an obvious identification of $\mathcal{F}/A_N$ with the algebra $\mathcal{F}_N = \mathcal{F}\big|_N$ (see the proof of Proposition 4.12) for which the quotient map $\varphi \colon \mathcal{F} \to \mathcal{F}\big|_{A_N}$ becomes the restriction homomorphism $p \colon \mathcal{F} \to \mathcal{F}_N$:

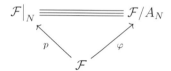

4.15. Example. If S^1 is the circle from Example 4.13, then A_{S^1} is the principal ideal in $C^\infty(\mathbb{R}^2)$ generated by the function $r_1^2 + r_2^2 - 1$.
◀ Let $f \in A_{S^1}$. Let us prove that

$$f(r_1, r_2) = g(r_1, r_2) \cdot (r_1^2 + r_2^2 - 1)$$

for a suitable function $g \in C^\infty(\mathbb{R}^2)$. Since the algebra $C^\infty(\mathbb{R}^2)$ is complete, if suffices to construct g in a neighborhood of S^1, say, in $\mathbb{R}^2 \backslash \{0\}$. To this end, introduce the following auxiliary functions:

$$u(t, r_1, r_2) = t + \frac{1 - t}{\sqrt{r_1^2 + r_2^2}},$$

$$h(t, r_1, r_2) = f\left(r_1 \cdot u(t, r_1, r_2), \ r_2 \cdot u(t, r_1, r_2)\right).$$

Then, taking into account the fact that

$$\frac{\partial u}{\partial t} = 1 - \frac{1}{\sqrt{r_1^2 + r_2^2}} = \frac{r_1^2 + r_2^2 - 1}{r_1^2 + r_2^2 + \sqrt{r_1^2 + r_2^2}},$$

$$h(0, r_1, r_2) = 0 \quad (\text{since } f \in A_{S^1}),$$

$$h(1, r_1, r_2) = f(r_1, r_2),$$

we obtain

$$f(r_1, r_2) = \int\limits_0^1 \frac{\partial h}{\partial t}(t, r_1, r_2)dt$$

$$= \frac{\int\limits_0^1 \left(r_1 \frac{\partial f}{\partial r_1}(r_1 u, r_2 u) + r_2 \frac{\partial f}{\partial r_2}(r_1 u, r_2 u) \right) dt}{r_1^2 + r_2^2 + \sqrt{r_1^2 + r_2^2}} \cdot \left(r_1^2 + r_2^2 - 1 \right).$$

The first factor in the last expression is the desired function $g(r_1, r_2)$. ▶

4.16 Exercise. Let $A = C^\infty (\mathbb{R}^2)$. Show that the algebra $A/ \left(y^2 - x^3 \right) A$ is not smooth.

A traditional purely algebraic approach to solving such problems is based on the following two facts:

1. If an algebra is smooth, then its localization at a maximum ideal is isomorphic to the algebra of germs of smooth functions on some $\mathbb{R}^k$ (at the origin), see Example III on page 156;

2. The formal completion of the above local algebra is isomorphic to the algebra of formal power series.

The task becomes much simpler if one uses certain elementary tools of differential calculus over commutative algebras (see Exercise 9.34).

4.17. Lemmas. It is not difficult to generalize the statements in Sections 2.3–2.7 from $\mathbb{R}^n$ to arbitrary manifolds. In particular, for any arbitrary algebra $\mathcal{F}$ of smooth functions on a manifold M, the following statements hold:

(i) *For any open set $U \subset M$ there exists a function $f \in \mathcal{F}$ such that*

$$\begin{cases} f(x) > 0 & for\ all\ x \in U, \\ f(x) = 0 & if\ x \notin U. \end{cases}$$

(ii) *For any two nonintersecting closed subsets $A, B \subset M$ there exists a function $f \in \mathcal{F}$ such that*

$$\begin{cases} f(x) = 0 & for\ all\ x \in A, \\ f(x) = 1 & for\ all\ x \in B, \\ 0 < f(x) < 1 & otherwise. \end{cases}$$

(The reader possibly established a weaker statement when working through Proposition 4.12.)

(iii) *There exists a function $f \in \mathcal{F}$ all of whose level surfaces are compact.*

4.18. In proving Lemma 4.17, the following statement, called the "partition of unity lemma," may be useful: *If $\{U_\alpha\}$ is a locally finite open covering of the space $M = |\mathcal{F}|$, then there exist functions $f_\alpha \in \mathcal{F}$ such that $f_\alpha(x) = 0$ if $x \in M \smallsetminus U_\alpha$ and*

$$\sum_\alpha f_\alpha(x) \equiv 1.$$

(A *locally finite covering* $\{U_\alpha\}$ is a covering such that for any $x \in M$ there exists a neighborhood $U \subset M$ of x that intersects only a finite number of sets U_α.)

We suggest that the reader try to prove this statement first in the particular case where the covering $\{U_\alpha\}$ is supplied with isomorphisms $\mathcal{F}|_{U_\alpha} \cong C^\infty(\mathbb{R}^n)$ (or with diffeomorphisms $U_\alpha \cong \mathbb{R}^n$).

4.19. Suppose $\mathcal{F}$ is the algebra of smooth functions on the manifold M. Consider the *action of a group* on this smooth manifold, i.e., a family Γ of automorphisms $\gamma \colon \mathcal{F} \to \mathcal{F}$ such that

(i) $\gamma_1, \gamma_2 \in \Gamma \Rightarrow \gamma_1 \circ \gamma_2 \in \Gamma$,

(ii) $\gamma \in \Gamma \Rightarrow \gamma^{-1} \in \Gamma$.

Suppose $\mathcal{F}^\Gamma \subset \mathcal{F}$ is the subalgebra of invariant functions for this action, i.e.,

$$\mathcal{F}^\Gamma = \{f \in \mathcal{F} \mid \gamma(f) = f \quad \text{for all } \gamma \in \Gamma\}.$$

4.20 Lemma. *Suppose $\mathcal{F}^\Gamma$ is the subalgebra of invariant functions of a group action of Γ on $M = |\mathcal{F}|$. If $\widetilde{\mathcal{F}^\Gamma}$ (see Section 3.4) contains a function all of whose level surfaces are compact, then the algebra $\mathcal{F}^\Gamma$ is geometrical.*

◀ The proof will be a repetition, word for word, of 3.16, if we note that the operations applied there preserve Γ-invariance.   ▶

In particular, if the set $|\mathcal{F}^\Gamma|$ is compact, then the algebra $\mathcal{F}^\Gamma$ is always geometrical.

4.21. In order to learn to visualize the algebra $\mathcal{F}^\Gamma$ of Γ-invariant functions on a manifold $|\mathcal{F}| = M$, let us consider the orbit $\mathcal{O}_a = \{|\gamma|(a) \mid \gamma \in \Gamma\}$ for each point $a \in M$. Denote the set of all orbits by N.

Elements of $\mathcal{F}^\Gamma$ can be understood as functions on N. Indeed, if $b = |\gamma|(a) \in \mathcal{O}_a$ and $f \in \mathcal{F}$ is a Γ-invariant function, then $f(b) = f(a)$. In other words, each "point" of the set N (i.e., each orbit) determines an $\mathbb{R}$-point of the algebra $\mathcal{F}^\Gamma$, so that we have the natural map

$$N \to |\mathcal{F}^\Gamma|, \quad \mathcal{O}_a \mapsto (f \mapsto f(a)).$$

This map will be bijective if the two following conditions hold:

(i) Any homomorphism $a \colon \mathcal{F}^\Gamma \to \mathbb{R}$ can be extended to a homomorphism $\tilde{a} \colon \mathcal{F} \to \mathbb{R}$ (surjectivity).

(ii) If $b \notin \mathcal{O}_a$, then there exists an $f \in \mathcal{F}^\Gamma$ such that $f(b) \neq f(a)$ (injectivity).

These two conditions are satisfied, for example, if the group Γ is finite. We leave their verification in the case of Examples 4.24 and 4.25 below as an exercise for the reader.

4.22 Proposition. *The algebra $\mathcal{F}^\Gamma$ of Γ-invariant functions on a smooth manifold $M = |\mathcal{F}|$ is complete if conditions* (i), (ii) *of 4.21 and the assumptions of* Lemma 4.20 *hold.*

◀ Each real-valued function $f \colon |\mathcal{F}^\Gamma| \to \mathbb{R}$ determines, by means of the projection $M \to N = |\mathcal{F}^\Gamma|$, $a \mapsto \mathcal{O}_a$, the function $\tilde{f} \colon M \to \mathbb{R}$. If f coincides in a neighborhood of each point $b \in \mathcal{O}_a \in |\mathcal{F}^\Gamma|$ with some function belonging to $\mathcal{F}^\Gamma$, then the Γ-invariant function $\tilde{f}$ coincides with some smooth function (from $\mathcal{F}$) in a neighborhood of each point $a \in |\mathcal{F}|$. Since $\mathcal{F}$ is complete, $\tilde{f} \in \mathcal{F}$, so that $f \in \mathcal{F}^\Gamma$. ▶

4.23. It is clear that the set of orbits N of a group action of Γ on the manifold $M = |\mathcal{F}|$ is the quotient set of M by an equivalence relation (the one identifying points within each orbit).

Definition. Assume that N coincides with $|\mathcal{F}^\Gamma|$ and the algebra $\mathcal{F}^\Gamma$ is smooth (or smooth with boundary); then we say that $\mathcal{F}^\Gamma$ is the algebra of smooth functions on the *quotient manifold of M by the group action of* Γ.

Following tradition, we shall often denote the quotient manifold by M/Γ, although this is sometimes a misleading notation (just as is the notation $|\mathcal{F}^\Gamma|$).

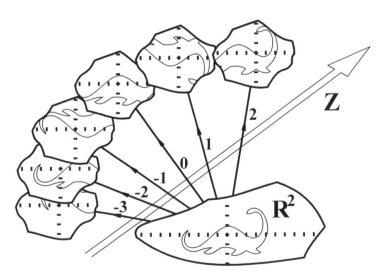

Figure 4.3. Group action of $\mathbb{Z}$ on $\mathbb{R}^2$ producing the Möbius band.

4.24. Examples. I. In Examples 4.5 and 4.7, I, we actually dealt with quotient manifolds of the line $\mathbb{R}$ and of the plane $\mathbb{R}^2$ by the discrete cyclic group $\mathbb{Z}$ of isometries (see Figure 4.3).

II. In Examples 4.7, II, and 4.8 we took the quotient manifolds with respect to the action of isometry groups with two generators. The reader may try to depict how the map $|i|: \mathbb{R}^2 \to |\mathcal{F}|$ winds the plane $\mathbb{R}^2$ about the Klein bottle (see Section 4.7, III).

III. Consider the action on $\mathcal{F} = C^\infty(\mathbb{R}^n)$ of the free Abelian group Γ with n generators $\gamma_1, \ldots, \gamma_n$, where γ_i is the parallel translation by the unit vector along the ith coordinate, i.e.,

$$\gamma_i(f)(r_1, \ldots, r_n) = f(r_1, \ldots, r_{i-1}, r_i + 1, r_{i+1}, \ldots, r_n)$$

for all $f \in \mathcal{F}$, $(r_1, \ldots, r_n) \in \mathbb{R}^n$. It is easy to see that $\mathcal{F}^\Gamma$ is the subalgebra of all functions in $C^\infty(\mathbb{R}^n)$ that are 1-periodic with respect to each variable. Generalizing the arguments carried out in Sections 4.5 and 4.7, the reader will easily check that the quotient of $\mathbb{R}^n$ with respect to this action of Γ will be a smooth manifold. This manifold is known as the *n-dimensional torus* and is denoted by T^n. (It is easy to see that $T^1 = S^1$; the most popular case, $n = 2$, was mentioned in the Introduction: It is the surface of the doughnut.)

4.25. Examples. I. Consider the $\mathbb{R}$-algebra $\mathcal{F} = C^\infty(\mathbb{R}^{n+1} \setminus \{0\})$. The multiplicative group $\mathbb{R}_+^{\cdot}$ of positive real numbers acts on $\mathcal{F}$ by the automorphism h_λ:

$$h_\lambda(f)(r_1, \ldots, r_{n+1}) = f(\lambda r_1, \ldots, \lambda r_{n+1})$$
$$\text{for all } f \in \mathcal{F}, \ (r_1, \ldots, r_{n+1}) \in \mathbb{R}^{n+1} \setminus \{0\}.$$

It turns out that the corresponding quotient set by this group action is a smooth manifold. Generalizing the argument of Section 4.13, show that the quotient manifold $\mathbb{R}^{n+1}/\mathbb{R}_+^{\cdot}$ can be identified with the closed submanifold in $\mathbb{R}^{n+1}$ whose points satisfy the equation $r_1^2 + \cdots + r_{n+1}^2 = 1$. This is the *n-dimensional sphere* S^n. Prove that the ideal $A_{S^n} = \{f \in \mathcal{F} \mid f(a) = 0 \ \forall a \in S^n\}$ is the principal ideal generated by the function

$$(r_1, \ldots, r_{n+1}) \mapsto r_1^2 + \cdots + r_{n+1}^2 - 1.$$

II. In the previous example replace the group $\mathbb{R}_+^{\cdot}$ by the multiplicative group $\mathbb{R}^{\cdot}$ of all nonzero real numbers (with the action described by the same formula). Prove that the quotient of $M = |\mathcal{F}|$ by this action is a smooth manifold. In the case $n = 2$ check that it can be identified with the projective plane (cf. Section 4.8 (ii)). In the general case this manifold is known as the *n-dimensional real projective space* and is denoted by $\mathbb{R}P^n$.

III. $\mathbb{R}P^n$ can also be obtained from S^n, $n \geqslant 1$, by taking the quotient with respect to the group action of $\mathbb{Z}_2 = \mathbb{R}^{\cdot}/\mathbb{R}_+^{\cdot}$. Geometrically, this quotient space can be visualized as obtained by "gluing together" all pairs of diametrically opposed points on the sphere.

Nevertheless, note that $\mathbb{R}P^1 \cong S^1$.

4.26 Exercises.   1. Suppose Γ is the automorphism group of the algebra $\mathcal{F} = C^\infty\left(\mathbb{R}^2\right)$ with one generator γ:

$$\gamma(f)(r_1, r_2) = f(-r_1, -r_2) \quad \text{for all } (r_1, r_2) \in \mathbb{R}^2, \ f \in \mathcal{F}.$$

Show that the algebra $\mathcal{F}^\Gamma$ of Γ-invariant functions is not smooth (nor smooth with boundary).

2. Suppose Γ is the rotation group of the plane about the origin:

$$|\gamma| = \begin{pmatrix} \cos\varphi & -\sin\varphi \\ \sin\varphi & \cos\varphi \end{pmatrix},$$

$$\gamma(f)(r_1, r_2) = f(r_1 \cos\varphi - r_2 \sin\varphi, r_1 \sin\varphi + r_2 \cos\varphi),$$

for all $(r_1, r_2) \in \mathbb{R}^2$, $f \in \mathcal{F} = C^\infty\left(\mathbb{R}^2\right)$.
Show that the space $\left|\mathcal{F}^\Gamma\right|$ is the closed half-line, so that $\mathcal{F}^\Gamma$ is an algebra of smooth functions on a manifold with boundary. Why does this algebra not coincide with the whole algebra of smooth functions on the half-line?

4.27. Remarks.

(i) For a number of reasons, the definition of a group action on a manifold given in Section 4.19 is not a very fortunate one; it should be regarded as preliminary. We shall give a satisfactory definition only in Section 6.10.

(ii) We also do not possess any meaningful criterion for the smoothness of the algebra $\mathcal{F}^\Gamma$ of Γ-invariant functions simple enough to mention here. The reader, however, will profit by proving the following: *If for any $a \in M$ there exists a neighborhood $U \subset M$, $U \ni a$, such that for all nontrivial $\gamma \in \Gamma$, $|\gamma|(U) \cap U = \varnothing$, then the algebra $\mathcal{F}^\Gamma$ is smooth.*

(iii) In the next six sections we are also anticipating a bit. Since these sections are relatively difficult, they may be omitted at first reading, which should then be continued from the beginning of Chapter 5.

4.28. If a physical system consists of independent parts, then it is natural to think of any state of the system as being a pair (a_1, a_2), where a_1 and a_2 are the corresponding states of the first and second part. If the states a_i are understood as points of the manifold M_i ($i = 1, 2$), the algebras $\mathcal{F}_i$ of smooth functions on the M_i being correctly defined and well known, it may be useful to define the manifold M of states of the entire system by using these algebras.

Exercise. Let $\mathbb{R}$-algebras $\mathcal{F}_1$ and $\mathcal{F}_2$ be geometrical. Show that their tensor product $\mathcal{F}_1 \otimes_\mathbb{R} \mathcal{F}_2$ is geometrical too.

Definition. The smooth envelope $\mathcal{F} = \overline{\mathcal{F}_1 \otimes_\mathbb{R} \mathcal{F}_2}$ of the tensor product $\mathcal{F}_1 \otimes_\mathbb{R} \mathcal{F}_2$ of geometrical $\mathbb{R}$-algebras (see Section 3.36) is said to be the *algebra of smooth functions of the Cartesian product of the smooth manifolds M_1 and M_2.*

In the next section this terminology will be justified.

4.29 Proposition. *If $\mathcal{F}$ is the algebra of smooth functions on the Cartesian product of the manifolds M_1 and M_2, then $|\mathcal{F}|$ is indeed homeomorphic to the Cartesian product of the topological spaces $M_1 = |\mathcal{F}_1|$ and $M_2 = |\mathcal{F}_2|$.*

◀ (We suggest that the reader return to this proof after having read Chapter 6.) Our goal is to identify the space $M_1 \times M_2$ with the set of $\mathbb{R}$-points of the algebra $\mathcal{F} = \overline{\mathcal{F}_1 \otimes_\mathbb{R} \mathcal{F}_2}$.

To each pair $(a_1, a_2) \in M_1 \times M_2$ let us assign the homomorphism $a_1 \otimes a_2$:

$$\mathcal{F}_1 \underset{\mathbb{R}}{\otimes} \mathcal{F}_2 \to \mathbb{R} \quad (f_1 \otimes f_2 \mapsto f_1(a_1) \cdot f_2(a_2)), \quad f_i \in \mathcal{F}_i \ (i = 1, 2).$$

Note that the algebra $\mathbb{R}$ is C^∞-closed (Section 3.32). Hence by the definition of smooth envelope, $a_1 \otimes a_2$ can be uniquely extended to the homomorphism

$$\overline{a_1 \otimes a_2} \colon \mathcal{F} = \overline{\mathcal{F}_1 \underset{\mathbb{R}}{\otimes} \mathcal{F}_2} \to \mathbb{R},$$

and we have constructed the map

$$\pi \colon M_1 \times M_2 \to |\mathcal{F}| \quad \left((a_1, a_2) \mapsto \overline{a_1 \otimes a_2}\right).$$

The map π is injective: If $\overline{a_1 \otimes a_2}$ coincides with $\overline{a_1' \otimes a_2'}$, then, using the fact that these homomorphisms coincide on elements of the form $f_1 \otimes 1$ and $1 \otimes f_2$, we immediately conclude that

$$f_i(a_i) = f_i(a_i') \text{ for all } f_i \in \mathcal{F}_i, \quad i = 1, 2.$$

Since the algebras $\mathcal{F}_i$ are geometrical, this implies $a_i = a_i'$ and hence the injectivity of π. Its surjectivity follows from elementary properties of tensor products. Thus π identifies $M_1 \times M_2$ and $|\mathcal{F}|$ as sets.

It remains to prove that π identifies the standard product topology in $M_1 \times M_2$ with the $\mathbb{R}$-algebra topology (Section 3.12) in $|\mathcal{F}|$. Consider the basis of the topology in $M_1 \times M_2$ consisting of the sets $U_1 \times U_2$ with

$$U_i = \{a \in M_i \mid \alpha_i < f_i(a) < \beta_i\}, \quad f_i \in \mathcal{F}_i, \ \alpha_i, \beta_i \in \mathbb{R}, \ i = 1, 2.$$

Then the sets

$$V_1 = U_1 \times M_2 = \{(a_1, a_2) \mid \alpha_1 < \overline{a_1 \otimes a_2}(f_1 \otimes 1) < \beta_1\},$$
$$V_2 = M_1 \times U_2 = \{(a_1, a_2) \mid \alpha_2 < \overline{a_1 \otimes a_2}(1 \otimes f_2) < \beta_2\},$$

are open in the topology induced from $|\mathcal{F}|$ by π in $M_1 \times M_2$. Therefore, $V_1 \cap V_2 = U_1 \times U_2$ is open as well.

Conversely, it follows from the construction of smooth envelopes (Section 3.36) that in order to obtain a basis of the topology in $|\mathcal{F}|$, we can take any subset of functions in $\mathcal{F}$ as long as the subalgebra generated by this subset has a smooth envelope coinciding with $\mathcal{F}$. In the given case it suffices to take the subset of functions of the form $f_1 \otimes f_2$, $f_i \in \mathcal{F}_i$, $i = 1, 2$.

Consider the basis open set

$$V = \{(a_1, a_2) \in M_1 \times M_2 = |\mathcal{F}| \mid \alpha < f_1(a_1)f_2(a_2) < \beta\}, \ \alpha, \beta \in \mathbb{R},$$

corresponding to such a function.

Figure 4.4. A basic open set in the Cartesian product.

The set of points on the plane $\mathbb{R}^2$ satisfying the inequality $\alpha < r_1 r_2 < \beta$ is open in the sense that, together with any of its points, it contains a rectangle $\{(r_1, r_2) \mid \alpha_1 < r_1 < \beta_1, \ \alpha_2 < r_2 < \beta_2\}$ (see Figure 4.4). Hence V is the union of sets of the form

$$\{(a_1, a_2) \in M_1 \times M_2 \mid \alpha_1 < f_1(a_1) < \beta, \ \alpha_2 < f_2(a_2) < \beta_2\}$$

and is therefore open in $M_1 \times M_2$.   ▶

4.30 Example–Lemma. *The smooth envelope of the $\mathbb{R}$-algebra*

$$C^\infty\left(\mathbb{R}^k\right) \otimes_{\mathbb{R}} C^\infty\left(\mathbb{R}^l\right)$$

is isomorphic to the $\mathbb{R}$-algebra $C^\infty\left(\mathbb{R}^{k+l}\right)$.

◀ Consider the $\mathbb{R}$-algebra homomorphism

$$i \colon C^\infty\left(\mathbb{R}^k\right) \otimes_{\mathbb{R}} C^\infty\left(\mathbb{R}^l\right) \to C^\infty\left(\mathbb{R}^{k+l}\right),$$
$$i(f \otimes g)(r_1, \ldots, r_{k+l}) = f(r_1, \ldots, r_k) \cdot g(r_{k+1}, \ldots, r_{k+l}).$$

We shall show that i satisfies the definition of smooth envelope (Section 3.36).

Indeed, suppose

$$\Phi \colon C^\infty\left(\mathbb{R}^k\right) \otimes_{\mathbb{R}} C^\infty\left(\mathbb{R}^l\right) \to \mathcal{F}$$

is a homomorphism to a C^∞-closed (Section 3.32) $\mathbb{R}$-algebra $\mathcal{F}$. A homomorphism

$$\Phi' \colon C^\infty\left(\mathbb{R}^{k+l}\right) \to \mathcal{F}$$

is a prolongation of Φ (i.e., $\Phi = \Phi' \circ i$) if and only if for all $(r_1, \ldots, r_k) \in \mathbb{R}^k$, $(s_1, \ldots, s_l) \in \mathbb{R}^l$, $g \in C^\infty\left(\mathbb{R}^{k+l}\right)$, we have

$$\Phi'(g) = g(\Phi(r_1 \otimes 1), \ldots, \Phi(r_k \otimes 1), \Phi(1 \otimes s_1), \ldots, \Phi(1 \otimes s_l)).$$

(Here, on the right-hand side, we regard r_i and s_j as functions on $\mathbb{R}^k$ and $\mathbb{R}^l$; the right-hand side is then well defined because $\mathcal{F}$ is C^∞-closed.)

Since the last formula is well defined for any $g \in C^\infty\left(\mathbb{R}^{k+l}\right)$, the required homomorphism exists and is unique. By the uniqueness theorem in Section 3.37, the lemma follows. ▶

4.31 Proposition. *If $\mathcal{F}_1$, $\mathcal{F}_2$ are smooth $\mathbb{R}$-algebras, then so is the algebra $\mathcal{F} = \overline{\mathcal{F}_1 \otimes_\mathbb{R} \mathcal{F}_2}$ (see Section 4.28).*

◀ Since the proof of this proposition repeats the one given in Section 4.29, we only indicate the main ideas, leaving the details to the industrious reader.

Suppose $a_i \in M_i = |\mathcal{F}|$ and let $U_i \ni a_i$ be neighborhoods such that $\mathcal{F}_i|_{U_i} \cong C^\infty\left(\mathbb{R}_i^n\right)$, $i = 1, 2$. We would like to establish the isomorphism

$$\mathcal{F}|_{U_1 \times U_2} \cong C^\infty\left(\mathbb{R}^{n_1 + n_2}\right).$$

By 4.30, it suffices to show that

$$\mathcal{F}|_{U_1 \times U_2} \cong \overline{\mathcal{F}_1|_{U_1} \otimes_\mathbb{R} \mathcal{F}_2|_{U_2}}.$$

But as before, there is a homomorphism

$$i \colon \mathcal{F}_1|_{U_1} \otimes_\mathbb{R} \mathcal{F}_2|_{U_2} \to \mathcal{F}|_{U_1 \times U_2},$$

which, as it turns out, satisfies the definition of a smooth envelope. ▶

Exercises.   1. Let $\mathcal{F}_1$ be a smooth algebra with boundary (see Section 4.2) and let $\mathcal{F}_2$ be a smooth algebra. Mimicking the proof of the above proposition, show that the algebra $\mathcal{F} = \overline{\mathcal{F}_1 \otimes_\mathbb{R} \mathcal{F}_2}$ is smooth with boundary.

2. Does the previous assertion remain valid if $\mathcal{F}_2$ is also a smooth algebra with boundary?

4.32. Example. I. The cylinder in Example 4.5, II, is the Cartesian product of S^1 and $\mathbb{R}$.

II. The n-dimensional torus T^n (see 4.24, III) is the Cartesian product of T^{n-1} and S^1. In particular, $T^2 = S^1 \times S^1$.

4.33 Exercise. The reader sufficiently versed in topology will profit a great deal by proving that the Möbius band (Example 4.7, I) is *not* the Cartesian product of $\mathbb{R}$ and S^1.

5

Charts and Atlases

5.1. In this chapter the intuitive idea of "introducing local coordinates" is elaborated into a formal mathematical definition of a differentiable manifold. The definition, of course, turns out to be equivalent to the algebraic one given in the previous chapter, as will be proved in Chapter 7.

The coordinate approach is more traditional, and is certainly more appropriate for practical applications (when something must be computed). However, it is less suitable for developing the theory, since it requires tedious verifications of the fact that the notions and constructions introduced in the theory by means of coordinates are *well defined*, i.e., independent of the specific choice of local coordinates.

In the coordinate approach, a manifold structure on a set is defined by a family of compatible charts constituting a smooth atlas, much in the same way as the geopolitical structure on the Earth's surface is described by the charts of a geographical atlas. The words in italics above will be given mathematical definition in subsequent sections.

5.2. A *chart* (U, x) on the set M is a bijective map $x \colon U \to \mathbb{R}^n$ of a subset $U \subset M$ onto an open set $x(U)$ of Euclidean space $\mathbb{R}^n$. The integer $n > 0$ is the *dimension* of the chart.

Examples. I. If U is an open set in $\mathbb{R}^n$, then the identity map defines a chart (U, id) on the set $\mathbb{R}^n$.

II. If T^2 is the configuration space of the plane *double pendulum* (Figure 5.1), then (U, s), where

$$U = \left\{ (\varphi, \psi) \in T^2 \mid -\frac{\pi}{4} < \varphi, \psi < \frac{\pi}{4} \right\}$$

© Springer Nature Switzerland AG 2020
J. Nestruev, *Smooth Manifolds and Observables*, Graduate Texts
in Mathematics 220, https://doi.org/10.1007/978-3-030-45650-4_5

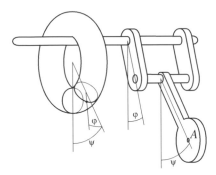

Figure 5.1. Double pendulum.

and s denotes the map

$$U \ni (\varphi, \psi) \xmapsto{\;s\;} (\sin \varphi, \sin \psi) \in \mathbb{R}^2,$$

is a chart on T^2. (Note that by assigning to each position of the pendulum the coordinates of its end point $A \in \mathbb{R}^2$, we do not obtain a chart, since this assignment is not bijective.)

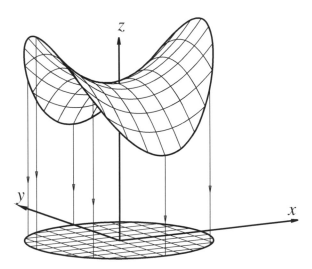

Figure 5.2. Saddle surface.

III. If S is the saddle surface $z = 1 + x^2 - y^2$, then the vertical projection

$$S \ni (x, y, z) \xmapsto{\;\mathrm{pr}\;} (x, y) \in \mathbb{R}^2$$

of a neighborhood $U \subset S$ of the point $(0,0,1)$ defines a chart (U, pr) on S (Figure 5.2).

5.3. Given a chart (U, x) and a point $a \in U$, note that $x(a)$ is a point of $\mathbb{R}^n$; i.e., we have $x(a) = (r_1, \ldots, r_n) \in \mathbb{R}^n$; the number r_i is called the *ith coordinate* of a, and the corresponding function (sending each $a \in U$ to its *i*th coordinate) is the *ith coordinate function* (in the chart (U, x)); it is denoted by $x_i \colon U \to \mathbb{R}$. A chart is entirely determined by its coordinate functions; in the literature, the expression *local coordinates* is often used to mean "chart" in this sense.

5.4. Two charts (U, x), (V, y) on the same set M are called *compatible* if the *change of coordinate map*, i.e.,

$$y \circ x^{-1} \colon x(U \cap V) \to y(U \cap V),$$

is a diffeomorphism of open subsets of the space $\mathbb{R}^n$ (see Figure 5.3) or if $U \cap V = \varnothing$. The compatibility relation is reflexive, symmetric, and transitive (because of appropriate properties of diffeomorphisms in $\mathbb{R}^n$) so that the family of all charts on a given set M splits into equivalence classes (sets of compatible charts).

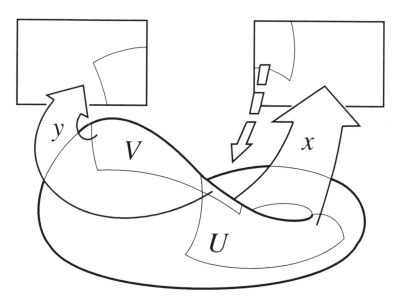

Figure 5.3. Compatible charts.

Examples. I. All the charts (U, id), where U is an open subset of $\mathbb{R}^n$ and n is fixed, are compatible.

II. The chart (U, s) on the double pendulum T^2 described in 5.2, II, is not compatible with the chart (U, c) defined by

$$U \ni (\varphi, \psi) \overset{c}{\longmapsto} (\sin \varphi, g(\psi)),$$

where $g(\psi)$ equals $\sin\psi$ for negative ψ and $1 - \cos\psi$ for nonnegative ψ. This is because the change of coordinates map $c \circ s^{-1}$ fails to be smooth at the point $(0,0) \in \mathbb{R}^2$.

5.5. A family $\mathcal{A}$ of compatible charts $x_\alpha \colon U_\alpha \to \mathbb{R}^n$ on the set M (where $n \geqslant 0$ is fixed and α ranges over some index set J) is said to be an *atlas* on M if the U_α cover M, i.e., $\bigcup_{\alpha \in J} U_\alpha = M$. The integer n is the *dimension* of $\mathcal{A}$. An atlas is *maximal* if it is not contained in any other atlas. Obviously, *any atlas is contained in a unique maximal atlas*, namely the one consisting of all charts compatible with any of the charts of the given atlas.

Two atlases are said to be *compatible* if any chart of one of them is compatible with any chart of the other. The last condition is equivalent to the fact that the union of these atlases is also an atlas. Note that any two compatible atlases, together with their union, are contained in the same maximal atlas.

This compatibility relation is reflexive, symmetric, and transitive (by the corresponding properties of diffeomorphisms in $\mathbb{R}^n$). Consequently, the family of all atlases on a given set M splits into equivalence classes, and any such a class contains a unique maximal atlas.

Examples. I. $\mathbb{R}^n$ has an atlas consisting of a single chart: $(\mathbb{R}^n, \mathrm{id})$.

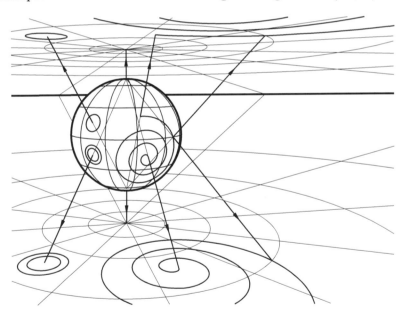

Figure 5.4. Stereographic projections.

II. The sphere S^2 has an atlas consisting of two charts (e.g., stereographic projections from the north and south poles; see Figure 5.4).

III. The double pendulum (see Example 5.2, II) also has two-chart atlases (try to find one).

5.6. The reader is perhaps wondering why we are not defining manifolds as sets supplied with an atlas. Well, we won't be. Because if we did, this excessively general definition would put us under the obligation to bestow the noble title of manifold upon certain ungainly objects. Such as

I. *The discrete line.* This is the set of points of $\mathbb{R}$ with the discrete atlas consisting of all charts of the form $(\{r\}, v)$, where $r \in \mathbb{R}$ and $v(r) = 0 \in \mathbb{R}^0$.

II. *The long line.* This is the disjoint union

$$\mathcal{R} = \coprod_\alpha \mathbb{R}_\alpha$$

of copies $\mathbb{R}_\alpha$ of $\mathbb{R}$, indexed by an ordered uncountable set of indices α (we can take α to range over $\mathbb{R}$ itself). The set $\mathcal{R}$ has a natural order and a natural topology (induced from $\mathbb{R}$ and disjoint union). It also has a natural atlas $\mathcal{A} = \{(\mathbb{R}_\alpha, \mathrm{id}_\alpha) \colon \alpha \in \mathbb{R}\}$, where $\mathrm{id}_\alpha \colon \mathbb{R}_\alpha \to \mathbb{R}^1$ is the identification of each copy $\mathbb{R}_\alpha$ with the original prototype $\mathbb{R}^1 = \mathbb{R}$.

III. *The line with a double point.* This is the line $\mathbb{R}$ to which a point θ is added, while the atlas consists of two charts (U, x) and (V, y), where

$$U = \mathbb{R}, \qquad\qquad x = \mathrm{id},$$
$$V = \{\theta\} \cup \mathbb{R} \smallsetminus \{0\}, \qquad y(\theta) = 0, \quad y(r) = r, \ r \in \mathbb{R} \smallsetminus \{0\}.$$

In other words, the line with a double point can be obtained if two copies of the line $\mathbb{R}$ are identified at all points with the same coordinates except for zero.

5.7. The following definitions are needed to exclude pathological atlases (of the types described in Section 5.6) from our considerations. We say that an atlas $\mathcal{A}$ on M satisfies the *countability condition* if it consists of a finite or a countable number of charts or if all its charts are compatible with those of such an atlas. An atlas $\mathcal{A}$ on M satisfies the *Hausdorff condition* if for any two points $a, b \in M$ there exist nonintersecting charts (U, x), (V, y) containing these points ($a \in U$, $b \in V$, $U \cap V = \varnothing$) and compatible with the charts of $\mathcal{A}$.

Clearly, the discrete and long lines (see Examples 5.6, I, and II) do not satisfy the countability condition, while the line with a double point, 5.6, III, fails to meet the Hausdorff condition.

5.8. Coordinate definition of manifolds. A set supplied with a maximal atlas $\mathcal{A}_{\max} = \{(U_\alpha, x_\alpha)\}$, $x_\alpha \colon U_\alpha \to \mathbb{R}^n$, $n \geqslant 0$, satisfying the countability and Hausdorff conditions is called an *n-dimensional differentiable* (or *smooth*) *manifold*. It will be proved in Chapter 7 that this definition is equivalent to the algebraic one given in Section 4.1.

In order to determine a specific manifold $(M, \mathcal{A}_{\max})$, we shall often indicate some smaller atlas $\mathcal{A} \subset \mathcal{A}_{\max}$, since $\mathcal{A}_{\max}$ is uniquely determined by any of its subatlases (see 5.5). In that case we denote our manifold by $(M, \mathcal{A})$ and say that $\mathcal{A}$ is a *smooth atlas* on M. Note that the adjective "smooth" implicitly includes the compatibility, Hausdorff, and countability conditions.

5.9. We now show that *any smooth atlas* $\mathcal{A} = \{(U_\alpha, x_\alpha)\}$ *on the manifold M determines a topological structure on the set M*, carried over from the Euclidean topology in the sets $x_\alpha(U_\alpha) \subset \mathbb{R}^n$ by the maps x_α^{-1}. To be more precise, a base of open sets in the space M is constituted by all the sets $x_\alpha^{-1}(B_\beta)$, where the B_β's are all the open Euclidean balls contained in all the $x_\alpha(U_\alpha)$. It follows immediately from the definitions that M then becomes a *Hausdorff topological space with countable base*. When the set M is supplied with an atlas, this topological structure is always understood. For example, when we say that the manifold M is *compact* or *connected*, we mean that it is a compact (connected) topological space with respect to the topology described above.

It is easy to see that *any chart (U, x) is a homeomorphism* of an open subset $U \subset M$ (in this topology) onto an open subset $x(U) \subset \mathbb{R}^n$.

5.10. Examples of manifolds from geometry. I. The *sphere*

$$S^n = \{\vec{x} \colon |\vec{x}| = 1\} \subset \mathbb{R}^{n+1}$$

has a two-chart atlas given by stereographic projections (similar to the two-dimensional case; see 5.5, II). The sphere S^n is the simplest compact connected n-dimensional manifold.

II. The *hyperboloid* $x^2 + y^2 - z^2 = 1$ in $\mathbb{R}^3$ has a simple four-chart atlas $(U_\pm, p_{zy})$, $(V_\pm, p_{zx})$, where

$$U_\pm = \{(x, y, z) \mid x = \pm\sqrt{1 + z^2 - y^2},\ \pm x > 0\},$$
$$V_\pm = \{(x, y, z) \mid y = \pm\sqrt{1 + z^2 - x^2},\ \pm y > 0\},$$

and the maps p_{zy}, p_{zx} are the projections on the corresponding planes. The corresponding two-dimensional manifold is connected but not compact.

III. *The projective space* $\mathbb{R}P^n$ is the set of all straight lines passing through the origin O of $\mathbb{R}^{n+1}$. For each such line l consider the following chart (U_l, p_l). The set U_l consists of all lines forming an angle of less than (say) $30°$ with l (Figure 5.5). To define p_l, choose a basis in the n-plane $l^\perp \ni O$ perpendicular to l and fix one of the half-spaces $\mathbb{R}^{n+1} \setminus l^\perp$; to every line $l' \in U_l$ let the map p_l assign the coordinates in $l^\perp$ of the projection on $l^\perp$ of the unit vector pointing into the chosen half-space and determining l'. The set of all such charts (U_l, p_l) constitutes a smooth atlas, endowing $\mathbb{R}P^n$ with the structure of a compact connected n-dimensional manifold.

IV. The *Grassmann manifold* $G_{n,m}$ is the set of all m-dimensional planes in $\mathbb{R}^n$ passing through the origin O. To construct a chart, let us choose in $\mathbb{R}^n$ a Cartesian system $(x_1, x_2, \ldots, x_n)$ and take for U the set of all

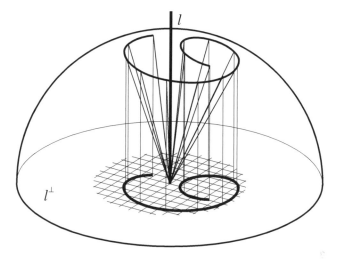

Figure 5.5. Construction of a chart on $\mathbb{R}P^n$.

m-dimensional planes given in these coordinates by the system of equations

$$x_{m+i} = \sum_{j=1}^{m} a_{ij}x_j, \qquad i = 1, \ldots, n - m.$$

Let also the map $p\colon U \to \mathbb{R}^{m(n-m)}$ take such a plane to the set of coefficients a_{ij} appearing in the system above. Choosing different Cartesian systems, one can construct different charts covering together the entire space $G_{n,m}$. However, to cover it, it suffices to use only one Cartesian system and change the order of coordinates in it. Compatibility of these charts follows from the smooth dependency of solutions of a linear system on the system's coefficients. Thus we obtain an $m(n - m)$-dimensional manifold that generalizes the previous example, namely $G_{n,1} = \mathbb{R}P^{n-1}$.

One can consider planes not only in $\mathbb{R}^n$ but also in any finite-dimensional vector space V. The manifold obtained in this case is denoted by $G_{V,m}$.

Exercise. How many connected charts do you need to obtain an atlas for the Klein bottle (cf. 4.7, III)? Prove that two is enough.

5.11. Examples of manifolds from algebra. I. The *general linear group* $\mathrm{GL}(n)$ of all linear isomorphisms of $\mathbb{R}^n$ has a one-chart atlas of dimension n^2 obtained by assigning to each $g \in \mathrm{GL}(n)$ the n^2 entries of its matrix written column by column in the form of a single column vector with n^2 components. The corresponding manifold is not compact and not connected.

II. The *special orthogonal group* $\mathrm{SO}(n)$ of all positive orthogonal matrices possesses a smooth atlas of $(n(n-1)/2)$-dimensional charts. Its construction is left to the reader, who might profit by referring to Example 5.10, III.

5.12. Examples of manifolds from mechanics. I. The configuration space of the double pendulum (see 5.2, II, and 5.4, II), as the reader must have guessed by now, is the two-dimensional *torus* $T^2 = S^1 \times S^1$.

II. The configuration space of a *thin uniform disk whose center is fixed by a hinge* (allowing it be inclined at all angles and directions in three-space) possesses a natural two-dimensional atlas. The reader is urged to find such an atlas and compare it with that of the projective plane $\mathbb{R}P^2$. He/she will also appreciate that if one side of the disk is painted, then the corresponding atlas will be that of S^2.

III. The configuration space (recall Section 1.1) of a *solid freely rotating in the space about a fixed point* is a three-dimensional compact connected manifold. The reader is asked to find an atlas for it and compare it with $\mathbb{R}P^3$ and SO(3).

5.13 Exercise. Consider the following manifolds (already discussed informally in Chapter 1):

1. The projective space $\mathbb{R}P^3$.

2. The sphere S^3 with antipodal points identified.

3. The disk D^3 with antipodal points of its boundary $\partial D^3 = S^2$ identified.

4. The special orthogonal group SO(3).

5. The configuration space of a solid freely rotating about a fixed point in $\mathbb{R}^3$.

Show that they are all diffeomorphic by

1. Constructing atlases and diffeomorphisms between the charts.

2. Constructing isomorphisms of the corresponding smooth $\mathbb{R}$-algebras.

5.14. Previously (see Section 4.2), we introduced the notion of manifold with boundary algebraically. Now we give the corresponding coordinate definition.

This definition is just the same as that of an ordinary manifold (see Section 5.8), except that the notion of chart must be modified (by substituting $\mathbb{R}^n_H$ for $\mathbb{R}^n$). Namely, a *chart with boundary* (U, x) on the set M is a bijective map $x \colon U \to \mathbb{R}^n_H$ of a subset $U \subset M$ onto an open subset $x(U)$ of the Euclidean half-space

$$\mathbb{R}^n_H = \{(r_1, \dots, r_n) \in \mathbb{R}^n \mid r_n \geqslant 0\}.$$

Note that a chart in $\mathbb{R}^n$ is a particular case of a chart in $\mathbb{R}^n_H$, since $x(U)$ may not intersect the "boundary"

$$\{(r_1, \dots, r_n) \in \mathbb{R}^n \mid r_n = 0\}$$

of the half-space.

All further definitions (those of dimension, compatibility, atlases, etc.) remain the same, except that $\mathbb{R}_H^n$ must be substituted for $\mathbb{R}^n$ in the appropriate places. Repeating these definitions in this modified form, we obtain the coordinate definition of a *manifold with boundary*.

The industrious reader will gain by actually carrying out these repetitions in detail; this is a good way to check that he/she has in fact mastered the main Definition 5.8.

5.15. If $\mathcal{A} = (U_\alpha, x_\alpha)$, $\alpha \in \mathcal{I}$, consists of charts with boundary, the set

$$\partial M = \left\{ m \in M \mid \exists \alpha \in \mathcal{I}, \; m = x_\alpha^{-1}(r_1, \ldots, r_n), \; r_n = 0 \right\}$$

of points mapped by coordinate maps on the boundary $(n-1)$-dimensional plane $\{r_n = 0\}$ of the half-space $\mathbb{R}_H^n$ is said to be the *boundary* of M.

When ∂M is empty, we recover the definition of ordinary manifold (the open half-space $\{r_n > 0\}$ being homeomorphic to $\mathbb{R}^n$).

Sometimes in the literature the term "manifold" is defined so as to include manifolds with boundary; in that case the expression *closed manifold* is used to mean manifold (with empty boundary).

Proposition. *The boundary ∂M of an n-dimensional manifold with boundary M has a natural $(n-1)$-dimensional manifold structure.*

◄ Hint of the proof. Take the intersection of the charts of M with the boundary $(n-1)$-plane of the half-space

$$\mathbb{R}^{n-1} = \{(r_1, \ldots, r_n) \in \mathbb{R}_H^n \mid r_n = 0\}$$

to get an atlas on ∂M.   ►

5.16. Examples of manifolds with boundary. I. The *n-dimensional disk* $D^n = \{x \mid \|x\| \leq 1\} \subset \mathbb{R}^n$ is a n-dimensional manifold with boundary $\partial D^n = S^{n-1}$.

II. If M is an n-dimensional manifold (without boundary) defined by its atlas $\mathcal{A}$, a manifold with boundary can be obtained from M by "removing an open disk from it." This means that we take any chart $(U, x) \in \mathcal{A}$, choose an open n-dimensional disk V in $x(U) \subset \mathbb{R}^n$, and consider the set $M \setminus x^{-1}(V)$, which has an obvious manifold-with-boundary structure; $M \setminus x^{-1}(V)$ is sometimes called a *punctured manifold*.

III. Figure 5.6 presents the beginning of a complete list of all two-dimensional manifolds (with boundary) whose boundary is the circle S^1. They are called the 2-disk, the punctured torus, the punctured orientable surface of genus 2, ..., the punctured orientable surface of genus k, ... (upper row), the Möbius band, the punctured Klein bottle,..., the punctured non-orientable surface of genus k, ... The term "orientable," which we do not discuss in the general case here, in the two-dimensional case means "does not contain a Möbius band."

Figure 5.6. Two-dimensional manifolds with S^1 as the boundary.

5.17. In the algebraic study of smooth manifolds, the fundamental concept was the $\mathbb{R}$-algebra of smooth functions $\mathcal{F}$. This $\mathbb{R}$-algebra can also be defined for a manifold M given by an atlas $\mathcal{A}$.

Definitions. A function $f \colon M \to \mathbb{R}$ on the manifold M with smooth atlas $\mathcal{A}$ is called *smooth* if for any chart $(U, x) \in \mathcal{A}$ the function $f \circ x^{-1} \colon x(U) \to \mathbb{R}$ defined on the open set $x(U) \subset \mathbb{R}^n$ is smooth (i.e., $f \circ x^{-1} \in C^\infty(x(U))$). The set of all smooth functions on M is denoted by $C^\infty(M)$. This set has an obvious $\mathbb{R}$-algebra structure, and will temporarily be called the $\mathbb{R}$-*algebra of smooth functions on M with respect to the atlas $\mathcal{A}$*.

It is easy to establish that $C^\infty(M)$ is the same for any atlas $\mathcal{A}'$ compatible with $\mathcal{A}$. Moreover, we shall see that $C^\infty(M)$ is the same $\mathbb{R}$-algebra as the one in the algebraic approach $(C^\infty(M) = \mathcal{F})$, but this will be proved only in Chapter 7).

Exercise.   1. Describe the smooth function algebra (in the sense of the above definition) of the configuration space for the double pendulum (see 5.12).

2. Same question for the case in which the φ-rod is shorter then the ψ-rod, so that the latter is blocked in its rotation when the point A hits the φ-axle (see Figure 5.1). Identify the corresponding smooth manifold.

5.18. In the equivalence proof carried out in Chapter 7, we shall need the following proposition.

Proposition. *If M is a manifold $\mathcal{A}$ its smooth atlas, and $C^\infty(M)$ the $\mathbb{R}$-algebra of smooth functions on M (with respect to $\mathcal{A}$), then there exists a function $f \in C^\infty(M)$ all of whose level surfaces (i.e. the sets $f^{-1}(\lambda)$, $\lambda \in \mathbb{R}$) are compact subsets of M.*

This proposition generalizes Proposition 2.7 and can be proved by using the latter and the partition of unity lemma, stated in the appropriate form for manifolds with atlases (compare with Section 4.18).

6
Smooth Maps

6.1. Suppose $\mathcal{F}_1$ is the algebra of smooth functions on a manifold M_1, and $\mathcal{F}_2$ is the one on another manifold M_2. The map $f\colon M_1 \to M_2$ is called *smooth* if $f = |\varphi|$, where $\varphi\colon \mathcal{F}_2 \to \mathcal{F}_1$ is an $\mathbb{R}$-algebra homomorphism.

Recall that $M_i = |\mathcal{F}_i|$, $i = 1, 2$, are the dual spaces to $\mathcal{F}_i$ (see Section 3.8), i.e., consist of all $\mathbb{R}$-algebra homomorphisms $x\colon \mathcal{F}_i \to \mathbb{R}$, $i = 1, 2$. Recall also that $|\varphi|\colon |\mathcal{F}_1| \to |\mathcal{F}_2|$ is the dual map defined in 3.19 as $|\varphi|\colon x \mapsto x \circ \varphi$ and that all homomorphisms (see Section 3.1) are unital: $\varphi(1) = 1$.

Exercise. Prove that φ is injective whenever f is surjective. Construct a counterexample to the converse statement (if you do not succeed, try again after reading Example 6.5 below).

6.2. Example. Suppose $\mathcal{F}$ is the algebra of smooth functions on the manifold M, and Γ is a group acting on M. The map

$$p = |i|\colon M \to M/\Gamma$$

dual to the inclusion $i\colon \mathcal{F}^\Gamma \to \mathcal{F}$ of the algebra of Γ-invariant functions (see 4.19) into $\mathcal{F}$ is of course smooth.

To be specific, consider the group $\Gamma = \mathbb{Z}$ acting on $\mathbb{R}$ by identifying points r_1 and r_2 whenever $r_1 - r_2 \in \mathbb{Z}$. Denote by S^1 the set of equivalence classes, and let $\mathcal{F}$ be, as in 4.5, the algebra of smooth 1-periodic functions on the line $\mathbb{R}$. The natural projection $p\colon \mathbb{R} \to S^1$ is a smooth map, since it coincides with $|i|$, where i is the inclusion $\mathcal{F} \subset C^\infty(\mathbb{R})$.

We say that the map p winds the line $\mathbb{R}$ around the circle S^1.

6.3. Example. Let $\mathcal{F}$ be the algebra of smooth functions on the manifold M, and $N \subset M$ a smooth submanifold (Definition 4.11) of M. In this case

© Springer Nature Switzerland AG 2020
J. Nestruev, *Smooth Manifolds and Observables*, Graduate Texts
in Mathematics 220, https://doi.org/10.1007/978-3-030-45650-4_6

the inclusion $N \hookrightarrow M$ is a smooth map, since it coincides with $|\rho|$, where

$$\rho \colon \mathcal{F} \to \mathcal{F}_N = \mathcal{F}\big|_N$$

is the restriction homomorphism.

To be specific, suppose that S^1 and $\mathcal{F}$ are the same as in Example 6.2 and let $\mathcal{F}_{\mathbb{C}}$ be the algebra of all real-valued functions of a complex variable $z = x + iy$ smoothly depending on the variables x and y. Consider the inclusion

$$i \colon S^1 \hookrightarrow \mathbb{C} \quad \left([r] \longmapsto e^{2\pi i r}\right),$$

where $[r] \in S^1$ is the equivalence class of the point $r \in \mathbb{R}$. Further, define the homomorphism

$$\beta \colon \mathcal{F}_{\mathbb{C}} \to \mathcal{F} \quad (\beta(f)(r) = f\left(e^{2\pi i r}\right),\ f \in \mathcal{F}_{\mathbb{C}},\ r \in \mathbb{R}).$$

The inclusion i coincides with $|\beta|$ and is, therefore, a smooth map of S^1 into $\mathbb{R}^2 = \mathbb{C}$.

Notice that we have already met this inclusion in another coordinate representation; see Example 4.13.

6.4. Examples. Suppose $\mathcal{F}$ is the algebra of smooth functions on the open Möbius band (4.7, I) and $\mathcal{F}(S^1)$ the algebra of 1-periodic smooth functions on the line ("the circle" 3.18).

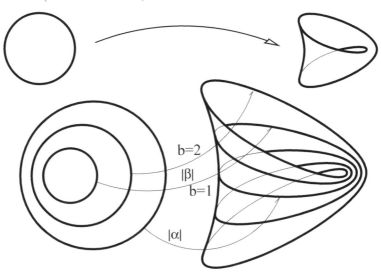

Figure 6.1. Maps from S^1 to the Möbius band.

I. Consider the homomorphisms

$$\alpha, \beta \colon \mathcal{F} \to \mathcal{F}(S^1) \quad (\alpha(f)(r) = f(r, 0),\ \beta(f)(r) = f(2r, b)),$$

where $f \in \mathcal{F}$, $r \in \mathbb{R}$, and $b \neq 0$ is any real number.

Exercise. Why does the formula $\gamma(f)(r) = f(r, b)$ not define a homomorphism $\gamma \colon \mathcal{F} \to \mathcal{F}(S^1)$?

The smooth maps $|\alpha|$ and $|\beta|$ are shown in Figure 6.1. Notice that the image of the map $|\beta|$ is "twice as long" as that of $|\alpha|$.

II. There is a remarkable smooth map of the Möbius band on the circle, namely the map $g = |\xi|$, where

$$\xi \colon \mathcal{F}\left(S^1\right) \to \mathcal{F} \quad (\xi(f)(r_1, r_2) = f(r_1)).$$

Note that $\alpha \circ \xi = \mathrm{id}_{\mathcal{F}(S^1)}$. The map $g = |\xi|$ may be visualized as "collapsing" the Möbius band to its central circle (Figure 6.2).

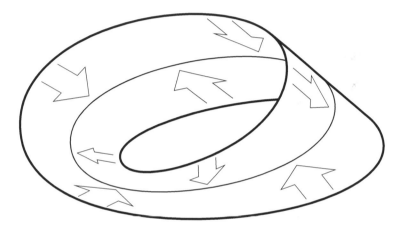

Figure 6.2. "Collapsing" the Möbius band.

6.5. Example Choose an irrational number $\lambda \in \mathbb{R}$ and consider the map

$$f \colon \mathbb{R}^1 \to \mathbb{R}^2, \quad r \mapsto (r, \lambda r).$$

Then $f = |\varphi|$, where

$$\varphi \colon C^\infty\left(\mathbb{R}^2\right) \to C^\infty\left(\mathbb{R}^1\right) \quad (\varphi(g)(r) = g(r, \lambda r))$$

for all $g \in C^\infty\left(\mathbb{R}^2\right)$, $r \in \mathbb{R}$.

Denote by $\overline{\varphi}$ the restriction of the homomorphism φ to the subalgebra of doubly periodic functions (see Example 4.24, III, with $n = 2$). The image of the smooth map $|\overline{\varphi}| \colon \mathbb{R}^1 \to T^2$ is everywhere dense in the torus T^2 and, therefore, the homomorphism $\overline{\varphi}$ is injective.

This example is interesting, since it shows that an algebra of functions "in several variables" may be isomorphic to a subalgebra of $C^\infty\left(\mathbb{R}^1\right)$.

When the number λ is taken to be the rational, the image of $|\overline{\varphi}|$ is compact. A particular case is shown in Figure 6.3. Try to guess what value of λ was taken there.

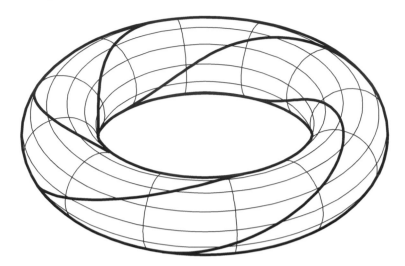

Figure 6.3. A map of $\mathbb{R}$ to the torus.

6.6. Now that smooth maps of manifolds have been introduced, manifolds no longer appear as unrelated, separate objects; they have been
brought together into something unified, called a *category*. Other examples of categories are groups and their homomorphisms, topological spaces
and continuous maps, $\mathbb{R}$-algebras and $\mathbb{R}$-algebra homomorphisms, and linear spaces and linear operators. As we pointed out in Section 3.38, we shall
not give any formal definitions from abstract category theory, but will often
"think categorically."

In particular, let us point out two fundamental properties of the *category
of smooth manifolds and maps*:

(i) if $a = |\alpha| \colon M_1 \to M_2$ and $b = |\beta| \colon M_2 \to M_3$ are smooth maps
corresponding to the $\mathbb{R}$-algebra homomorphisms

$$\alpha \colon \mathcal{F}_2 \to \mathcal{F}_1 \text{ and } \beta \colon \mathcal{F}_3 \to \mathcal{F}_2,$$

then $b \circ a \colon M_1 \to M_3$ is a smooth map (since it corresponds to the
composition $\alpha \circ \beta$, in inverse order, of the homomorphisms α and β);

(ii) the identity map

$$\mathrm{id} \colon M_1 = M \to M_2 = M$$

is smooth, since it corresponds to the identity homomorphism

$$\mathrm{id} \colon \mathcal{F}_2 = \mathcal{F} \to \mathcal{F}_1 = \mathcal{F}.$$

The other categories mentioned above possess similar properties.

Suppose we are given a collection of maps possessing properties (i) and (ii). Then a typical "categorical trick" is to "inverse all arrows"; i.e., whenever a map $A \to B$ belongs to our collection, assume that there is a map $B \to A$ in a new, dual collection of maps, in which composition of maps is written in inverse order. Then we obtain the *dual category*, also satisfying (i), (ii).

We have, in fact, been using that construction in passing from homomorphisms of smooth $\mathbb{R}$-algebras to smooth maps of manifolds. The contents of Chapters 4 and 6 may be summarized as follows: A smooth manifold is a smooth $\mathbb{R}$-algebra, understood as an object of the dual category.

6.7. Suppose M_1 and M_2 are manifolds. The smooth map $a\colon M_1 \to M_2$ is said to be a *diffeomorphism* if there exists a smooth map $b\colon M_2 \to M_1$ such that $b \circ a = \mathrm{id}_{M_1}$, $a \circ b = \mathrm{id}_{M_2}$.

The manifolds M_1 and M_2 are called *diffeomorphic* if there exists a diffeomorphism of one onto the other. Note that two manifolds are diffeomorphic if and only if their algebras of smooth functions are isomorphic.

The relation of being diffeomorphic is an equivalence relation and will be denoted by $\cong$. In Chapter 4, when we spoke of two manifolds being "the same" or "identical," we actually meant that they were diffeomorphic; indeed, from the point of view of the theory, diffeomorphic manifolds are the same manifold presented in different guises.

6.8. Examples. I. The argument in Example 4.13 can be understood as a proof of the fact that the two methods for constructing the circle (as the quotient space of $\mathbb{R}^1$ and as a submanifold of $\mathbb{R}^2$) result in diffeomorphic manifolds.

II. A linear operator $A\colon \mathbb{R}^n \to \mathbb{R}^n$ will be a diffeomorphism if and only if $\det A \neq 0$, i.e., if A is bijective.

III. Suppose $\mathcal{F}_1$ and $\mathcal{F}_2$ are the algebras of smooth functions on manifolds M_1 and M_2. In general, the bijectivity of the smooth map $|\alpha|\colon M_1 \to M_2$ (where $\alpha\colon \mathcal{F}_2 \to \mathcal{F}_1$ is an $\mathbb{R}$-algebra homomorphism) is not sufficient for this map to be a diffeomorphism. As an example, we can take the map

$$\varphi\colon \mathbb{R} \to \mathbb{R}, \quad r \mapsto r^3.$$

In order to establish that various specific bijective smooth maps are in fact diffeomorphisms, one often uses the implicit function theorem 6.22. We shall not dwell on this here.

6.9 Exercises.   1. Prove that the boundary of the closed Möbius band (4.7, II) is diffeomorphic to the circle S^1.

2. Following Example 4.13 and I above, construct a diffeomorphism between the two models of the sphere S^n described in 4.25, I.

6.10. We now return to the topics of Chapter 4 in order to give, as promised, more satisfactory definitions of group action, quotient manifolds, and Cartesian products. These definitions will be "categorical" in character: They will be based on smooth maps and diagrams.

Suppose $\mathcal{F}$ is the algebra of smooth functions on the manifold M, and Γ is some group of automorphisms of this $\mathbb{R}$-algebra. The smooth map $a\colon M \to N$ of M into the manifold N is called Γ-*invariant* if $a \circ |\gamma| = a$ for all $\gamma \in \Gamma$. Obviously, the quotient map q of M on the quotient manifold M/Γ is Γ-invariant. (Of course, this is true, provided that M/Γ is a smooth manifold, i.e., if the algebra $\mathcal{F}^{\Gamma}$ of invariant functions is a smooth $\mathbb{R}$-algebra; see Definition 4.1 and Section 4.23.) It turns out that the quotient map is the "universal" Γ-invariant smooth map.

Proposition. *Let* $\mathcal{F}(N)$ *be the algebra of smooth functions on the smooth manifold* N *and* $a\colon M \to N$ *be any* Γ-*invariant smooth map with respect to an action of* Γ *on* M *such that* M/Γ *is a smooth manifold. Then there exists a unique map* $b\colon M/\Gamma \to N$ *for which the following diagram is commutative:*

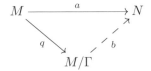

◀ We must prove the existence and uniqueness of an $\mathbb{R}$-algebra homomorphism $\beta\colon \mathcal{F}(N) \to \mathcal{F}^{\Gamma}$ for which the diagram

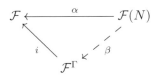

where $a = |\alpha|$ is commutative. Clearly, there is no more than one such β. It exists iff $\Im\alpha = \alpha(\mathcal{F}(N))$ consists of Γ-invariant elements. But for a Γ-invariant map a, this is always the case: For all $\gamma \in \Gamma$, $f \in \mathcal{F}(N)$ we have

$$\gamma(\alpha(f)) = \alpha(f) \circ |\gamma| = f \circ (a \circ |\gamma|) = f \circ a = \alpha(f). \quad \blacktriangleright$$

6.11. Remark. The universal property characterizing the quotient map $M \to M/\Gamma$ determines the quotient manifold M/Γ uniquely up to diffeomorphism.

The proof of this statement can be copied over from the uniqueness proof of smooth envelopes (Proposition 3.37), and we suggest that the reader carry it out. The underlying general principle for proofs of this type, which we do not wish to formalize here, is that "any universal property determines an object uniquely."

6.12. Now we return to Cartesian products (Definition 4.28). Let $\mathcal{F}_l$ be the algebra of smooth functions on the manifold M_l, $l = 1, 2$. The *projection* *maps*

$$p_l \colon M_1 \times M_2 \to M_l, \quad (a_1, a_2) \mapsto a_l, \quad l = 1, 2,$$

are smooth, since $p_l = |\pi_l|$, where the $\mathbb{R}$-algebra homomorphism π is the composition

$$\mathcal{F}_l \xrightarrow{i_l} \mathcal{F}_1 \otimes_{\mathbb{R}} \mathcal{F}_2 \xrightarrow{\sigma} \overline{\mathcal{F}_1 \otimes_{\mathbb{R}} \mathcal{F}_2} = \mathcal{F}(M_1 \times M_2);$$

here σ is the smooth envelope homomorphism (3.36), and

$$i_l(f) = \begin{cases} f \otimes 1, & l = 1, \\ 1 \otimes f, & l = 2, \end{cases} \quad \text{for all } f \in \mathcal{F}_l.$$

The pair of projection maps (p_1, p_2) possesses the following universal property.

Proposition. *For any smooth manifold N and any pair of smooth maps $f_l \colon N \to M_l$, $l = 1, 2$, there exists a unique smooth map $f \colon N \to M_1 \times M_2$ completing the commutative diagram*

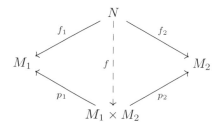

◀ Denote the algebra of smooth functions on N by $\mathcal{F}(N)$. Assume that $f_l = |\varphi_l|$, where $\varphi_l \colon \mathcal{F}_l \to \mathcal{F}(N)$, $l = 1, 2$, are the dual $\mathbb{R}$-algebra homomorphisms. Our proposition will be proved if we establish the existence and uniqueness of the following diagram:

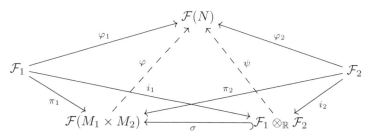

By the universal property of tensor products, there exists a unique homomorphism ψ shown in the diagram. Since the $\mathbb{R}$-algebra $\mathcal{F}(N)$ is smooth, while the smooth envelope homomorphism has the universal property stated in Proposition 3.37, the homomorphism φ is also well defined and unique.   ▶

6.13. Remark. Proposition 6.12 may be used to construct smooth maps from a third manifold M_3 to the product $M_1 \times M_2$ of two given ones. For example, a smooth map of a manifold N to $S^1 \times S^1$ is a pair of smooth maps from N to S^1, i.e., a pair of smooth functions on N defined modulo 1.

6.14. The remainder of this section is a discussion of the notions and constructions developed in 6.1–6.13 carried out in the "coordinate language" introduced in Chapter 5.

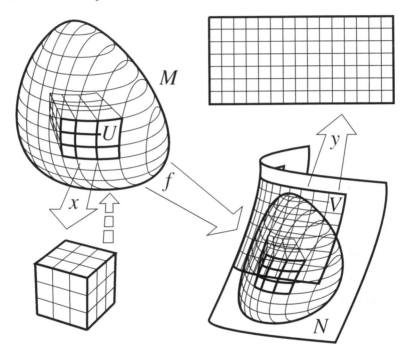

Figure 6.4. Smooth map in the "coordinate language."

Suppose (M, A) and (N, B) are manifolds with smooth atlases A and B (see Section 5.8). The map $f \colon M \to N$ is called *smooth* (or *differentiable*) at *the point* $a \in M$ if for some (hence for all) pairs of charts (U, x) and (V, y) compatible with the atlases A and B and covering the points a and $f(a)$, respectively, the map $y \circ f \circ x^{-1}$, defined in the neighborhood $x(f^{-1}(V) \cap U)$ of the point $x(a) \in \mathbb{R}^n$, is an infinitely differentiable map of domains in Euclidean space (see Figure 6.4). The map $f \colon M \to N$ is called *smooth* if it is smooth at each point.

In the coordinate language, a smooth bijective map $f \colon M \to N$ is called a *diffeomorphism* if the map f^{-1} is also smooth.

6.15. When working with specific maps of manifolds, we ordinarily use a coordinate representation. Actually, this means that we use the above-mentioned map $y \circ f \circ x^{-1}$, which, being a map of Euclidean spaces, is represented by certain functions

$$y_i = f_i(x_1, \ldots, x_m), \quad i = 1, \ldots, n,$$

where $m = \dim M$, $n = \dim N$. For a smooth map (at a point), all the functions f_i, $i = 1, \ldots, n$, will also be smooth (respectively, at a point).

It is easy to see that if a map is described by smooth functions in a family of pairs of charts compatible with the corresponding atlases (and the charts cover the entire manifold M), then this map will be described by smooth functions for any pair of charts compatible with the same atlases (of course, charts for which the composition $y \circ f \circ x^{-1}$ is undefined are not taken into account).

The equivalence of these coordinate definitions and the corresponding algebraic ones will be established in Chapter 7.

6.16. Examples. I. The map $a \colon T^2 \to \mathbb{R}^2$ that assigns to each position of the double pendulum (Example 5.2, III) its end point A (see Figure 5.1) is smooth.

◀ Choose some fixed position of the double pendulum; then all sufficiently close positions of the pendulum are characterized by two angles x, y, so that we have a chart

$$U \ni (\text{position}) \overset{\Phi}{\mapsto} (x, y)$$

that is compatible with the standard atlas of the torus. The manifold $\mathbb{R}^2$ can be covered by a single chart $\mathrm{id} \colon \mathbb{R}^2 \to \mathbb{R}^2$. We can say that *in the chosen local coordinates the map a is described by the formulas*

$$r_1 = R \cos x + r \cos y, \ r_2 = R \sin x + r \sin y.$$

The rigorous meaning of the words in italics is that the formulas actually describe the map $(\mathrm{id})^{-1} \circ a \circ \Phi^{-1}$; hence it follows from Definition 6.14 that a is smooth.   ▶

II. Choose a fixed unit vector $v \in S^{n-1}$ in $\mathbb{R}^n$ and consider the map

$$f_v \colon \mathrm{SO}(n) \to S^{n-1}, \quad A \mapsto A(v).$$

Verify that f_v is a smooth map by using the atlases described in Examples 5.10, I, and 5.11, II.

We can also consider the map

$$\varphi \colon \mathrm{SO}(n) \times S^{n-1} \to S^{n-1}, \quad (A, v) \mapsto A(v),$$

and attempt to prove its smoothness, working with an atlas on the product $\mathrm{SO}(n) \times S^{n-1}$. The reader will be wise not to take this attempt too seriously, but to prove smoothness of φ eventually by using the algebraic definitions.

6.17. More examples I. Let us consider four-dimensional Euclidean space $\mathbb{R}^4$ as the algebra of quaternions $\mathbb{H}$, denote by $V \cong \mathbb{R}^3$ the subspace of purely imaginary quaternions $r_1 i + r_2 j + r_3 k$, and introduce the map

$$(\mathbb{H} \smallsetminus \{0\}) \times V \to V, \quad (q, v) \mapsto qvq^{-1}.$$

It is easy to check that for each nonzero quaternion q, the matrix of the linear operator $v \mapsto qvq^{-1}$ in the coordinates r_1, r_2, r_3 is orthogonal, so that this is a map $\mathbb{H} \smallsetminus \{0\} \to \mathrm{SO}(3)$. Two quaternions q_1 and q_2 determine the same transformation of the space V if and only if $q_1 = \lambda q_2$ for some $\lambda \in \mathbb{R} \smallsetminus \{0\}$, so that we have obtained a bijective smooth map

$$\mathbb{R}P^3 \to \mathrm{SO}(3).$$

Recall that $\mathbb{R}P^3$ is precisely the quotient manifold of the punctured space $\mathbb{R}^4 \smallsetminus \{0\} = \mathbb{H} \smallsetminus \{0\}$ by the group $\mathbb{R}^{\cdot}$ of all homotheties with center at the origin (see Example 4.25, II) and that the projection $\mathbb{H} \smallsetminus \{0\} \to \mathbb{R}P^3$ has the universal property (see Proposition 6.10). It is not useless to try to establish the smoothness of this map by using the atlases 5.10, III, and 5.11, II. It is more difficult to prove the fact that it is a diffeomorphism (as mentioned in Section 1.6).

II. The composition

$$S^3 \to \mathbb{R}^4 \smallsetminus \{0\} \to \mathrm{SO}(3) \to S^2,$$

where the first arrow is the inclusion and the others are defined in Examples 6.16, II, and 6.17, I, is called the *Hopf map*. Notice that the composition of the first two arrows may be represented in the form

$$S^3 \to \mathbb{R}P^3 \xrightarrow{\cong} \mathrm{SO}(3),$$

where the map $S^3 \to \mathbb{R}P^3$ was described in Example 4.25, III, as the quotient map $S^3 \to S^3/\mathbb{Z}_2 \cong \mathbb{R}P^3$.

Exercise. Try to show that the inverse image of any point of S^2 under the Hopf map $h \colon S^3 \to S^2$ is a closed submanifold of S^3, diffeomorphic to S^1.

6.18. Example (of a smooth map of a manifold with boundary). Suppose D^3 is the 3-dimensional closed disk with center at the origin $O \in \mathbb{R}^3$ and of radius π. To each point $a \in D^3$ associate the rotation of $\mathbb{R}^3$ about the line joining a to the origin by the angle $\alpha = \|a\|$. Thus we obtain the map $g \colon D^3 \to \mathrm{SO}(3)$.

Clearly, diametrically opposed points on the boundary $\partial D^3 = S^2$ of D^3 determine the same rotation. Thus the manifold $\mathrm{SO}(3) \cong \mathbb{R}P^3$ of orthogonal transformations of $\mathbb{R}^3$ can be represented as the disk D^3 whose diametrically opposed boundary points have been glued together.

The map $g \colon D^3 \to \mathrm{SO}(3)$ is a smooth surjective map of a manifold with boundary onto a manifold (without boundary).

Exercise. Show that the image of the boundary $S^2 = \partial D^3$ under this map is a smooth closed submanifold in $\mathrm{SO}(3)$ diffeomorphic to the manifold $\mathbb{R}P^2$.

6.19 Exercises.   1. Write out the formulas for the orthogonal projection of the unit sphere $S^n \subset \mathbb{R}^{n+1}$ onto the hyperplane $\mathbb{R}^n \subset \mathbb{R}^{n+1}$ in the charts indicated in 5.10 (i) and verify that this projection is a smooth map.

2. Prove that $SO(4) \cong S^3 \times SO(3)$. *Hint*: $SO(3)$ may be understood as the set of orthonormed pairs $\{u, v\}$ in $\mathbb{R}^3$, $SO(4)$ as the set of orthonormed triples $\{u, v, w\}$ in $\mathbb{R}^4$; let $\mathbb{R}^4 = \mathbb{H}$ and $\mathbb{R}^3 = V \subset \mathbb{H}$ as in Example 6.17, I; investigate the map

$$S^3 \times SO(3) \to SO(4), \quad (u, \{v, w\}) \mapsto \{u, uv, uw\}.$$

6.20. The collection of all smooth maps (in the sense of the coordinate Definition 6.13) also possesses the two properties concerning composition and identity maps mentioned in Section 6.6 (which is not surprising, since the coordinate definition is equivalent to the algebraic one; see Section 6.1). The class of all smooth manifolds in the sense of Section 5.8 together with the family of all smooth (Section 6.14) maps constitutes the *category of coordinate manifolds*.

In what follows, we shall need some classical results of multidimensional local calculus, i.e., the theorems on implicit and inverse functions. We formulate them here in a form convenient for the subsequent exposition, without any proofs. The latter may be found in any advanced calculus course.

6.21 Inverse Function Theorem. *Let a smooth map*

$$f = (f_1, \ldots, f_n) \colon \mathbb{R}^n \to \mathbb{R}^n$$

possess a nondegenerate Jacobi matrix in a neighborhood of the origin $0 \in \mathbb{R}^n$:

$$\det f'(0) = \det \left(\frac{\partial f_i}{\partial x_j}(0) \right) \neq 0.$$

Then there exist open sets $U \ni 0$ and $V \ni f(0)$ such that the map

$$\varphi \overset{\text{def}}{=} f|_U \colon U \to V$$

possesses a smooth inverse $\varphi^{-1} \colon V \to U$. The Jacobi matrix of the latter at any point $y \in V$ can be computed by the formula

$$(\varphi^{-1})'(y) = \left(\varphi'(\varphi^{-1}(y)) \right)^{-1}. \quad \blacktriangleright$$

6.22 Implicit Function Theorem. *Let*

$$f = (f_1, \ldots, f_{n+m}) \colon \mathbb{R}^n \times \mathbb{R}^m \to \mathbb{R}^m$$

be a smooth map with $f(a,b) = 0 \in \mathbb{R}^m$ possessing a nondegenerate $(m \times m)$ matrix of partial derivatives with respect to the variables $x_{n+1}, \dots, x_{n+m}$:

$$\det\left(\frac{\partial f_i}{\partial x_{n+j}}(a)\right)_{1 \leqslant i,j \leqslant m} \neq 0.$$

Then there exist open sets U and V, $a \in U \subset \mathbb{R}^n$, $b \in V \subset \mathbb{R}^m$, and a smooth function $g\colon U \to V$ such that $f(x, g(x)) = 0$ for all $x \in U$. ▶

Finally, we shall need another classical theorem (the theorem on the linearization of a smooth map), which is an amalgam of the previous two results.

6.23 Theorem. *Let $f\colon \mathbb{R}^n \to \mathbb{R}^m$, $m \leqslant n$, be a smooth map such that $f(0) = 0$ and the rank of the matrix*

$$M = \left(\frac{\partial f_i}{\partial x_j}(0)\right)_{\substack{1 \leqslant i \leqslant m. \\ 1 \leqslant j \leqslant n}}$$

is equal to m. Then there exist neighborhoods U, $V \subset \mathbb{R}^n$ of 0 and a diffeomorphism $\varphi\colon U \to V$ such that

$$(f \circ \varphi)(x_1, \dots, x_n) = (x_{n-m+1}, \dots, x_n).$$ ▶

6.24. Remark. The three theorems above are "local": their assumptions are formulated in Euclidean spaces, while the conclusions are valid for neighborhoods (of Euclidean spaces). Nevertheless, they may be applied to the "global" case of manifolds by considering separate coordinate neighborhoods. By *uniqueness* of the inverse (and implicit) functions in the corresponding neighborhood, the functions constructed in these theorems coincide on the common part of the two neighborhoods. Therefore, we can "glue" them together over the entire manifold. Thus, for example, the conclusion of the implicit function theorem may be formulated as follows: *If $f\colon M \to N$ is a smooth map of manifolds, $m = \dim M > n = \dim N$, and the Jacobian of f has rank n at any point, then $f^{-1}(z) \subset M$ is a submanifold for any $z \in N$.*

6.25 Exercises. A plane hinge mechanism (see Section 1.14) is called *generic* if the configuration with all the hinges positioned along the same straight line is impossible (i.e., there is no linear relation with coefficients ± 1 among the lengths of the rods).

1. Prove that the configuration space of a generic hinge mechanism is a smooth manifold.

2. Show that the configuration space of a generic pentagon (Section 1.14) is diffeomorphic either to the sphere with no more than 4 handles, or to the disjoint union of two tori, or to the disjoint union of two spheres.

7

Equivalence of Coordinate and Algebraic Definitions

7.1. The aim of this chapter is to prove that the two definitions of smooth manifold (4.1 and 5.8) and of smooth maps (Sections 6.1 and 6.13) yield the same concepts, thus showing that the coordinate approach and the algebraic one are equivalent.

The equivalence of the two definitions of smooth manifold will be stated in the form of two theorems (7.2 and 7.7). The equivalence of the definitions of smooth maps is Theorem 7.16.

7.2 Theorem. *Suppose $\mathcal{F} = C^\infty(M)$ is the algebra of smooth functions on a manifold M defined by its smooth atlas A. Then $\mathcal{F}$ is a smooth $\mathbb{R}$-algebra (in the sense of Section 4.1), and the map*

$$\theta \colon M \to |\mathcal{F}|, \quad \big(\theta(p)\big)(f) = f(p),$$

is a homeomorphism.

◀ The proof will be in four steps. First we shall establish that the map $\theta \colon M \to |\mathcal{F}|$ is bijective (Section 7.3), then that it is a homeomorphism (Section 7.4); then we shall show that $C^\infty(M)$ is geometrical and complete (Section 7.5) and finally prove that $C^\infty(M)$ is smooth (Section 7.6).

7.3. *The map $\theta \colon M \to |\mathcal{F}|$ is bijective.*
◁ Injectivity is obvious: If $p, q \in M$ are distinct points, then there exists a function $f \in C^\infty(M)$ such that $f(p) \neq f(q)$ (e.g., any function positive in a small neighborhood of p not containing q and identically zero outside of it; see Proposition 2.4); for this function the values of the homomorphisms

© Springer Nature Switzerland AG 2020
J. Nestruev, *Smooth Manifolds and Observables*, Graduate Texts
in Mathematics 220, https://doi.org/10.1007/978-3-030-45650-4_7

$p, q \colon f \to \mathbb{R}$ differ, since

$$q(f) = f(q) \neq f(p) = p(f).$$

To prove surjectivity, suppose $p \colon \mathcal{F} \to \mathbb{R}$ is any homomorphism; let $f \in C^\infty(M)$ be any function with compact level surfaces (see Proposition 5.17) and $\lambda = p(f)$. Suppose that none of the points of the compact set $L = f^{-1}(\lambda)$ correspond to the homomorphism p. Then there exists a family of functions $\{f_x \mid x \in L\}$ such that $f_x(x) \neq p(f_x)$. Consider the covering of L by the open sets

$$U_x = \{q \in M \mid f_x(q) \neq p(f_x)\}$$

and choose a finite subcovering $U_{x_1}, \ldots, U_{x_m}$. Consider the function

$$g = (f - \lambda)^2 + \sum_{k=1}^{m} (f_{x_k} - p(f_{x_k}))^2.$$

This is a smooth function on M that vanishes nowhere and therefore possesses a smooth inverse $1/g$. Now we easily obtain the standard contradiction, familiar from the examples:

$$p(g) = (p(f) - \lambda)^2 + \sum_{k=1}^{m} (p(f_{x_k}) - p(f_{x_k}))^2 = 0,$$

$$1 = p(1) = p\left(g \cdot \frac{1}{g}\right) = p(g)p\left(\frac{1}{g}\right) = 0.$$

Hence the homomorphism p is given by some point of $L \subset M$, so that θ is surjective. $\triangleright$

7.4. *The map $\theta \colon M \to |\mathcal{F}|$ is a homeomorphism.*
$\triangleleft$ Let U be an open set in $|\mathcal{F}|$. Then, by Definition 3.12, the set U is the union of sets of the form $f^{-1}(V)$, where $V \subset \mathbb{R}$ is open. Since we have $f \in \mathcal{F} = C^\infty(M)$ is smooth (hence continuous), the sets $f^{-1}(V)$ are open in the topology of M, and so is U.

Conversely, for any open set U in the topology of M there exists a function $f \in \mathcal{F}$ such that $U = f^{-1}(\mathbb{R}_+)$. In fact, Lemma 4.17 (i) remains obviously valid if in its statement M is a manifold in the sense of Section 5.8. But $\mathbb{R}_+ \subset \mathbb{R}$ is open, so that U is open in the topology of $|\mathcal{F}|$. $\triangleright$

7.5. *The algebra $\mathcal{F} = C^\infty(M)$ is geometrical and complete.*
$\triangleleft$ The fact that $\mathcal{F}$ is geometrical is obvious, since the elements $f \in \mathcal{F}$ are real functions on the set M, so that only the identically zero function vanishes at all points of M. The fact that $\mathcal{F}$ is complete is also obvious: Any function $f \colon M \to \mathbb{R}$ that is smooth in a neighborhood of every point is smooth on the entire manifold, i.e., belongs to $\mathcal{F} = C^\infty(M)$ (see Section 5.17). $\triangleright$

7.6. *$C^\infty(M)$ is a smooth $\mathbb{R}$-algebra.*

◁ To prove this, we shall construct a countable atlas $A_2 = \{(U_k, x_k)\}$ such that $x_k(U_k) = \mathbb{R}^n$. We begin with an arbitrary countable atlas $A_0 = \{(V_l, y_l)\}$ compatible with the given atlas A. Each open set $y_l(V_l) \subset \mathbb{R}^n$ may be represented as the countable union of open balls in $\mathbb{R}^n$; i.e., $y_l(V_l) = \bigcup_{i=1}^{\infty} G_{li}$, where

$$G_{li} = \{r \in \mathbb{R}^n \mid \|r - a_{li}\| < r_{li}\}.$$

The family $A_1 = \left\{ \left(y_l^{-1}(G_{li}), y_l\big|_{y_l^{-1}(G_{li})} \right) \mid l, i \in \mathbb{N} \right\}$ is obviously a countable atlas on M, compatible with A_0 and hence with A. Now note that for any chart (U, x) on M, where $x(U)$ is an open ball in $\mathbb{R}^n$, we can construct the chart $(U, \eta \circ x)$, where $\eta \circ x(U) = \mathbb{R}^n$, by taking η to be any diffeomorphism of the ball $x(U)$ onto $\mathbb{R}^n$. For instance, if $x(U)$ is the ball of radius ρ and center a, we can take η to be the bijective map $\eta \colon \mathbb{R}^n \to x(U)$ given by the formula

$$\eta(s) = a + \frac{s\rho}{\sqrt{\rho^2 + \|s\|^2}}.$$

The map η has an inverse, namely

$$\eta^{-1}(r) = \frac{\rho(r - a)}{\sqrt{\rho^2 - \|r - a\|^2}},$$

and therefore the chart $(U, \eta \circ x)$ is compatible with (U, x).

Carrying out this construction for every chart in A_1, we obtain the required atlas $A_2 = \{(U_k, x_k)\}$.

Consider any chart $(U_k, x_k) \in A_2$. The set U_k is obviously open in the topological space M. Clearly, the restriction of the algebra $\mathcal{F} = C^\infty(M)$ to this set consists of all functions $f \colon U_k \to \mathbb{R}$ such that $f \circ x_k^{-1}$ is smooth on $\mathbb{R}^n$; hence the assignment $f \mapsto f \circ x_k^{-1}$ is an isomorphism of the $\mathbb{R}$-algebra $C^\infty(M)\big|_{U_k}$ onto $C^\infty(\mathbb{R}^n)$. This means that $\mathcal{F} = C^\infty(M)$ is smooth.   ▷

This concludes the proof of Theorem 7.2.   ◀

7.7 Theorem. *Suppose $\mathcal{F}$ is any smooth $\mathbb{R}$-algebra. Then there exists a smooth atlas A on the dual space $M = |\mathcal{F}|$ such that the map*

$$\mathcal{F} \to C^\infty(M), \quad f \mapsto (p \mapsto p(f)),$$

of the algebra $\mathcal{F}$ onto the algebra $C^\infty(M)$ of smooth functions on M with respect to A (see Section 5.17) is an isomorphism.

◀ The proof will require four steps. In the first one (Section 7.8) we construct a chart $x \colon U \to \mathbb{R}^n$ for each of the open sets U of a countable covering of $|\mathcal{F}|$, using a lemma proved in the second step (Section 7.9). In the third step (Section 7.10) we show that any two such charts are compatible. In the fourth and final step (Section 7.11) we show that $f \in \mathcal{F}$ iff $f \in C^\infty(M)$ (where f, an abstract element of $\mathcal{F}$, is identified with the function $f \colon |\mathcal{F}| \to \mathbb{R}, \; p \mapsto p(f)$).

7.8. Construction of a chart $x \colon U \to \mathbb{R}^n$.

◁ By the definition of smooth $\mathbb{R}$-algebras (Section 4.1), there is an open covering of $|\mathcal{F}|$ by open sets U for each of which there exist an isomorphism $i \colon \mathcal{F}|_U \to C^\infty(\mathbb{R}^n)$. Taking any such U, we shall construct the required chart $x \colon U \to \mathbb{R}^n$.

The composition

$$U \xrightarrow{\mu} \left|\mathcal{F}|_U\right| \xrightarrow{|\rho|} |\mathcal{F}|,$$

where μ is the inclusion (see Section 3.29) and $\rho \colon \mathcal{F} \to \mathcal{F}|_U$ the restriction map (Section 3.25), as can easily be checked, coincides with the inclusion $U \subset |\mathcal{F}|$. Since μ is a homeomorphism *onto* $\left|\mathcal{F}|_U\right|$ (this is proved in Lemma 7.9 below) and $|\rho| \circ \mu$ is the inclusion $U \subset |\mathcal{F}|$, it follows that $|\rho|$ must be a homeomorphism onto $U \subset |\mathcal{F}|$.

Now let $h \colon \mathcal{F} \to C^\infty(\mathbb{R}^n)$ be the composition

$$\mathcal{F} \xrightarrow{\rho} \mathcal{F}|_U \xrightarrow{i} C^\infty(\mathbb{R}^n), \quad h = i \circ \rho.$$

Consider the dual map $|h| = |\rho| \circ |i|$. Since $|C^\infty(\mathbb{R}^n)| = \mathbb{R}^n$ (by Example 3.16) and i is an isomorphism, it follows from Section 3.20 that $|i|$ is a homeomorphism $|i| \colon \mathbb{R}^n \to \left|\mathcal{F}|_U\right|$. But $|\rho|$ is also a homeomorphism (onto $U \subset |\mathcal{F}|$), and so the composition $|h| = |\rho| \circ |i|$ is a homeomorphism $|h| \colon \mathbb{R}^n \to U$, where $U \subset |\mathcal{F}|$ is open. Hence we obtain the required chart by putting $x = |h|^{-1} \colon U \to \mathbb{R}^n$.   ▷

7.9 Lemma. *The inclusion map* $\mu \colon U \to \left|\mathcal{F}|_U\right|$ *is a homeomorphism onto* $\left|\mathcal{F}|_U\right|$.

◁ Suppose μ is not surjective; i.e., assume that there exists a point $a \in \left|\mathcal{F}|_U\right| \setminus \mu(U)$; set $\overline{a} = |i|^{-1}(a)$, where $i \colon \mathcal{F}|_U \to C^\infty(\mathbb{R}^n)$ is the isomorphism appearing in Section 7.8. We consider two cases, depending on whether a belongs to the closure of $\mu(U)$ or not.

First case: $a \in \overline{\mu(U)}$. Consider the function $f \colon x \mapsto 1/\|x - \overline{a}\|$ defined on $\mathbb{R}^n \setminus \{\overline{a}\}$ and the function $g = f \circ |i|^{-1} \circ \mu$ defined on U. We claim that $g \in \mathcal{F}|_U$. Indeed, any point $r \in \mathbb{R}^n \setminus \{\overline{a}\}$ possesses a neighborhood on which f coincides with a smooth function defined on all of $\mathbb{R}^n$. Taking the inverse images of such neighborhoods under the map $|i|^{-1} \circ \mu$, we see that any point $q \in U$ possesses a neighborhood on which g coincides with a function from $\mathcal{F}|_U$. By the definition of $\mathcal{F}|_U$ (Section 3.23), this means that g locally coincides with functions belonging to $\mathcal{F}$; hence $g \in \mathcal{F}|_U$, as claimed. But now if we consider the function $i(g)$, which is a smooth function (on the entire space $\mathbb{R}^n$) coinciding with f on the set $|i|^{-1}(\mu(U))$ whose closure contains $\overline{a}$, we obtain a contradiction (since f "becomes infinite" at $\overline{a}$).

Second case: $a \notin \overline{\mu(U)}$. Consider two smooth functions on $\mathbb{R}^n$: the identically zero one, and a function f_0 that vanishes on the closed set $|i|^{-1}(\overline{\mu(U)})$ and equals 1 at $\overline{a}$ (f_0 exists by Corollary 2.5). These are distinct functions, so that their pullbacks by i on $\mathcal{F}\big|_U$ are different elements of this algebra, which is impossible, since both pullbacks vanish on U.

This proves the surjectivity of μ. The fact that it is a homeomorphism follows from Proposition 3.29. $\triangleright$

7.10. *The charts constructed in* Section 7.8 *are compatible.*
$\triangleleft$ Suppose $x\colon U \to \mathbb{R}^n$ and $y\colon V \to \mathbb{R}^n$ are two such charts, while $i\colon \mathcal{F}\big|_U \to C^\infty(\mathbb{R}^n)$ and $j\colon \mathcal{F}\big|_V \to C^\infty(\mathbb{R}^n)$ are the corresponding $\mathbb{R}$-algebra isomorphisms. Let $W = U \cap V \neq \varnothing$ (the case $W = \varnothing$ is trivial). By Proposition 3.25 we have the isomorphisms

$$ i\big|_W \colon \mathcal{F}\big|_W \to C^\infty(x(W)), \quad j\big|_W \colon \mathcal{F}\big|_W \to C^\infty(y(W)), $$

which give us the isomorphism $t\colon C^\infty(x(W)) \to C^\infty(y(W))$; this, by Proposition 3.16, shows that $|t|\colon y(W) \to x(W)$, the change of coordinate map, is a diffeomorphism. $\triangleright$

7.11. Final step. $\triangleleft$ Suppose $f \in \mathcal{F}$. To prove that f is smooth in the sense of $C^\infty(M)$ (see Section 5.16), we must show that the function $f \circ x^{-1}$ is a smooth function on $\mathbb{R}^n$ for each of the charts $x\colon U \to \mathbb{R}^n$ constructed in 7.8. But we have (see 7.8)

$$ f \circ x^{-1} = f \circ |h| = f \circ |\rho| \circ |i| = i(\rho(f)) \in C^\infty(\mathbb{R}^n), $$

since i is the isomorphism $i\colon \mathcal{F}\big|_U \to C^\infty(\mathbb{R}^n)$. Thus $f \circ x^{-1} \in C^\infty(\mathbb{R}^n)$.

Conversely, let $f \in C^\infty(M)$. This means that for any chart (U, x) the function $f \circ x^{-1} = i(f \circ |\rho|)$ belongs to $C^\infty(\mathbb{R}^n)$ and is the image (by i) of an element of $\mathcal{F}\big|_U$ (namely $f \circ |\rho|$). Thus f locally coincides with elements of $\mathcal{F}\big|_U$ and hence of $\mathcal{F}$. Since $\mathcal{F}$ is complete, $f \in \mathcal{F}$. $\triangleright$ $\blacktriangleright$

7.12. Thus we have established the equivalence of the two definitions of smooth manifold: the algebraic one (Definition 4.1) and the coordinate one (Definition 5.8). Theorems similar to 7.2 and 7.7 are valid for manifolds with boundary. The proofs are similar (with obvious modification here and there). The reader who wishes to check that he or she has mastered the contents of Sections 7.1–7.11 will benefit by carrying them out in detail.

7.13 Definition. A *smooth set* is a pair $(W, C^\infty(W))$, where W is a closed subset $W \subset M$ of a smooth manifold M, and $C^\infty(W)$ is the algebra of smooth functions on W defined as follows:

$$ C^\infty(W) \overset{\text{def}}{=} \left\{ f\big|_W \mid f \in C^\infty(M) \right\}. $$

Exercise. Prove that theorems similar to 7.2 and 7.7 are valid for smooth sets as well.

7.14 Exercise. Describe the algebras $C^\infty(W)$ for the following cases:

1. $W = \mathbf{K}$ is the coordinate cross on the plane:
$$ \mathbf{K} = \left\{ (x, y) \subset \mathbb{R}^2 \mid xy = 0 \right\}. $$

2. $W \subset \mathbb{R}^2$ is given by the equation $y = \sqrt{|x|}$.

3. W is a triangle in $\mathbb{R}^2$: $W = T_1 \cup T_2 \cup T_3$, where

$$T_1 = \{(x,y) \mid 0 \leqslant y \leqslant 1,\ x = 0\},$$
$$T_2 = \{(x,y) \mid 0 \leqslant x \leqslant 1,\ y = 0\},$$
$$T_3 = \{(x,y) \mid x + y = 1,\ x, y \geqslant 0\}.$$

4. W is the triangle described in Problem 3 together with its interior part: $W = \{(x,y)\} \mid x + y \leqslant 1,\ x, y \geqslant 0\}$.

5. W is the cone $x^2 + y^2 = z^2$ in $\mathbb{R}^3$.

6. $W = W_i$, $i = 1, 2, 3$, is one of the three one-dimensional pairwise homeomorphic polyhedra depicted in Figure 7.1 (W_1 lives in $\mathbb{R}^2$, while W_2 and W_3 are in $\mathbb{R}^3$). In Chapter 9 the reader will find Exercise 9.36, 5, in which it will be required to prove that the algebras $C^\infty(W_i)$, $i = 1, 2, 3$, are mutually non-isomorphic.

7. $W \subset \mathbb{R}^2$ is the closure of the graph of the function $y = \sin \frac{1}{x}$.

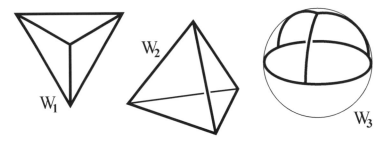

Figure 7.1. 1-skeleton of the tetrahedron.

Of course, there are various alternative descriptions of the algebras in question. For instance, perhaps the most direct and constructive way to represent smooth functions on the triangle (see Problem 3 above) is by means of triples (f_1, f_2, f_3), $f_i \in C^\infty([0,1])$, such that

$$f_1(1) = f_2(0),\ f_2(1) = f_3(0),\ f_3(1) = f_1(0).$$

Try to give a similar description of smooth functions on the cross and on the polyhedra mentioned in Problems 1 and 6, respectively.

7.15. Smooth sets sometimes appear implicitly in various mathematical problems. We illustrate this in the following exercises.

Exercises.   1. Show that the configuration space of a hinge mechanism can be viewed as a smooth set.

2. Determine which of the smooth sets corresponding to quadrilaterals and pentagons listed in Exercise 1 from Section 1.14 are not smooth manifolds and describe their smooth function algebras (cf. Exercise 6.25).

3. Prove that the smooth set corresponding to a nonrigid nongeneric pentagon (see Exercise 6.25), e.g., $(2; 2, 1, 1, 2)$, is not a smooth manifold and describe its singular points (a pentagon is called rigid if its configuration space consists of one point, e.g., $(4; 1, 1, 1, 1)$).

Remark. The smooth set corresponding to the quadrilaterals $(5; 3, 3, 1)$ is "diffeomorphic" to a pair of tangent circles at a point, say z. Smooth functions on this set can be viewed as pairs (f_1, f_2), where $f_1, f_2 \in C^\infty(S^1)$, such that $f_1(z) = f_2(z)$ and $f_1'(z) = f_2'(z)$. Try to give similar descriptions for smooth functions on smooth sets appearing in Exercises 2 and 3.

7.16. The coincidence of the two definitions of a smooth map given in 6.1 and 6.14 is guaranteed by the following theorem.

Theorem. *Let M and N be manifolds with smooth atlases A and B and smooth function algebras $\mathcal{F}_M$ and $\mathcal{F}_N$, respectively. A map $\varphi\colon M \to N$ is smooth with regard to A and B (Section 6.14) if and only if $\varphi^*(\mathcal{F}_N) \subset \mathcal{F}_M$, where*

$$\varphi^*\colon \mathcal{F}_N \to \mathcal{F}_M, \quad f \mapsto f \circ \varphi.$$

The proof is given below, with Sections 7.17 and 7.18 corresponding to the "only if" and the "if" parts of the theorem, respectively.

7.17. *If $\varphi\colon M \to N$ is smooth, then $\varphi^*(\mathcal{F}_N) \subset \mathcal{F}_M$.*
◁ Suppose $f \in \mathcal{F}_N$ and $a \in M$. Choose a chart (V, y) in a neighborhood of $\varphi(a)$ and a chart (U, x) in a neighborhood of a, the charts (V, y) and (U, x) being compatible with B and A, respectively, and satisfying $\varphi(U) \subset V$. Then

$$\varphi^*(f) \circ x^{-1} = (f \circ \varphi) \circ x^{-1} = \left(f \circ y^{-1} \right) \circ \left(y \circ \varphi \circ x^{-1} \right)$$

is smooth as the composition of two smooth maps of Euclidean domains. Thus locally the map $\varphi^*(f)$ coincides with a map locally coinciding with an element of $\mathcal{F}_M$. Since $\mathcal{F}_M$ is complete (see 7.5), it follows that $\varphi^*(f) \in \mathcal{F}_M$. ▷

7.18. *If $\varphi^*(\mathcal{F}_N) \subset \mathcal{F}_M$, then $\varphi\colon M \to N$ is smooth.*
◁ Choose arbitrary charts $(U, x) \in A$ and $(V, y) \in B$ such that $\varphi(U) \subset V$. We must prove, according to Definition 6.13, that the local coordinates of the point $\varphi(a)$ are smooth functions of the local coordinates of $a \in U$. In other words, the functions $y^i \circ \varphi = \varphi^*\left(y^i\right)$ must be smooth. For every function y^i, let us choose a function $f_i \in \mathcal{F}_N$ such that $y^i = f_i|_V$. By the assumption $\varphi^*(\mathcal{F}_N) \subset \mathcal{F}_M$, the functions $\varphi^*(f_i) = f_i \circ \varphi$ are smooth, and thus the functions $y^i \circ \varphi|_U = \varphi^*(f_i)|_U$ are smooth as well. ▷

7.19. We have proved that two categories, the category of manifolds (as smooth atlases) with smooth maps in the sense of 6.13 and the category of manifolds (as smooth $\mathbb{R}$-algebras) with smooth maps in the sense of 6.1, are equivalent. From now on we shall not differentiate between the two categories. The notation $(M, \mathcal{F})$ or $(M, C^\infty(M))$ will be used to denote a manifold; the identifications $M = |\mathcal{F}|$ and $\mathcal{F} = C^\infty(M)$ will always be implied.

The reader acquainted with the notion of complex manifold has probably noticed already that, in general, such a manifold does not coincide with the complex spectrum of its algebra of holomorphic functions. For example, as is well known from the elementary theory of functions of a complex variable, all holomorphic functions on the Riemann sphere (and on any compact complex manifold) reduce to constants. For this reason, it might seem that the "spectral approach" adopted in this book is less universal than the standard one, based on charts and atlases. Nevertheless, the observability principle forces us to understand complex manifolds as smooth ones, but equipped with an additional (complex) structure. In other words, complex manifolds within this approach are understood as solutions of certain differential equations, while complex charts appear as local solutions of these equations. This viewpoint, going back to Riemann, has many advantages, despite its apparent lack of simplicity. For example, it can be generalized to any commutative algebra by the methods of the "algebraic" differential calculus described below in Chapters 9 and 12.

8
Points, Spectra, and Ghosts

8.1. Spectra of commutative $\Bbbk$-algebras. In the previous chapters, we looked at commutative $\mathbb{R}$-algebras from the "physical" point of view, interpreting them as observation instruments. Namely, a commutative $\mathbb{R}$-algebra A allows to observe (or see) the topological space $|A|$. Moreover, we noted that the very construction of this space is in no way related to the specifics of the field $\mathbb{R}$ and is of a more general algebraic nature. Indeed, let A be a commutative unitary ($=$ with unit) algebra over an arbitrary field $\Bbbk$. (In what follows, if this is required by the context, the unit of the algebra A is denoted by 1_A and, if the converse is not stated explicitly, the term "commutative algebra" means commutative algebra with unit.)

Let us put

$$|A| \stackrel{\text{def}}{=} \{h\colon A \to \Bbbk \mid h \in \text{Hom}_{\Bbbk}(A, \Bbbk), \quad h \neq 0\}.$$

The set $|A|$ defined in this way is called the $\Bbbk$-*spectrum* or the *dual space* of the algebra A. It is endowed with the natural topology, whose base of open sets consists of sets of the form

$$U_a \stackrel{\text{def}}{=} \{h \in |A| \mid h(a) \neq 0\}, \quad a \in A.$$

The obvious relation $U_{ab} = U_a \cap U_b$ shows that any finite intersection of sets of the base is also a set of the base. In algebraic geometry, such a topology is known as the *Zariski topology*, and we shall use this term in our more general context. Similarly, using geometrical terminology, we will refer to the homomorphism $h \in |A|$ as a *point of the spectrum* of the algebra A.

© Springer Nature Switzerland AG 2020
J. Nestruev, *Smooth Manifolds and Observables*, Graduate Texts
in Mathematics 220, https://doi.org/10.1007/978-3-030-45650-4_8

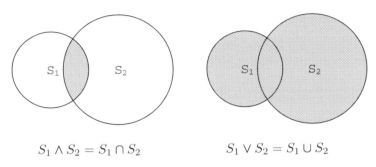

$$S_1 \wedge S_2 = S_1 \cap S_2 \qquad\qquad S_1 \vee S_2 = S_1 \cup S_2$$

Figure 8.1. Conjunction and disjunction.

8.2. Spectra of Boolean algebras. The fact that commutative algebras over fields other than $\mathbb{R}$ can actually serve as "observation instruments" can be seen in other branches of mathematics far removed from physics. The first of such examples is, of course, algebraic geometry, in which spectra of a certain class of commutative algebras (see below) are studied. Another remarkable example, appearing in mathematical logic, is that of Boolean algebras, which we begin to discuss here and will continue studying further on.

Recall that a *Boolean algebra* B is a commutative algebra over the two-element field $\mathbb{F}_2$, all of whose elements are *idempotent*, i.e., satisfy the condition $b^2 = b$. A very visual example of a Boolean algebra is the algebra of all subsets of a given set S. In that case, the logical operations of *conjunction* and *disjunction* correspond to the operations of intersection and union of subsets:

$$S_1 \wedge S_2 = S_1 \cap S_2, \quad S_1 \vee S_2 = S_1 \cup S_2, \quad S_i \subset S.$$

As to the operations that endow the collection of subsets of S with a commutative algebra structure are intersection (multiplication) and *symmetric difference* (addition):

$$S_1 \cdot S_2 \overset{\text{def}}{=} S_1 \cap S_2, \quad S_1 + S_2 \overset{\text{def}}{=} (S_1 \cup S_2) \setminus (S_1 \cap S_2), \quad S_i \subset S.$$

Exercise. Show that the set of all subsets of a set S is a commutative algebra over $\mathbb{F}_2$ with respect to the above defined operations of multiplication and addition. The zero and the unit of this algebra are the empty set and the entire set S, respectively.

In the following proposition, we list some main properties of Boolean algebras.

Proposition. *Let B be a Boolean algebra. Then*

(a) *If the element $b \in B$ is invertible, then $b = 1_B$;*

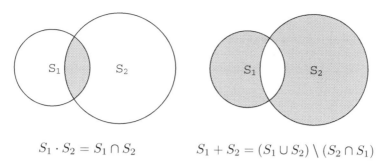

$$S_1 \cdot S_2 = S_1 \cap S_2 \qquad S_1 + S_2 = (S_1 \cup S_2) \setminus (S_2 \cap S_1)$$

Figure 8.2. Product and sum.

(b) *If $I \subset B$ is a maximal ideal, then the algebra B/I is isomorphic to $\mathbb{F}_2$;*

(c) *Every element $b \in B$ different from 1_B is contained in a certain maximal ideal. In particular, $|B| \neq \emptyset$;*

(d) *The intersection of all the nontrivial maximal ideals of B is trivial.*

(e) *The open sets U_b, $b \in B$ from the base of the Zariski topology of the spectrum $|B|$ are also closed;*

(f) *The topological space $|B|$ is Hausdorff, see Section 5.7.*

◀

(a)–(b) The proof of the first two points is left to the reader as an exercise.

(c) The principal ideal $b \cdot B$ is proper, because it does not contain the element $1_B - b \neq 0$. Indeed, assuming that $1_B - b = bb'$, we come to a contradiction: $0 = b(1_B - b) = b^2 b' = bb' = 1_B - b$. But $b \cdot B$, as any proper ideal, is contained in some maximal ideal I. Hence $b \in b \cdot B \subset I$.

(d) Assume $\mathcal{I}$ is the intersection of all the maximal ideals of the algebra B, $0 \neq b \in \mathcal{I}$, and I is the maximal ideal containing $1_B - b$. But I does not contain b, or else it would contain 1_B and, therefore, would coincide with B. But since $\mathcal{I} \subset I$, it follows that $b \notin \mathcal{I}$, contradicting the assumption.

(e) Obviously,

$$U_b = \{h \in |B| \mid h(b) \neq 0\} = \{h \in |B| \mid h(b) = 1\}$$
$$= \{h \in |B| \mid h(b - 1_B) = 0\} \quad (8.1)$$

This shows that U_b is the complement of the open set U_{b-1_B} and is therefore closed.

(f) Let $h_1, h_2 \in |B|$ and $h_1 \neq h_2$. The latter means that there exists an element $b \in B$ such that $h_1(b) \neq h_2(b)$. This implies $h_1(b) = 1$, $h_2(b) = 0$ or $h_1(b) = 0$, $h_2(b) = 1$. In the first of those two cases, this is equivalent to $h_1 \in U_a$ and $h_2 \in U_{1_B - b}$, in the second case, to $h_2 \in U_a$ and $h_1 \in U_{1_B - b}$. But, as we saw above, the open sets U_b and $U_{1_B - b}$ don't intersect.   ▶

8.3. Homomorphisms and maps of $\Bbbk$-spectra. As we have previously seen, smooth maps of smooth manifolds and homomorphisms of their algebras are in a natural bijective correspondence. Actually, this is a particular case of the following more general and fundamental algebraic construction.

Let $H \colon A \to B$ be a homomorphism of commutative $\Bbbk$-algebras. If $h \in |B|$, then the homomorphism $h \circ H \colon A \to \Bbbk$ is a point of the spectrum of the algebra A. Thus we obtain the following map of $\Bbbk$-spectra

$$|H| \colon |B| \to |A|, \quad h \mapsto h \circ H.$$

It possesses the following properties.

Proposition. (a) *The map $|H|$ is continuous.*

(b) *If the homomorphism H is surjective, then the map $|H|$ is injective.*

(c) *Let $A_1 \xrightarrow{H_2} A_2 \xrightarrow{H_1} A_3$ be a homomorphism of commutative $\Bbbk$-algebras. Then $|H_2 \circ H_1| = |H_1| \circ |H_2|$.*

◄ (a) It suffices to verify that the sets $|H|^{-1}(U_a)$, $a \in A$, are open in $|B|$. But

$$h \in |H|^{-1}(U_a) \Leftrightarrow h \circ H \in U_a \Leftrightarrow h(H(a)) \neq 0 \Leftrightarrow h \in U_{H(a)},$$

i.e., $|H|^{-1}(U_a) = U_{H(a)}$.

(b) If $h_1, h_2 \in |B|$ $h_1 \neq h_2$, then there exists an element $b \in B$ such that $h_1(b) \neq h_2(b)$. Suppose further that $b = H(a)$. In view of the subjectivity of H, such an element $a \in A$ exists and so

$$(|H|(h_1))(a) = h_1(H(a)) = h_1(b) \neq h_2(b) = h_2(H(a)) = (|H|(h_2))(a),$$

i.e., $|H|(h_1) \neq |H|(h_2)$.

(c) This statement is obviously equivalent to the associativity of the composition of maps: $h \circ (H_2 \circ H_1) = (h \circ H_2) \circ H_1$, $h \in |A_3|$.   ►

Exercises.   1. Show that the injectivity of the homomorphism H does not imply, in general, the surjectivity of the map $|H|$.

2. Is the assertion converse to item (b) of Proposition 8.3 true?

3. Show that the surjectivity of $|H|$ does not imply, in general, the injectivity of the homomorphism H.

4. Let $A = \mathbb{Z}_m[x]$, where m is prime. Show that any continuous map S of the set $|A|$ into itself has the form $S = |H|$, where $H \colon A \to A$ is some algebra homomorphism.

8.4. Geometrization of algebras. It is remarkable that elements of an algebra A can be represented by functions on its spectrum. Namely, to each element $a \in A$, let us assign the $\Bbbk$-valued function

$$f_a \colon |A| \to \Bbbk, \quad f_a(h) \stackrel{\text{def}}{=} h(a), \quad h \in |A|.$$

and denote by $\Gamma(A)$ the set of all such functions. If $a, b \in A$, $\lambda \in \Bbbk$, then

$$f_{a+b} = f_a + f_b, \quad f_{ab} = f_a f_b, \quad f_{\lambda a} = \lambda f_a, \quad f_{1_A} = 1_{\Gamma(A)},$$

This is equivalent to the map

$$\Gamma = \Gamma_A \colon A \to \Gamma(A), \quad a \mapsto f_a,$$

being a $\Bbbk$-algebra homomorphism. This map is obviously surjective, but not necessarily injective. If $f_a \equiv 0$, then $a \in \bigcap_{h \in |A|} \operatorname{Ker} h$ and conversely. In other words,

$$\operatorname{Ker} \Gamma_A = \operatorname{Inv}(A) \overset{\text{def}}{=} \bigcap_{h \in |A|} \operatorname{Ker} h.$$

In what follows, it will be convenient for us to use the shorter notation $\mu_h = \operatorname{Ker} h$. Obviously, $\mu_z = \mu_{h_z}$ if $A = C^\infty(M)$ and $z \in M$.

The function f_a is the geometrical realization of the element a of the abstract algebra A. Accordingly, it is natural to call the map Γ_A the *geometrization of the algebra* A. It follows from the above that the zero function corresponds to elements $a \in \operatorname{Inv}(A)$. In that sense all such elements are *invisible* or *non-observable* .

Definition. A commutative $\Bbbk$-algebra A is said be *geometrical*, if Γ_A is an isomorphism or, equivalently, if $\operatorname{Inv}(A) = 0$.

Exercises.   1. Let m be prime, let $A = \mathbb{F}_m[x_1, \ldots x_n]$ be the polynomial algebra over the m-element field $\mathbb{F}_m$. Show that $|A|$ consists of m^n points and describe $\operatorname{Inv}(A)$ and $\Gamma(A)$.

2. Let $A = \Bbbk(x_1, \ldots, x_n)$ be the algebra of power series over a field $\Bbbk$ of zero characteristic. Show that $|A|$ is a singleton and describe $\operatorname{Inv}(A)$ and $\Gamma(A)$.

The next proposition shows that the homomorphism Γ_A "kills" the invisible elements of the algebra A without changing its spectrum.

Proposition. *For any commutative algebra* A, *the algebra* $\Gamma(A)$ *is geometrical and* $|\Gamma(A)| = |A|$.

◄ Since the homomorphism Γ_A is surjective, the map $|\Gamma_A|$ is injective (Proposition 8.3, (b)). Further, $h \in |A|$ implies $\operatorname{Inv}(A) \subset \operatorname{Ker} h$ and so h can be represented as the following composition of homomorphisms

$$A \xrightarrow{\Gamma_A} \Gamma(A) = \frac{A}{\operatorname{Inv}(A)} \xrightarrow{h'} \Bbbk.$$

This shows that $h = |\Gamma_A|(h')$.   ►

8.5. "Boolean observability." Now we can readily describe what can be "seen" by means of Boolean algebras. The main fact here is

Proposition. *Boolean algebras are geometrical. In particular, the geometrization homomorphism* $\Gamma_B \colon B \to \Gamma(B)$ *is an isomorphism for any Boolean algebra* B.

◀ If I is a maximal ideal of the algebra B, then the quotient algebra B/I is canonically isomorphic to the field $\mathbb{F}_2$ (Proposition 8.2, (b)). Therefore, the canonical projection $B \to B/I$ may be identified with a certain element of the spectrum $|B|$. On the other hand, if $h \in |B|$, then $\operatorname{Ker} h$ is a maximal ideal, because the range of the homomorphism h is a field. Thus the correspondence $h \mapsto \operatorname{Ker} h$ is bijective and so $\operatorname{Inv}(\mathrm{B}) = \bigcap_{h \in |B|} \operatorname{Ker} h$ coincides with the intersection of all the maximal ideals of the algebra B, and this intersection is trivial (Proposition 8.2, (d)). ▶

Thus any Boolean algebra B may be identified with its geometrization $\Gamma(B)$. Moreover, since the elements of the algebra $\Gamma(B)$ are functions $f_b \colon |B| \to \mathbb{F}_2$, $b \in B$, each such function uniquely determines the domain where it takes the value $1_{\mathbb{F}_2}$. For the function f_b, such a domain will obviously be U_b. Hence the assignment

$$B \ni b \overset{U}{\mapsto} U_b \subset |B|$$

is bijective. This assignment, as can be readily seen, does the following.

Theorem.

$$f_{b_1} + f_{b_2} \mapsto (U_{b_1} \cup U_{b_2}) \setminus (U_{b_1} \cap U_{b_2}) \quad (\textit{symmetric difference}).$$
$$f_{b_1} f_{b_2} \mapsto U_{b_1} \cap U_{b_2} \quad (\textit{intersection}).$$
$$f_{b_1} + f_{b_2} + f_{b_1} f_{b_2} \mapsto U_{b_1} \cup U_{b_2} \quad (\textit{union}). \quad ▶$$

The first two of these maps show that the canonical base $\{U_b\}_{b \in |B|}$ in the Zariski topology in $|B|$, supplied with the operations of symmetric difference and intersection, is an algebra isomorphic to B (see Exercise 8.2). But since the Boolean operations of conjunction ($\wedge$) and disjunction ($\vee$) in the algebra of subsets of the given set are actually the operations of intersection and union, it follows that the last two of the three maps defined above show that the Boolean operations in B itself are the following

$$b_1 \wedge b_2 = b_1 b_2, \quad b_1 \vee b_2 = b_1 + b_2 + b_1 b_2.$$

This also explains how Venn diagrams arise (see pages 86–87), as geometrical representations of Boolean operations.

The geometrical description of Boolean algebras given above, except for the abstract characteristic of their spectra, is basically the well-known *Theorem of Stone*. The interpretation of Boolean algebras as a means of observation makes the proof of that theorem simpler and more transparent. What we actually "observe" in this approach will be discussed in the Appendix.

8.6. Invisible elements and homomorphisms. The behavior of "invisible" elements under homomorphisms of $\Bbbk$-algebras is described by the following proposition.

Proposition. *Suppose* $H\colon A \to B$ *is a homomorphism of* $\Bbbk$-*algebras and* $h \in |B|$. *Then*

(a) $H\big(\operatorname{Ker}|H|(h)\big) \subset \operatorname{Ker} h;$

(b) $H(\operatorname{Inv}(A)) \subset \operatorname{Inv}(B);$

(c) *if the map* $|H|$ *is surjective, then* $\operatorname{Ker} H \subset \operatorname{Inv}(A);$

(d) *if the algebra* A *is geometrical and the map* $|H|$ *is surjective, then the homomorphism* H *is injective.*

◀ (a) Since $\operatorname{Ker}\big(|H|(h)\big) = \operatorname{Ker}(h \circ H)$, we have

$$\operatorname{Ker} H \subset \operatorname{Ker}\big(|H|(h)\big) = H^{-1}(\operatorname{Ker} h). \tag{8.2}$$

This implies the inclusion (a).

(b) We have

$$\operatorname{Inv}(A) = \bigcap_{h' \in |A|} \operatorname{Ker} h' \subset \bigcap_{h' \in \Im|H|} \operatorname{Ker} h' = \bigcap_{h \in |B|} \operatorname{Ker}\big(|H|(h)\big).$$

Therefore, in view of (8.2),

$$H\left(\operatorname{Inv}(A)\right) \subset H\Big(\bigcap_{h \in |B|} \operatorname{Ker}\,|H|(h)\Big)$$

$$= H\Big(\bigcap_{h \in |B|} H^{-1}(\operatorname{Ker} h)\Big) = \bigcap_{h \in |B|} \operatorname{Ker} h = \operatorname{Inv}(B).$$

(c) From (8.2) we have $\operatorname{Ker} H \subset \bigcap_{h \in |B|} \operatorname{Ker}\big(|H|(h)\big)$. On the other hand, $\bigcap_{h \in |B|} \operatorname{Ker}\big(|H|(h)\big) = \operatorname{Inv}(A)$, because $|H|$ is surjective.

(d) This follows immediately from (c), since in the given situation we have $\operatorname{Inv}(A) = 0$. ▶

Note that for a geometrical algebra A, assertion (d) of the previous proposition is dual to assertion (b) of Proposition 8.1. For nongeometrical algebras A this is, generally speaking, not true. For example, the homomorphism $H = \Gamma_A$ induces a bijection of the spectra (Proposition 8.4), but is not injective if the ideal $\operatorname{Inv}(A)$ is nontrivial.

8.7. What is a point? Making more visible, more geometrical, any algebraic, analytic, or some other formal mathematical context is not only natural, but can be very useful. When trying to establish some fact, it is necessary to organize in a definite way the chain of logical arguments and calculations. For such a task, the geometrical picture, if it is available, may prove very helpful. In this connection, the story behind the proof of Fermat's Last Theorem is rather instructive. Attempts to prove it by means of various purely algebraic tricks were unsuccessful for over 350 years. The solution was found on the basis of methods of contemporary algebraic geometry, by reformulating the problem in terms of *rational points* on algebraic curves.

In many cases the necessary visibility can be achieved by linking the specific objects under consideration to some appropriate spaces. Since spaces consist of points, the very notion of point becomes crucial. Above, considering commutative algebras, we found the appropriate notion of point by analyzing the observation mechanism in classical physics. However, this should not be regarded as the universal solution. To show this, we shall introduce some other approaches below, so as to give the reader a more complete picture.

In this connection, here are some remarks.

(i) Although the "readings of measuring devices" (see Section 1.9) that we have used to motivate our constructions are usually real numbers, it is often necessary to consider measurements of a more general nature (complex numbers, matrices, residues modulo some positive integer, etc.). A striking example is the complex phase method used in the elementary theory of electricity.

(ii) If the solution of some problem (in physics or mathematics) reduces to the solution of algebraic equations with coefficients in a ring A (or in a field A which is not algebraically closed), then as a rule, it is useful to seek the solution in an extension $B \supset A$ of the ring, rather than in the ring A itself. The simplest example is the use of complex roots of a polynomial with real coefficients. A less trivial example is given by the so-called Pauli operators, which arise in the quantum mechanics of electrons as matrix solutions σ_1, σ_2, σ_3 of the system of equations

$$\sigma_1^2 = \sigma_2^2 = \sigma_3^2 = -1, \quad \sigma_2\sigma_3 - \sigma_3\sigma_2 = \sigma_1,$$
$$\sigma_1\sigma_2 - \sigma_2\sigma_1 = \sigma_3, \quad \sigma_1\sigma_3 - \sigma_3\sigma_1 = -\sigma_2.$$

(iii) Mathematicians like to avoid exceptions and strive for the aesthetic unification of any theory; this often leads them to invent a language in which the exceptions turn out to be part of the general rule. Thus parallel lines "intersect at infinity," imaginary points and points at infinity appear in many situations, in particular in connection with $\mathbb{R}$-algebras (see Example 1.18). This trick of giving intuitive geometrical meaning to different algebraic situations is particularly fruitful in algebraic geometry, whose ideas will be used extensively here.

8.8. Motivating example. If $f \in \mathbb{R}[x]$ is an irreducible polynomial of degree 2, then there obviously exists no $\mathbb{R}$-point of the algebra $\mathbb{R}[x]$ that would be a root of this polynomial; i.e., there is no homomorphism $\alpha\colon \mathbb{R}[x] \to \mathbb{R}$ such that $\alpha(f) = 0$.

However, there exist exactly two homomorphisms into the field of complex numbers, α, $\overline{\alpha}\colon \mathbb{R}[x] \to \mathbb{C}$, such that $\alpha(f) = \overline{\alpha}(f) = 0$. This will be proved below (see Theorem 8.11), but we suggest that the reader try to prove this now as an exercise. Homomorphisms such as α and $\overline{\alpha}$ should be viewed as complex points for the algebra $\mathbb{R}[x]$.

8.9. Suppose $\mathbb{k}$ is an arbitrary field, $K \supset \mathbb{k}$ a ring without zero divisors, and $\mathcal{F}$ a commutative $\mathbb{k}$-algebra with unit. In view of Example 8.8, it is natural to define a point of the algebra $\mathcal{F}$ "over the ring K" as an epimorphism of $\mathbb{k}$-algebras $a\colon \mathcal{F} \to K$. It is no less natural to consider two points

$$a_i\colon \mathcal{F} \to K_i \supset \mathbb{k}, \quad i = 1, 2,$$

identical (equivalent) if there exists a $\mathbb{k}$-algebra isomorphism $i\colon K_1 \to K_2$ such that $a_2 = i \circ a_1$. Now we can give the following definition.

Definition. An equivalence class of $\mathbb{k}$-algebra epimorphisms

$$a\colon \mathcal{F} \to K \supset \mathbb{k}$$

is called a *K-point* of the $\mathbb{k}$-algebra $\mathcal{F}$. The ring K is then referred to as the *domain* of the point a.

Further, we shall often speak of points of the algebra without specifying their domain. The reader should keep in mind that each point of the $\mathbb{k}$-algebra $\mathcal{F}$ has its own domain, and different points may have different (or identical) domains.

8.10. From the viewpoint of our physical interpretation (in terms of measurements), the isomorphism i appearing in Definition 8.9 can be construed as an equivalent change of the "system of observations," which, of course, should not influence the collection of points (states) determined by the algebra $\mathcal{F}$.

Definition 8.9 also possesses a purely algebraic motivation, which will appear in the next section.

8.11 Theorem. *Points of a commutative $\mathbb{k}$-algebra $\mathcal{F}$ (understood in the sense of Definition 8.9) correspond bijectively to prime ideals of the algebra $\mathcal{F}$, the correspondence being given by the map*

$$(a\colon \mathcal{F} \to K) \mapsto \operatorname{Ker} a \subset \mathcal{F}.$$

◀ The fact that the ideal $\operatorname{Ker} a$ is prime and depends only on the equivalence class of the homomorphism a and that different equivalence classes correspond to different ideals is obvious.

To prove surjectivity, take an arbitrary prime ideal $p \subset \mathcal{F}$. Then the ring $\mathcal{F}/p$ has no zero divisors, and the quotient map $q\colon \mathcal{F} \to \mathcal{F}/p$ is a point of the algebra $\mathcal{F}$ for which $\operatorname{Ker} q = p$. ▶

The set of prime ideals of a $\mathbb{k}$-algebra $\mathcal{F}$ is called the *prime spectrum* of $\mathcal{F}$ and is denoted by $\operatorname{Spec} \mathcal{F}$. According to the theorem, $\operatorname{Spec} \mathcal{F}$ may be viewed as constituting the set of all points of the $\mathbb{k}$-algebra $\mathcal{F}$. Obviously, when $\Bbbk = \mathbb{R}$, we have $\operatorname{Spec} \mathcal{F} \supset |\mathcal{F}|$, so that elements of $\operatorname{Spec} \mathcal{F}$ generalize the notion of $\mathbb{R}$-point; i.e., $\mathbb{R}$-points in the sense of Section 3.4 are "points over the field $\mathbb{R}$" in the sense of Section 8.9.

8.12. Examples (based on Theorem 8.11). I. A smooth manifold has no points with values $z \in \mathbb{C}$ with $\operatorname{Im} z \neq 0$.

◄ Suppose $a \colon C^\infty(M) \to \mathbb{C}$ is a $\mathbb{C}$-point and $a(f) = i \in \mathbb{C}$. Therefore, we have $a\left(1 + f^2\right) = 0$, and hence $1 + f^2 \in \operatorname{Ker} a$. On the other hand, the element $1 + f^2$ is obviously invertible in $C^\infty(M)$ and as such does not belong to any proper ideal of $C^\infty(M)$.   ►

II. The set of all points of the polynomial algebra $\mathbb{R}[x]$ can be identified with the complex half-plane $\{z \mid \Im z \geqslant 0\}$ to which a point ω, corresponding to the ideal $\{0\} \subset \mathbb{R}[x]$, has been added.

◄ Since $\mathbb{R}[x]$ is a principal ideal domain, any prime ideal in it is of the form $\mathbb{R}[x]f$, where f is an irreducible polynomial. But then either $f \equiv 0$ or $f = a(x - b)$ or $f = a(x - c)(x - \bar{c})$, where a, $b \in \mathbb{R}$, while c and $\bar{c}$ are conjugate complex numbers.   ►

Note that we have proved the statement mentioned at the end of Section 8.8.

Exercises.   1. Let $\Bbbk$ be a field, and X a formal variable; describe $\operatorname{Spec} \Bbbk[X]$.

2. Show that $\mu_z^\infty \overset{\text{def}}{=} \bigcap_k \mu_z^k$, $z \in M$, is a prime ideal in $C^\infty(M)$.

8.13. We have learned to assign a set of points, $\operatorname{Spec} \mathcal{F}$, to every $\Bbbk$-algebra $\mathcal{F}$. It is only natural to try to supply this set with a topology. In a similar situation, working with $|\mathcal{F}|$, we introduced the topology induced from the topology in $\mathbb{R}$, but in the case of an arbitrary $\Bbbk$-algebra $\mathcal{F}$ we no longer have a fixed field $\mathbb{R}$ with a trustworthy topology. A new idea is needed to find a topology in $\operatorname{Spec} \mathcal{F}$.

Suppose C is a closed set in $\mathbb{R}^n$, then there exists a function $f \in C^\infty(\mathbb{R}^n)$ such that $f(r) = 0 \iff r \in C$. The same is true for a closed set B in a manifold M: There exists a function $f \in \mathcal{F} = C^\infty(M)$ such that $f(p) = 0 \iff p \in B$ (Section 4.17 (i)). This circumstance will be the basis for introducing a topology in $\operatorname{Spec} \mathcal{F}$.

First note that any element $f \in \mathcal{F}$ may be viewed as a function on $\operatorname{Spec} \mathcal{F}$ ranging in $\mathcal{F} \bmod p$ (which may have different algebraic structures). Namely, for any prime ideal $p \in \operatorname{Spec} \mathcal{F}$, we put

$$f(p) = f \bmod p.$$

For any subset $E \subset \mathcal{F}$, denote by $V(E) \subset \operatorname{Spec} \mathcal{F}$ the set of all prime ideals containing E. In other words, if $p \in V(E)$, then for any function $f \in E$ we have $f(p) = 0$. We can now define the *Zariski topology* in the prime spectrum $\operatorname{Spec} \mathcal{F}$ of an arbitrary $\Bbbk$-algebra $\mathcal{F}$ as the topology whose basis of closed sets is the collection $\{V(E) \mid E \subset \mathcal{F}\}$.

8.14. The definition of the Zariski topology in $\operatorname{Spec}\mathcal{F}$ can also be given in terms of the closure operation. For any $M \subset \operatorname{Spec}\mathcal{F}$, consider the ideal $I_M = \bigcap_{p\in M} p$ and define the closure $\overline{M}$ of M as

$$\overline{M} = \{p \in \operatorname{Spec}\mathcal{F} \mid p \supset I_M\}.$$

This definition has a clear algebraic meaning: If the element $f \in \mathcal{F}$ vanishes on M (i.e., $p \in M \Rightarrow f(p) = 0 \iff f \in p$), then it also vanishes on $\overline{M}$; conversely, if any f that vanishes on M is also zero on some point $p \in \operatorname{Spec}\mathcal{F}$, then $p \in \overline{M}$. (The fact that this construction gives the same topology as 8.13 follows directly from the two definitions.)

The reader has perhaps wondered what price we shall have to pay for the extreme generality of this construction. It turns out that the Zariski topology in $\operatorname{Spec}\mathcal{F}$ is non-Hausdorff. In particular, $\operatorname{Spec}\mathcal{F}$ contains non-closed points, which will be considered in the next section. Note, however, that every point of $|\mathcal{F}|$ is closed.

8.15 Exercises.   1. Prove that a one-point set $\{p\} \subset \operatorname{Spec}\mathcal{F}$ is closed in the Zariski topology iff the ideal p is maximal.

2. Describe the Zariski topology of $|\mathbb{R}[X]|$. Compare the Zariski topologies of $|\mathbb{R}[X]| = \mathbb{R}$ and of $|C^\infty(\mathbb{R})| = \mathbb{R}$.

3. Describe the Zariski topology of $\operatorname{Spec}\mathbb{R}[X]$.

8.16 Proposition. *The closure of a point $q \in \operatorname{Spec}\mathcal{F}$ coincides with the set*

$$V(q) = \{p \in \operatorname{Spec}\mathcal{F} \mid q \subset p\}$$

(in other words, $\overline{\{q\}}$ is the set of all common zeros of all the functions from the ideal q).

◀ By Definition 8.13, the set $V(q)$ is closed. On the other hand, $E \subset \mathcal{F}$, and $q \in V(E) \Rightarrow q \supset E \Rightarrow V(q) \subset V(E)$.   ▶

8.17. Corollaries and examples. I. If a $\Bbbk$-algebra has no zero divisors, then $\{0\} \in \operatorname{Spec}\mathcal{F}$ and $\overline{\{0\}} = \operatorname{Spec}\mathcal{F}$. For this reason, the ideal $\{0\}$ is said to be the *common point* of the set $\operatorname{Spec}\mathcal{F}$.

II. Suppose $f \in \Bbbk[x_1,\ldots,x_n] = \mathcal{F}$ is an irreducible polynomial in n variables over the algebraically closed field $\Bbbk$. Consider the prime ideal $p = \Bbbk[x_1,\ldots,x_n]\cdot f$. The closure of the point p in the prime spectrum $\operatorname{Spec}\mathcal{F}$ of the polynomial algebra $\mathcal{F}$ contains all the maximal ideals corresponding to the points of the hypersurface

$$H_f = \{(r_1,\ldots,r_n) \in \Bbbk^n \mid f(r_1,\ldots,r_n) = 0\}.$$

The point p is therefore called the *common point* of this hypersurface. (To each point $(r_1,\ldots,r_n) \in \Bbbk^n$ there corresponds the maximal ideal of $\mathcal{F}$ generated by all monomials $x_1 - r_1,\ldots,x_n - r_n$.)

III. By Proposition 8.16, the closure of the point p in Example III above contains all the prime ideals containing p. From the point of view of "maximal ideal topology," each such ideal determines a surface of "lesser dimension" contained in the hypersurface H_f and is the common point of this lesser surface.

8.18 Theorem. *The prime spectrum* $\operatorname{Spec}\mathcal{F}$ *of any* $\Bbbk$-*algebra* $\mathcal{F}$ *is compact.*

Let us restate the theorem in terms of closed sets: *if* $\{M_\alpha\}_{\alpha\in A}$ *is a family of closed sets such that* $\bigcap_{\alpha\in A} M_\alpha = \varnothing$, *then we can choose a finite subfamily* $M_{\alpha_1},\dots,M_{\alpha_N}$ *with empty intersection.* We shall prove the theorem in this (equivalent) form.

◀ Without loss of generality, we can assume that $M_\alpha = V(E_\alpha)$, where $E_\alpha \subset \mathcal{F}$ is some subset. Denote by $[E_\alpha] \subset \mathcal{F}$ the ideal generated by E_α and note that

$$\varnothing = \bigcap_{\alpha\in A} M_\alpha = \bigcap_{\alpha\in A} V(E_\alpha) = V\left(\sum_{\alpha\in A}[E_\alpha]\right).$$

But this means that the ideal $\sum_{\alpha\in A}[E_\alpha]$ is not contained in any prime ideal and hence in any maximal ideal of the algebra $\mathcal{F}$. In other words,

$$\sum_{\alpha\in A}[E_\alpha] = \mathcal{F} \ni 1.$$

Therefore, we can find $\alpha_1,\dots,\alpha_N \in A$ and $f_i \in [E_{\alpha_i}]$, $i = 1,\dots,N$, such that $\sum_{i=1}^N f_i = 1$. But in this case

$$\bigcap_{i=1}^N M_{\alpha_i} = V\left(\sum_{i=1}^N [E_{\alpha_i}]\right) = V(\mathcal{F}) = \varnothing. \quad ▶$$

8.19. Let us now investigate how prime spectra behave under homomorphisms of their $\Bbbk$-algebras. Suppose $\mathcal{F}_1$, $\mathcal{F}_2$ are $\Bbbk$-algebras and $\alpha\colon \mathcal{F}_2 \to \mathcal{F}_1$ is a $\Bbbk$-algebra homomorphism. We claim that *if* $p \subset \mathcal{F}_1$ *is a prime ideal, then so is* $\alpha^{-1}(p)$.

◀ If $q_1 \in \alpha^{-1}(p)$, then for any $q_2 \in \mathcal{F}_2$, we have

$$\alpha(q_1q_2) \in \alpha(\alpha^{-1}(p)q_2) = p\alpha(q_2) \subset p.$$

Thus $q_1q_2 \in \alpha^{-1}(p)$, and $\alpha^{-1}(p)$ is an ideal. To show that it is prime, let $q_1,\, q_2 \in \mathcal{F}$ and $q_1q_2 \in \alpha^{-1}(p)$. Then $\alpha(q_1)\alpha(q_2) \in p$, and since p is prime, at least one of the elements $\alpha(q_1)$, $\alpha(q_2)$, say the first, belongs to p. Then we see that

$$q_1 \in \alpha^{-1}(\alpha(q_1)) \subset \alpha^{-1}(p);$$

i.e., the ideal $\alpha^{-1}(p)$ is prime.   ▶

Now, to every $\Bbbk$-algebra homomorphism $\alpha\colon \mathcal{F}_2 \to \mathcal{F}_1$, we can assign the map of prime spectra

$$|\alpha|\colon \ \mathrm{Spec}\,\mathcal{F}_1 \to \mathrm{Spec}\,\mathcal{F}_2, \quad p \mapsto \alpha^{-1}(p).$$

8.20 Proposition. *For any $\Bbbk$-algebra homomorphism $\alpha\colon \mathcal{F}_2 \to \mathcal{F}_1$ the corresponding prime spectra map $|a|\colon \ \mathrm{Spec}\,\mathcal{F}_1 \to \mathrm{Spec}\,\mathcal{F}_2$ is continuous in the Zariski topology.*

◀ The proof is a straightforward verification of definitions; we leave it to the reader. ▶

8.21. Now, copying a similar definition for smooth manifolds (see Section 6.1), we can give the following definition:

Definition. A map $\beta\colon \ \mathrm{Spec}\,\mathcal{F}_1 \to \mathrm{Spec}\,\mathcal{F}_2$ is said to be *smooth* if there exists a $\Bbbk$-algebra homomorphism $\alpha\colon \mathcal{F}_2 \to \mathcal{F}_1$ such that $\beta = |\alpha|$.

Thus the reader has met with another example of a category: the category of $\Bbbk$-algebra prime spectra, in which the morphisms are smooth maps of spectra.

8.22. Suppose the ideal $p \in \mathrm{Spec}\,\mathcal{F}$ of the $\Bbbk$-algebra $\mathcal{F}$ satisfies $\mathcal{F}/p = \Bbbk$. Then it is easy to show that p is a maximal ideal. We also suggest that the reader work out the following exercises.

Exercises.   1. The ideal p is maximal iff $\mathcal{F}/p$ is a field.

   2. If $\mathcal{F}$ is finitely generated and p is maximal, then $\mathcal{F}/p$ is a finite algebraic extension of $\Bbbk$.

8.23. Definition. The *maximal spectrum* $\mathrm{Spm}\,\mathcal{F}$ of the $\Bbbk$-algebra $\mathcal{F}$ is the set of maximal ideals of $\mathcal{F}$.

The previous section, as well as Sections 8.15–8.16, where we studied the closure of one-point sets in $\mathrm{Spec}\,\mathcal{F}$, suggests that instead of $\mathrm{Spec}\,\mathcal{F}$ we should have been studying $\mathrm{Spm}\,\mathcal{F}$. Indeed, we would thus have avoided such pathology as non-closed points, and the set of domains of points would be more manageable. However, this is not quite reasonable because of the fact that *unlike prime ideals, maximal ideals may have inverse images that are no longer maximal* (see Example 8.24). Therefore, generally there is no map of maximal spectra that could naturally be assigned to a homomorphism of the corresponding algebras. In other words, the correspondence $A \mapsto \mathrm{Spm}\,A$ does not define a functor from the category of K-algebras to the category of topological spaces.

8.24. Example. Suppose $\mathcal{F}_2 = \Bbbk[x_1, x_2]$ is the algebra of polynomials in two variables, $\mathcal{F}_1 = \Bbbk(x_1)$ is the field of rational functions, and $\alpha\colon \mathcal{F}_2 \to \mathcal{F}_1$ is the composition of the quotient epimorphism

$$\mathcal{F}_2 = \Bbbk[x_1, x_2] \to \Bbbk[x_1, x_2]/(x_2) = \Bbbk[x_1]$$

and the inclusion $\Bbbk[x_1] \hookrightarrow \Bbbk(x_1) = \mathcal{F}_1$.

The ideal $\{0\} \subset \Bbbk(x_1)$ is a maximal one; however, $\alpha^{-1}(\{0\}) = \mathcal{F}_2 \cdot x_2$ is a prime ideal, but not a maximal one. This establishes the statement in italics in the previous section.

8.25 Exercises.   1. Let M be a compact manifold. Prove that any maximal ideal in $C^\infty(M)$ is of the form μ_z, $z \in M$ (see Exercise 2 from Section 8.12).

2. Show that for any noncompact manifold M there exist maximal ideals in $C^\infty(M)$ different from the μ_z's.

3. Show that any such maximal ideal, regarded as a $C^\infty(M)$-module or an algebra, has an infinite number of generators.

8.26. In the remainder of this section, we shall discuss certain aspects of prime spectra that distinguish them from smooth manifolds.

As we saw in Example 8.12, II, a simple spectrum may contain, besides "visible" points, certain points whose geometrical interpretation is not obvious (e.g., the prime ideal ω in that example).

We shall begin with a few examples showing that "points" of that type may already appear in the maximal spectrum $\operatorname{Spm} \mathcal{F}$ of a $\Bbbk$-algebra.

8.27. Ghosts. The reader who has studied Example 8.12, I, in detail has undoubtedly noticed that the maximal ideals of the algebra of smooth functions on a compact manifold correspond bijectively to ordinary $\mathbb{R}$-points of this algebra (i.e., to ordinary points of the manifold). For noncompact manifolds this is no longer the case.

Indeed, suppose I_c is the ideal of all functions with compact support on a noncompact manifold. None of the maximal ideals containing I_c is the kernel of any $\mathbb{R}$-point (the proof is left to the reader as an exercise). Such maximal ideals correspond to "points" of noncompact manifolds we have called *ghosts*. In Sections 8.29–8.31, we shall see how such ghosts can actually materialize.

8.28. Before continuing, the following remark is called for. Suppose $\mathcal{F}$ is the $\mathbb{R}$-algebra of smooth functions on the compact manifold M. As was pointed out above, $M = \operatorname{Spm} \mathcal{F}$, but the reader should not think that $M = \operatorname{Spec} \mathcal{F}$. To that end, the reader should work out the following exercise.

Exercise. A function $f \in \mathcal{F}$ is called *flat* at the point $a \in M$ if it vanishes with all its derivatives at that point. The set of all functions that are flat at the given point $a \in M$ constitutes an ideal. Prove that this ideal is prime (i.e., is a point of $\operatorname{Spec} \mathcal{F}$ that corresponds to no point of M).

8.29. Example (compactifications of the line)**.** Let $\mathcal{F}$ be the $\mathbb{R}$-algebra of smooth functions on the line $\mathbb{R}$, and let I_c, as above, be its ideal of functions with compact support. One can try to describe the maximal ideals that contain I_c, but most of these ideals—ghosts—have no reasonable constructive description.

Nevertheless, there are two very nice sets of functions that contain I_c, namely the sets $\overline{\mu}_{+\infty}$ and $\overline{\mu}_{-\infty}$:

$$\overline{\mu}_{\pm\infty} = \left\{ f \in \mathcal{F} \mid \lim_{r \to \pm\infty} f(r) = 0 \right\}.$$

These sets, unfortunately, are not ideals (in poetic language we can say that "were they ideals, they would be maximal ones"). This unpleasant circumstance can be overcome as follows. Put

$$\mathcal{F}_1 = \left\{ f \in \mathcal{F} \mid \forall k \geqslant 0 \ \lim_{r \to +\infty} \frac{d^k f}{dr^k} \ \text{and} \ \lim_{r \to -\infty} \frac{d^k f}{dr^k} \ \text{exist} \right\}$$

and define $\mu_{\pm\infty}$ by putting $\mu_{\pm\infty} = \mathcal{F}_1 \cap \overline{\mu}_{\pm\infty}$. Then,

(i) $\mu_{+\infty}$ and $\mu_{-\infty}$ are maximal ideals of the algebra $\mathcal{F}_1$ containing the ideal I_c.

(ii) If $M = [0,1] \subset \mathbb{R}^1$, and the algebra $\mathcal{F}_M$ consists of the restrictions of all the functions from $\mathcal{F}$ to M, then the manifold with boundary M is diffeomorphic to $|\mathcal{F}_1|$.

(iii) Under this identification, the ideal I_c becomes the ideal of functions that vanish at the end points of the closed interval $[0,1]$, while the ideals $\mu_{-\infty}$ and $\mu_{+\infty}$ become the points 0 and 1, respectively.

(iv) This implies that all the maximal ideals of the algebra $\mathcal{F}_1$, except $\mu_{+\infty}$ and $\mu_{-\infty}$, correspond to ordinary points of the line $\mathbb{R}^1$; as for the ideals $\mu_{\pm\infty}$, they are "ghosts," which are adjoined to the line in order to make it compact.

(v) Prove that the algebra $\mathcal{F}_1$ is not isomorphic to the algebra $\mathcal{F}_M$. Try to find another subalgebra of $\mathcal{F}$, different from $\mathcal{F}_1$, isomorphic to $\mathcal{F}_M$.

8.30. Another example. Among the numerous other methods of compactifying the line, we consider only one more. Let

$$\mathcal{F}_2 = \left\{ f \in \mathcal{F}_1 \mid \forall k \geqslant 0 \ \lim_{r \to +\infty} \frac{d^k f}{dr^k} = \lim_{r \to -\infty} \frac{d^k f}{dr^k} \in \mathbb{R} \right\}$$

and put

$$\mu_\infty = \mu_{+\infty} \cap \mathcal{F}_2 \quad (= \mu_{-\infty} \cap \mathcal{F}_2).$$

It is easy to prove that the algebra $\mathcal{F}_2$ is isomorphic to $C^\infty(S^1)$, the algebra of smooth functions on the circle. This can be visualized by imagining that the "infinite end points" of the line are glued together by means of the "ghost" corresponding to the ideal μ_∞ (see Figure 8.3).

8.31. Further examples. I. Denote by $\mathcal{F}$ the algebra of complex-valued functions of a complex variable, defined, holomorphic, and bounded in the domain $|z| < 1$. Among the maximal ideals of the algebra $\mathcal{F}$, we know, of

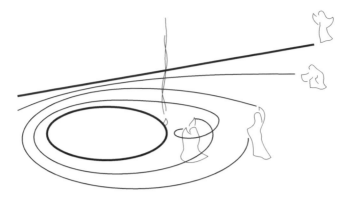

Figure 8.3. Gluing "infinite end points."

course, the ideals corresponding to points a, $|a| < 1$, namely

$$\mu_a = \{f \in \mathcal{F} \mid f(a) = 0\}.$$

It is also possible to define μ_a when $|a| = 1$ by putting

$$\mu_a = \{f \in \mathcal{F} \mid \lim_{z \to a} f(z) = 0\}.$$

Prove that the maximal spectrum of the algebra $\mathcal{F}$ consists of all the ideals μ_a, $|a| \leqslant 1$.

II. Find the maximal spectrum of the algebra $\mathcal{F}_2$ of complex-valued functions of two complex variables, analytic and bounded in the open polydisk

$$\{(z_1, z_2) \mid |z_1| < 1, \; |z_2| < 1\}.$$

9

Differential Calculus as Part of Commutative Algebra

9.1. The formal approach to observation procedures in classical physics, described in the previous chapters, leads to rather important conclusions. Let us first note that *the differential calculus is the natural language of classical physics*. On the other hand, *all information about some classical physical system is encoded in the corresponding algebra of observables*. From this it follows that the *differential calculus needed to describe physical problems is a part of commutative algebra*.

The basic aim of this chapter is to explain how to obtain a purely algebraic definition of differential operators using general facts known from the classical calculus. It is very important that the constructions obtained below can be used for arbitrary, not necessarily smooth, algebras. But perhaps the key point is somewhat different.

We shall see that the differential calculus is a formal consequence of arithmetical operations. This unexpected and beautiful fact plays an important role not only in mathematics itself. It also allows us to reconsider some paradigms reflecting the relationship of mathematics to the natural sciences and, above all, with physics and mechanics. By including observability in our considerations, we ensure that mathematics may be regarded as a branch of the natural sciences.

9.2. Let us start with the simplest notion of calculus, namely that of the derivative, which is the formal mathematical counterpart of velocity. From elementary mechanics we know that velocity is a vector, i.e., a directed segment. This point of view is hardly satisfactory, since it is completely unclear what a *directed segment* is in the case of an abstract (*curved*)

© Springer Nature Switzerland AG 2020
J. Nestruev, *Smooth Manifolds and Observables*, Graduate Texts
in Mathematics 220, https://doi.org/10.1007/978-3-030-45650-4_9

manifold. For this reason, the founders of differential geometry defined the tangent vector as a quantity described in a given coordinate system as an n-tuple of numbers; when passing from one coordinate system to another, this n-tuple is to be transformed in a prescribed manner. This approach is also unsatisfactory, because it describes vectors in local coordinates and does not explain what they are in essence.

In many modern textbooks on differential geometry one can find another definition of a tangent vector, which does not use local coordinates: A tangent vector is an equivalence class of smooth curves tangent to each other at a given point of the manifold under consideration. But try to find the sum of such classes or multiply a class by a number (i.e., try to introduce the structure of a linear space), and you will immediately see that this definition is inconvenient to work with.

The principal reason why these definitions of a tangent vector are unsatisfactory is their descriptive nature: They say nothing about the functional role of this notion in the differential calculus. This role can be understood if algebras of observables are used.

Let A be an algebra of observables. Then, by definition, $M = |A|$ is the "manifold" of states for the corresponding system, and a particular state is an element $h \in |A|$. Therefore, a time evolution of the system's state is described by a family h_t. Consequently, the velocity of evolution at a moment t_0 is

$$\Delta_h \overset{\text{def}}{=} \frac{dh_t}{dt}\bigg|_{t=t_0} : A \to \mathbb{R}, \qquad (9.1)$$

where, by definition,

$$\frac{dh_t}{dt} = \lim_{\Delta t \to 0} \frac{1}{\Delta t}(h_{t+\Delta t} - h_t),$$

whatever the meaning of the limit above may be. In other words, the motion that we conceive is in fact the change in time in the readings given by the measuring devices, while the velocity of motion is the velocity of these changes.

The translation of the above to geometrical language is accomplished by the correspondence $M \ni z \leftrightarrow h = h_z \in |A|$, where $A = C^\infty(M)$ and $h_z(f) = f(z)$ for any $f \in C^\infty(M)$. After this translation, the family $\{h_t\}$ becomes a curve $z(t)$ on the manifold M. This curve is such that $h_t = h_{z(t)}$, while the derivative

$$\frac{dh_t}{dt}$$

is a tangent vector to $z(t)$ and consequently to the manifold M. Formula (9.1) now acquires the form

$$\Delta_z = \frac{dh_{z(t)}}{dt}\bigg|_{t=t_0} : A \to \mathbb{R}, \qquad (9.2)$$

where $z = z(t_0)$, and obviously $\Delta_z \leftrightarrow \Delta_h$.

Thus, we arrive at the following interpretation:

| velocity of changes of system states | $\longleftrightarrow$ | tangent vector to the manifold M |

where the system is described by the algebra A.

The words *tangent vector* were used intuitively above. Our aim now is to define it rigorously, solely in terms of the algebra of observables A. The above interpretation allows us to do this in a natural way. It remains only to understand the mathematical nature of the operator Δ_h (or Δ_z), informally defined by formula (9.1) (respectively by (9.2)): Not all maps from A to $\mathbb{R}$ (for example, h) may be appropriately called velocities of state change.

9.3. The algebra of observables is a combination of two structures: that of a vector space and that of a multiplicative structure. The interaction of the operator Δ_h (or Δ_z) with the first structure is obvious: It is $\mathbb{R}$-linear, i.e.,

$$\Delta_h \left(\sum_{j=1}^{k} \lambda_j f_j \right) = \sum_{j=1}^{k} \lambda_j \Delta_h(f_j), \quad \lambda_j \in \mathbb{R}, \ f_j \in C^\infty(U), \tag{9.3}$$

where U is a coordinate neighborhood.

Exercise. Check this.

To understand how Δ_h interacts with the multiplicative structure of the algebra $C^\infty(M)$, one needs to compute the action of this operator on the product of two observables. We have

$$\Delta_h(fg) = \frac{dh_t(fg)}{dt}\bigg|_{t=t_0} = \frac{d(h_t(f)h_t(g))}{dt}\bigg|_{t=t_0}$$

$$= \frac{dh_t(f)}{dt}\bigg|_{t=t_0} h_{t_0}(g) + h_{t_0}(f)\frac{dh_t(g)}{dt}\bigg|_{t=t_0}$$

$$= \Delta_h(f)h(g) + h(f)\Delta_h(g);$$

i.e., Δ_h satisfies the following *Leibniz rule*:

$$\Delta_h(fg) = \Delta_h(f)h(g) + h(f)\Delta_h(g). \tag{9.4}$$

In geometrical form, this rule can be written as

$$\Delta_z(fg) = \Delta_z(f)g(z) + f(z)\Delta_z(g). \tag{9.5}$$

Thus, the rules (9.3) and (9.4) completely govern the interrelations between the operator Δ_h and the basic structures in the algebra of observables. Therefore, we have a good reason for giving the following definition:

9.4. Definition. A map

$$\xi \colon C^\infty(M) \to \mathbb{R}$$

is said to be a *tangent vector to a manifold* M at a point $z \in M$ if it satisfies the two following conditions:

(1) $\mathbb{R}$-linearity:

$$\xi\left(\sum_{j=1}^{k} \lambda_j f_j\right) = \sum_{j=1}^{k} \lambda_j \xi(f_j), \quad \lambda_j \in \mathbb{R}, \ f_j \in C^\infty(M).$$

(2) The local Leibniz rule (or the Leibniz rule at a point z):

$$\xi(fg) = f(z)\xi(g) + g(z)\xi(f), \quad f, g \in C^\infty(M).$$

Obviously, if we now define the sum of two tangent vectors and the multiplication of a tangent vector by a real number using the rules

$$(\xi + \xi')(f) = \xi(f) + \xi'(f),$$
$$(\lambda\xi)(f) = \lambda\xi(f), \quad \lambda \in \mathbb{R},$$

then in both cases the result will be an $\mathbb{R}$-linear operator satisfying the Leibniz rule; i.e., we shall obtain a tangent vector again. In other words, the set $T_z M$ of all tangent vectors at a point $z \in M$ possesses a natural structure of a vector space over $\mathbb{R}$. This space is called the *tangent space of the manifold* M *at* z.

Remark. The zero vector $0_z \in T_z M$ is just the zero map from $C^\infty(M)$ to $\mathbb{R}$ and as such coincides with $0_{z'}$ for any other point z'. But it is natural to distinguish between vectors 0_z and $0_{z'}$, $z \neq z'$, since they are tangent to M at two different points (are subject to different Leibniz rules). For a formally satisfactory explanation of this distinction see Section 9.52.

9.5. Let us now describe the operators $\xi \in T_z M$ in local coordinates. Let U be a domain in $\mathbb{R}^n$ and fix a local coordinate system $x_1, \ldots, x_n$. Assume that

$$z = (z_1, \ldots, z_n), \quad y = y(\Delta t) = (z_1 + \alpha_1 \Delta t, \ldots, z_n + \alpha_n \Delta t), \quad \alpha_i \in \mathbb{R},$$

in this system. Then obviously $z = y(0)$ and

$$\Delta_z(f) = \lim_{\Delta t \to 0} \frac{f(y(\Delta t)) - f(z)}{\Delta t} = \left.\frac{df(y(t))}{dt}\right|_{t=0} = \sum_{i=1}^{n} \alpha_i \left.\frac{\partial f}{\partial x_i}\right|_z.$$

This is nothing but the derivation of the function f in the direction $\alpha = (\alpha_1, \ldots, \alpha_n)$. Hence, the operator Δ_z is described by an n-tuple of real numbers. This observation explains the hidden meaning of the classical descriptive definition of tangent vectors.

The arguments of Section 9.3 and 9.5 use the operator Δ_z, which was not defined in conceptually correct way, and so these arguments are not rigorous either. Nevertheless, they make it possible to define tangent vectors in terms of the algebra of observables (see Definition 9.4). Using this definition as a starting point, we can now compare our approach with the usual one.

9.6 Tangent Vector Theorem. *Let M be a smooth manifold, $z \in M$, and let $x_1, \ldots, x_n$ be a local coordinate system in a neighborhood $U \ni z$. Then, in this coordinate system, any tangent vector $\xi \in T_z M$ can be represented in the form*

$$\xi = \sum_{i=1}^{n} \alpha_i \frac{\partial}{\partial x_i}\Big|_z, \quad \alpha_i \in \mathbb{R}.$$

In other words, the notions of tangent vector and of differentiation in a given direction coincide.

◀ The proof of this theorem consists of several steps. The first of them is proposed to the reader.

9.7 Lemma–Exercise. *Let $f = \mathrm{const} \in \mathbb{R}$; then $\xi(f) = 0$.*   ▷

9.8 Lemma. *Tangent vectors are local operators, i.e., if two functions $f, g \in \mathcal{F}$ coincide on an open set $U \ni z$, then for any tangent vector $\xi \in T_z M$ the equality $\xi(f) = \xi(g)$ holds.*

◁ To prove this statement, it suffices to check that if the equality $f|_U = 0$ holds in some neighborhood $U \ni z$, then $\xi(f) = 0$. Indeed, in this case, by Corollary 2.5, there exists a function $h \in C^\infty(M)$ such that $h(z) = 0$ and $h|_{M \setminus U} = 1$. Consequently, $f = hf$, and, by the Leibniz rule,

$$\xi(f) = \xi(hf) = f(z)\xi(h) + h(z)\xi(f) = 0.   ▷$$

9.9 Lemma. *The spaces $T_z U$ and $T_z M$ are naturally isomorphic for any open neighborhood $U \ni z$.*

◁ The embedding $i\colon U \subset M$ of the open set U into the manifold M induces the map $d_z i\colon T_z U \to T_z M$ as follows. Let $\xi \in T_z U$ and $f \in C^\infty(M)$; set $d_z i(\xi)(f) = \xi(f|_U)$. (Check that $d_z i(\xi)$ is indeed a tangent vector.) Obviously, the map $d_z i$ is $\mathbb{R}$-linear.

Let us construct the inverse map. To this end, note first that for any function $g \in C^\infty(U)$ one can find a function $f \in C^\infty(M)$ coinciding with g in some neighborhood in U of the point z. Indeed, consider a function $h \in C^\infty(U)$ vanishing outside some compact neighborhood $V_0 \subset U$ and equal to 1 in a neighborhood $V_1 \subset V_0$ of the point z (see Section 2.5). Then for $f \in C^\infty(M)$ we can take a function vanishing outside U and coinciding with gh in U. Now define an A-linear map $\pi_U\colon T_z M \to T_z U$ by setting $\pi_U(\eta)(g) = \eta(f)$. It follows from Lemma 9.8 that the value $\eta(f)$ does not depend on the choice of the function f; i.e., the homomorphism π_U is well defined. It is now easy to see that $d_z i \circ \pi_U = \mathrm{id}$ and $\pi_U \circ d_z i = \mathrm{id}$. ▷

9.10. From the lemma above it follows that we may confine ourselves to the case $M = U \subset \mathbb{R}^n$. Moreover, we may assume the domain U to be *star-shaped* with respect to the point z (i.e., such that $y \in U$ implies $[z, y] \subset U$, the domain U contains the entire closed interval $[z, y]$). By Corollary 2.9, any smooth function f in a star-shaped neighborhood of z can be represented in the form

$$f(x) = f(z) + \sum_{i=1}^{n}(x_i - z_i)\frac{\partial f}{\partial x_i}(z) + \sum_{i,j=1}^{n}(x_i - z_i)(x_j - z_j)g_{ij}(x).$$

Applying the tangent vector ξ to the last equality and using the Leibniz rule, we immediately see that for any derivation $\xi \in T_z M$,

$$\xi(f) = \sum_{i=1}^{n}\alpha_i\frac{\partial f}{\partial x_i}(z),$$

where $\alpha_i = \xi(x_i - z_i) = \xi(x_i)$, which concludes the proof of the Tangent Vector theorem. ▶

9.11. From the Tangent Vector Theorem it follows that $T_z M$ is an n-dimensional vector space over $\mathbb{R}$. In fact, by this theorem, the tangent vectors

$$\frac{\partial}{\partial x_1}\Big|_z, \quad \ldots, \quad \frac{\partial}{\partial x_n}\Big|_z$$

generate the space $T_z M$ for any coordinate neighborhood U of the point z. Let

$$\xi = \sum_{j=1}^{n}\alpha_j\frac{\partial}{\partial x_j}$$

be a linear combination of the vectors $\partial/\partial x_j$. Since obviously $\alpha_j = \xi(x_j)$, the vector ξ does not vanish if at least one of the coefficients α_j is not zero. Hence, the vectors $\partial/\partial x_j$ are linearly independent and form a basis of the tangent space $T_z U$. The isomorphism $d_z i \colon T_z U \to T_z M$ constructed during the proof of the theorem now shows that $\dim T_z M = n$. Below we shall identify the vectors $\partial/\partial x_j$ forming a basis of $T_z U$ with the vectors $d_z i\,(\partial/\partial x_j)$ forming a basis of $T_z M$.

It follows from the above that the dimension of the tangent space $T_z M$ is equal to the number of local coordinates in any chart containing z. In other words, it is equal to the dimension of the manifold M.

9.12. Let $y_1, \ldots, y_n$ be another local coordinate system in a neighborhood of the point z. Then in the corresponding basis of the tangent space $\partial/\partial y_1\Big|_z, \ldots, \partial/\partial y_n\Big|_z$, one has

$$\xi = \sum_{k=1}^{n}\beta_k\frac{\partial}{\partial y_k}\Big|_z.$$

Further, in view of the classical *chain rule* (see any advanced calculus textbook),

$$\xi = \sum_{i=1}^{n} \alpha_i \frac{\partial}{\partial x_i}\Big|_z = \sum_{i=1}^{n} \alpha_i \sum_{k=1}^{n} \frac{\partial y_k}{\partial x_i} \frac{\partial}{\partial y_k}\Big|_z = \sum_{k=1}^{n} \left(\sum_{i=1}^{n} \alpha_i \frac{\partial y_k}{\partial x_i} \right) \frac{\partial}{\partial y_k}\Big|_z,$$

i.e.,

$$\beta_k = \sum_{i=1}^{n} \alpha_i \frac{\partial y_k}{\partial x_i}, \qquad k = 1, \ldots, n.$$

The matrix that transforms the basis

$$\frac{\partial}{\partial x_1}\Big|_z, \quad \cdots \quad \frac{\partial}{\partial x_n}\Big|_z,$$

corresponding to the local coordinates $x_1, \ldots, x_n$, to the basis

$$\frac{\partial}{\partial y_1}\Big|_z, \quad \cdots \quad \frac{\partial}{\partial y_n}\Big|_z,$$

corresponding to the coordinates $y_1, \ldots, y_n$, is the *Jacobi matrix*

$$J_z = \begin{pmatrix} \dfrac{\partial y_1}{\partial x_1} & \cdots & \dfrac{\partial y_1}{\partial x_n} \\ \vdots & \ddots & \vdots \\ \dfrac{\partial y_n}{\partial x_1} & \cdots & \dfrac{\partial y_n}{\partial x_n} \end{pmatrix}_z.$$

The subscript z indicates that the elements of the Jacobi matrix are computed at the point z. We see that the *coordinate change rules* obtained here for tangent vectors are in agreement with the approach accepted in the tensor calculus.

9.13. The differential of a smooth map. It is quite natural that any smooth map of manifolds generates a map of tangent vectors. (Try to check this yourself by looking at Figure 9.1 on the next page and continuing the informal arguments of Section 9.2.) A rigorous construction is as follows. Let $\varphi \colon M \to N$ be a smooth map and $\xi \in T_z M$. Then the map $\eta = \xi \circ \varphi^* \colon C^\infty(N) \to \mathbb{R}$ is a tangent vector to the manifold N at the point $\varphi(z)$.

In fact, its $\mathbb{R}$-linearity is obvious. In addition, for any $f, g \in C^\infty(N)$, we have

$$\eta(fg) = \xi(\varphi^*(fg)) = \xi(\varphi^*(f)\varphi^*(g))$$
$$= \xi(\varphi^*(f))(\varphi^*(g)(z)) + (\varphi^*(f)(z))\xi(\varphi^*(g))$$
$$= \eta(f)g(\varphi(z)) + f(\varphi(z))\eta(g).$$

Definition. The map

$$d_z\varphi \colon T_z(M) \to T_z(N), \quad \xi \mapsto \xi \circ \varphi^*, \quad \xi \in T_z M,$$

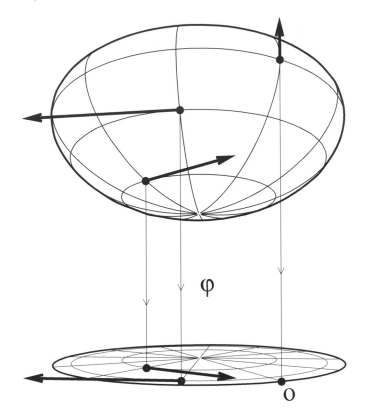

Figure 9.1. Mapping tangent vectors.

is called the *differential of the map* φ at the point $z \in M$.

Obviously, the differential $d_z\varphi$ is a linear map.

9.14 Exercise. Prove that if $\psi\colon N \to L$ is another smooth map, then

$$d_z(\psi \circ \varphi) = d_{\varphi(z)}\psi \circ d_z\varphi. \qquad (9.6)$$

Prove also that if $N = L$ and $\psi = \mathrm{id}_N$, then $d_z\psi = \mathrm{id}_{T_z N}$.

Formula (9.6), applied to $\psi = \varphi^{-1}$, shows that

$$d_{\varphi(z)}\varphi^{-1} = (d_z\varphi)^{-1}.$$

In particular, $d_z\varphi$ is an isomorphism whenever φ is a diffeomorphism.

We can now return to the discussion of Section 4.10 and prove the following statement:

9.15 Proposition. *The algebras $C^\infty(M)$ and $C^\infty(N)$ are not isomorphic if $\dim M \neq \dim N$. In particular, the algebras $C^\infty(\mathbb{R}^n)$ and $C^\infty(\mathbb{R}^m)$ are not isomorphic if $m \neq n$.*

◀ Indeed, let $\Phi\colon C^\infty(N) \to C^\infty(M)$ be an isomorphism. In that case, $\varphi = |\Phi|\colon M \to N$ is a diffeomorphism. As was observed earlier, the differential

$$d_z\varphi\colon T_zM \to T_{\varphi(z)}N$$

is then an isomorphism for all $z \in M$. Therefore,

$$\dim M = \dim T_zM = \dim T_{\varphi(z)}N = \dim N. \quad ▶$$

9.16. Let us now describe $d_z\varphi$ in coordinates. As in Section 6.15, choose local charts $(x_1, \ldots, x_n)$ and $(y_1, \ldots, y_m)$ in M and N containing the points z and $\varphi(z)$, respectively. Let $\varphi^*(y_i) = \varphi_i(x_1, \ldots, x_n)$ be the functions describing the map φ in coordinates. Then, for $g \in C^\infty(N)$, we have

$$\left[d_z(\varphi)\left(\frac{\partial}{\partial x_i}\Big|_z \right) \right](g) = \left[\frac{\partial}{\partial x_i}\Big|_z \circ \varphi^* \right](g)$$

$$= \frac{\partial g(\varphi_1(x_1, \ldots, x_n), \ldots, \varphi_m(x_1, \ldots, x_n))}{\partial x_i}\Big|_z$$

$$= \sum_{j=1}^{m} \frac{\partial g(\varphi_1(x_1, \ldots, x_n), \ldots))}{\partial y_j}\Big|_z \frac{\partial \varphi_j(x_1, \ldots, x_n)}{\partial x_i}\Big|_z$$

$$= \sum_{j=1}^{m} \frac{\partial \varphi_j(x_1, \ldots, x_n)}{\partial x_i}\Big|_z \left[\varphi * \left(\frac{\partial g}{\partial y_j} \right)(z) \right]$$

$$= \sum_{j=1}^{m} \frac{\partial \varphi_j(x_1, \ldots, x_n)}{\partial x_i}\Big|_z \frac{\partial g}{\partial y_j}\varphi(z) = \sum_{j=1}^{m} \frac{\partial \varphi_j}{\partial x_i}(z)\frac{\partial g}{\partial y_j}(\varphi(z))$$

$$= \left[\sum_{j=1}^{m} \frac{\partial \varphi_j}{\partial x_i}(z)\frac{\partial}{\partial y_j}\Big|_{\varphi(z)} \right](g).$$

In other words,

$$d_z(\varphi)\left(\frac{\partial}{\partial x_i}\Big|_z \right) = \sum_{j=1}^{m} \frac{\partial \varphi_j}{\partial x_i}(z)\frac{\partial}{\partial y_j}\Big|_{\varphi(z)}; \tag{9.7}$$

i.e., the matrix of the linear map $d_z\varphi$ in the bases

$$\left\{ \frac{\partial}{\partial x_i}\Big|_z \right\} \subset T_z(M), \quad \left\{ \frac{\partial}{\partial y_j}\Big|_{\varphi(z)} \right\} \subset T_{\varphi(z)}(N),$$

respectively, is the Jacobi matrix $\left\| \partial \varphi_i/\partial x_j \right\|_z$ of φ at the point z. The subscript z indicates here that all the derivatives in this matrix are taken at the point z. Thus, the coordinate representation for the differential $d_z\varphi$

is of the form

$$
\begin{pmatrix} \beta_1 \\ \vdots \\ \vdots \\ \beta_m \end{pmatrix} = \begin{pmatrix} \dfrac{\partial \varphi_1}{\partial x_1} & \cdots & \dfrac{\partial \varphi_1}{\partial x_n} \\ \vdots & \ddots & \vdots \\ \dfrac{\partial \varphi_m}{\partial x_1} & \cdots & \dfrac{\partial \varphi_m}{\partial x_n} \end{pmatrix}_z \cdot \begin{pmatrix} \alpha_1 \\ \vdots \\ \vdots \\ \alpha_n \end{pmatrix}, \tag{9.8}
$$

where $(\alpha_1, \ldots, \alpha_n)$ and $(\beta_1, \ldots, \beta_m)$ are the coordinates of the vector $v \in T_z(M)$ and of its image $d_z\varphi(v)$ in the bases

$$
\left\{ \frac{\partial}{\partial x_i}\Big|_z \right\}, \quad \left\{ \frac{\partial}{\partial y_j}\Big|_{\varphi(z)} \right\},
$$

respectively.

9.17. The tangent manifold. As we know from elementary mechanics, any particular state of a mechanical system S is determined by its position (configuration) and instantaneous velocity. If $M = M_S$ is the configuration space (see Sections 1.1, 5.12) of this system, then, as follows from the arguments of Section 9.2, the notion of a tangent vector to the manifold M_S is identical to the concept of the system's state. More precisely, if we consider a tangent vector $\xi \in T_z M$, then z is the position of the system, while ξ is its instantaneous velocity. Thus, the set of all system states (pay attention to Remark 9.4) is

$$
TM \overset{\text{def}}{=} \bigcup_{z \in M} T_z M.
$$

We will equip this set with a smooth manifold structure in a natural way below. The object obtained is called the *tangent manifold* of the manifold M. Since the evolution of the system is uniquely determined by its initial state, differential equations describing possible evolutions should be equations on the manifold TM.

Besides mechanics, tangent manifolds naturally arise in various branches of mathematics, and first of all in differential geometry.

9.18. To introduce a smooth manifold structure on TM, we shall need the following simple facts:

I. Any smooth map $\Phi \colon M \to N$ generates the map of sets

$$
T\Phi \colon TM \to TN,
$$

taking a tangent vector $\xi \in T_z M$ to $d_z\Phi(\xi) \in T_{\Phi(z)}N$. By (9.6),

$$
T(\Phi \circ \Psi) = T\Phi \circ T\Psi.
$$

II. If $W \subset \mathbb{R}^n$ is an open domain of the arithmetical space $\mathbb{R}^n$, then TW is naturally identified with the domain $W \times \mathbb{R}^n \subset \mathbb{R}^n \times \mathbb{R}^n = \mathbb{R}^{2n}$.

Namely, if $z \in W$, $z = (z_1, \dots, z_n)$ and $\xi \in T_zW$, then

$$\xi \Longleftrightarrow (z_1, \dots, z_n, \alpha_1, \dots, \alpha_n) \quad \text{if} \quad \xi = \sum_{i=1}^{n} \alpha_i \frac{\partial}{\partial x_i}.$$

III. A natural map of sets

$$\pi_T = \pi_{TM} \colon TM \to M, \quad T_zM \ni \xi \mapsto z \in M,$$

is defined, taking any tangent vector to its point of application.

IV. If $U \subset M$ is an open subset, then

$$\pi_T^{-1}(U) = \bigcup_{z \in U} T_zM = \bigcup_{z \in U} T_zU = TU.$$

9.19. Let us now note that any chart (U, x) of the manifold M, by the above, generates the map

$$Tx \colon TU \to TW \subset \mathbb{R}^{2n}, \quad \text{where} \quad W = x(U).$$

This map is obviously a bijection. Using 9.18, IV, we can identify TU with $\pi_T^{-1}(U)$ and obtain a $2n$-dimensional chart $\left(\pi_T^{-1}(U), Tx\right)$ in TM. In other words, coordinate functions $\{x_i, q_j\}$ associated with Tx are such that x_i for $(z, \xi) \in TU$ is the ith coordinate of the point z, while q_j is the jth component of the coordinate presentation for the vector ξ in the basis $\partial/\partial x_i|_z$. Charts of this kind are called *special*. If the charts (U, x) and (U', y) are compatible on M, the corresponding special charts $\left(\pi_T^{-1}(U), Tx\right)$ and $\left(\pi_T^{-1}(U'), Ty\right)$ are also compatible. In fact, the analytical form for the coordinate change

$$(Ty) \circ (Tx)^{-1} \colon T(W) \to T(W'), \quad W' = y(U'),$$

is (see (9.12))

$$y_i = y_i(x), \quad \beta_j = \sum_{k=1}^{n} \alpha_k \frac{\partial y_j}{\partial x_k}(x),$$

and consequently the map under consideration has the Jacobi matrix

$$\begin{pmatrix} J & * \\ 0 & J \end{pmatrix},$$

where J is the Jacobi matrix for the coordinate change $y \circ x^{-1}$, while the asterisk denotes an $(n \times n)$–matrix. Hence, $(Ty) \circ (Tx)^{-1}$ is a diffeomorphism of open sets in $\mathbb{R}^{2n}$.

Let $A = \{(U_k, x_k)\}$ be an atlas on M. Then, by the above, we see that $TA = \{(\pi_T^{-1}(U_k), Tx_k)\}$ is a *special* atlas on TM. If two atlases A_1 and A_2 are compatible on M, then the corresponding special atlas TA_1 is compatible with TA_2 as well. If A is a countable and Hausdorff atlas, then TA enjoys the same properties. For all these reasons, the smooth manifold structure on the set TM determined by the atlas TA does not depend on the choice of the atlas A on M. The set TM equipped with the described smooth structure is called the *tangent manifold* of the manifold M. Let us also note that the map

$$\pi_T \colon TM \to M$$

described in Section 9.18, III, is smooth. It is called the *tangent fiber bundle* or simply *tangent bundle* of the manifold M. This correspondence defines a functor.[1] We use here the words *fiber bundle* for the first time. The exact definition will be given in Section 11.10; Chapters 11 and 12 are completely devoted to the study of this notion. In the case under consideration, these words mean that the tangent spaces $T_z M$ (*fibers of the projection π_T*) are identical (i.e., diffeomorphic to each other) and "fiber" the tangent manifold TM. Moreover, these fibers are mutually isomorphic vector spaces. Fiber bundles of this type are called *vector bundles* and studied in detail in Chapter 12.

Exercise. Prove that the map $d\Phi \colon TM \to TN$, corresponding to the smooth map $\Phi \colon M \to N$, is smooth and the following diagram

$$
\begin{array}{ccc}
TM & \xrightarrow{\ d\Phi\ } & TN \\
{\scriptstyle \pi_T}\Big\downarrow & & \Big\downarrow{\scriptstyle \pi_T} \\
M & \xrightarrow[\ \Phi\ ]{} & N
\end{array}
$$

is commutative.

9.20. Sticking to the observability principle, it would be much more attractive to associate with any (smooth) algebra A an algebra TA such that $|TA| = T|A|$. That is why we have to confine ourselves to the example of the *cotangent manifold* considered in the next sections. The question of the algebraic definition of cotangent manifolds is delayed to Chapter 10.

9.21. Besides the description of a mechanical system in terms of position–velocity, there exists another, often more convenient, description in terms of position–momentum. The fundamental relation

$$p = mv,$$

which ties the velocity and the momentum of a mass point, shows that momenta are linear functionals on the space of velocities. In other words, the momentum of a system S in a position $z \in M = M_S$ is a linear functional on the tangent space $T_z M$; i.e., it is an element of the dual

---

[1] The reader will recall that a covariant functor Φ from a category $\mathfrak{M}$ to a category $\mathcal{N}$ is given by assignments $\Phi \colon \mathrm{Ob}\, M \to \mathrm{Ob}\, N$ and $\Phi_* \colon \mathrm{Mor}(P, Q) \to \mathrm{Mor}(\Phi(P), \Phi(Q))$, such that $\Phi_*(\xi \circ \eta) = \Phi_*(\xi) \circ \Phi_*(\eta)$ and $\Phi_*(\mathrm{id}_{\mathfrak{M}}) = \mathrm{id}_{\mathcal{N}}$. Contravariant functors are defined similarly (reverse arrows and raise indices).

space

$$T_z^* M \overset{\text{def}}{=} \text{Hom}_{\mathbb{R}}(T_z M, \mathbb{R}).$$

The space $T_z^* M$ is called the *cotangent space* to M at the point z, and its elements are called *tangent covectors* to the manifold M at the point z. So, the momenta of a mechanical system S are tangent covectors to the configuration manifold $M = M_S$.

These and many other considerations lead us to the notion of *cotangent manifold*. To give a formal definition of the latter, let us consider the set

$$T^* M \overset{\text{def}}{=} \bigcup_{z \in M} T_z^* M$$

together with the natural projection

$$\pi_T^* = \pi_{T^* M} : T^* M \to M, \quad T_z^* M \ni \theta \mapsto z \in M.$$

9.22. A natural smooth manifold structure on $T^* M$ can be defined using the scheme applied already to TM. It is extremely helpful here to understand tangent covectors as differentials of functions on M.

Namely, let $z \in M$ and let f be a function from $C^\infty(M)$. Let us define a function $d_z f$ on $T_z M$ by setting

$$d_z f(\xi) \overset{\text{def}}{=} \xi(f), \quad \xi \in T_z M.$$

Exercise. Prove the following statements:

1. If $f = \text{const}$, then $d_z f = 0$.

2. $d_z(f + g) = d_z f + d_z g$,

3. $d_z(fg) = f(z)d_z g + g(z)d_z f$.

By the definition of the linear space structure on $T_z M$ (see Section 9.4), the function $d_z f$ is linear. Therefore, $d_z f$ is a tangent covector at the point z, called the *differential of the function f* at z. If (U, x) is a chart containing the point z, then

$$d_z x_i \left(\frac{\partial}{\partial x_j} \Big|_z \right) = \frac{\partial x_i}{\partial x_j}(z) = \delta_i^j,$$

where δ_i^j is the Kronecker symbol. This shows that $(d_z x_1, \dots, d_z x_n)$ is the basis of the space $T_z^* M$ dual to the basis

$$\frac{\partial}{\partial x_1}\Big|_z, \dots, \frac{\partial}{\partial x_n}\Big|_z$$

in $T_z M$. Therefore, any covector $\theta \in T_z^* M$ is uniquely written in the form

$$\theta = \sum_{i=1}^n p_i d_z x_i, \quad p_i \in \mathbb{R}.$$

It is useful to note that

$$p_i = \theta\Big(\frac{\partial}{\partial x_i}\Big|_z\Big).$$

In particular, if $\theta = d_z f$, then

$$\theta\Big(\frac{\partial}{\partial x_i}\Big|_z\Big) = \frac{\partial f}{\partial x_i}(z) \quad \text{and hence} \quad d_z f = \sum_{i=1}^{n} \frac{\partial f}{\partial x_i}(z)d_z x_i.$$

This formula justifies the adopted terminology and shows that any covector θ can be represented in the form $\theta = d_z f$. Thus, tangent covectors at a given point are exhausted by differentials of functions at this point.

9.23. Any smooth map $\Phi\colon M \to N$ generates the linear map

$$d_z\Phi^*\colon T^*_{\Phi(z)}N \to T^*_z M$$

dual to the linear map

$$d_z\Phi\colon T_z M \to T_{\Phi(z)}N.$$

If $\xi \in T_z M$ and $g \in C^\infty(N)$, then, by definition,

$$d_z\Phi^*(d_{\Phi(z)}g)(\xi) = d_{\Phi(z)}g(d_z\Phi(\xi))$$
$$= \big(d_z\Phi(\xi)\big)(g) = \xi\big(\Phi^*(g)\big) = \big(d_z\Phi^*(g)\big)(\xi).$$

This means that

$$d_z\Phi^*(d_{\Phi(z)}g) = d_z\Phi^*(g).$$

Note also that, in the notation of Section 9.16, the matrix of the map $d_z\Phi^*$ in the bases $\{d_z x_i\}$ and $\{d_{\Phi(z)}y_j\}$ in $T^*_z M$ and $T^*_{\Phi(z)}N$, respectively, is the transposed Jacobi matrix $J_z = \|\partial y_i/\partial x_j(z)\|$.

9.24. The construction of special charts on T^*M is accomplished in a way similar to that used above for TM. Instead of properties I–IV from Section 9.18, the following facts should be used:

I. Any diffeomorphism $\Phi\colon M \to N$ generates the bijection

$$\Phi_*\colon T^*M \to T^*N, \quad T^*_z M \ni \theta \mapsto (d_z\Phi^*)^{-1}(\theta).$$

II. If $W \subset \mathbb{R}^n$ is an open domain, then

$$T^*W = W \times \mathbb{R}^n \subset \mathbb{R}^n \times \mathbb{R}^n = \mathbb{R}^{2n},$$

and the identification $T^*W = W \times \mathbb{R}^n$ follows the rule

$$T^*W \ni \theta \Longleftrightarrow (z_1,\ldots,z_n,p_1,\ldots,p_n),$$

where $z = (z_1,\ldots,z_n)$, $\theta = \sum_{i=1}^{n} p_i d_z x_i$ and $(x_1,\ldots,x_n)$ are the standard coordinates in $\mathbb{R}^n$.

Now let (U,x) be a chart on M and $W = x(U) \subset \mathbb{R}^n$. By duality, the natural identification of $T_z U$ with $T_z M$ allows the identification of $T^*_z U$

with $T_z^* M$. In turn, this leads to the identification of $T^* U$ with $\pi_{T^*}^{-1}(U)$. Now using II, we obtain a *special chart* $(\pi_{T^*}^{-1}(U), T^* x)$ on $T^* M$. Here $T^* x$ denotes the system of coordinate functions $\{x_i, p_j\}$, where x_i for (z, θ) taken from $T^* U$ is the ith coordinate of the point z, while p_j is the jth component of the decomposition of the vector θ in the basis dx_i.

If $A = \{(U_k, x_k)\}$ is an atlas on M, then $T^* A \overset{\text{def}}{=} \{(\pi_{T^*}^{-1}(U_k), T^* x_k)\}$ is an atlas (of dimension $2n$) on $T^* M$. If two atlases A_1 and A_2 are compatible on M, then so are the atlases $T^* A_1$ and $T^* A_2$. For this reason, the atlas $T^* A$ determines a smooth manifold structure on $T^* M$ independent of the choice of a particular atlas A.

The $2n$-dimensional manifold $T^* M$ thus obtained is called the *cotangent manifold* of the manifold M. Note also that the map

$$\pi_{T^*} : T^* M \to M$$

is smooth. It is called the *cotangent bundle* of M.

9.25. Any function $f \in C^\infty(M)$ generates the smooth map

$$s_{df} : M \to T^* M, \quad s_{df}(z) = d_z(f).$$

This map is characterized by the fact that any point $z \in M$ is taken to a point in the fiber of the cotangent bundle $\pi_{T^*}^{-1}(z) = T_z^* M$ over z. Such maps are called *sections*. This notion will be discussed in more detail in subsequent chapters; see Sections 11.12 and 12.7.

Exercise. Describe s_{df} in special local coordinates.

9.26. Any map $\Phi \colon M \to N$ generates a family $d_z \Phi^* \colon T_{\Phi(z)}^* N \to T_z^* M$ of maps taking cotangent spaces of points in M to those in N. Unfortunately, it does not allow one, in general, to construct a map of cotangent manifolds $T^* N \to T^* M$ that reduces to $d_z \Phi^*$ when restricted to $T_{\Phi(z)}^* M$.

If $\dim M = \dim N = n$ and the map $\Phi \colon M \to N$ is *regular* at all points of the manifold, i.e., all differentials are isomorphisms, then one can define a smooth map $T^* M \to T^* N$ *covering* Φ and reducing to $(d_z \Phi^*)^{-1}$ when restricted to $T_z^* M$. The map Φ_* considered in Section 9.24 is its particular case.

9.27. Note that the cotangent space $T_z^* M$ can be defined in a purely algebraic way, as it is done in algebraic geometry. Let μ_z be the ideal consisting of all functions vanishing at the point z:

$$\mu_z \overset{\text{def}}{=} \{f \in C^\infty(M) \mid f(z) = 0\}.$$

Proposition. *There exists a natural isomorphism between* $T_z^* M$ *and the quotient* μ_z / μ_z^2.

◀ Consider the quotient algebra

$$J_z^1 M \overset{\text{def}}{=} C^\infty(M) / \mu_z^2$$

and the map

$$\bar{d}_z \colon J_z^1 M \to T_z^* M, \quad \bar{d}_z([f]) = d_z f,$$

where $f \in C^\infty(M)$ and $[f] = f \bmod \mu_z^2$. Since $d_z f = 0$ for $z \in \mu_z^2$ (see the exercise from Section 9.22), this map is well defined. Obviously, it is $\mathbb{R}$-linear and surjective (see Section 9.22) because any covector θ can be presented as the differential of some function, $\theta = d_z f$.

The decomposition of the algebra $C^\infty(M)$ into the direct sum of linear spaces

$$C^\infty(M) = \mathbb{R} \oplus \mu_z, \quad f = f(z) + (f - f(z)),$$

gives the direct sum decomposition

$$J_z^1 M = \mathbb{R} \oplus \mu_z / \mu_z^2. \tag{9.9}$$

By the exercise from Section 9.22, the map d_z annihilates the first summand. On the other hand, Hadamard's lemma (Lemma 2.8) shows that $f \in \mu_z$ and $d_z f = 0$ imply $f \in \mu_z^2$. Therefore, the restriction $\bar{d}_z$ to μ_z / μ_z^2 is an isomorphism.   ▶

Corollary. $\dim J_z^1 M = n + 1$.

◀ Indeed, the above proposition allows us to rewrite equality (9.9) in the form

$$J_z^1 M = \mathbb{R} \oplus T_z^* M.   ▶$$

9.28. The quotient algebra $J_z^1 M$ was useful in the proof of Proposition 9.27; as we shall see later, this algebra is one of the most important constructions of the differential calculus. It is called the *algebra of first-order jets* (or of *1-jets*) at the point $z \in M$ for the algebra of smooth functions $C^\infty(M)$.

The union

$$J^1 M = \bigcup_{z \in M} J_z^1 M$$

can be endowed with a natural smooth manifold structure in a similar way as was done above for $T^* M$.

Exercise. Develop the corresponding constructions in detail. Describe the special coordinates in $J^1 M$.

The manifold $J^1 M$ is called the *manifold of first-order jets* for the manifold M.

Similarly to tangent and cotangent manifolds, $J^1 M$ is fibered over M by means of the natural map

$$\pi_{J^1} \colon J^1 M \to M, \quad \pi_{J^1}([f]_z^1) = z,$$

where $[f]_z^1$ denotes the image of the function f under the quotient map $C^\infty(M) \to J_z^1 M$. Similar to π_T and π_{T^*} (see Sections 9.19 and 9.24, respectively), the map π_{J^1} is also a vector bundle over M. Its fibers are of dimension $(n+1)$. For any function $f \in C^\infty(M)$, one can consider the smooth map

$$s_{j_1 f} \colon M \to J^1 M, \quad s_{j_1 f}(z) = [f]_z^1,$$

which is a section of the bundle π_{J^1}.

The map

$$\pi_{J^1, T^*} \colon J^1 M \to T^* M, \quad \pi_{J^1, T^*}([f]_z^1) = d_z f,$$

which relates the manifold of 1-jets in the natural way to the cotangent manifold, is a one-dimensional vector bundle over $T^* M$.

A remarkable feature of the manifold $J^1 M$ is that it allows us to construct an exhaustive theory of first-order partial differential equations in one unknown. In this theory, differential equations are interpreted as submanifolds in $J^1 M$.

The Tangent Vector Theorem allowed us to make the first step in understanding the differential calculus as a part of commutative algebra. The next step is to define tangent vectors to the spectrum of an arbitrary commutative algebra.

Let A be an arbitrary unital commutative K-algebra. Denote by $|A|$ its K-spectrum, i.e., the set of all (unital) K-homomorphisms from A to K.

9.29. Definition. A map $\xi \colon A \to K$ is called a *tangent vector*, or a *derivation* at a point $h \in |A|$, if it

(i) is K-linear, i.e.,

$$\xi\left(\sum_{j=1}^k \lambda_j f_j\right) = \sum_{j=1}^k \lambda_j \xi(f_j), \qquad \lambda_j \in K, \quad f_j \in A;$$

(ii) satisfies the Leibniz rule at h, i.e.,

$$\xi(fg) = f(h)\xi(f) + g(h)\xi(f), \qquad f, g \in A.$$

This definition, in the case $K = \mathbb{R}$ and $A = C^\infty(M)$, coincides with the definition of tangent vectors to the manifold M $(= |A|)$ at a point $z \in M$. To understand this fact, it suffices to recall the identification $M = |C^\infty(M)|$ and to treat z as the homomorphism $h_z \colon f \mapsto f(z)$, $f \in C^\infty(M)$. The set of all tangent vectors at a given point is naturally endowed with a K-module structure (or that of a vector space over K when K is a field):

1. $(\xi_1 + \xi_2)(a) \stackrel{\text{def}}{=} \xi_1(a) + \xi_2(a)$, $a \in A$;

2. $(k\xi)(a) \stackrel{\text{def}}{=} k\xi(a)$, $k \in K, a \in A$.

Let us denote this K-module by $T_h A$. If $K = \mathbb{R}$ and $A = C^\infty(M)$, then, under the above identification of points $z \in M$ with K-homomorphisms h_z, the space $T_z M$ will coincide with $T_{h_z} A$.

Remark. In algebraic geometry, one considers various spectra of algebras, maximal, primitive, etc. Treating the symbol h in the previous definition in an adequate way, the reader will easily define tangent vectors for points of all these spectra.

9.30. Cotangent spaces of commutative algebra spectra. Proposition 9.27, revealing the purely algebraic nature of cotangent bundles, shows how to define the *cotangent space of the spectrum* $|A|$ for an arbitrary commutative K-algebra A at some point $h \in |A|$. Namely, set

$$T_h^* A \stackrel{\text{def}}{=} \mu_h / \mu_h^2, \qquad (9.10)$$

where μ_h is the kernel of the K-algebra homomorphism $h \colon A \to K$. By definition, $T_h^* A$ is a K-module. Its role is illustrated by the following proposition:

Proposition. *For any K-algebra A, the natural surjection of K-modules*

$$\nu_h \colon \operatorname{Hom}_K(T_h^* A, K) \to T_h A \qquad (9.11)$$

is defined. If K is a field, then ν_h is an isomorphism.

◄ Let us first note that any K-linear map $\varphi \colon T_h^* A \to K$ determines the tangent vector

$$\xi_\varphi \in T_h A, \quad \xi_\varphi(a) = \varphi([a - h(a) \cdot 1_A]),$$

where $[b] = b \bmod \mu_h^2$ (check it). The correspondence $\varphi \mapsto \xi_\varphi$ in an obvious way determines the K-module homomorphism

$$\nu_h \colon \operatorname{Hom}_K(T_h^* A, K) \to T_h A.$$

The map ν_h is a surjection. In fact, let $\xi \in T_h A$. Consider the K-linear map

$$\varphi_\xi \colon T_h^* A \to K, \qquad \varphi_\xi([a]) = \xi(a), \quad a \in \mu_h.$$

The Leibniz rule implies $\xi(\mu_h^2) \subset \mu_h$, and thus the map φ_ξ is well defined. Obviously, $\nu_h(\varphi_\xi) = \xi$.

If K is a field, then ν_h is also an injection. In fact, now let $a, b \in \mu_h$ and $[a] \neq [b]$. Since K is a field, one can always find a linear function φ defined on the vector K-space $T_h^* A$ and satisfying $\varphi([a]) \neq \varphi([b])$, i.e., $\xi_\varphi(a) \neq \xi_\varphi(b)$. ►

9.31. To find an algebraic counterpart for the concept of the differential of a smooth map, let us note that to any (unital) K-algebra homomorphism $F \colon A_1 \to A_2$ there corresponds a map of K-spectra, namely

$$|F| \colon |A_2| \to |A_1|, \quad |A_2| \ni h \mapsto h \circ F \in |A_1|.$$

If, in addition, $\xi \in T_h(A_1)$, then the map

$$d_h|F|(\xi) \stackrel{\text{def}}{=} \xi \circ F : A_1 \to K$$

is a tangent vector to the space $|A_1|$ at the point $|F|(h) = h \circ F$. Thus we obtain the K-linear map

$$d_h|F| : T_h(A_1) \to T_{h \circ F}(A_2)$$

(prove this fact). If $F = \varphi^*$, where $\varphi : M_2 \to M_1$ is a smooth map and $A_i = C^\infty(M_i)$, $i = 1, 2$, then the differentials $d_h \varphi$ and $d_h|F|$ coincide.

9.32 Exercises.   1. Prove that $T_{\mathrm{id}_K} K = 0$, where $\mathrm{id}_K : K \to K$, is the only point of the K-spectrum for K.

2. Let $i : K \to A$, $k \mapsto k \cdot 1_A$, be the canonical embedding and let $\xi \in T_h A$. Prove that $\xi|_{\Im i} = 0$; i.e., any derivation at a given point takes constants to zero.

3. Let $F : A_1 \to A_2$ be a K-algebra epimorphism. Prove that the map $d_h|F|$ is a monomorphism for any point $h \in |A_2|$.

4. Let $C^0(M)$ be the algebra of all continuous functions on M. Prove that $T_z(C^0(M)) = 0$ for any point $z \in M$.

9.33. The advantages of the algebraic approach to the differential calculus can already be shown at this point, though so far we have succeeded only in giving the definition of tangent vectors. For example, we can define tangent spaces to manifolds with singularities and furthermore—to arbitrary smooth sets (see Section 7.13)—and obtain the simplest invariants of singular points. Some examples will be given below. The following statement, whose proof can be literally carried over from Section 9.9, will be quite useful in analyzing these examples. Below we use the notation of Sections 3.23–3.25.

Proposition. *Suppose $\mathcal{F}$ is an arbitrary geometrical $\mathbb{R}$-algebra and let $U \subset |\mathcal{F}|$ be an open subset. Then the restriction homomorphism*

$$\rho_U : \mathcal{F} \to \mathcal{F}_U \stackrel{\text{def}}{=} \mathcal{F}|_U$$

induces an isomorphism

$$d_h(\rho_U) : T_h(\mathcal{F}_U) \to T_{\rho_U \circ h}(\mathcal{F}), \quad h \in |\mathcal{F}_U|. \quad \blacktriangleright$$

A similar construction is valid for arbitrary K-algebras.

9.34 Exercises.   1. Let $W = \left\{ (x, y) \in \mathbb{R}^2 \mid y^2 = x^3 \right\}$ be the semicubical parabola. Show that $T_z W$ is two-dimensional for $z = (0, 0)$ and one-dimensional otherwise.

An obvious consequence of this fact is that the algebra

$$C^\infty(W) = C^\infty\left(\mathbb{R}^2\right) / \left(y^2 - x^3\right) C^\infty\left(\mathbb{R}^2\right)$$

is not smooth (cf. 4.16).

2. Give an example of a smooth set whose tangent spaces are all one-dimensional except for single point in which it is 3-dimensional.

9.35. Example. Suppose $\mathbf{K}$ is the coordinate cross on the plane (see Section 7.14, 1), $\mathbf{K} = \{(x, y) \subset \mathbb{R}^2 \mid xy = 0\}$, and $\mathcal{F} = C^\infty(\mathbf{K}) = C^\infty(\mathbb{R}^2)\big|_{\mathbf{K}}$ is the algebra of smooth functions on $\mathbf{K}$. Let us describe $T_z C^\infty(\mathbf{K})$ for all points $z \in \mathbf{K}$. Elements of the algebra $C^\infty(\mathbf{K})$ may be understood as pairs $(f(x), g(y))$ of smooth functions on the line satisfying $f(0) = g(0)$. In other words,

$$C^\infty(\mathbf{K}) = \{(f(x), g(y)) \mid f(0) = g(0)\}.$$

Note that for any nonsingular point on the cross, i.e., for a point of the form $(x, 0)$, with $x \neq 0$, or $(0, y)$, with $y \neq 0$, the tangent space is one-dimensional. Let us consider, say,

$$z = (x, 0), \ x \neq 0, \ \text{and} \ U = \{(x', 0) \mid xx' > 0\}.$$

Then U is open in $\mathbf{K} = |\mathcal{F}|$, and consequently, by Proposition 9.33, we have $T_z \mathcal{F} = T_z \mathcal{F}\big|_U$. It remains to note that $\mathcal{F}\big|_U = C^\infty(\mathbb{R}^1_+) = C^\infty(\mathbb{R}^1)$. For a basis vector in the space $T_{(x,0)}$, we can take the operator

$$\frac{d}{dx}\Big|_{(x,0)}, \qquad \frac{d}{dx}\Big|_{(x,0)} (f(x), g(y)) = \frac{df}{dx}(x),$$

while for tangent spaces of the form $T_{(0,y)}$, we can take the operator

$$\frac{d}{dy}\Big|_{(0,y)}, \qquad \frac{d}{dy}\Big|_{(0,y)} (f(x), g(y)) = \frac{dg}{dy}(y).$$

Now consider the point $(0, 0)$. Obviously, the operators

$$\frac{d}{dx}\Big|_{(0,0)} \quad \text{and} \quad \frac{d}{dy}\Big|_{(0,0)}$$

will be tangent vectors at this point. They are linearly independent, and hence the space $T_{(0,0)}$ is at least two-dimensional. Since the natural restriction map $\tau\colon C^\infty(\mathbb{R}^2) \to C^\infty(\mathbf{K})$ is an epimorphism, then by Problem 3 of Section 9.32, the kernel of the map

$$d_{(0,0)}\tau\colon T_{(0,0)}(C^\infty(\mathbf{K})) \to T_{(0,0)}(C^\infty(\mathbb{R}^2)) = \mathbb{R}^2$$

is trivial. Therefore, the tangent space $T_{(0,0)}(C^\infty(\mathbf{K}))$ is isomorphic to $\mathbb{R}^2$.

Thus, the property of the point $(0, 0) \in \mathbf{K}$ to be singular manifests itself, in particular, in the fact that the dimension of the tangent space at this point is greater than for "normal" ones.

Let us stress that the standard coordinate approach does not allow one to define tangent vectors at the point $(0, 0)$. But if one tried to understand a tangent vector as an equivalence class of curves, then there would be no linear space structure in the set of such tangent vectors to $\mathbf{K}$ at this point.

9.36 Exercise. Let $W \subset \mathbb{R}^n$ be a smooth set. Recall that by definition, $C^\infty(W) = \{f|_W \mid f \in C^\infty(\mathbb{R}^n)\}$. Describe $T_z W$ for all points $z \in W$ in the following cases (see Exercise 7.14):

1. $W \subset \mathbb{R}^2$ is given by the equation $y = \sqrt{|x|}$.

2. W is the triangle in $\mathbb{R}^2$: $W = W_1 \cup W_2 \cup W_3$, where

$$W_1 = \{(x,y) \mid 0 \leqslant y \leqslant 1,\ x = 0\},$$
$$W_2 = \{(x,y) \mid 0 \leqslant x \leqslant 1,\ y = 0\},$$
$$W_3 = \{(x,y) \mid x + y = 1,\ x, y \geqslant 0\}.$$

3. W is the triangle from the previous problem together with the interior domain: $W = \{(x,y) \mid x + y \leqslant 1,\ x, y \geqslant 0\}$.

4. W is the cone $x^2 + y^2 = z^2$ in $\mathbb{R}^3$.

5. $W = W_i$, $i = 1, 2, 3$, is one of the one-dimensional homeomorphic polyhedra shown in Figure 7.1. Explain why the algebras $C^\infty(W_i)$, $i = 1, 2, 3$, are pairwise nonisomorphic.

6. $W \subset \mathbb{R}^2$ is the closure of the graph of the function $y = \sin 1/x$.

9.37 Exercise. Let $\dim T_z W = 0$, where W is a smooth set. Prove that z is an isolated point of W. This is no longer true for arbitrary algebras. Find an example of an algebra $\mathcal{F}$ such that $|\mathcal{F}| \simeq \mathbb{R}$, but $\dim T_z \mathcal{F} = 0$ for all $z \in |\mathcal{F}|$.

Can you construct another algebra $\mathcal{F}$ with $|\mathcal{F}| \simeq \mathbb{R}$ and $\dim T_z \mathcal{F} = 2$ for some $z \in |\mathcal{F}|$? And the same for all $z \in |\mathcal{F}|$?

9.38. More complicated objects of the differential calculus, which can be constructed using tangent vectors, are vector fields. Vivid geometrical images of vector fields are provided by numerous fields of forces in mechanics and physics, velocity fields of continuous media, etc. A "field" of arrows on a meteorological map may be considered as the velocity field of moving air masses.

Let us try to formalize this notion in the same spirit as was done for tangent vectors. The first step in this direction is obvious: a vector field on a manifold M is a family of tangent vectors $\{X_z\}_{z \in M}$, where $X_z \in T_z M$. In terms of the algebra of observables $C^\infty(M)$, this means that we are dealing with the family of operators

$$X_z \colon C^\infty(M) \to \mathbb{R}, \quad z \in M.$$

In particular, such a family of operators assigns to each function f from $C^\infty(M)$ the set of numbers $\{X_z(f),\ z \in M\}$, which a physicist would call a *scalar field*, while a mathematician would just call it a function on M. Denoting this function by $X(f)$, we obtain by definition

$$X(f)(z) \stackrel{\text{def}}{=} X_z(f), \quad z \in M.$$

In this notation, it becomes clear that the words *vector field* must be understood as a sort of operation on the algebra $C^\infty(M)$:

$$X \colon C^\infty(M) \to ?\,,$$

where the question mark means some set of functions on M. A natural way to formalize the idea of smoothness of a vector field X is to set $? = C^\infty(M)$: $X(f) \in C^\infty(M)$ for any function $f \in C^\infty(M)$. Thus, a smooth vector field X on M is an operator acting on $C^\infty(M)$:

$$X \colon C^\infty(M) \to C^\infty(M).$$

By the $\mathbb{R}$-linearity of the maps X_z, of which the operator X "consists," this operator is also $\mathbb{R}$-linear. Moreover, the Leibniz rule for a tangent vector X_z at a point z implies

$$X(fg)(z) = X_z(fg) = X_z(f)g(z) + f(z)X_z(g)$$
$$= \big(X(f)(z)\big)g(z) + f(z)\big(X(g)(z)\big) = [X(f)g + fX(g)](z);$$

i.e., the operator X satisfies the Leibniz rule

$$X(fg) = X(f)g + fX(g), \quad f, g \in C^\infty(M). \tag{9.12}$$

The above motivates the following definition:

Definition. An $\mathbb{R}$-linear operator $X \colon C^\infty(M) \to C^\infty(M)$ satisfying the Leibniz rule (9.12) is called a *smooth vector field* on the manifold M.

Everywhere below the word "smooth" is omitted, since we shall deal with smooth vector fields only.

9.39. The above definition of a vector field was formulated in terms of the base algebra $C^\infty(M)$ and completely satisfies the principle of observability. Moreover, we can now a posteriori justify the use of the words *vector field* in this definition: We can associate with any vector field X the family of tangent vectors $\{X_z \in T_zM\}_{z \in M}$. Namely, setting

$$X_z(f) = X(f)(z), \quad z \in M, \tag{9.13}$$

we easily see that the maps $X_z \colon C^\infty(M) \to \mathbb{R}$ thus defined are $\mathbb{R}$-linear and satisfy the Leibniz rule at any point z, i.e., they are tangent vectors at z.

Exercise. Prove this fact.

9.40 Proposition. (Locality of vector fields.) *Let X be a vector field on M. If functions $f, g \in C^\infty(M)$ coincide on an open set $U \subset M$, then the functions $X(f), X(g)$ also coincide on U.*

◀ In fact, by Lemma 9.8, one has $X_z(f) = X_z(g)$ for all $z \in U$. Hence, $X(f)(z) = X(g)(z)$ for all $z \in U$.   ▶

The interpretation of a vector field as a family of tangent vectors allows one to consider the section of the tangent bundle

$$s_X \colon M \to TM, \quad z \mapsto X_z \in T_zM \subset TM,$$

related to this field.

Exercise. Prove that s_X is a smooth section, and vice versa, any (smooth) section of the tangent bundle is of the form s_X. Therefore, vector fields on M can be understood as sections of the tangent bundle.

9.41. Equality (9.13) shows that the family of tangent vectors $\{X_z\}_{z \in M}$ generated by X determines this vector field uniquely. Using this fact, we can easily understand how vector fields are described in terms of local coordinates. In fact, if (U, x) is a chart on M and $z \in U$, then X_z, as a tangent vector at the point $z \in U$, can be presented in the form

$$X_z = \sum_{i=1}^{n} \alpha_i(z) \frac{\partial}{\partial x_i}\bigg|_z . \tag{9.14}$$

The notation $\alpha_i(z)$ underlines the fact that the coordinates of the vector X_z depend on a point $z \in U$; i.e., they are functions on U. By (9.13) and (9.14), we have

$$X(f)(z) = X_z(f) = \sum_{i=1}^{n} \alpha_i(z) \frac{\partial f}{\partial x_i}(z) = \left(\left(\sum_{i=1}^{n} \alpha_i \frac{\partial}{\partial x_i} \right)(f) \right)(z),$$

and consequently

$$X(f) = \sum_{i=1}^{n} \alpha_i \frac{\partial}{\partial x_i}(f).$$

Therefore,

$$X = \sum_{i=1}^{n} \alpha_i \frac{\partial}{\partial x_i}.$$

Note that all functions α_i belong to $C^\infty(U)$. In fact, let $z \in U$. Consider a function $\tilde{x}_i \in C^\infty(M)$ coinciding with x_i in a neighborhood V of the point z. By definition, $X(\tilde{x}_i) \in C^\infty(M)$. Further, $X(\tilde{x}_i)\big|_V = \alpha_i$, since $\tilde{x}_i\big|_V = x_i\big|_V$, and consequently $\alpha_i \in C^\infty(V)$. Since $z \in U$ is an arbitrary point, we see that α_i is a smooth function on U.

9.42. Transformation of vector fields. Let X be a vector field on M, and $\varphi \colon M \to N$ a smooth map. The differential $d_z\varphi$ takes any vector X_z to the tangent vector $Y_{\varphi(z)} = d_z\varphi(X_z) \in T_{\varphi(z)}N$. In general, the family of tangent vectors $\{Y_{\varphi(z)}\}_{z \in M}$ does not constitute a vector field on N. In fact, if $u \in N \setminus \varphi(M)$, then the vector Y_u is undefined, while a point $u \in \varphi(M)$ may have several inverse images, and thus the vector Y_u may be defined ambiguously (see Figure 9.2).

So, as a rule, there are no maps of vector fields corresponding to maps of manifolds. But diffeomorphisms are exceptions from the general rule, and if φ is a diffeomorphism, then its action on a vector field X can be defined

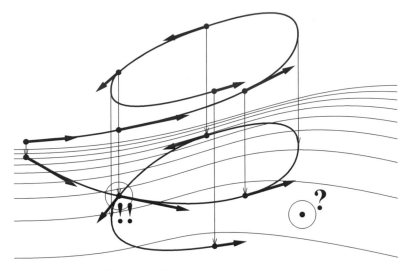

Figure 9.2. Trying to map a vector field.

by the formula

$$Y = (\varphi^{-1})^* \circ X \circ \varphi^*.$$

Exercise.   1. Prove that Y is really a vector field.

2. Prove that $Y_{\varphi(z)} = d_z \varphi(X_z)$.

Below (see Section 9.47) it will be shown that the image of a vector field may be defined in a reasonable way, provided that the notion of vector field can be adequately generalized.

9.43. The definition of a vector field given above is a particular case of the general algebraic notion of derivation, which is as follows. Let A be a commutative K-algebra.

Definition. A K-linear map $\Delta \colon A \to A$ is called a *derivation* of the algebra A if it satisfies the Leibniz rule

$$\Delta(ab) = a\Delta(b) + b\Delta(a) \quad \forall\, a, b \in A.$$

Let us denote the set of all derivations of A by $D(A)$. Let $\Delta, \nabla \in D(A)$ and $a \in A$. Then obviously, $\Delta + \nabla \in D(A)$ and $a\Delta \in D(A)$. These operations endow $D(A)$ with a natural A-module structure.

Any derivation of the K-algebra A can be understood as a vector field on $|A|$. To see this, it suffices to carry over formula (9.13) to the algebraic setting. Let $\Delta \in D(A)$ and $h \in |A|$. Put

$$\Delta_h = h \circ \Delta \colon A \to K.$$

Then, obviously, the operator Δ_h, being the composition of two K-linear operators, is also K-linear, and

$$\Delta_h(ab) = h(\Delta(ab)) = h(\Delta(a)b + a\Delta(b))$$
$$= h(\Delta(a))h(b) + h(a)h(\Delta(b)) = \Delta_h(a)h(b) + h(a)\Delta_h(b).$$

Thus $\Delta_h \in T_h(A)$. In what follows, for brevity we shall write $D(M)$ instead of $D(C^\infty(M))$.

Let us note that the definition of a tangent vector Δ_h written in the form

$$\Delta_h(f) = h(\Delta(f)), \quad f \in A,$$

is identical to (9.13) if $K = \mathbb{R}$, $A = C^\infty(M)$, while $h = h_z \in M = |A|$, is, as usual, understood as a homomorphism taking f to $f(z)$. If the algebra A is geometrical, then the system $\{\Delta_h\}_{h \in |A|}$ of vectors tangent to $|A|$ determines the "vector field" Δ uniquely.

9.44. Just as in the case of tangent vectors (see Sections 9.33–9.36), one can construct the theory of vector fields for geometrical objects of a much more general nature than smooth manifolds. For example, using Section 9.33, it is possible to obtain a theory of vector fields on arbitrary closed subsets of smooth manifolds, just as is done for the manifolds themselves. In the examples below we use the notation introduced for smooth manifolds. The locality of vector fields (Proposition 9.40) is valid in this more general situation and is proved in the same way.

9.45. Example. Let us describe vector fields on the cross $\mathbf{K}$ using the notation of Section 9.35, where we studied tangent vectors. By A_x and A_y we denote the algebras of smooth functions on the line with fixed coordinate functions x and y, respectively. The natural embeddings

$$i_x \colon A_x \to C^\infty(\mathbf{K}), \qquad\qquad f(x) \mapsto (f(x), f(0)),$$
$$i_y \colon A_y \to C^\infty(\mathbf{K}), \qquad\qquad g(y) \mapsto (g(0), g(y)),$$

are defined together with the projections

$$\pi_x \colon C^\infty(\mathbf{K}) \to A_x, \qquad\qquad (f(x), g(y)) \mapsto f(x),$$
$$\pi_y \colon C^\infty(\mathbf{K}) \to A_y, \qquad\qquad (f(x), g(y)) \mapsto g(y).$$

Obviously, $\pi_x \circ i_x = \mathrm{id}$ and $\pi_y \circ i_y = \mathrm{id}$. Therefore, if $\Delta \in D(\mathbf{K})$, then $\Delta^x = \pi_x \circ \Delta \circ i_x \in D(A_x)$ and $\Delta^y = \pi_y \circ \Delta \circ i_y \in D(A_y)$.

Let us show that

$$\Delta(f(x), g(y)) = (\Delta^x(f), \Delta^y(g)) \tag{9.15}$$

and the fields Δ^x, Δ^y vanish at the point 0.
◀ Indeed, let

$$\varphi = i_x(f) = (f(x), c) \in C^\infty(\mathbf{K}), \quad c = f(0).$$

Consider the point $z = (0, y) \in \mathbf{K}$, $y \neq 0$. Then a sufficiently small neighborhood U of the point z is an interval and $\varphi|_U \equiv c$. By the locality of tangent vectors, one has $\Delta_z(\varphi) = 0$. In other words, $\Delta(\varphi)(z) = 0$ for all points of the y-axis except for the point $(0,0)$. By continuity, $\Delta(\varphi)$ is zero identically on the whole axis. Therefore, $i_x(\pi_x(\Delta(\varphi))) = \Delta(\varphi)$, or $i_x(\Delta^x(f)) = \Delta(i_x(f))$. The last equality means that

$$i_x \circ \Delta^x = \Delta \circ i_x.$$

In addition, $\Delta^x(f)(0) = 0$, since $\Delta(\varphi)(0,0) = 0$. Consequently, the vector field $\Delta^x \in D(A_x)$ vanishes at the point $(0,0)$. In a similar way,

$$i_y \circ \Delta^y = \Delta \circ i_y,$$

and the vector field $\Delta^y \in D(A_y)$ also vanishes at $(0,0)$.

Note now that

$$(f(x), g(y)) = i_x(f) + i_y(g) - (c, c), \quad c = f(0) = g(0),$$

and $\Delta((c,c)) = 0$, since functions of the form (c,c) are constants in the algebra $C^\infty(\mathbf{K})$. From this we eventually obtain that

$$\Delta((f(x), g(y))) = \Delta(i_x(f)) + \Delta(i_y(g))$$
$$= i_x(\Delta^x(f)) + i_y(\Delta^y(g)) = (\Delta^x(f), \Delta^y(g)),$$

where the last equality is a consequence of

$$\Delta^x(f)(0) = \Delta^y(g)(0) = 0. \quad \blacktriangleright$$

Obviously, the inverse statement is also valid: Any pair of vector fields $\Delta^x \in D(A_x)$, $\Delta^y \in D(A_y)$ determines a vector field Δ on $\mathbf{K}$ by formula (9.15), provided that these fields vanish at the point $(0,0)$.

Exercise. Describe the A-modules of vector fields for the following algebras:

1. For all algebras considered in Exercise 9.36.

2. For the $\mathbb{R}$-algebra $C^m(\mathbb{R}^1)$, $m \geqslant 1$, of m-times differentiable functions on the line. (Hint: Start from the case $m = 0$.)

3. For the algebra $A = K[X]/X^{l+1}K[X]$ of truncated polynomials, where $K = \mathbb{R}$ or $\mathbb{Z}/m\mathbb{Z}$.

4. For a Boolean algebra, i.e., a commutative algebra over the field $\mathbb{F}_2 = \mathbb{Z}/2\mathbb{Z}$ whose elements satisfy the relation $a^2 = a$.

9.46. Vector fields on submanifolds. In the above considerations, we formalized two geometrical images: a manifold at one point of which an arrow "grows" and a manifold on which arrows "grow" at all points. Clearly, an intermediate situation also exists: arrows may grow at points of some submanifold (or, more generally, of a closed subset). Examples of this kind are velocity fields on a moving thread or on an oscillating membrane. Arguments similar to those that have led us to the definition of vector fields on a manifold M lead to the desired formalization in this case also.

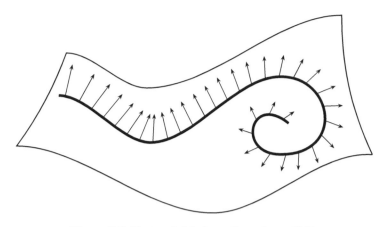

Figure 9.3. Vector field along the submanifold.

Let $N \subset M$ be a submanifold of a manifold M.

Definition. An $\mathbb{R}$-linear map

$$X \colon C^\infty(M) \to C^\infty(N)$$

is said to be a *tangent* (to M) *vector field along* N if

$$X(fg) = X(f)g\big|_N + f\big|_N X(g). \qquad (9.16)$$

The sum of two vector fields along N is obviously a vector field along N. One can also define multiplication of such fields by functions from $C^\infty(M)$:

$$(fX)(g) \stackrel{\text{def}}{=} f\big|_N X(g), \quad f, g \in C^\infty(M).$$

The set $D(M, N)$ of all tangent fields along N on the manifold M is a $C^\infty(M)$-module with respect to these operations. If $z \in N$, then the following analog of formula (9.13),

$$X_z(f) = X(f)(z), \qquad (9.17)$$

determines a tangent vector to the manifold M at a point z, corresponding to the vector field X along the submanifold N. Note that (9.17) makes no sense when $z \notin N$, and the definition above really introduces a field of vectors along the submanifold N.

9.47. Vector fields along maps. Relation (9.16) may be also rewritten in the form

$$X(fg) = X(f)i^*(g) + i^*(f)X(g),$$

where $i \colon N \hookrightarrow M$ denotes the inclusion map. After this, it becomes clear that it still makes sense, provided that i is an arbitrary map of N to M.

Definition. Let $\varphi\colon N \to M$ be a smooth map of manifolds. An $\mathbb{R}$-linear map

$$X\colon C^\infty(M) \to C^\infty(N)$$

is said to be a *tangent* (to M) *vector field along the map* φ if

$$X(fg) = X(f)\varphi^*(g) + \varphi^*(f)X(g) \quad \forall f, g \in C^\infty(M). \tag{9.18}$$

The set $D_\varphi(M)$ of all vector fields along a given map φ is a $C^\infty(M)$-module if the multiplication of a field $X \in D_\varphi(M)$ by $f \in C^\infty(M)$ is defined by the rule $(fX)(g) \overset{\text{def}}{=} \varphi^*(f)X(g)$. The $C^\infty(M)$-module $D_\varphi(M)$ also becomes a $C^\infty(N)$-module if the multiplication of its elements by elements of the algebra $C^\infty(N)$ is defined by

$$(fX)(g) \overset{\text{def}}{=} fX(g), \quad f \in C^\infty(N),\ g \in C^\infty(M),\ X \in D_\varphi(M).$$

Exercise. Check that formula (9.17), in the context under consideration, allows one to assign to any point $z \in M$ a tangent vector to M at the point $\varphi(z)$.

Any vector field $X \in D_\varphi(M)$ may be understood as an *infinitesimal deformation* of the map φ. In fact, since $X_z \in T_{\varphi(z)}M$, this vector can be naturally understood as an *infinitesimal shift* of the image of z under the map φ.

Vector fields along maps are also often called *relative* vector fields.

Example. Let $\varphi\colon N \to M$ be an arbitrary smooth map, $X \in D(N)$, and $Y \in D(M)$. Then $X \circ \varphi^*$ and $\varphi^* \circ Y$ are vector fields along the map φ. It is appropriate to interpret the relative field $X \circ \varphi^*$ as the *image of the field X under the map* φ (cf. Section 9.42).

9.48. An important example of a relative vector field is the *universal vector field* on M. It is constructed in the following way.

Consider the tangent bundle $\pi_T\colon TM \to M$ (see Section 9.17). Let ξ be a tangent vector to M, also understood as a point of the manifold TM. The universal vector field Z on M is defined as the following vector field along the map π_T:

$$Z(f)(\xi) = \xi(f), \quad f \in C^\infty(M).$$

9.49 Exercises.   1. Show that $Z(f) \in C^\infty(TM)$. (Hint: Use special local coordinates on TM.)

2. Check that Z is indeed a vector field along π_T.

3. Let $X \in D(M)$. Prove that

$$X = s_X^* \circ Z, \tag{9.19}$$

where $s_X\colon M \to TM$, $z \mapsto X_z \in T_zM$, is the section of the tangent bundle corresponding to the vector field X. Formula (9.19) explains why the field Z is called universal: Any vector field M can be obtained from Z by using the appropriate section.

9.50. The only difference between the two definitions of vector fields discussed above is the interpretation of the Leibniz rule, i.e., the rule for differentiation of products. Let us rewrite it without specifying the range of the map X:

$$X(fg) = fX(g) + gX(f). \tag{9.20}$$

This formula makes sense when the product of the objects $X(g), X(f)$ and the functions f, g is defined, or in the other words, when the range of X is a module over its domain. For this reason, the following definition exhausts everything discussed above in relation to tangent vectors and vector fields.

Definition. Let A be a commutative K-algebra and let P be an arbitrary A-module. A K-linear map $\Delta \colon A \to P$ is called a *derivation* of the algebra A with values in P if it satisfies the Leibniz rule (9.20), i.e.,

$$\Delta(fg) = f\Delta(g) + g\Delta(f) \quad \forall\, f, g \in A.$$

The set $D(P)$ of all derivations of the algebra A with values in P carries a natural A-module structure.

9.51. Let $X \in D(P)$, and $h \colon P \to Q$ be an A-module homomorphism. Then $h \circ X \in D(Q)$. (Check this.)

Moreover, the map

$$D(h) \colon D(P) \to D(Q), \quad D(P) \ni X \mapsto h \circ X \in D(Q),$$

is obviously an A-module homomorphism, and

$$D(\mathrm{id}_P) = \mathrm{id}_{D(P)},$$
$$D(h_1 \circ h_2) = D(h_1) \circ D(h_2).$$

This means that the correspondence $P \mapsto D(P)$ is a functor in the category of A-modules and their homomorphisms. This functor is one of the basic ones of the differential calculus. Some others will be discussed below. A complete and systematic description of the algebra of these functors together with specific features of its realization for concrete commutative algebras is the object of differential calculus in its modern meaning. Therefore, it may be asserted that the construction of the differential calculus, started by Newton and Leibniz, is not finished yet; it must be accomplished in the future.

9.52. Let us show how to specify Definition 9.50 in order to obtain the definitions of tangent vectors and various vector fields considered above. We shall also describe the procedures that assign to a vector field a tangent vector at a fixed point. In all our considerations here, we assume that $K = \mathbb{R}$.

I. The tangent vector to the manifold M at a point z:

$$A = C^\infty(M), \quad P = A/\mu_z = \mathbb{R},$$

where μ_z is an ideal of the point z.

II. A vector field on the manifold M:

$$A = P = C^\infty(M).$$

The tangent vector $X_z \in T_z M$ is assigned to a vector field $X \in D(M)$ and a point $z \in M$ in the following way:

$$D(A) \ni X \mapsto X_z = h \circ X \in D(A/\mu_z),$$

where $h\colon A \to A/\mu_z$ is the natural projection. In other words, we have $X_z = D(h)(X)$.

III. A vector field along a submanifold (without boundary) $N \subset M$:

$$A = C^\infty(M), P = A/\mu_N, \quad \text{where } \mu_N = \{f \in C^\infty(M) \mid f|_N = 0\}.$$

Let us consider the natural isomorphism $C^\infty(N) = A/\mu_N$. If we are given $X \in D(M, N) = D(P)$ and $z \in N$, then $\mu_z \supset \mu_N$, and the natural projection

$$h\colon P = A/\mu_N \to A/\mu_z$$

is defined. Also we have $X_z = h \circ X = D(h)(X)$.

IV. A vector field along a map $\varphi\colon N \to M$:

$$A = C^\infty(M), \quad P = C^\infty(N), \quad \varphi = |F|,$$

where $F\colon A \to P$ is an $\mathbb{R}$-algebra homomorphism. Note also that the A-module structure in the algebra $P = C^\infty(N)$ is defined by the rule

$$(f, g) \mapsto F(f)g = \varphi^*(f)g, \quad f \in C^\infty(M), \ g \in C^\infty(N).$$

Let $X \in D_\varphi(M)$. If $z \in N$ and $h\colon P \to Q = P/\mu_z$ is the natural projection, then $X_z = h \circ X$.

By the way, at this point let us answer a question that naturally arises: What is a continuous vector field on M? It is an element of the $C^\infty(M)$-module $D(C^0(M))$, where $C^0(M)$ is the algebra of continuous functions on M equipped with a natural $C^\infty(M)$-module structure. We stress that $C^0(M)$ is regarded as $C^\infty(M)$-module, we consider derivations from $C^\infty(M)$ to $C^0(M)$ and not from $C^0(M)$ to $C^0(M)$. Vector fields of class C^m are defined in a similar way.

9.53. To conclude our discussion of geometrical and algebraic problems related to the notion of vector field, let us note that the module $D(M)$ of vector fields (or more generally, derivations of the algebra A) carries another important algebraic structure. Namely, $D(A)$ is a *Lie algebra* due to the following $\mathbb{R}$-linear skew-symmetric operation satisfying the Jacobi identity.

Proposition. *The commutator* $[X, Y] \overset{\text{def}}{=} X \circ Y - Y \circ X$ *of two derivations* $X, Y \in D(A)$ *is again a derivation.*

◀ Indeed,

$$
\begin{aligned}
[X, Y](fg) &= (X \circ Y - Y \circ X)(fg) \\
&= X(fY(g) + gY(f)) - Y(fX(g) + gX(f)) \\
&= X(f)Y(g) + fX(Y(g)) + X(g)Y(f) + gX(Y(f)) \\
&\quad - Y(f)X(g) - fY(X(g)) - Y(g)X(f) - gY(X(f)) \\
&= fX(Y(g)) - fY(X(g)) + gX(Y(f)) - gY(X(f)) \\
&= f[X, Y](g) + g[X, Y](f). \quad \blacktriangleright
\end{aligned}
$$

Since the commutator is obviously skew-symmetric, it remains only to check the Jacobi identity.

9.54 Proposition. *Let* $X, Y, Z \in D(A)$. *Then*

$$
[X, [Y, Z]] = [[X, Y], Z] + [Y, [X, Z]].
$$

◀ Indeed,

$$
\begin{aligned}
[[X, Y], Z] + [Y, [X, Z]] &= [[X, Y], Z] - [[X, Z], Y] \\
&= [X \circ Y - Y \circ X, Z] - [X \circ Z - Z \circ X, Y] \\
&= X \circ Y \circ Z - Y \circ X \circ Z - Z \circ X \circ Y + Z \circ Y \circ X \\
&\quad - X \circ Z \circ Y + Z \circ X \circ Y + Y \circ X \circ Z - Y \circ Z \circ X \\
&\quad - X \circ Y \circ Z - Z \circ Y \circ X - X \circ Z \circ Y + Y \circ Z \circ X \\
&= X \circ (Y \circ Z - Z \circ Y) - (Y \circ Z - Z \circ Y) \circ X \\
&= X \circ [Y, Z] - [Y, Z] \circ X = [X, [Y, Z]]. \quad \blacktriangleright
\end{aligned}
$$

9.55. The local coordinate description of vector fields given in Section 9.41 shows that they are (scalar) first-order differential operators. On the other hand, a first-order (scalar) differential operator Δ of general form on the manifold M can be locally written as

$$
\Delta = \sum_{i=1}^{n} \alpha_i \frac{\partial}{\partial x_i} + \beta, \quad \alpha_i, \beta \in C^\infty(U).
$$

Let us note that its free term β has an invariant meaning: $\beta = \Delta(1)$. Therefore, we can assert that Δ is a first-order differential operator if and only if $\Delta - \Delta(1)$ is a derivation. This gives us a coordinate-free definition of (scalar) linear first-order differential operators. But being insufficiently "clever," it does not allow us to guess a similar definition for operators of higher orders. Let us trim this definition: Note that the Leibniz rule for the derivation $\Delta - \Delta(1)$ is equivalent to the following equality:

$$\Delta(fg) - f\Delta(g) = g\Delta(f) - fg\Delta(1), \quad \text{or} \quad [\Delta, f](g) = g[\Delta, f](1).$$

Thus, we have obtained the following result:

9.56 Proposition. *An $\mathbb{R}$-linear map*

$$\Delta \colon C^\infty(M) \to C^\infty(M)$$

is a first-order differential operator if and only if

$$[[\Delta, f], g] = 0 \quad \forall f, g \in C^\infty(M). \tag{9.21}$$

Let us note that (9.21) is equivalent to the fact that the commutator $[\Delta, f]$ is a $C^\infty(M)$-homomorphism for any $f \in A$.

Looking at the last equality, the reader has probably understood already how to define a differential operator of any order over an arbitrary commutative algebra A. Before stating this definition, let us observe that the expression $[[\Delta, f], g]$ is not manifestly symmetric with respect to f and g, while in fact f and g enter this expression symmetrically:

$$[[\Delta, f], g] = \Delta \circ fg + fg\Delta - g\Delta \circ f - f\Delta \circ g = [[\Delta, g], f].$$

Therefore, we shall change our notation and for any element $f \in A$ introduce the map

$$\delta_f \colon \mathrm{Hom}_K(A, A) \to \mathrm{Hom}_K(A, A), \quad \delta_f(\Delta) \stackrel{\mathrm{def}}{=} [\Delta, f].$$

By the above, the operators δ_f and δ_g commute, and condition (9.21) acquires the form

$$(\delta_g \circ \delta_f)(\Delta) = 0 \quad \forall f, g \in A.$$

We can now give the following fundamental definition:

9.57. Definition. Let A be a K-algebra; then a K-homomorphism $\Delta \colon A \to A$ is called a *linear differential operator of order $\leqslant l$* with values in A if for any $f_0, \ldots, f_l \in A$ we have the identity

$$(\delta_{f_0} \circ \cdots \circ \delta_{f_l})(\Delta) = 0, \tag{9.22}$$

where $\delta_f(\Delta) = \Delta \circ f - f \circ \Delta$.

Let us denote the set of all differential operators of order $\leqslant l$ acting from A to A by $\mathrm{Diff}_l\, A$. Like $D(A)$, this set is stable with respect to summation and multiplication by elements of the algebra A. Therefore, it is naturally endowed with an A-module structure. Moreover, another A-module structure can be introduced in it, by defining the action of an element $f \in A$ on an operator Δ as the composition $\Delta \circ f$. This structure is called the *right* one, and the action of an element $f \in A$ on an operator Δ will be denoted[2] by $f^> \Delta$ instead of $\Delta \circ f$. The set $\mathrm{Diff}_l\, A$ endowed with the module structure with respect to the right multiplication will be denoted by $\mathrm{Diff}_l^>\, A$, the one with the left multiplication by $\mathrm{Diff}_l^<\, A$ or simply $\mathrm{Diff}_l\, A$. The two multiplicative structures in $\mathrm{Diff}_l\, A$ commute and thus determine a bimodule structure, denoted by $\mathrm{Diff}_l^\diamondsuit\, A$.

9.58 Exercises.   1. Prove the last statement. Namely, check that the set $\mathrm{Diff}_l\, A$ is stable with respect to right multiplication and that the left and right multiplications commute in $\mathrm{Diff}_l\, A$.

   2. Check whether the set $D(A)$ is stable with respect to right multiplication.

To deduce some natural and useful properties of differential operators, we shall need the following notation. Let $\mathrm{I}^{(n)} = (1, 2, \ldots, n)$ be the ordered set of the first n integers, and $I = (i_1, \ldots, i_l)$, $l \leqslant n$, $J = (j_1, \ldots, j_r)$, $r \leqslant n$, ordered subsets of $\mathrm{I}^{(n)}$. Let us set by definition $|I| = l$, $|J| = r$, $a_I = (a_{i_1}, \ldots, a_{i_l})$, $a^I = a_{i_1} \cdot \ldots \cdot a_{i_l}$, and $\delta_{a_{\mathrm{I}^{(n)}}} = \delta_{a_{i_1}} \circ \cdots \circ \delta_{a_{i_l}}$. The multiindex obtained by ordering the union $I \bigcup J$ will be denoted by $I + J$.

Exercise. Let A be a K-algebra, and Δ, ∇ K-linear maps from A to A. Then

$$\delta_{a_{\mathrm{I}^{(n)}}}(\Delta \circ \nabla) = \sum_{I+J=\mathrm{I}^{(n)}} \delta_{a_I}(\Delta) \circ \delta_{a_J}(\nabla), \quad a_i \in A, \qquad (9.23)$$

$$\delta_{a_{\mathrm{I}^{(n)}}}(\Delta)(b) = \sum_{I+J=\mathrm{I}^{(n)}} (-1)^{|I|} a^I \Delta(a^J b), \quad a_i, b \in A. \qquad (9.24)$$

For the case $\Delta \in \mathrm{Diff}_m\, A$, $m < n$, the left-hand side of the last equality vanishes by Section 9.57, and the equality can be rewritten in the following form:

$$\Delta(a_{\mathrm{I}^{(n)}} b) = - \sum_{I+J=\mathrm{I}^{(n)}, |I|>0} (-1)^{|I|} a^I \Delta(a^J b). \qquad (9.25)$$

These formulas allow one to readily prove the following two important statements:

9.59 Proposition. *Let ∇ and Δ be linear differential operators of orders $\leqslant l$ and $\leqslant m$, respectively. Then their composition $\Delta \circ \nabla$ is a linear differential operator of order $\leqslant l + m$.*

◄ Indeed, let us set $n = m+l+1$ in formula (9.23). Then each monomial on the right-hand side of the equality thus obtained will vanish by the definition of differential operators: Either $|I| \geqslant m + 1$ and therefore $\delta_{a_I}(\Delta) = 0$, or $|J| \geqslant l + 1$ and, respectively, $\delta_{a_J}(\nabla) = 0$.   ►

---

[2] Sometimes the notation $a^+\Delta$ is also used for right structure and $a\Delta$, for the left one.

9.60 Proposition. *Let $\nu \subset A$ be an arbitrary ideal, $a \in \nu^k$, $\Delta \in \mathrm{Diff}_n A$, and $n < k$. Then $\Delta(a) \in \nu^{k-n}$.*

◀ To prove the proposition, it suffices to confine ourselves to the case $a = a_1 \cdots a_k$, $a_i \in \nu$. Let $k = n + 1$. Consider equality (9.25) with $b = 1$. Then every summand on the right-hand side will contain at least one element $a_i \in \nu$ and consequently will belong to ν itself. The passage from $k = n + r$ to $k = n + r + 1$ is accomplished as follows. Let us use formula (9.25) again. Each of the summands on the right-hand side is of the form

$$a_{i_1} \cdots a_{i_m} \Delta(a_{j_1} \cdots a_{j_{k-m}}). \tag{9.26}$$

Note that $a_{i_1} \cdots a_{i_m} \in \nu^m$, $a_{j_1} \cdots a_{j_{k-m}} \in \nu^{k-m}$. If $m \geqslant k - n$, then the monomial 9.26 obviously belongs to ν^{k-n}. Otherwise, if $k - m > n$, we see that $\Delta(a_{j_1} \cdots a_{j_{k-m}}) \in \nu^{k-m-n}$ by the induction hypothesis, and the monomial (9.26) as a whole belongs to ν^{k-n}. ▶

9.61 Corollary. *If functions f and g coincide in some neighborhood $U \ni z$, then for any differential operator Δ (in the sense of Definition 9.57) of the equality $\Delta(f)(z) = \Delta(g)(z)$ is valid. In other words, differential operators are local.*

◀ Indeed, let Δ be an operator of order $\leqslant l$. Since $f - g \in \mu_z^{l+1}$ for any l, then by Proposition 9.60 one has $\Delta(f - g) \in \mu_z$. ▶

Now let us prove that for algebras of smooth functions, Definition 9.57 *coincides with the usual definition of linear differential operator.*

This corollary allows us to obtain, for any differential operator $\Delta \in \mathrm{Diff}_l C^\infty(M)$, its well-defined restriction $\Delta|_U : C^\infty(U) \to C^\infty(U)$ to any open domain $U \subset M$ by setting

$$\Delta|_U (f)(z) = \Delta(g)(z), \quad f \in C^\infty(U), \ g \in C^\infty(M), \ z \in U,$$

where g is an arbitrary function coinciding with f in some neighborhood of the point z. This definition implies $\Delta|_U (f|_U) = \Delta(f)|_U$ for $f \in C^\infty(M)$. Obviously, any operator is uniquely determined by its restrictions on charts of an arbitrary atlas.

Now we can prove the following important statement:

9.62 Theorem. *Let $\Delta \in \mathrm{Diff}_l(C^\infty(M))$, and let $x_1, \ldots, x_n$ be local coordinates in a neighborhood $U \subset M$. Then the operator $\Delta|_U$ can be presented in the form (for the notation, see Section 2.8)*

$$\Delta|_U = \sum_{|\sigma|=0}^{l} \alpha_\sigma \frac{\partial^{|\sigma|}}{\partial x^\sigma}, \quad \alpha_\sigma \in C^\infty(U).$$

◀ Let $z \in U$ and $f \in C^\infty(M)$. Consider an arbitrary star-shaped neighborhood $U_z \subset U$ of the point z and, using Section 2.9, present the function f

in this neighborhood in the form

$$f = \sum_{|\sigma|=0}^{l} \frac{\partial^{|\sigma|} f}{\partial x^{\sigma}}(z) \left(\frac{(x-z)^{\sigma}}{\sigma!} \right) + h(x),$$

where $h(x) \in \mu_z^{l+1}$ and $(x-z)^{\sigma} \stackrel{\text{def}}{=} (x_1 - a_1)^{\sigma_1} \cdots (x_n - a_n)^{\sigma_n}$. Therefore,

$$\Delta(f)(z) = \Delta|_U (f|_U)(z) = \sum_{|\sigma|=0}^{l} \frac{\partial^{|\sigma|} f}{\partial x^{\sigma}}(z) \alpha_{\sigma}(z),$$

where

$$\alpha_{\sigma}(x) \stackrel{\text{def}}{=} \Delta|_U \left(\frac{(x-z)^{\sigma}}{\sigma!} \right) \quad \text{and} \quad \alpha_{\sigma}(z) = \Delta|_U \left(\frac{(x-z)^{\sigma}}{\sigma!} \right) \Big|_z.$$

It remains to note that the functions $\alpha_{\sigma}(x)$ are smooth by construction.
▶

To understand how the algebraic Definition 9.57 of differential operators works for the case in which the algebra A is not the smooth function algebra on a smooth manifold, let us do the following exercise:

9.63 Exercises.   1. Describe the modules of differential operators for the algebra $C^{\infty}(\mathbf{K})$. (See Examples 9.35 and 9.45.)

2. Do the same for the algebra of truncated polynomials

$$A = K[X]/X^n K[X], \quad \text{where} \quad K = \mathbb{R}, \; K = \mathbb{Z}_m.$$

(See Exercise 3 from Section 9.45.)

3. In the classical situation $A = C^{\infty}(\mathbb{R})$, any differential operator of order > 1 may be represented as the sum of compositions of first-order operators. May one assert the same thing for the algebras from the previous exercises?

9.64. Jets of order l at a point. Let us formulate an important consequence of Proposition 9.60. Recall that μ_z^{l+1} denotes the $(l+1)$st power of the ideal μ_z consisting of all functions on M vanishing at the point z.

Corollary. Let $\Delta \in \text{Diff}_l \, C^{\infty}(M)$, $f, g \in C^{\infty}(M)$, and $z \in M$. Then $\Delta(f)(z) = \Delta(g)(z)$ if $f = g \mod \mu_z^{l+1}$.

◀ Indeed, in this case $f - g \in \mu_z^{l+1}$, and consequently, $\Delta(f - g) \in \mu_z$, i.e., $\Delta(f - g)(z) = 0$. ▶

It will be useful to consider this fact after introducing the vector *space of lth order jets*, or *l-jets*, of (smooth) functions on M at some point z (cf. Section 9.27):

$$J_z^l M \stackrel{\text{def}}{=} C^{\infty}(M)/\mu_z^{l+1}.$$

The image of the function f under the natural projection

$$C^\infty(M) \to C^\infty(M)/\mu_z^{l+1} = J_z^l M$$

is called its *jet of order* l (or l-*jet*) at the point z and is denoted by $[f]_z^l$. In these terms, the condition $f = g \mod \mu_z^{l+1}$ means that $[f]_z^l = [g]_z^l$, while the previous corollary asserts that $\Delta(f)(z) = \Delta(g)(z)$ if $[f]_z^l = [g]_z^l$. In other words, the map

$$h_{\Delta,z}: J_z^l M \to \mathbb{R}, \quad [f]_z^l \mapsto \Delta(f)(z),$$

is well defined. It is obviously $\mathbb{R}$-linear. Its importance is explained by the fact that it completely determines the operator Δ at the point z.

Exercise. The map $h_{\Delta,z}$ is a linear function on the space $J_z^l M$. Find a basis of the space $J_z^l M$ in which the components of this function are the numbers $\alpha_\sigma(z)$ appearing in Theorem 9.62.

9.65. The manifold of jets. The family $\{h_{\Delta,z}\}_{z \in M}$ of linear functionals uniquely determines the operator Δ, since

$$\Delta(f)(z) = h_{\Delta,z}([f]_z^l). \tag{9.27}$$

Therefore, it makes sense to construct a new object combining the separate maps $h_{\Delta,z}$ into a single whole. To do this, one first needs to join their domains, in the same way as was done in Section 9.28 for $l = 1$:

$$J^l M = \bigcup_{z \in M} J_z^l M.$$

The set $J^l M$ is equipped with a smooth manifold structure by a procedure similar to that used for TM and T^*M. The details of this construction will be described in Section 11.11. This smooth manifold is called the *manifold of jets of order* l (or of l-*jets*) of the manifold M.

The map

$$\pi_{J^l} = \pi_{J^l M}: J^l M \to M, \quad J^l M \supset J_z^l M \ni \theta \mapsto z \in M,$$

fibers the manifold $J^l M$ over M. By this definition, $\pi_{J^l}^{-1}(z) = J_z^l M$. Moreover, to any function $f \in C^\infty(M)$, we can assign the section

$$s_{j_l(f)}: M \to J^l M, \quad z \mapsto [f]_z^l \in J_z^l M \subset J^l M,$$

of this bundle. This section is called the l-*jet of* f.

Any operator $\Delta \in \mathrm{Diff}_l\, C^\infty(M)$ determines the map

$$h_\Delta: J^l M \to M \times \mathbb{R}, \quad J_z^l M \ni \theta \mapsto (z, h_{\Delta,z}(\theta)).$$

Let $\pi_\mathbb{R}: M \times \mathbb{R} \to \mathbb{R}$ be the canonical projection. Then, by (9.27), $\pi_\mathbb{R}(h_\Delta([f]_z^l)) = \Delta(f)(z)$, and consequently,

$$\Delta(f) = \pi_\mathbb{R} \circ h_\Delta \circ s_{j_l(f)}, \tag{9.28}$$

where $\Delta(f)$ is understood as a smooth map from M to $\mathbb{R}$. This relation shows that all the information on the operator Δ is encoded in the map of smooth manifolds h_Δ. It will be shown in Section 14.15 that $s_{j_l} : f \mapsto s_{j_l(f)}$ is a differential operator of order l, whose range of values is the set of sections of the bundle π_{J^l}. It is natural to call this operator the *universal differential operator* of order l, since all concrete operators are obtained by composing this operator with maps from $J^l M$ to $M \times \mathbb{R}$. The specifics of maps of the form h_Δ acting from $J^l M$ to $M \times \mathbb{R}$ can be described in the following way.

Consider the projection

$$\pi \colon M \times \mathbb{R} \to M, \quad (z, \lambda) \mapsto z.$$

It is a trivial bundle over M with fiber $\mathbb{R}$ (see Section 12.2). Then, as is easily seen, maps of the form h_Δ are morphisms of the vector bundle π_{J^l} to the vector bundle π (see Section 12.4), and in particular, they take the fiber $\pi_{J^l}^{-1}(z) = J_z^l M$ to the fiber $\pi^{-1}(z) = \mathbb{R}$. This map of fibers obviously coincides with $h_{\Delta, z}$.

All these facts reveal the fundamental role of vector bundles in the differential calculus. For this and many other reasons (some of them will appear in our subsequent exposition), the theory of vector bundles is a necessary part of the differential calculus over smooth manifolds. This theory will be considered in detail in Chapter 12.

Due to the universality of the operator s_{j_l} expressed by formula (9.28), the manifolds $J^l M$ and their natural generalizations constitute an important part of the foundations of the modern theory of partial differential equations. The universal property of this operator is also revealed by the fact that the module of sections of the bundle

$$\pi_{J^l} \colon J^l M \to M$$

(see Section 12.7) is the *representing object* for the functor Diff_l of the differential calculus in the category of geometrical $C^\infty(M)$-modules (see Section 12.44).

9.66. It was shown above for the case in which $A = C^\infty(M)$ is the algebra of smooth functions on the manifold M that Definition 9.57 is equivalent to the usual definition of a linear differential operator acting on functions and taking its values in functions on M (i.e., to the definition of scalar differential operators). In fact, the more general case, that of matrix differential operators, can also be described in purely algebraic terms. Let us recall that such an operator Δ is usually defined as a matrix composed of differential operators,

$$\Delta = \begin{pmatrix} \Delta_{1,1} & \cdots & \Delta_{1,m} \\ \vdots & \ddots & \vdots \\ \Delta_{k,1} & \cdots & \Delta_{k,m} \end{pmatrix},$$

where $\Delta_{i,j}$ are scalar differential operators of order $\leqslant l$, while the action of this operator on a vector function $\bar{f} = (f_1, \ldots, f_n)$ is defined in the following natural way:

$$\begin{pmatrix} \Delta_{1,1} & \cdots & \Delta_{1,m} \\ \vdots & \ddots & \vdots \\ \Delta_{k,1} & \cdots & \Delta_{k,m} \end{pmatrix} \begin{pmatrix} f_1 \\ \vdots \\ f_m \end{pmatrix} = \begin{pmatrix} \Delta_{1,1}(f_1) + \ldots + \Delta_{1,m}(f_m) \\ \vdots \\ \Delta_{k,1}(f_1) + \ldots + \Delta_{k,m}(f_m) \end{pmatrix}.$$

In Chapter 12, it will be shown that vector functions of the above type can naturally be considered as sections of an m-dimensional vector bundle over M and that the category of all vector bundles over M is equivalent to the category of projective modules over the algebra $C^\infty(M)$. This fact, together with the observations of Section 9.50, leads one to believe that differential operators of general nature should be maps connecting modules over some base algebra A. It is remarkable that to define a general differential operator it suffices simply to repeat the scalar Definition 9.57. The only thing that matters here is that for any K-linear map of A-modules $\Delta\colon P \to Q$ and any $a \in A$ one can define the commutator

$$\delta_a(\Delta) \overset{\text{def}}{=} [\Delta, a]\colon P \to Q,$$

where the element $a \in A$ is understood as the operator of multiplication by a applied to elements of the corresponding A-module. In other words,

$$\delta_a(\Delta)(p) = \Delta(ap) - a\Delta(p), \quad p \in P.$$

So, one can hope that the following purely algebraic definition reduces to the usual notion of a (matrix) differential operator in the "standard" situation.

9.67. Definition. Let A be an arbitrary commutative K-algebra, and let P and Q be A-modules. A K-homomorphism $\Delta\colon P \to Q$ is called a *linear differential operator of order* $\leqslant l$ acting from P to Q if for any $a_0, \ldots, a_l \in A$ one has

$$(\delta_{a_0} \circ \cdots \circ \delta_{a_l})(\Delta) = 0. \tag{9.29}$$

The fact that under an adequate specialization ($K = \mathbb{R}$, $A = C^\infty(M)$ with projective A-modules P and Q) the definition given above coincides with the usual one will be proved in Chapter 12, after we have established relations between vector bundles and projective modules.

Let us denote the set of all differential operators of order $\leqslant l$ acting from P to Q by $\mathrm{Diff}_l(P, Q)$. This set is stable with respect to summation and to the ordinary (left) multiplication by elements of the algebra A:

$$(a\Delta)(p) \overset{\text{def}}{=} a \cdot \Delta(p), \quad a \in A, \ p \in P.$$

Therefore, it possesses a natural left A-module structure. One can also introduce another A-module structure, defining the action of an element $a \in A$ on the operator Δ as the composition $\Delta \circ a$. This structure is called *right*, and the action of $a \in A$ to Δ is denoted by $a^> \Delta$ instead of $\Delta \circ a$. The set $\mathrm{Diff}_l(P, Q)$, as a module with respect to the right multiplication, will be denoted by $\mathrm{Diff}_l^>(P, Q)$. Two multiplicative structures in $\mathrm{Diff}_l(P, Q)$ commute and thus determine the structure of a bimodule, denoted by $\mathrm{Diff}_l^\diamond(P, Q)$. For the sake of brevity, we use the notation $\mathrm{Diff}_l^\diamond Q$ for $\mathrm{Diff}_l^\diamond(A, Q)$.

If $h \colon P \to Q$ is an A-module homomorphism, then the correspondence $\Delta \mapsto h \circ \Delta$, $\Delta \in \mathrm{Diff}_l P$, determines a homomorphism of the A-module $\mathrm{Diff}_l P$ to the A-module $\mathrm{Diff}_l Q$. Therefore, the correspondence $P \mapsto \mathrm{Diff}_l P$ is a functor on the category of A-modules. Let us denote this functor by Diff_l. We obtain another example of a functor of the differential calculus (see Section 9.51). Such functors are defined for all commutative unital algebras. If we choose an A-module P, we obtain an example of what may be called a *relative* functor of differential calculus, $\mathrm{Diff}_l(P, \cdot) \colon Q \mapsto \mathrm{Diff}_l(P, Q)$.

Formulas (9.23)–(9.25) and Proposition 9.59 are proved in the general situation exactly in the same way as for scalar operators. As to Proposition 9.60, its analog in the general case is the following.

Proposition. *Let $I \subset A$ be an ideal, let P, Q be A-modules, $p \in I^k P$, $\Delta \in \mathrm{Diff}_n(P, Q)$, and $n < k$. Then $\Delta(p) \in I^{k-n} Q$.*

The proof is the same as in the scalar case.
Exercises.   1. Check that

$$\mathrm{Diff}_0(P, Q) = \mathrm{Diff}_0^>(P, Q) = \mathrm{Hom}_K(P, Q).$$

2. Consider the maps $i^>$ and $i_<$ of A-modules that are the identities on the underlying sets:

$$i^> \colon \mathrm{Diff}_l(P, Q) \to \mathrm{Diff}_l^>(P, Q), \quad i^>(f) = f,$$
$$i_< \colon \mathrm{Diff}_l^>(P, Q) \to \mathrm{Diff}_l(P, Q), \quad i_<(f) = f.$$

Prove that these maps are differential operators of order $\leqslant l$.

9.68. Let us note that any differential operator of order $\leqslant l$ is an operator of order $\leqslant m$ as well, provided $l \leqslant m$. Therefore, we have a natural bimodule embedding $\mathrm{Diff}_l^\diamond(P, Q) \subset \mathrm{Diff}_m^\diamond(P, Q)$. Let us denote the direct limit of the sequence of embeddings

$$\mathrm{Diff}_0^\diamond(P, Q) \subset \cdots \subset \mathrm{Diff}_l^\diamond(P, Q) \subset \mathrm{Diff}_{l+1}^\diamond(P, Q) \subset \cdots$$

by $\mathrm{Diff}^\diamond(P, Q)$.

As we saw above, the composition of two differential operators, if it is defined, is again a differential operator. Therefore, the bimodule $\mathrm{Diff}^\diamond(P, P)$ becomes a left (noncommutative) A-algebra with respect to the operation

$$a(\Delta \nabla) = (a\Delta)\nabla.$$

Moreover, $\mathrm{Diff}^\diamond(P, Q)$ can be regarded as a right $\mathrm{Diff}^\diamond(P, P)$- and left $\mathrm{Diff}^\diamond(Q, Q)$-module.

10

Symbols and the Hamiltonian Formalism

10.1. We now have everything needed to answer the question stated in Section 9.20: *What is the algebra whose spectrum is the cotangent manifold T^*M?* Let us start with necessary algebraic definitions.

Let K be a field and let A be a K-algebra. The embedding of A-modules $\mathrm{Diff}_{k-1} A \subset \mathrm{Diff}_k A$ (see Section 9.68) allows us to define the quotient module

$$\mathcal{S}_k(A) \stackrel{\mathrm{def}}{=} \mathrm{Diff}_k A / \mathrm{Diff}_{k-1} A,$$

which is called the *module of symbols of order k* (or the module of *k-symbols*). The coset of an operator $\Delta \in \mathrm{Diff}_k A$ modulo $\mathrm{Diff}_{k-1} A$ will be denoted by $\mathrm{smbl}_k \Delta$ and called the *symbol* of Δ. Let us define the *algebra of symbols* $\mathcal{S}_*(A)$ for the algebra A by setting

$$\mathcal{S}_*(A) = \bigoplus_{n=0}^{\infty} \mathcal{S}_n(A).$$

The operation of multiplication in $\mathcal{S}_*(A)$ is induced by the composition of differential operators. To be more precise, for two elements

$$\mathrm{smbl}_l \Delta \in \mathcal{S}_l(A), \quad \mathrm{smbl}_k \nabla \in \mathcal{S}_k(A)$$

let us set by definition

$$\mathrm{smbl}_l \Delta \cdot \mathrm{smbl}_k \nabla \stackrel{\mathrm{def}}{=} \mathrm{smbl}_{k+l}(\Delta \circ \nabla) \in \mathcal{S}_{l+k}(A).$$

This operation is well defined, since the result does not depend on the choice of representatives in the cosets $\mathrm{smbl}_l \Delta$ and $\mathrm{smbl}_k \nabla$. Indeed, if, say,

© Springer Nature Switzerland AG 2020
J. Nestruev, *Smooth Manifolds and Observables*, Graduate Texts
in Mathematics 220, https://doi.org/10.1007/978-3-030-45650-4_10

$\mathrm{smbl}_l \, \Delta = \mathrm{smbl}_l \, \Delta'$, then $\Delta - \Delta' \in \mathrm{Diff}_{l-1} \, A$ and consequently

$$(\Delta - \Delta') \circ \nabla \in \mathrm{Diff}_{l+k-1} \, A.$$

Proposition. $\mathcal{S}_*(A)$ *is a commutative algebra.*

◀ We must check that

$$\Delta \circ \nabla - \nabla \circ \Delta = [\Delta, \nabla] \in \mathrm{Diff}_{l+k-1} \, A$$

if $\Delta \in \mathrm{Diff}_l \, A$ and $\nabla \in \mathrm{Diff}_k \, A$. Let us use induction on $l + k$. In the case $l + k = 0$, i.e., for $l = k = 0$, the statement is obvious, since scalar differential operators of order zero are the operators of multiplication by elements of the algebra A, and this algebra is commutative. The induction step from $l + k < n$ to $l + k = n$ is based on the following formula, which is a particular case of (9.23):

$$\delta_a(\Delta \circ \nabla - \nabla \circ \Delta) = \delta_a(\Delta) \circ \nabla + \Delta \circ \delta_a(\nabla) - \delta_a(\nabla) \circ \Delta - \nabla \circ \delta_a(\Delta)$$
$$= [\delta_a(\Delta), \nabla] + [\Delta, \delta_a(\nabla)].$$

The orders of the operators $\delta_a(\Delta)$ and $\delta_a(\nabla)$ are $l-1$ and $k-1$, respectively. By the induction hypothesis, the last expression is an operator of order $\leqslant k+l-2$. Hence, the order of the operator $[\Delta, \nabla]$ does not exceed $l+k-1$. ▶

Let us note that $\mathcal{S}_0(A) = A$ is a subalgebra of the algebra $\mathcal{S}_*(A)$, and the operations of left (right) multiplication of differential operators by elements of the algebra A reduce to the left (right) multiplication by elements of this subalgebra. By the commutativity of the algebra $\mathcal{S}_*(A)$, these multiplication operations coincide.

10.2. Now let $\mathrm{smbl}_l \, \Delta \in \mathcal{S}_l(A)$ and $\mathrm{smbl}_k \, \nabla \in \mathcal{S}_k(A)$. Then, by the last proposition, $[\Delta, \nabla] \in \mathrm{Diff}_{l+k-1} \, A$. One can assign to the pair $(\mathrm{smbl}_l \, \Delta, \mathrm{smbl}_k \, \nabla)$ the element

$$\{\mathrm{smbl}_l \, \Delta, \mathrm{smbl}_k \, \nabla\} \overset{\text{def}}{=} \mathrm{smbl}_{k+l-1}[\Delta, \nabla] \in \mathcal{S}_{k+l-1}(A),$$

which is well defined, i.e., does not depend on the choice of representatives in the cosets $\mathrm{smbl}_l \, \Delta$ and $\mathrm{smbl}_k \, \nabla$ (this is proved exactly in the same way as we proved that the multiplication in $\mathcal{S}_*(A)$ is well defined). The operation $\{\cdot, \cdot\}$ is K-linear and skew-symmetric. It satisfies the Jacobi identity, since the commutator of linear differential operators satisfies this identity. Thus, $\mathcal{S}_*(A)$ is a *Lie algebra* with respect to this operation. If $\mathfrak{s}_1, \mathfrak{s}_2 \in \mathcal{S}_1(A)$, then $\{\mathfrak{s}_1, \mathfrak{s}_2\} \in \mathcal{S}_1(A)$ as well. In other words, $\mathcal{S}_1(A) \subset \mathcal{S}_*(A)$ is a *Lie subalgebra* of the Lie algebra of symbols $S_*(A)$.

Exercises.   1. Let $\mathfrak{s} = \mathrm{smbl}_1 \, \Delta$. Prove that the correspondence

$$\mathfrak{s} \leftrightarrow \Delta - \Delta(1) \in D(A)$$

is well defined and establishes an isomorphism between the Lie algebras $\mathcal{S}_1(A)$ and $D(A)$.

2. Fix an arbitrary element $\mathfrak{s} \in \mathcal{S}_*(A)$. Show that the map

$$\{\mathfrak{s}, \cdot\} \colon \mathcal{S}_*(A) \to \mathcal{S}_*(A), \quad \mathfrak{s}_1 \mapsto \{\mathfrak{s}, \mathfrak{s}_1\},$$

is a derivation of the algebra $\mathcal{S}_*(A)$.

3. Let $a_1, \ldots, a_n \in A$ and $\Delta \in \mathrm{Diff}^n A$. Show that the map $\mathrm{smbl}_n(\Delta) \colon A \otimes \ldots \otimes A \to A$ given by

$$\big(\mathrm{smbl}_n(\Delta)\big)(a_1, \ldots, a_n) \mapsto \delta_{a_1, \ldots, a_n}(\Delta)$$

is well defined, depends only on the symbol of the operator Δ, and is linear in each argument. Further, instead of $\big(\mathrm{smbl}_n(\Delta)\big)(a_1, \ldots, a_n)$, we write $[a_1, \ldots, a_n]_{\mathrm{smbl}_n(\Delta)}$.

10.3. Assume that the ring K is an algebra over the field of rational numbers $\mathbb{Q}$. Then any element $a \in A$ determines a K-algebra homomorphism

$$\Xi_a \colon \mathcal{S}_*(A) \to A, \quad \mathrm{smbl}_k(\Delta) \mapsto \frac{[\delta_a^k(\Delta)]}{k!}(1).$$

Let us check this fact. Note first that $\delta_a^k(\nabla) = 0$ if $\nabla \in \mathrm{Diff}_{k-1} A$. Therefore, the map Ξ_a is well defined. Its K-linearity is obvious. Further, $\Xi_a\big|_{\mathcal{S}_0(A)} \colon \mathcal{S}_0(A) = A \to A$ is the identity map, and hence $\Xi_a(1_{\mathcal{S}_*(A)}) = 1_A$ (unitarity!). Finally, if $\Delta \in \mathrm{Diff}_k A$, $\nabla \in \mathrm{Diff}_l A$, then from (9.23) it follows that

$$\delta_a^{k+l}(\Delta \circ \nabla) = \binom{k+l}{k} \delta_a^k(\Delta) \circ \delta_a^l(\nabla). \tag{10.1}$$

Since $\delta_a^k(\Delta) \in \mathrm{Diff}_0 A = A$ and $\delta_a^l(\nabla) \in \mathrm{Diff}_0 A = A$ are operators of multiplication by the elements $[\delta_a^k(\Delta)](1)$ and $[\delta_a^l(\nabla)](1)$ of the algebra A, the multiplicativity of the map Ξ_a is a direct consequence of (10.1).

Proposition. Let $I \subset A$ be an ideal, and $\Pi \colon A \to A/I$ the natural projection. Then $\Pi \circ \Xi_a = 0$ if $a \in I^2$.

◀ From formula (9.24) it follows that

$$[\delta_a^k(\Delta)](1) = \sum_{i=0}^k \binom{k}{i} a^i \Delta \big(a^{k-i}\big).$$

Therefore, if $a \in I^2$, then

$$[\delta_a^k(\Delta)](1) = \Delta\big(a^k\big) \quad \mathrm{mod}\ I^2.$$

If, in addition, $\Delta \in \mathrm{Diff}_k A$, then, by Proposition 9.60, $\Delta\big(a^k\big) \in I^k$, $k \geqslant 1$, and consequently $\Pi(\Xi_a(\mathrm{smbl}_k \Delta)) = 0$. For $k = 0$, the assertion is obvious. ▶

10.4. We can now describe the K-spectrum $|\mathcal{S}_*(A)|$ of the algebra $\mathcal{S}_*(A)$. Let $h \in |A|$. Then the composition

$$\gamma_{h,a} \overset{\text{def}}{=} h \circ \Xi_a : \mathcal{S}_*(A) \to K$$

is a K-algebra homomorphism and thus is a point of the spectrum $|\mathcal{S}_*(A)|$.

Corollary. *Let $\mu_h = \operatorname{Ker} h$ and*

$$a - h(a) \cdot 1_A = b - h(b) \cdot 1_A \quad \mathrm{mod}\ \mu_h^2.$$

Then $\gamma_{h,a} = \gamma_{h,b}$.

◀ Note first that $\delta_{\lambda \cdot 1_A} = 0$ for $\lambda \in K$. Therefore,

$$\Xi_a = \Xi_{a-h(a)\cdot 1_A}, \quad \Xi_b = \Xi_{b-h(b)\cdot 1_A},$$

and we may confine ourselves to the case $h(a) = h(b) = 0$, which is equivalent to $a, b \in \mu_h$. Further, by our assumptions, we have $a' = a - b \in \mu_h^2$. Since

$$\delta_b^k(\Delta) = \delta_{a+a'}^k(\Delta) = \sum_{s=0}^{k} \binom{k}{s} \delta_{a'}^s (\delta_a^{k-s}(\Delta)),$$

by setting

$$\Delta_s = \frac{k!}{(k-s)!} \delta_a^{k-s}(\Delta)$$

and assuming that $\Delta \in \operatorname{Diff}_k A$, we find that

$$
\begin{aligned}
\Xi_b(\operatorname{smbl}_k \Delta) &= \frac{1}{k!} \left[\delta_{a+a'}^k(\Delta) \right](1) \\
&= \frac{1}{k!} \left[\delta_a^k(\Delta) \right](1) + \sum_{s=1}^{k} \frac{1}{s!} \left[\delta_{a'}^s(\Delta_s) \right](1) \\
&= \Xi_a(\operatorname{smbl}_k \Delta) + \sum_{s=1}^{k} \Xi_{a'}(\operatorname{smbl}_s(\Delta_s)).
\end{aligned}
$$

Therefore, Proposition 10.3 implies that $h \circ \Xi_b = h \circ \Xi_a$. ▶

Let us recall that by definition (see Section 9.30) the cotangent space to the spectrum $|A|$ at a point $h \in |A|$ is the quotient module $T_h^*(A) \overset{\text{def}}{=} \mu_h/\mu_h^2$. Corollary 10.4 makes it possible to construct a map

$$i_h : T_h^*(A) \to |\mathcal{S}_*(A)|,$$

by setting $i_h([a]) = \gamma_{h,a}$, $a \in \mu_h$. In other words, to any "cotangent vector" to the "manifold" $|A|$ there corresponds a point of the "manifold" $|\mathcal{S}_*(A)|$. Let us study this correspondence in more detail.

10.5 Proposition. *Let K be a field and assume that any tangent vector $\xi \in T_h A$ can be continued to a "vector field" $X \in D(A)$; i.e., for any*

$\xi \in T_h A$ *there exists a derivation* $X \in D(A)$ *such that* $\xi = h \circ X$. *Then the map* i_h *is injective.*

◀ By (9.11), to any K-linear map $\varphi \colon T_h^* A \to K$ there corresponds a tangent vector $\xi_\varphi = \nu_h(\varphi) \in T_h A$. Let now $a, b \in \mu_h$ and $[a] \neq [b]$, where $[g] = g$ mod μ_h^2. Since K is a field, by Section 9.30, ν_h is an isomorphism and consequently $\xi_\varphi(a) \neq \xi_\varphi(b)$. Let us continue the tangent vector ξ_φ to a vector field $X \in D(A)$. Then the above inequality can be interpreted as $h(X(a)) \neq h(X(b))$. Now identifying $D(A)$ with $\mathcal{S}_1(A)$ (see Exercise 1 from Section 10.2), we see that $X(a) = \Xi_a(X)$, $X(b) = \Xi_b(X)$, and thus the last inequality can be rewritten in the form

$$(h \circ \Xi_a)(X) \neq (h \circ \Xi_b)(X).$$

Hence we have $\gamma_{h,a} \neq \gamma_{h,b}$, which is equivalent to the desired inequality $i_h([a]) \neq i_h([b])$.   ▶

Obviously, the assumptions of the proposition proved above hold for the algebra $A = C^\infty(M)$. Therefore, setting $i_z = i_{h_z}$ for a point $z \in M$, we obtain the following:

10.6 Corollary. *The map* $i_z \colon T_z^* M \to |\mathcal{S}_*(C^\infty(M))|$ *is injective.*   ▶

Combining the maps i_z for all points $z \in M$, we obtain the embedding

$$i \colon T^* M \to |\mathcal{S}_*(C^\infty(M))|, \quad i|_{T_z M} = i_z. \tag{10.2}$$

10.7. Let us discuss some other facts useful in our subsequent study of the K-spectrum of the algebra $\mathcal{S}_*(A)$.

Suppose that $\widetilde{h} \in |\mathcal{S}_*(A)|$. Let us identify A with $\mathcal{S}_0(A)$. Then obviously $h \stackrel{\text{def}}{=} \widetilde{h}\big|_A \in |A|$.

Exercise. Show that $\widetilde{h} \in \Im i_h$ implies $\widetilde{h}\big|_A = h$. Check that the projection $\pi_{T^*} \colon T^* M \to M$ is the geometrical analog of the map $\widetilde{h} \mapsto h$ in the case $A = C^\infty(M)$. (In other words, if $\widetilde{h} = h_\theta$, $\theta \in T^* M$, then $h = h_z$, where $z = \pi_{T^*}(\theta)$.)

Note now that if $a \in A$ and $X \in D(A) = \mathcal{S}_1(A)$, then $\widetilde{h}(aX) = h(a)\widetilde{h}(X)$. In particular, $\widetilde{h}(aX) = 0$ if $a \in \mu_h$. Therefore, the map of K-modules

$$\widetilde{\widetilde{h}} \colon D(A)/\mu_h D(A) \to K, \quad \widetilde{\widetilde{h}}(X \bmod \mu_h D(A)) = \widetilde{h}(X),$$

is well defined. On the other hand, we have the natural map

$$\tau_h \colon D(A)/\mu_h D(A) \to T_h A, \quad X \bmod \mu_h D(A) \mapsto h \circ X.$$

Lemma. *If* $A = C^\infty(M)$, *then the map* τ_h *defined above is an isomorphism of vector spaces over* $\mathbb{R}$.

◀ From the spectrum theorem, Theorem 7.7, it follows that $h = h_z$ for some point $z \in M$, and consequently $h \circ X = X_z$ (see Section 9.52, II).

Since any tangent vector $\xi \in T_h A = T_z M$ can obviously be continued to a vector field on M, τ_h is a surjective map. Now the injectivity of τ_h means that $X_z = 0$ implies $X \in \mu_h D(A) = \mu_z D(M)$. This implication is easily proved using the following fact: if

$$X = \sum_{i=1}^{n} \alpha_i(x) \frac{\partial}{\partial x_i}$$

in a local coordinate system, then $\alpha_i(z) = 0$, i.e., the coefficients α_i belong to μ_z.   ▶

The following fact may be used to rigorously complete the proof of the above lemma.

Exercise. Let $X \in D(M)$ be such that $X_z = 0$, $z \in M$. Prove that

$$X = \sum_{i=1}^{n} f_i X_i + Y, \ f_i \in C^\infty(M), \ Y, X_i \in D(M),$$

where f_i (respectively, X_i) coincides with α_i (respectively, $\partial/\partial x_i$), $i = 1, \dots, n$, in a neighborhood of z, while Y vanishes in this neighborhood.

We can now completely describe the $\mathbb{R}$-spectrum of the algebra of symbols $\mathcal{S}_*(C^\infty(M))$.

10.8 Theorem. *The map* $i \colon T^*M \to |\mathcal{S}_*(C^\infty(M))|$ *is an isomorphism; i.e.,* T^*M *is the* $\mathbb{R}$*-spectrum of the algebra* $\mathcal{S}_*(C^\infty(M))$.

◀ The injectivity of the map i was proved in Corollary 10.6. Let us prove its surjectivity. Suppose that, in the notation of Section 10.7, we have $A = C^\infty(M)$, $\widetilde{h} \in |\mathcal{S}_*(A)|$, $h = \widetilde{h}\big|_A$, and let $z \in M$ be a point such that $h = h_z$. Then $T_h A = T_z M$, and by Lemma 10.7 the map $\widetilde{h}$ can be understood as an $\mathbb{R}$-linear map from $T_z M$ to $\mathbb{R}$, i.e., as a covector $d_z f \in T^*M$ (see Section 9.22). By the definition of $\widetilde{h}$, we see that

$$\gamma_{h_z, f}(X) = (h_z \circ \Xi_f)(X) = (h_z \circ X)(f) = X_z(f)$$

$$= d_z f(X_z) = \widetilde{h}(X_z) = \widetilde{h}(X).$$

Thus, $\widetilde{h}\big|_{\mathcal{S}_1(A)} = \gamma_{h_z, f}\big|_{\mathcal{S}_1(A)}$. The following lemma, whose assumptions hold for the algebra $C^\infty(M)$ because the partition of unity lemma (4.18) applies, shows that $\widetilde{h}$ lies in the image of the map i.   ▶

10.9 Lemma. *Let a* K*-algebra* A *be such that for all natural numbers* l *any differential operator of order* $\leqslant l$ *is representable as the sum of monomials of the form* $X_1 \circ \cdots \circ X_s$, *where* $X_i \in D(A)$, $s \leqslant l$. *In this case, if*

$$\widetilde{h}_1, \widetilde{h}_2 \in |\mathcal{S}_*(A)|, \ \widetilde{h}_1\big|_A = \widetilde{h}_2\big|_A \ and \ \widetilde{h}_1\big|_{\mathcal{S}_1(A)} = \widetilde{h}_2\big|_{\mathcal{S}_1(A)},$$

then $\widetilde{h}_1 = \widetilde{h}_2$.

◄ Passing to symbols of differential operators, we see that the algebra $\mathcal{S}_*(A)$ is generated by its submodule $\mathcal{S}_1(A) = D(A)$; i.e., any symbol $s = \mathrm{smbl}_k(\delta)$ can be represented as the sum of monomials $s_1 \cdots s_k$, where $s_i = \mathrm{smbl}_1 X_j$, $X_j \in D(A)$. Since $\tilde{h}_i$ is an algebra homomorphism, we have $h_i(s_1 \cdots s_k) = h_i(s_1) \cdots h_i(s_k)$ and the required equality $\tilde{h}_1(s) = \tilde{h}_2(s)$ follows from the fact that by our assumptions, $\tilde{h}_1(s) = \tilde{h}_2(s)$ for all symbols $s \in D(A) = \mathcal{S}_1(A)$.   ►

10.10 Exercise. Describe the $\mathbb{R}$-spectrum of the algebra of symbols $\mathcal{S}_*(C^\infty(\mathbf{K}))$ on the cross. Use a reasonable modification of the constructions that allowed us to describe the spectrum of the algebra $\mathcal{S}_*(C^\infty(M))$. By Proposition 10.3, this spectrum can be naturally interpreted as the cotangent space for the cross.

10.11. The algebra of symbols in coordinates. Theorem 10.8 allows one to understand elements of the algebra $\mathcal{S}_*(C^\infty(M))$ as functions on T^*M. Let us describe this interpretation in special coordinates (see Section 9.24).

Let U be a local chart on M. Then, by Section 9.24, T^*U is a local chart on T^*M. The localization of differential operators defined on M to the domain U naturally generates the corresponding localization of the algebra $\mathcal{S}_*(C^\infty(M))$. This localization clearly coincides with $\mathcal{S}_*(C^\infty(U))$. Therefore, we can restrict ourselves to the interpretation of its elements as functions on T^*U. We shall use the notation of Section 10.8.

Let $\theta = d_z f \in T^*U$, $f \in C^\infty(U)$, and $\Delta \in \mathrm{Diff}_k C^\infty(M)$. Denote by $\mathfrak{s} = \mathfrak{s}_\Delta$ the function on T^*M corresponding to the symbol $\mathrm{smbl}_k \Delta$. Then, by definition,

$$\mathfrak{s}(\theta) = \gamma_{h_z,f}(\mathrm{smbl}_k(\Delta)).$$

Let us identify, as above, a vector field $X \in D(U)$ with its symbol. It was noted in Section 10.8 that $\gamma_{h_z,f}(X) = X_z(f)$. Therefore,

$$\mathfrak{s}(\theta) = X_z(f).$$

In particular, if $X = \partial/\partial x_i$, then $\mathfrak{s}(\theta) = \partial f/\partial x_i(z)$.

Recall that by the definition of special coordinates (x, p) in T^*U, coordinate $p_i(\theta)$ is the ith component of the covector θ in the basis $\{d_z x_i\}$ in T_z^*M. Since $\theta = d_z f$ in our case, $p_i(\theta) = \partial f/\partial x_i(z)$ and consequently,

$$\mathfrak{s}_{\partial/\partial x_i} = p_i.$$

Note further that the definition

$$\mathrm{smbl}_k(\Delta) \cdot \mathrm{smbl}_l(\nabla) = \mathrm{smbl}_{k+l}(\Delta \circ \nabla),$$

for $\Delta \in \mathrm{Diff}_k\, C^\infty(M)$, $\nabla \in \mathrm{Diff}_l\, C^\infty(M)$, implies $\mathfrak{s}_{(\Delta \circ \nabla)} = \mathfrak{s}_\Delta \cdot \mathfrak{s}_\nabla$. Therefore, if $\Delta = \sum_{|\sigma| \leq h} a_\sigma \partial^{|\sigma|}/\partial x^\sigma$, then

$$\mathfrak{s}_\Delta = \sum_{|\sigma|=k} a_\sigma p^\sigma, \quad \text{where } p^\sigma = p_1^{i_1} \cdots p_n^{i_n} \text{ if } \sigma = (i_1, \ldots, i_n).$$

Thus the algebra $\mathcal{S}_*(C^\infty(U))$ is isomorphic to the algebra of fiberwise polynomials in the variables $p_1, \ldots, p_n$ with coefficients in the algebra $C^\infty(U)$. From this we immediately obtain the following:

10.12 Proposition. *The algebra $\mathcal{S}_*(C^\infty(M))$ is isomorphic to the subalgebra of the algebra $C^\infty(T^*M)$ consisting of functions whose restrictions to the fibers T_z^*M of the cotangent bundle are polynomials. The algebra $C^\infty(T^*M)$ is isomorphic to the smooth closure of the algebra $\mathcal{S}_*(C^\infty(M))$.*

Exercise. 1. Prove that the restriction of $s_{df}^* \colon C^\infty(T^*M) \to C^\infty(M)$ to the subalgebra $\mathcal{S}_*(C^\infty(M))$ coincides with the map

$$\Xi_f \colon C^\infty(T^*M) \to C^\infty(M)$$

(see Section 10.3).

2. Let A be a K-algebra, $h \in |A|$, and $\mathcal{S}_+(A) = \sum_{i>0} \mathcal{S}_i(A)$. Then the map

$$\bar{h} \colon \mathcal{S}_*(A) \to K, \quad \bar{h}\big|_A = h, \quad \bar{h}\big|_{\mathcal{S}_+(A)} = 0,$$

is a K-algebra homomorphism, i.e., $\bar{h} \in |\mathcal{S}_*(A)|$. Show that when $A = C^\infty(M)$, the map $|A| \to |\mathcal{S}_*(A)|$, $h \mapsto \bar{h}$, coincides with s_{df} for $f \equiv 0$ (the canonical embedding of M into T^*M).

3. Find an analog of the map s_{df} for arbitrary K-algebras.

4. Describe the algebra of symbols $\mathcal{S}_*(C^\infty(\mathbf{K}))$ and realize it as the algebra of functions on the spectrum $|\mathcal{S}_*(C^\infty(\mathbf{K}))|$.

10.13. The Hamiltonian formalism in T^*M and $|\mathcal{S}_*(A)|$. Now consider the case $A = C^\infty(M)$; let us describe, in special coordinates, the bracket $\{\cdot, \cdot\}$ introduced in Section 10.2. By the skew-symmetry of this bracket and because of the relation

$$\{\mathfrak{s}, \mathfrak{s}_1\mathfrak{s}_2\} = \{\mathfrak{s}, \mathfrak{s}_1\}\mathfrak{s}_2 + \mathfrak{s}_1\{\mathfrak{s}, \mathfrak{s}_2\} \tag{10.3}$$

(see the exercise from Section 10.2), it suffices to compute this bracket for the coordinate functions. Using the notation of the previous section, we have by definition

$$\{\mathfrak{s}_\Delta, \mathfrak{s}_\nabla\} = \mathfrak{s}_{[\Delta, \nabla]}. \tag{10.4}$$

Since $[f, g] = [g, f] = 0$ for $f, g \in C^\infty(U)$,

$$\left[\frac{\partial}{\partial x_i}, \frac{\partial}{\partial x_j}\right] = 0, \quad \left[\frac{\partial}{\partial x_i}, f\right] = \frac{\partial f}{\partial x_i},$$

it follows by (10.4) that

$$\{f(x), g(x)\} = 0, \quad \{p_i, p_j\} = 0, \quad \{p_i, f(x)\} = \frac{\partial f(x)}{\partial x_i}. \tag{10.5}$$

Further, applying (10.3) to $F = f(x)p^\sigma$, $G = g(x)p^\tau$, we obtain

$$\begin{aligned}
\{F, G\} &= \{f(x)p^\sigma, g(x)p^\tau\} = \{f(x), g(x)\}p^\sigma p^\tau + \{f(x), p^\tau\}g(x)p^\sigma \\
&\quad + \{p^\sigma, g(x)\}f(x)p^\tau + \{p^\sigma, p^\tau\}f(x)g(x) \\
&= \{p^\sigma, g(x)\}f(x)p^\tau - \{p^\tau, f(x)\}g(x)p^\sigma,
\end{aligned}$$

where $p^\sigma = p_{i_1} \ldots p_{i_n}$, $p^\tau = p_{j_1} \ldots p_{i_n}$. By the last equality in (10.5),

$$\{p^\sigma, g(x)\} = \sum_{i=1}^{n} \frac{\partial p^\sigma}{\partial p_i} \frac{\partial g}{\partial x_i}, \quad \{p^\tau, f(x)\} = \sum_{i=1}^{n} \frac{\partial p^\tau}{\partial p_i} \frac{\partial f}{\partial x_i}.$$

As a result of these computations, we finally obtain the formula:

$$\{F, G\} = \sum_{i=1}^{n} \left(\frac{\partial F}{\partial p_i} \frac{\partial G}{\partial x_i} - \frac{\partial G}{\partial p_i} \frac{\partial F}{\partial x_i} \right), \tag{10.6}$$

which is the *standard Poisson bracket* on T^*M. Moreover, this formula shows that the derivation $X_F \overset{\text{def}}{=} \{F, \cdot\}$ of the algebra of symbols (which is, geometrically, a vector field on its $\mathbb{R}$-spectrum, i.e., on T^*M) is of the form

$$X_F = \sum_{i=1}^{n} \left(\frac{\partial F}{\partial p_i} \frac{\partial}{\partial x_i} - \frac{\partial F}{\partial x_i} \frac{\partial}{\partial p_i} \right). \tag{10.7}$$

Thus X_F is the *Hamiltonian vector field* on T^*M with Hamiltonian F. These facts justify calling the bracket $\{\cdot, \cdot\}$ on the algebra of symbols $\mathcal{S}_*(A)$ of an arbitrary K-algebra A the *Poisson bracket*, while the derivations $\{\mathfrak{s}, \cdot\}$, $\mathfrak{s} \in \mathcal{S}_*(A)$, are naturally called *Hamiltonian vector fields* on $|\mathcal{S}_*(A)|$. This is another evidence in favor of treating differential calculus as a part of commutative algebra (this treatment being a consequence of the *observability principle*). The reader can now enjoy constructing Hamiltonian mechanics on smooth sets or over arithmetic fields.

Exercise. 1. Let $F = F(x, p) \in C^\infty(T^*M)$. Check that finding functions satisfying the condition $s^*_{df}(F) = 0$, $f \in C^\infty(M)$, is equivalent to solving the Hamilton–Jacobi equations. Find the analog of these equations for an arbitrary K-algebra A.

2. Find the analog of the Hamilton–Jacobi equations on the cross $\mathbf{K}$.

10.14. Thus we see that the differential calculus is a consequence of the classical observability principle and is developed simply and naturally if one keeps this fact in mind. The commutativity of the algebra of observables is a formalization of the fundamental idea of classical physics: the independence of observations. A more sophisticated realization of this idea by means of commutative graded algebras (traditionally called *superalgebras*) does not involve overcoming any additional difficulties. All definitions and constructions of the differential calculus are carried over to this case and to any other case in which commutativity can be treated in a reasonable way. This is done in detail in Chapter 21.

As is known, in quantum physics, one has to reject the principle of independence of observations. Nevertheless, it would not be correct to try to quantize the differential calculus in order to describe quantum phenomena by a simple change from commutative algebras to noncommutative ones. The reader will see this by trying systematically to carry over the constructions of this chapter to noncommutative algebras. The failure of such attempts becomes really catastrophic when one tries to repeat subtler and deeper constructions in the noncommutative situation. These and many other reasons show that it is hardly possible to obtain a mathematically adequate quantum principle of observability by using the language of noncommutative algebra (or noncommutative geometry).

There are serious reasons to believe now that this aim can be reached in a natural way by using the language of the *secondary differential calculus*, which is a sort of synthesis of the usual (= primary) differential calculus with homological algebra. In any case, we have no doubt that this calculus is the natural language for the geometry of nonlinear partial differential equations.

11

Smooth Bundles

11.1. The inner structure of points. The concept of observability developed in this book assumes that a point is an elementary object that can be individualized with the help of a given set of instruments, i.e., a given algebra of observables A. We know, however, that points of the physical manifold where we live may have *inner parameters*, such as temperature, color, and humidity. To give a precise mathematical meaning to this phrase, we need the notion of fibration or fiber space, which is the main protagonist of the present chapter.

A priori, there are two possibilities: the concept of inner structure can be either *relative* or *absolute*. The inner structure is relative if it can be expressed within the classical framework by simply adding new instruments. The mathematical meaning of this construction is that the algebra of observables A is extended to a bigger algebra B by means of the inclusion $i\colon A \hookrightarrow B$. The inner structure of a point $z \in |A|$ is then described by its inverse image $|i|^{-1}(z) \subset |B|$ under the map of $\mathbb{R}$-spectra $|i|\colon |B| \to |A|$.

Example. The 3-dimensional world $\mathbb{R}^3$ can be made colored if to the set of instruments measuring a point's coordinates, we add one more instrument measuring the color, i.e., the frequency of electromagnetic waves. In algebraic language, this means that we pass from the algebra $A = C^\infty\left(\mathbb{R}^3\right)$ to the algebra $B = A \otimes_{\mathbb{R}} C$, where C is the algebra of smooth functions on the "manifold of colors," which we identify naturally with $\mathbb{R}^1_+$. The inclusion $i\colon A \hookrightarrow B$ is defined by the rule

$$A \ni a \mapsto a \otimes 1_C \in A \otimes_{\mathbb{R}} C.$$

© Springer Nature Switzerland AG 2020
J. Nestruev, *Smooth Manifolds and Observables*, Graduate Texts
in Mathematics 220, https://doi.org/10.1007/978-3-030-45650-4_11

Returning to the general case, let us note that $|i|^{-1}(z) = |B_z|$, where $B_z = B/(\mu_z \cdot B)$. The inner structure of a point $z \in |A|$ is thus observable by means of the algebra B_z. The assumption that all points $z \in |A|$ have the same inner structure means that all algebras of additional observables B_z are the same, i.e., isomorphic to each other. If this condition is fulfilled in a certain regular manner (see Section 11.9), then the map $|i| \colon |B| \to |A|$ is referred to as a *locally trivial smooth bundle*. In the above example, this condition holds, and all algebras B_z are isomorphic to C.

At first glance, relative inner structures do not add anything new to the classical scheme of observability, because any such structure can be reduced to a standard one through an appropriate extension of the algebra A. This approach, however, is not convenient when the manifold $M = |A|$ is considered as a display that shows the points with different inner structures. For example, this is the case for real physical space. Moreover, the problem ceases to be a question of mere convenience if the inner structures of the points displayed on M are *absolute* in the sense that they cannot be described by the above classical approach. In particular, this is true of quantum phenomena that have survived many unsuccessful attempts of explanation based on so-called "hidden parameters."

Unless otherwise specified, all algebras in this chapter are assumed to be smooth.

11.2. Fibrations as extension of algebras. Before going on to bundles, we shall introduce the more general notion of a fibration. In algebraic terms it can be defined as follows.

Definition. A *smooth fibration* is an injective homomorphism of smooth algebras $i \colon A \hookrightarrow B$. The manifold $|A|$ is called the *base* of the fibration i, the manifold $|B|$ is its *total space*, while the map $|i| \colon |B| \to |A|$ is referred to as the *projection* of the fibration.

11.3. Examples. I. Product fibration. Let A and C be smooth algebras. Set $B = \overline{A \otimes_{\mathbb{R}} C}$ (we recall that the bar stands for the smooth envelope of an algebra; see Section 3.36) and define the inclusion $i \colon A \hookrightarrow B$ by the rule $a \mapsto a \otimes 1$. A concrete example of this construction is the fibration of the torus over the circle, defined as the extension $i \colon A \hookrightarrow B$, where

$$B = \{ g \in C^\infty(\mathbb{R}^2) \mid g(x+1, y) = g(x, y+1) = g(x, y) \}$$

is the algebra of twice periodic functions in two variables and A is the subalgebra of B consisting of all functions that do not depend on y. Indeed, if

$$A = \{ f \in C^\infty(\mathbb{R}) \mid f(x+1) = f(x) \}, \tag{11.1}$$

$$C = \{ f \in C^\infty(\mathbb{R}) \mid f(y+1) = f(y) \}, \tag{11.2}$$

then it is readily verified that $\overline{A \otimes C} \cong B$.

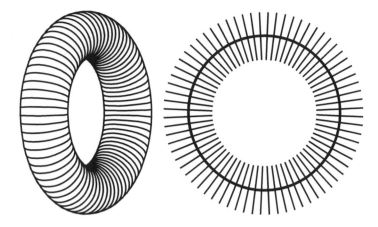

Figure 11.1. Product fibration.

II. **The Klein bottle fibered over the circle** (see Figure 11.2). This is the inclusion of the algebra of smooth periodic functions on the real line

$$A = \{f \in C^\infty(\mathbb{R}) \mid f(x+1) = f(x)\}$$

into the algebra

$$B = \{g \in C^\infty(\mathbb{R}^2) \mid g(x+1, y) = -g(x, y+1) = g(x, y)\}$$

by the rule $f \mapsto g \colon g(x, y) = f(x)$.

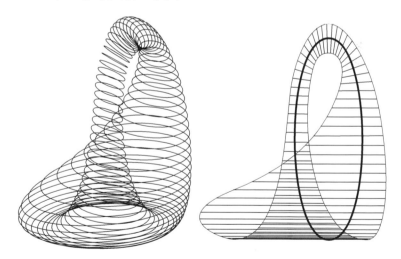

Figure 11.2. Nontrivial fibration.

III. **The two-sheeted covering of the circle.** This is the map of the algebra

$$A = \{f \in C^\infty(\mathbb{R}) \mid f(x+1) = f(x)\}$$

into itself defined by the formula $f \mapsto g \colon g(x) = f(2x)$.

IV. **One-sheeted fibration of the line over the circle.** Consider $B = C^\infty(\mathbb{R})$ and let $A \subset B$ consist of all functions $f \in B$ for which the function $x \mapsto f(1/x)$ and all of its derivatives have finite limits as $x \to 0$. One can see that $A \cong C^\infty(S^1)$, and under a proper choice of this isomorphism, the inclusion $A \hookrightarrow B$ corresponds to the map

$$\mathbb{R} \to S^1 = \left\{(x,y) \mid x^2 + y^2 = 1\right\} \colon \quad t \mapsto \left(\frac{1-t^2}{1+t^2}, \frac{2t}{1+t^2}\right).$$

The image of this map is the entire circle with one point removed.

V. **The projection of the tangent space TM of a manifold M onto M.** In Section 9.18, we described the map $\pi_T \colon TM \to M$, given by $(z, \xi) \mapsto z$. The corresponding homomorphism of smooth algebras $C^\infty(M) \to C^\infty(TM)$ is injective; therefore, π_T can be regarded as the projection of the fibration.

The most important class of fibrations consists of *bundles* (see Sections 11.9 and 11.10), defined as locally trivial fibrations. In the previous list, examples I–III, V possess this property, while example IV does not. To give a precise definition of local triviality, we need the notion of fiber and the procedure of localization.

11.4. Fiber of a fibration. Geometrically, the fiber of a given fibration $i \colon A \to B$ over the point $z \in |A|$ is the inverse image of z under the projection $|i| \colon |B| \to |A|$.

In examples I–II, the fiber over any point is a circle; in example III it is two points; in example IV, depending on the choice of the point z, the fiber is either empty or consists of one point. Finally, in example V, the fiber $T_z M$ is isomorphic to the linear space $\mathbb{R}^n$. Note that in examples I and II both the base space and the fiber of both fibrations are the same, whereas the total spaces are different.

An algebraic definition of the *fiber over a point* $a \in |A|$ can be given as follows: it is the quotient algebra of B over the ideal generated by the set $i(\mu_a)$, where $\mu_a \subset A$ is the maximal ideal of the point a.

11.5. The category of fibrations. By definition, a morphism of a fibration $i_1 \colon A \to B_1$ into a fibration $i_2 \colon A \to B_2$ is an algebra homomorphism $\varphi \colon B_2 \to B_1$ making the following diagram commutative

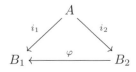

An equivalent definition in terms of spectra (see Sections 3.4 and 8.11) reads that the diagram

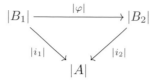

commutes. This means that the map $|\varphi|$ takes the fibers of one fibration into the fibers of another: $|\varphi|(|i_1|^{-1}(a)) \subset |i_2|^{-1}(a)$, or, equivalently, that $\varphi(i_2(\mu_a) \cdot B_2) \subset i_1(\mu_a) \cdot B_1$ for any point $a \in |A|$.

The totality of all fibrations over a smooth algebra $A = C^\infty(M)$ together with all morphisms between them constitutes the *category of fibrations over* M.

If the homomorphism φ is an isomorphism, or, which is the same thing, the map $|\varphi|$ is a diffeomorphism, then the fibrations i_1 and i_2 are said to be *equivalent*. In the case where the homomorphism φ is surjective, which corresponds to a proper embedding of manifolds $|B_1| \to |B_2|$, the fibration i_1 is called a *subfibration* of the fibration $|i_2|$.

The simplest example of a fibration with base M and fiber F is provided by the direct product $M \times F$ with the natural projection on the first factor, or, in algebraic terms, the natural inclusion of the algebra $A = C^\infty(M)$ into the smooth envelope of the tensor product $\overline{A \otimes C}$, where $C = C^\infty(F)$.

A fibration equivalent (in the category of fibrations) to a fibration of this kind is referred to as a *trivial fibration*. Roughly speaking, a *bundle* is a fibration that is *locally* isomorphic to a trivial fibration, see Definition 11.9. This phrase will become an exact definition after we have explained the meaning of the word "locally."

11.6. Localization. The aim of this section is to describe an algebraic construction that allows one to define the restriction of an algebra to an open set (see Section 3.23) in purely algebraic terms.

Let A be a commutative ring with unit and let $S \subset A$ be a *multiplicative set*, i.e., a subset of A, containing 1, not containing 0, and closed with respect to multiplication. In the set of all pairs (a, s), where $a \in A$, $s \in S$, we introduce the equivalence relation

$$(a_1, s_1) \sim (a_2, s_2) \stackrel{\text{def}}{\Longleftrightarrow} \exists s \in S \colon s(a_1 s_2 - a_2 s_1) = 0.$$

The equivalence class of a pair (a, s) is denoted by $\frac{a}{s}$ (or a/s) and called a *formal fraction*; we denote the set of all such classes by $S^{-1}A$. The sum and product of formal fractions are defined by the ordinary formulas

$$\frac{a_1}{s_1} \cdot \frac{a_2}{s_2} = \frac{a_1 a_2}{s_1 s_2}, \qquad \frac{a_1}{s_1} + \frac{a_2}{s_2} = \frac{a_1 s_2 + a_2 s_1}{s_1 s_2}.$$

The resulting ring $S^{-1}A$ is referred to as the *localization of the ring A over the multiplicative system S*. We leave the straightforward checks to the reader.

There is a *canonical homomorphism* $\iota\colon A \to S^{-1}A$ defined by $\iota(a) = a/1$. In general, ι is neither injective nor surjective.

Suppose now that P is a module over A. In the same way as above, in the set of pairs (p, s) with $p \in P$, $s \in S$, we can introduce the equivalence relation

$$(p_1, s_1) \sim (p_2, s_2) \overset{\text{def}}{\Longleftrightarrow} \exists s \in S\colon s(s_2 p_1 - s_1 p_2) = 0.$$

A *formal fraction* $\frac{p}{s}$ (or p/s) is the equivalence class of the pair (p, s). The set of all such classes, denoted by $S^{-1}P$, is referred to as the *localization of P over S*.

Exercises.   1. Introduce the addition of two elements of $S^{-1}P$ and the multiplication of an element of $S^{-1}P$ by an element of $S^{-1}A$ as follows:

$$\frac{p_1}{s_1} + \frac{p_2}{s_2} = \frac{s_2 p_1 + s_1 p_2}{s_1 s_2}, \qquad \frac{a_1}{s_1} \cdot \frac{p_2}{s_2} = \frac{a_1 p_2}{s_1 s_2}.$$

Verify that these operations are well defined and turn the set $S^{-1}P$ into an $(S^{-1}A)$-module.

2. Let $\varphi\colon P \to Q$ be a homomorphism of A-modules. Prove that the map $S^{-1}(\varphi)\colon S^{-1}P \to S^{-1}Q$, given by the formula

$$S^{-1}(\varphi)\left(\frac{p}{s}\right) \overset{\text{def}}{=} \frac{\varphi(p)}{s}, \quad p \in P,\ s \in S,$$

is well defined and represents a homomorphism of $(S^{-1}A)$-modules.

Summarizing, we can say that for a given multiplicative set $S \subset A$ we have defined a functor S^{-1} from the category of A-modules to the category of $(S^{-1}A)$-modules.

Examples. I. If A has no zero divisors and $S = A \setminus \{0\}$, then $S^{-1}A$ is the quotient field of A.

II. Let $A = \mathbb{Z}$ and let S be the set of all nonnegative powers of 10. Then $S^{-1}A$ consists of all rational numbers that have a finite decimal representation.

III. If M is a smooth manifold, $A = C^\infty(M)$, $x \in M$, and $S = A \setminus \mu_x$, where $\mu_x = \{f \in C^\infty(M) | f(x) = 0\}$, then $S^{-1}A$ is the *ring of germs of smooth functions on M at the point x* (readers who are familiar with the notion of germ may prove this fact as an exercise; others may take it as the definition and try to understand its geometrical meaning).

11.7 Proposition. *Let U be an open subset of the manifold M,*

$$A = C^\infty(M), \quad \text{and} \quad S = \{f \in A \mid f(x) \neq 0\ \forall x \in U\}.$$

Then $S^{-1}A$ is isomorphic to $C^\infty(U)$, and the canonical homomorphism $\iota\colon C^\infty(M) \to C^\infty(U)$ coincides with the restriction $\iota(f) = f\big|_U$.

◀ To prove this fact, consider the map $\alpha\colon S^{-1}A \to C^\infty(U)$ given by

$$\alpha\left(\frac{f}{s}\right)(x) = \frac{f(x)}{s(x)}, \qquad f \in A,\ s \in S,\ x \in U,$$

which converts a formal fraction into the ordinary quotient of two functions. This map is well defined, because the functions $s \in S$ do not vanish anywhere in U. Suppose that $\alpha(f_1/s_1) = \alpha(f_2/s_2)$. Then the function $f_1 s_2 - f_2 s_1$ is identically zero on U. By Lemma II of Section 4.17, there is a function $s \in S$ that is identically zero outside of U. The product of these two functions is identically zero on all of M, i.e., $s(f_1 s_2 - f_2 s_1) = 0$ as an element of the algebra A. By definition, the two formal fractions f_1/s_1 and f_2/s_2 are equal. This proves that α is injective. The fact that it is surjective follows from the next lemma. ▶

Exercise. Give an example of a geometrical algebra A that is not smooth and of an open subset $U \subset |A|$ such that the algebras $S^{-1}A$ and $A|_U$ are not isomorphic.

11.8 Lemma. *Suppose that U is an open subset of the manifold M and $f \in C^\infty(U)$. Then there exists a function $g \in C^\infty(M)$, having no zeros on U, such that the product fg can be smoothly extended to the entire manifold M and therefore $f = \alpha(fg/g)$.*

◀ A rigorous proof of this fact can be obtained by the techniques of partition of unity described in Chapter 2. We leave the details to the reader. ▶

We are now in a position to give a precise algebraic definition of a bundle.

Recall that a homomorphism of K-algebras $\varphi\colon A \to B$ gives rise to the operation of the *change of rings*: any B-module R can be regarded as an A-module with multiplication $a \cdot r \overset{\text{def}}{=} \varphi(a)r$, $a \in A$, $r \in R$. In particular, the algebra B itself can be regarded as an A-module. Therefore, for a given multiplicative set $S \subset A$, the $(S^{-1}A)$-module $S^{-1}B$ is defined.

11.9. Definition. Let A, B, and F be smooth algebras. An injective homomorphism $i\colon A \to B$ is called a *bundle* $|B|$ *over* $|A|$ *with fiber* $|F|$ if every point $z \in |A|$ has an open neighborhood $U_z \subset |A|$ over which the localization of i is equivalent to the product fibration

$$S_z^{-1}A \to \overline{S_z^{-1}A \otimes F},$$

where $S_z \subset A$ is the multiplicative system of U_z, i.e., the set of all elements of A whose values at the points of U_z are nonzero. More exactly, there must exist an algebra isomorphism $p\colon S_z^{-1}B \to \overline{S_z^{-1}A \otimes F}$ that makes the

following triangle commutative

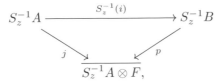

$$S_z^{-1}A \xrightarrow{ S_z^{-1}(i) } S_z^{-1}B$$

with j and p into $S_z^{-1}A \otimes F,$

where j is the map taking every element a to $a \otimes 1$. (Recall once again that the bar over the notation of an algebra means that we take its smooth envelope.)

A bundle π is *trivial* if and only if the previous condition (*local triviality axiom*) is fulfilled for $U_z = M$.

The geometrical definition of a bundle, which follows, is simpler. The equivalence of the two definitions is quite obvious.

11.10. Definition. Let E and M be smooth manifolds. A smooth map $\pi\colon E \to M$ is said to be a *fiber bundle*, or *bundle* for short, if there exists a manifold F such that the following condition holds: any point $x \in M$ has a neighborhood $U \subset M$ for which there exists a *trivializing* diffeomorphism $\varphi\colon \pi^{-1}(U) \to U \times F$ that closes the commutative diagram

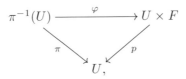

$$\pi^{-1}(U) \xrightarrow{\varphi} U \times F$$

with π and p into $U,$

where p is the projection of the product on the first factor.

Under these conditions, M, F, and E are referred to as the *base*, the *fiber*, and the *total space* of the bundle π, respectively. The dimension of the bundle is by definition equal to the dimension of the fiber. The whole thing is conventionally written as $E \xrightarrow{F} M$. The total space of the bundle π is usually denoted by E_π.

The fiber bundle π *over a point* $x \in M$ is the set $\pi_x = \pi^{-1}(x)$. The fiber over any point, equipped with the structure of the submanifold of E, is diffeomorphic to the fixed "outer" fiber F.

11.11. Some more examples. In the examples that follow, we give only the geometrical construction of the bundle, leaving it to the reader to describe the corresponding extension of smooth algebras.

I. **The open Möbius band** is a *line bundle* over the circle: $M \xrightarrow{\mathbb{R}} S^1$. If M is viewed as the band $[0,1] \times \mathbb{R} \subset \mathbb{R}^2$ with the points $(0,y)$ and $(1,-y)$ identified for any $y \in \mathbb{R}$, and the circle S^1 is viewed as the segment $[0,1]$ with identified end points 0 and 1, then the projection $\pi\colon M \to S^1$ is simply $\pi(x,y) = x$. A visual representation of this bundle is given in Figure 6.2, where the Möbius band embedded into $\mathbb{R}^3$ squeezes to its midline; the fibers are the segments perpendicular to the midline.

Let us check local triviality in this example. If $a \in S^1$ is an inner point of the segment $[0, 1]$, then for the neighborhood U we can take the interval $]0, 1[$. If $a = 0$, then we put $U = \{x \in S^1 \mid x \neq 1/2\}$ and define the diffeomorphism $\varphi \colon \pi^{-1}(U) \to U \times \mathbb{R}$ as follows:

$$\varphi(x, y) = \begin{cases} (x, y), & \text{if } x < \frac{1}{2}, \\ (x, -y), & \text{if } x > \frac{1}{2}. \end{cases}$$

II. **The bundle of the line over the circle** (cf. Section 6.2). Representing the circle S^1 as the set of complex numbers of modulus 1, we define the map $\exp \colon \mathbb{R}^1 \to S^1$ by $t \mapsto e^{it}$. Any neighborhood of $x \in S^1$, but not the whole circle S^1, can be taken as the neighborhood U in the condition of local triviality.

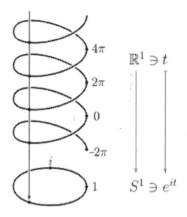

The fiber in this example is the zero-dimensional manifold $\mathbb{Z}$. Bundles with zero-dimensional fibers are called *coverings*.

III. Another example of a covering is provided by the map $S^n \to \mathbb{R}P^n$ that assigns to a point $x \in S^n \subset \mathbb{R}^n$ the line in $\mathbb{R}^{n+1}$ passing through that point and the origin (see the definitions in Section 5.10). This bundle is trivial over the complement to any hyperplane $\mathbb{R}P^n \setminus \mathbb{R}P^{n-1}$. Its fiber consists of two points.

Figure 11.3. $\exp \colon \mathbb{R}^1 \to S^1$

IV. **The bundle of unit tangent vectors to the sphere** $\pi \colon T_1 S^2 \to S^2$. The space of this bundle

$$T_1 S^2 = \{(x, y) \in \mathbb{R}^3 \times \mathbb{R}^3 \mid |x| = 1, |y| = 1, x \perp y\}$$

is a submanifold of $\mathbb{R}^6$. Making the orthogonal group act on a fixed unit vector to the sphere, we obtain a diffeomorphism $T_1 S^2 \cong SO(3)$. The total space of the bundle under study thus provides another (fifth!) realization of the manifold considered at the beginning of the book in Examples 1.1–1.4. The fiber is the circle S^1.

To prove local triviality, we shall show that this bundle is trivial over any open hemisphere S^2. Indeed, if U is a hemisphere and S^1 is its boundary, then we can identify the points of U and S^1 with the vectors drawn from the center of the sphere and set $\varphi(x, z) = x \times z$ (cross product of vectors) for $x \in U$ and $z \in S^1$. The two vectors $x \in U$ and $z \in S^1$ are never collinear; hence the map $\varphi \colon U \times S^1 \to \pi^{-1}(U)$ is a diffeomorphism.

V. The composition of the maps $S^3 \to \mathbb{R}P^3$ and $\mathbb{R}P^3 \to S^2$ defined in Examples III and IV above is a bundle of S^3 over S^2 with fiber S^1, called the *Hopf fibration*, compare with Section 6.17, II. We leave it to the reader to check its local triviality.

VI. **Tautological bundle over a Grassmannian.** Suppose $G_{n,k}$ is the Grassmann manifold (see Example IV in Section 5.10) whose points are k-dimensional linear subspaces of the n-dimensional space $\mathbb{R}^n$; let $E_{n,k}$ be the set of all pairs (x, L) such that $x \in L \in G_{n,k}$; $E_{n,k}$ is viewed as a submanifold in $\mathbb{R}^n \times G_{n,k}$. The correspondence $(x, L) \mapsto L$ defines a fibration

$$\Theta = \Theta_{n,k} : E_{n,k} \to G_{n,k},$$

called the *tautological bundle*.

Exercise. Show that the tautological bundle $E_{n,1} \to G_{n,1} = \mathbb{R}P^{n-1}$ for $k = 1$ can be described as the projection

$$\pi \colon \mathbb{R}P^n \setminus \{L_0\} \to \mathbb{R}P^{n-1},$$

where L_0 is the $(n+1)$th coordinate axis in $\mathbb{R}^{n+1}$ (= "vertical" line), and π assigns to each "slanted" line its projection to $\mathbb{R}^n \cong \{x_{n+1} = 0\}$. In particular, if $k = 1$ and $n = 2$, we obtain the fibration of the Möbius band over the circle from Example III.

Let us prove that the tautological fibration is a bundle, i.e., possesses the property of local triviality. We shall use the open covering of the manifold $G_{n,k}$ by the family of open sets

$$U_I, \quad \text{where } I = \{i_1, \dots, i_k\}, \; 1 \leqslant i_1 < \cdots < i_k \leqslant n.$$

By definition, the neighborhood U_I consists of all k-planes in $\mathbb{R}^n$ that do not degenerate under the projection on $\mathbb{R}^k_I$ along $\mathbb{R}^{n-k}_{\overline{I}}$, where $\overline{I} = \{1, \dots, n\} \setminus I$ and the symbol $\mathbb{R}^m_J$, $J = \{j_1, \dots, j_m\}$, stands for the m-plane in $\mathbb{R}^n$ spanned by the basic vectors numbered $j_1, \dots, j_m$. If $x \in L$ and $L \in U_I$, then to the pair $(x, L) \in E_{n,k}$ we assign the pair $(\overline{x}, L)$, where $\overline{x} \in \mathbb{R}^k_I \cong \mathbb{R}^k$ is the projection of x along $\mathbb{R}^{n-k}_{\overline{I}}$ onto $\mathbb{R}^k_I$. This assignment is a trivializing diffeomorphism for the tautological fibration over the set U_I. This is why $\Theta_{n,k}$ is in fact a bundle.

All bundles listed above are nontrivial. For example, the Möbius band is nonorientable and therefore not diffeomorphic to the cylinder $S^1 \times \mathbb{R}$. The nontriviality of the bundle of unit tangent vectors (Example IV) can be proved by using the basic facts about fundamental groups. In fact, the manifolds $\mathbb{R}P^3$ and $S^2 \times S^1$ are not diffeomorphic, because their fundamental groups are different: $\pi_1(\mathbb{R}P^3) = \mathbb{Z}_2$, $\pi_1(S^2 \times S^1) = \mathbb{Z}$. The same argument shows the nontriviality of the Hopf fibration (Example V).

VII. **The tangent bundle** $\pi_T \colon TM \to M$ (see Section 9.19). If (U, x) is a chart on M, then the corresponding trivializing diffeomorphism

$$U \times \mathbb{R}^n \to \pi_T^{-1}(U)$$

is the composition of natural identifications

$$U \times \mathbb{R}^n \longleftrightarrow TU \quad \text{and} \quad TU \longleftrightarrow \pi_T^{-1}(U),$$

described in Section 9.18 II and IV:

$$(z,q) \mapsto \sum_{i=1}^{n} q_i \frac{\partial}{\partial x_i}\Big|_z \in T_z M \subset TU, \text{ where } q = (q_1, \ldots, q_n).$$

Depending on the manifold M, its tangent bundle can be either trivial or nontrivial. A manifold with a trivial tangent bundle is called *parallelizable*. For example, any Lie group is parallelizable.

Exercise. Among the examples of manifolds considered earlier in this book, find some that are parallelizable and some that are not.

VIII. **The cotangent bundle** $\pi_{T^*} \colon T^*M \to M$ (Section 9.24). Just as in the previous case, the required trivialization $U \times \mathbb{R}^n \to \pi_{T^*}^{-1}(U)$ can be obtained from the identifications

$$U \times \mathbb{R}^n \longleftrightarrow T^*U \quad \text{and} \quad T^*U \longleftrightarrow \pi_{T^*}^{-1}(U),$$

described in Section 9.24:

$$(z,p) \mapsto \sum_{i=1}^{n} p_i d_z x_i \in T_z^* M \subset T^*U, \quad \text{where } p = (p_1, \ldots, p_n).$$

IX. **The bundle of l-jets of functions** $\pi_{J^l} \colon J^l M \to M$. In the chart (U, x) on M, the trivializing map $U \times \mathbb{R}^N \to \pi_{J^l}^{-1}(U) = J^l(U)$, where N is the total number of different derivatives of order $\leq l$,

$$\left(z, \mathbf{p}^l\right) \mapsto [f_{\mathbf{p}^l}]_z^l \in J_z^l M \subset J^l U, \quad \text{where } f_{\mathbf{p}^l} = \sum_{\sigma \leq l} \frac{1}{\sigma!} p_\sigma (x - z)^\sigma,$$

and $\mathbf{p}^l = (p_\sigma)$ is the vector with components p_σ, $|\sigma| \leq l$, arranged in the lexicographic order of the subscripts. Formula (2.4) shows that l-jets of functions $f_{\mathbf{p}^l}$ at the point z exhaust $J_z M$. The functions x_i, $i = 1, \ldots, n$, and p_σ, $|\sigma| \leq l$, constitute a local coordinate system in $\pi_{J^l}^{-1}(U)$, and such special charts form an atlas of $J^l M$. See Section 14.20.

Exercise. Find the value of N defined above.

11.12. Sections. A *section* of a bundle $\pi \colon E \to M$ is a smooth map $s \colon M \to E$ that assigns to every point $x \in M$ an element of the fiber over this point: $s(x) \in \pi_x$. In other words, the map s is such that $\pi \circ s = \mathrm{id}_M$. The set of all sections of the bundle π is denoted by $\Gamma(\pi)$.

In algebraic language, a section of the bundle $i \colon A \hookrightarrow B$ is represented by an algebra homomorphism $\sigma \colon B \to A$, left inverse to i, i.e., such that $\sigma \circ i = \mathrm{id}_A$.

Examples. I. The bundle $\pi\colon \mathbb{R}^1 \to S^1$, described in Example 11.13, I, has no sections. Indeed, let $f\colon S^1 \to \mathbb{R}^1$ be a smooth map and $\pi \circ f = \mathrm{id}_{S^1}$. The last equality implies that the restriction $\pi|_{f(S^1)}\colon f(S^1) \to S^1$ is a diffeomorphism of the set $f(S^1) \subset \mathbb{R}^1$ onto S^1. However, the image of a continuous map $f\colon S^1 \to \mathbb{R}^1$ is a certain segment $[a,b] \subset \mathbb{R}$ and thus cannot be homeomorphic to the circle.

II. The bundle $T_1 S^2 \to S^2$ (Example 11.11, IV) has no sections. This fact is sometimes referred to as the *hedgehog theorem*. Indeed, the existence of a section $f\colon S^2 \to T_1 S^2$ would lead to the following construction of a diffeomorphism $\varphi\colon S^2 \times S^1 \to T_1 S^2$: put $\varphi(x,\alpha)$ equal to the vector, obtained from $f(x)$ by a rotation through angle α in a fixed direction (say, counterclockwise, if the sphere is viewed from the outside).

III. The sections of the trivial bundle over M with fiber F are in one-to-one correspondence with smooth maps from M to F.

IV. The sections of the tangent bundle over M are naturally interpreted as vector fields on M (see Section 9.40). In a similar way, the sections of the cotangent bundle over M are naturally associated with first-order differential forms, or 1-forms. This notion is introduced and discussed below in Sections 14.4–14.8. A similar situation takes place for jet bundles as well; see Sections 9.65, 14.14–14.15.

Exercise. Describe the sections s_X, s_{df}, and $s_{j_l(f)}$ of the tangent, cotangent, and l-jet bundles (see 9.40, 9.25, and 9.65, respectively) in terms of special local coordinates.

V. This example concerns some remarkable sections of the tautological bundle over the Grassmannian $\Theta_{n,k}\colon E_{n,k} \to G_{n,k}$ (see Example VI in Section 11.11). Let $\mathfrak{M}_{k,n}$ be the space of $(k \times n)$ matrices of rank k. If

$$J = (j_1, \ldots, j_k),\ 1 \le j_1 < \cdots < j_k \le n,$$

and $\mathcal{M} \in \mathfrak{M}_{k,n}$, then $\mathcal{M}_J$ denotes the $(k \times k)$ matrix formed by the columns of $\mathcal{M}$ with numbers $j_1, \ldots, j_k$. At least one of the minors $|\mathcal{M}_J|$ in the matrix $\mathcal{M}$ is different from zero; therefore, $\sum_J |\mathcal{M}_J|^2 > 0$.

Fix a multiindex $I = (i_1, \ldots, i_k)$ and consider the following function on $\mathfrak{M}_{k,n}$:

$$\nu_I(\mathcal{M}) = \frac{|\mathcal{M}_I|}{\sum_J |\mathcal{M}_J|^2}.$$

Obviously, $\nu_I \in C^\infty(\mathfrak{M}_{k,n})$, and for any $g \in \mathrm{GL}(k,\mathbb{R})$, we have

$$\nu_I(g\mathcal{M}) = |g|^{-1}\nu_I(\mathcal{M}). \tag{11.3}$$

Further, let $\widetilde{\mathcal{M}}_I$ be the adjoint to the matrix $\mathcal{M}_I$, i.e., the one formed by the minors of order $k-1$ of $\mathcal{M}_I$. Denote by $\mathrm{Mat}_{k,n}$ the space of all $(k \times n)$ matrices over $\mathbb{R}$. The map

$$m_I\colon \mathfrak{M}_{k,n} \to \mathrm{Mat}_{k,n}, \quad m_I(\mathcal{M}) = \nu_I(\mathcal{M})\widetilde{\mathcal{M}}_I \mathcal{M}, \tag{11.4}$$

is $GL(k, \mathbb{R})$-*equivariant*, i.e., satisfies

$$m_I(g\mathcal{M}) = m_I(\mathcal{M}), \quad g \in GL(k, \mathbb{R}). \tag{11.5}$$

It is evidently smooth. Furthermore, $m_I(\mathcal{M}) \in \mathfrak{M}_{k,n}$ if $|\mathcal{M}_I| \neq 0$; otherwise, $m_I(\mathcal{M}) = 0$.

Consider the natural projection

$$\mu \colon \mathfrak{M}_{k,n} \to G_{n,k},$$

where $\mu(\mathcal{M})$ is the subspace of $\mathbb{R}^n$ spanned by the rows of the matrix $\mathcal{M}$. The two conditions: $\mu(\mathcal{M}') = \mu(\mathcal{M})$ and $\mathcal{M}' = g(\mathcal{M})$ for some $g \in GL(k, \mathbb{R})$ are equivalent. Therefore, $\mu(m_I(\mathcal{M})) = \mu(\mathcal{M})$ if $|\mathcal{M}_I| \neq 0$. On the other hand, the last condition means that $\mu_I(\mathcal{M}) \in U_I$ (see Example VI in Section 11.11).

We are now in a position to define the section

$$s_{I,i} \colon G_{n,k} \to E_{n,k}$$

of the tautological bundle $\Theta_{n,k}$ by setting

$$s_{I,i}(L) = (i\text{th line of the matrix } m_I(\mathcal{M}), L),$$

where $\mathcal{M}$ is any matrix such that $m_I(\mathcal{M}) = L$. By virtue of (11.5), this construction is well defined. Formula (11.4) ensures that the sections $s_{I,i}$ are smooth.

11.13. Subbundles. A bundle $\eta \colon E_\eta \to M$ is said to be a *subbundle* of the bundle $\pi \colon E_\pi \to M$ (notation: $\eta \subset \pi$) if

  (i) the total space E_η is a submanifold of E_π;

  (ii) the map η is the restriction of π on E_η;

  (iii) for any point $x \in M$ the fiber η_x is a submanifold of the fiber π_x.

Exercise. Give an algebraic definition of subbundles.

Examples. I. The tautological bundle over the Grassmannian (Example VI from 11.11) is a subbundle of the trivial bundle $\mathbb{R}^n \times G_{n,k} \to G_{n,k}$.

II. The bundle of unit tangent vectors of the sphere (Example IV from 11.13), which is a subbundle of the tangent bundle of the sphere, has no proper subbundles.

11.14. Whitney sum. Given two bundles η and ζ over one and the same manifold M, one can construct a new bundle π whose fiber over an arbitrary point $x \in M$ is the Cartesian product of the fibers of η and ζ:

$$\pi_x = \eta_x \times \zeta_x.$$

This bundle π is called the *direct sum*, or *Whitney sum*, or *Whitney product*, of the bundles η and ζ and denoted by $\eta \oplus \zeta$.

To give this construction an exact meaning, we must explain how the individual fibers are put together to make a smooth manifold. As always, there are two ways to do this.

The *algebraic* definition of the Whitney sum reads as follows: If the two given bundles correspond to algebra extensions $i : A \hookrightarrow B$ and $j : A \hookrightarrow C$, then their Whitney sum is represented by the homomorphism

$$i \otimes j : A \hookrightarrow \overline{B \otimes_A C}$$

that takes every element a to $a(1 \otimes 1) = i(a) \otimes 1 + 1 \otimes j(a)$.

Note that the tensor product of the algebras B and C is taken over the algebra A, not over the ground ring. This reflects the fact that it is the fibers that get multiplied in this construction, not the total spaces of the bundles.

The *geometric* construction of the Whitney sum consists in the following. The total space of the bundle $\eta \oplus \zeta$ is defined as

$$E_{\eta \oplus \zeta} = \{(y, z) \in E_\eta \times E_\zeta \mid \eta(y) = \zeta(z)\},$$

and the projection as the map that takes the pair (y, z) to the point $\eta(y) = \zeta(z)$.

Exercise. Check that these definitions of the Whitney sum are equivalent and the fiber of the resulting bundle over an arbitrary point x is in a natural bijection with the manifold $\eta_x \times \zeta_x$.

There is one more useful description of Whitney sum. The map

$$\eta \times \zeta \colon E_\eta \times E_\zeta \to M \times M, \quad (e_1, e_2) \mapsto (\eta(e_1), \zeta(e_2)),$$

is a bundle with fiber $\eta_u \times \zeta_v$ over the point (u, v), where $u, v \in M$. The diagonal $M_\Delta = \{(z, z) \mid z \in M\} \subset M \times M$ is a submanifold in $M \times M$, identified with M via the map $z \mapsto (z, z)$. The total space $E_{\eta \oplus \zeta}$ and the projection $\eta \oplus \zeta$ are identified with the manifold $(\eta \times \zeta)^{-1}(M_\Delta)$ and the map $\eta \times \zeta|_{(\eta \times \zeta)^{-1}(M_\Delta)}$, respectively. The restrictions of

$$p_\eta \colon E_\eta \times E_\zeta \to E_\eta \quad \text{and} \quad p_\zeta \colon E_\eta \times E_\zeta \to E_\zeta$$

to the submanifold $(\eta \times \zeta)^{-1}(M_\Delta)$ give rise to smooth surjective maps

$$p_\eta \colon E_{\eta \oplus \zeta} \to E_\eta \quad \text{and} \quad p_\zeta \colon E_{\eta \oplus \zeta} \to E_\zeta.$$

If $s \in \Gamma(\eta \oplus \zeta)$, then $s_\eta \overset{\text{def}}{=} p_\eta \circ s \in \Gamma(\eta)$, $s_\zeta \overset{\text{def}}{=} p_\eta \circ s \in \Gamma(\zeta)$, and $s(z) = (s_\eta(z), s_\zeta(z)) \in \eta_z \times \zeta_z$. This establishes a natural bijection between the sets of sections

$$\Gamma(\eta \oplus \zeta) = \Gamma(\eta) \times \Gamma(\zeta). \tag{11.6}$$

11.15. Examples of direct sums. I. Denote by $\alpha \colon S_1^1 \cup S_2^1 \to S^1$, where S_i^1 are copies of S^1, $i = 1, 2$, and by $\beta \colon S^1 \to S^1$ the trivial and the nontrivial two-sheeted coverings of the circle, respectively. Let A be the algebra of smooth functions on the circle, i.e., the algebra of smooth periodic functions on the line. In algebraic terms, the map α is described by the injection $i\colon A \to A \oplus A$, $f \mapsto (f, f)$, while β corresponds to the map $j\colon A \to A$ taking $f(x)$ to $f(2x)$. Then

    1. $\alpha \oplus \alpha$ is a trivial four-sheeted covering of the circle;

    2. $\alpha \oplus \beta \cong \beta \oplus \beta$ is a four-sheeted covering of the circle, whose total space consists of two connected components, each of which represents a nontrivial two-sheeted covering.

To understand this fact geometrically, it is sufficient to sketch the behavior of the four points of the fiber after one complete turn of the base circle.

Exercise. Prove these facts algebraically, considering the tensor products of algebras A_1 and A_2 over A, where $A_1 = A_2 = A = C^\infty(S^1)$ and A_1, A_2 are equipped with the A-module structure induced by the inclusions i and j.

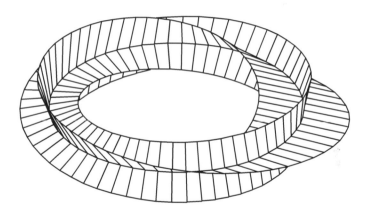

Figure 11.4. Direct sum of two Möbius bands.

II. The direct sum of two Möbius bands, considered as bundles over the circle, is trivial. A visual proof of this fact is shown in Figure 11.4. Let us represent the Möbius bundle as a subbundle of the trivial bundle over the circle with fiber $\mathbb{R}^2$. Then the lines perpendicular to its fibers constitute another Möbius bundle.

III. The bundle of 1-jets (see Section 9.28) is the direct sum of the trivial 1-dimensional bundle $M \times \mathbb{R} \to M$ and the cotangent bundle π_{T^*} (see Section 9.24).

11.16. Pullback bundle. Given a bundle $\pi\colon E_\pi \to M$ and a smooth map $f\colon N \to M$, we can attach a copy of the fiber $\pi_{f(y)}$ to every point $y \in N$. The union of all these fibers constitutes the total space of the *bundle*, $f^*(\pi)$, *induced from π by means of the map f or the pullback of π by f.*

There are two ways to turn this intuitive picture into a precise definition. Geometrically, the total space of the pullback bundle is defined as

$$E_{f^*(\pi)} \stackrel{\text{def}}{=} \{(y,z) \mid y \in N, z \in E_\pi, \pi(z) = f(y)\}.$$

The projection $f^*(\pi)$ acts as follows: $f^*(\pi)(y,z) = y$. Let us check its local triviality.

For a point $b \in N$, we set $a = f(b)$ and choose a neighborhood U of the point a in M such that π is trivial over U. Let $\psi\colon \pi^{-1}(U) \to U \times F$ be the trivializing diffeomorphism and let $\overline{\psi}$ be its composition with the projection $U \times F \to F$. Set $\chi(y,z) = (y, \overline{\psi}(z))$. Then

$$\chi\colon (f^*(\pi))^{-1}\left(f^{-1}(U)\right) \to f^{-1}(U) \times F$$

is the required diffeomorphism.

The *restriction of the bundle π to a submanifold* $N \subset M$ is a particular case of pullback bundle. It is defined as follows:

$$\pi\big|_N \stackrel{\text{def}}{=} \pi\big|_{\pi^{-1}(N)} \colon \pi^{-1}(N) \to N. \tag{11.7}$$

Exercise. Check that $\pi\big|_N = i^*(\pi)$, where $i\colon N \hookrightarrow M$ is the inclusion map.

The algebraic definition of the pullback bundle can be stated as follows. Let $i\colon A \hookrightarrow B$ be a bundle, understood as an algebra extension, and let $\varphi\colon A \to A_1$ be the algebra homomorphism corresponding to the smooth map $|\varphi|\colon |A_1| \to |A|$. Consider the algebras A_1 and B as A-modules with multiplication defined via i and φ. Then the *pullback bundle* $|\varphi|^*(i)$ is the natural homomorphism $A_1 \to \overline{A_1 \otimes_A B}$.

Remark. The commutative diagram

$$
\begin{array}{ccc}
A & \xrightarrow{\;\varphi\;} & A_1 \\
{\scriptstyle i}\big\downarrow & & \big\downarrow{\scriptstyle |\varphi|^*(i)} \\
B & \longrightarrow & \overline{A_1 \otimes_A B}
\end{array}
$$

shows that the notion of pullback bundle is a generalization of the Whitney sum.

Exercises. 1. Prove the equivalence of the geometrical and the algebraic definitions of the pullback bundle.

2. Show that a vector field along a map of manifolds $\varphi\colon N \to M$ (see Section 9.47) can be interpreted as a section of the induced bundle $\varphi^*(\pi_{TM})$ in the same way as an ordinary vector field is interpreted as a section of the tangent bundle (see Section 9.40).

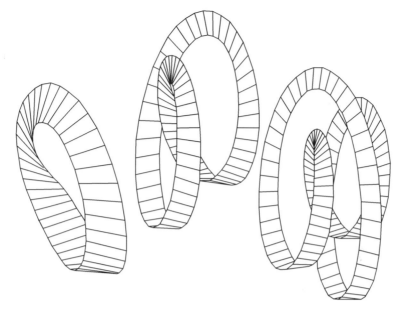

Figure 11.5. Bundles $f_n^*(\mu)$ for $n = 1,\ 2,\ 3$.

11.17. Examples. I. Let μ be the Möbius band bundle over the circle (11.11, III) and $f_n \colon S^1 \to S^1$ the n-sheeted covering of the circle (representing $S^1 = \{z \in \mathbb{C} \mid |z| = 1\}$, one can set $f_n(z) = z^n$, $n \in \mathbb{Z}$). Then

$$
f_n^*(\mu) = \begin{cases} \mu, & n \text{ odd,} \\ \mathbb{I}_{S^1}, & n \text{ even.} \end{cases}
$$

(See Figure 11.5; here and below, $\mathbb{I}_M$ denotes the trivial bundle over M with fiber $\mathbb{R}$.)

II. **Triviality criterion in terms of pullback bundles.** *A bundle is trivial if and only if it is equivalent to the pullback bundle obtained from a bundle over one point.*

Exercise. Prove this fact.

11.18. Define the *canonical morphism* $\varkappa \colon E_{f^*(\pi)} \to E_\pi$ by $\varkappa(y, z) = z$. The map $\varkappa$ is included into the commutative diagram

$$
\begin{array}{ccc}
E_{f^*(\pi)} & \xrightarrow{\ \varkappa\ } & E_\pi \\[2pt]
{\scriptstyle f^*(\pi)}\big\downarrow & & \big\downarrow{\scriptstyle \pi} \\[2pt]
N & \xrightarrow[\ f\ ]{} & M
\end{array}
$$

and therefore is an *f-morphism* from the bundle $f^*(\pi)$ to the bundle π.

More generally, given two bundles π over M and η over N and a smooth map $f \colon N \to M$, then an *f-morphism*, or a *morphism over f*, from η

into π is a smooth map $\psi\colon E_\eta \to E_\pi$ that makes the following diagram commutative:

$$
\begin{array}{ccc}
E_\eta & \xrightarrow{\ \psi\ } & E_\pi \\
\eta \downarrow & & \downarrow \pi \\
N & \xrightarrow[f]{} & M.
\end{array}
$$

The notion of f-morphism generalizes the notion of a morphism of bundles over M: the latter is nothing but a morphism over the map id_M.

Example. The map $T\Phi\colon TM \to TN$, arising from a map $\Phi\colon M \to N$ (Section 9.18), is a Φ-morphism from π_{TM} into π_{TN}.

The pair $(f^*(\pi), \eta)$ has the following universal property: for any bundle η over M and any f-morphism $\psi\colon \eta \to \pi$ there exists a unique smooth map χ that makes commutative the diagram

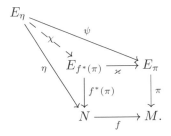

The proof is easy. For an arbitrary $y \in E_\eta$, both projections of $\chi(y)$ onto N and E_π are uniquely defined due to the commutativity of the diagram. This implies the uniqueness of χ. Existence follows from the explicit formula $\chi(y) = (\eta(y), \psi(y))$.

There is a natural map $\widehat{f}\colon \Gamma(\pi) \to \Gamma(f^*(\pi))$ called the *lifting of sections*. By definition, for any $s \in \Gamma(\pi)$ the value of the section $\widehat{f}(s)$ at the point $y \in N$ is equal to the value of s at $f(y)$. The precise formula is

$$
\widehat{f}(s)(y) = (y, s(f(y))) \in E_{f^*(\pi)}.
$$

The section $\widehat{f}(s)$ is called the *lift* of s along f.

11.19. Regular morphisms. Working with manifolds whose points have one and the same inner structure, it is natural to introduce the class of morphisms that preserve this structure. More specifically, an f-morphism ψ is said to be *regular* if for any point $z \in N$ the map of fibers $\psi_z\colon \eta_z \to \pi_{f(z)}$ is a diffeomorphism.

Proposition. *Let $\psi\colon \eta \to \pi$ be a regular morphism of bundles over the map $f\colon N \to M$. Then the canonical morphism $\chi\colon \eta \to f^*(\pi)$ defined in the previous section is an equivalence of bundles over N.*

◄ For any $z \in N$, the fiber map $\chi_z \colon \eta_z \to f^*(\pi)_z$ is identified with the map $\psi \colon \eta_z \to \pi_{f(z)}$ by the construction of χ, and therefore is a diffeomorphism. As a consequence, the map $\chi \colon E_\eta \to E_{f^*(\pi)}$ is a diffeomorphism, too. ►

The proposition shows that the class of bundles related to a given bundle π by means of regular morphisms is exhausted by the bundles pulled back from π. This observation leads to the tempting idea of building, for a given type of fibers, a *universal bundle* such that any bundle with this fiber could be the pullback of the universal bundle via a suitable smooth map. After an appropriate concretization, this idea can be implemented. Here is an example.

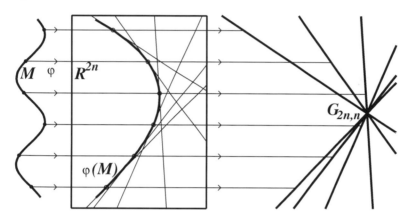

Figure 11.6. The Gauss map.

11.20. Example (Gauss map). Let M be an n-dimensional manifold. By the Whitney immersion theorem, M can be immersed into $\mathbb{R}^{2n}$, i.e., there exists a map $\varphi \colon M \to \mathbb{R}^{2n}$ such that for any point $z \in M$ the differential $d_z\varphi \colon T_z M \to T_{\varphi(z)}\mathbb{R}^{2n}$ is injective. Denote by r_a, $a \in \mathbb{R}^{2n}$, the linear shift in $\mathbb{R}^{2n}$ through the vector $-a$:

$$\mathbb{R}^{2n} \ni v \mapsto v - a \in \mathbb{R}^{2n}.$$

Let $G_{2n,n}$ be the Grassmann manifold of n-dimensional linear subspaces in $T_O\mathbb{R}^{2n}$, where $O = (0, \ldots, 0) \in \mathbb{R}^{2n}$. The *Gauss map* $g \colon M \to G_{2n,n}$ takes every point $z \in M$ to the image of the corresponding tangent space $T_z M$ under the map

$$d_z(r_{\varphi(z)} \circ \varphi) \colon T_z M \to T_O\mathbb{R}^{2n}$$

(see Figure 11.6). The map g is covered by the morphism of bundles $\gamma \colon \pi_{TM} \to \Theta_{2n,n}$, where $\Theta_{2n,n} \colon E_{2n,n} \to G_{2n,n}$ is the tautological bundle described in Example VI of Section 11.11. Indeed, if $\xi \in T_z M$, then

$$\gamma(\xi) = \big(d_z(r_{\varphi(z)} \circ \varphi)(\xi), g(z)\big) \in E_{2n,n}.$$

Therefore, by Proposition 11.19, the tangent bundles of all n-dimensional manifolds can be obtained as pullbacks from the tautological bundle over the Grassmannian $G_{2n,n}$. This fact plays an important role in the study of manifolds. In particular, it lies at the foundation of the theory of characteristic classes.

12

Vector Bundles and Projective Modules

12.1. We have seen in the previous chapter that for a given bundle, the fiber over a point of the base space describes the inner structure of this point. The fiber may have a certain mathematical structure. For example, the fibers of the tangent bundle have the natural structure of a linear space, and this structure has an evident physical meaning. Indeed, if the manifold M is the configuration space of a mechanical system (see Section 9.22), then for a fixed point $a \in M$, the tangent vector is interpreted as the velocity vector of the system having configuration a. Bundles of this kind, where fibers are vector spaces, are called *vector bundles* (see Section 12.2 for an exact definition). They form an interesting and important class of bundles. In particular, besides tangent bundles, this class contains cotangent bundles and jet bundles, which are fundamental objects of study in the geometrical theory of differential equations.

Vector bundles have a simpler algebraic description than bundles of general type. It turns out that under certain natural regularity conditions the extensions of algebras $A \hookrightarrow B$ that correspond to vector bundles are of the form $A \hookrightarrow S(P)$, where $S(P)$ is the symmetric algebra of a certain A-module P, appropriately completed. The study of vector bundles over a manifold M is thus reduced to the study of a certain class of modules over the algebra $A = C^\infty(M)$.

We begin with the geometrical definition of a vector bundle. After an investigation of the basic properties of vector bundles, we shall prove the fundamental theorem on the equivalence of the notions of a vector bundle over M and a finitely generated projective module over A, and then explain how symmetric algebras of modules appear in this context.

© Springer Nature Switzerland AG 2020
J. Nestruev, *Smooth Manifolds and Observables*, Graduate Texts
in Mathematics 220, https://doi.org/10.1007/978-3-030-45650-4_12

12.2. Geometrical definition of a vector bundle. A fiber bundle $\pi\colon E \to M$ with fiber V is said to be a *vector bundle* if

(1) V is a vector space (over $\mathbb{R}$);

(2) For any point $x \in M$, the fiber π_x is a vector space;

(3) The vector property of local triviality holds: for any point $x \in M$ there exists a neighborhood $U \subset M$, $x \in U$, and a trivializing diffeomorphism $\varphi\colon \pi^{-1}(U) \to U \times V$, linear on every fiber, i.e., such that all maps $\varphi_y\colon \pi_y \to V$, $y \in U$, are linear.

The *dimension* or *rank* of a vector bundle is the dimension of its fiber. Zero-dimensional vector bundles are called *zero bundles* and denoted by $\mathbb{O}_M$. Trivial one-dimensional bundles are called *unit bundles* and denoted by $\mathbb{I}_M$.

12.3. Adapted coordinates in vector bundles. Suppose (U, x) is a chart on M satisfying Condition (3) of Definition 12.2, while

$$\varphi\colon \pi^{-1}(U) \to U \times V$$

is the corresponding trivializing diffeomorphism, and $\xi \in V$. The map

$$s_\xi^\varphi \colon U \to \pi^{-1}(U), \quad U \ni x \mapsto \varphi^{-1}(x, \xi) \in \pi^{-1}(U),$$

is a section of the fibration $\pi|_U$. If $v_1, \ldots, v_m$ is a basis of the vector space V, then the sections $e_j = s_{v_j}^\varphi$, $1 \leqslant j \leqslant m$, have the property that at every point $z \in U$ the vectors

$$e_1(z) = \varphi^{-1}(z, v_1), \ldots, e_m(z) = \varphi^{-1}(z, v_m)$$

form a basis of the space $\pi_U^{-1}(z) \cong V$. Below (see Section 12.7), we shall show that the totality of all sections of a vector bundle π over M has a natural $C^\infty(M)$-module structure induced by the linear structure in the fibers π_z. In this sense we can say that in the chosen coordinate chart U the module of sections of the bundle $\pi|_U$ is free and $e_j = s_{v_j}^\varphi$, $1 \leqslant j \leqslant m$, is its free basis.

Now suppose that $z \in U$ and $(x_1, \ldots, x_n)$ are coordinate functions on U. A point $y \in \pi_z \subset \pi^{-1}(U)$ is defined by the set of $n + m$ numbers $(x_1, \ldots, x_n, u^1, \ldots, u^m)$, where $(x_1, \ldots, x_n)$ are coordinates of the point z and $(u^1, \ldots, u^m)$ are coordinates of the point y with respect to the basis $e_1|_z, \ldots, e_m|_z$. The functions $(x_1, \ldots, x_n, u^1, \ldots, u^m)$ form a coordinate system on $\pi^{-1}(U)$, called the *adapted* coordinates of the bundle π.

Accordingly, the chart $(\pi^{-1}(U), x_1, \ldots, x_n, u^1, \ldots, u^m)$ on the manifold E is referred to as an *adapted* chart. Finally, an atlas made up of adapted charts is also called adapted. It is readily verified that if two charts (U, x) and (U', x') on M are compatible, then the corresponding adapted charts on E are compatible, too. (The reader is invited to check this fact as an exercise.) In particular, this means that every atlas on the manifold M gives rise to an adapted atlas on the total space E.

12.4. Morphisms of vector bundles. A *morphism of vector bundles* $\alpha\colon \pi \to \eta$ *over* M is a bundle morphism $\alpha\colon E_\pi \to E_\eta$, which is fiberwise linear (i.e., the map α_z, the restriction of α to the fiber over z, is $\mathbb{R}$-linear for any point $z \in M$ of the base space). The set of all morphisms from π to η is denoted by $\mathrm{Mor}(\pi, \eta)$, and the category of vector bundles thus arising will be denoted by VB_M.

For the local study of morphisms, the following point of view is convenient: a morphism of trivial vector bundles over M is the same thing as an operator-valued function on the manifold M. An exact statement of this observation is contained in the obvious lemma that follows.

12.5 Lemma. *Let* $\pi\colon M \times V \to M$ *and* $\eta\colon M \times W \to M$ *be trivial vector bundles. To every fiberwise linear map* $\varphi\colon M \times V \to M \times W$ *one can assign a family of linear operators* $\widetilde{\varphi}\colon M \to \mathrm{Hom}(V, W)$ *by setting the value of the operator* $\widetilde{\varphi}(x)$ *on the vector* $v \in V$ *equal to the* W-*component of the element* $\varphi(x, v) \in M \times W$. *Then the following conditions are equivalent:*

(a) *the map* φ *is smooth (i.e.,* φ *is a bundle morphism);*

(b) *the map* $\widetilde{\varphi}$ *is smooth (the space* $\mathrm{Hom}(V, W)$ *is endowed with the structure of a smooth manifold, because it is a finite-dimensional real vector space).*

12.6. Examples of vector bundles. I. The Möbius band fibered over the circle (Example 11.11, III) can be viewed as a vector bundle if its fibers are regarded as one-dimensional linear spaces. Representing the algebra of functions on the Möbius band as the subalgebra $B \subset C^\infty(\mathbb{R}^2)$ distinguished by the condition $f(x + 1, y) = f(x, -y)$, we can define this bundle by the inclusion of algebras $A \hookrightarrow B$ that takes a function f to the function g, $g(x, y) = f(x)$; here $A = \{f \in C^\infty(\mathbb{R}) \mid f(x+1) = f(x)\}$ is the algebra of functions on the circle.

II. The tangent bundle $\pi_T\colon TM \to M$ (see Section 9.19). The trivializing diffeomorphisms described in Section 11.11, VII, are fiberwise linear. Therefore, the tangent bundle is a vector bundle.

III. The cotangent bundle $\pi_{T^*}\colon T^*M \to M$ (see Section 9.24). As in the previous example, the trivializing diffeomorphisms of Section 11.11, VIII, are obviously linear.

IV. The trivializations described in Section 11.11, IX, are also fiberwise linear. Therefore, the bundle of l-jets $\pi_{J^l}\colon J^l M \to M$ is a vector bundle.

Note that in the last three examples, the special coordinate systems defined in 9.19, 9.24, and 11.11, IX, respectively, are adapted.

Exercise. Check whether the maps (see 9.28)
$$\pi_{l,m}\colon J^l M \to J^m M, \quad [f]_z^l \mapsto [f]_z^m, \quad l \geqslant m,$$
$$\tau_l\colon J^l M \to T^* M, \quad [f]_z^l \mapsto d_z(f), \quad l \geqslant 1,$$
are vector bundles.

12.7. Module of sections. A remarkable property of vector bundles is that their sets of sections possess a module structure over the algebra of smooth functions. The resulting interrelation between the bundles and modules is of fundamental importance.

Note first of all that the set of sections of any vector bundle is nonempty: It always contains the *zero section* s_0. By definition, the value of s_0 at any point $z \in M$ is the zero of the vector space π_z.

Using the linear structure in the fibers π_x, one can introduce two operations in the set of sections of a vector bundle: addition and multiplication by a function on the manifold,

$$(s_1 + s_2)(z) = s_1(z) + s_2(z), \qquad (fs)(z) = f(z)s(z),$$

for any sections s, s_1, s_2, any smooth function $f \in C^\infty(M)$, and any point $z \in M$. The definition immediately implies that the sum of two sections and the product of a section and a smooth function are again (smooth) sections. These operations turn the totality of all smooth sections of a vector bundle π into a $C^\infty(M)$-module, called the *module of sections* and denoted by $\Gamma(\pi)$.

The next lemma clarifies the relationship between the global and the pointwise approaches to the sections of a vector bundle. As before, we denote by μ_z the maximal ideal of the algebra $C^\infty(M)$, defined by

$$\mu_z = \{f \in C^\infty(M) \,|\, f(z) = 0\}$$

and called the *ideal of the point z*.

12.8 Lemma. *Let π be a vector bundle over a manifold M and $z \in M$. Then*

(a) *for any point $y \in \pi_z$ there is a section $s \in \Gamma(\pi)$ such that $s(z) = y$;*

(b) *if $s \in \Gamma(\pi)$ and $s(z) = 0$, then there exist functions $f_i \in \mu_z$ and sections $s_i \in \Gamma(\pi)$ such that s can be written as a finite sum $s = \sum f_i s_i$.*

◀ (a) For a trivial bundle, the assertion evidently holds. Therefore, by local triviality, there is a neighborhood U of z and a section $s\big|_U \in \Gamma(\pi\big|_U)$ satisfying $s\big|_U(z) = y$. In order to obtain a global (i.e., defined over all M) section of the bundle π possessing the same property, it remains to multiply $s\big|_U$ by a smooth function whose support is contained in U and that has value 1 at the point z (see Chapter 2).

(b) First suppose that the bundle is trivial. In this case (see Section 12.3) there are sections $e_1, \dots, e_m \in \Gamma(\pi)$ whose values at every point $z \in M$ form a basis of the linear space π_z. A given section s can be expanded over the basis:

$$s = \sum_{i=1}^{m} f_i e_i.$$

The equality $s(z) = 0$ implies that $f_i(z) = 0$ for all $i = 1, \ldots, m$; therefore, $f_i \in \mu_z$, as required.

In the case of an arbitrary bundle, there is a neighborhood U of the point z, sections $e_i \in \Gamma\left(\pi|_U\right)$, and functions $f_i \in C^\infty(U)$ satisfying the equality $s|_U = \sum_{i=1}^m f_i e_i$. Choose a function $f \in C^\infty(M)$ such that $\operatorname{supp} f \subset U$ and $f(z) = 1$. Extending the function $f f_i \in C^\infty(U)$ (respectively, the section $f e_i \in \Gamma\left(\pi|_U\right)$) as the identical zero function outside of U, we obtain a smooth function (respectively, section) on the entire manifold M. Keeping the notation $f f_i$ and $f e_i$ for the extensions, we can write

$$f^2 s = \sum_{i=1}^m (f f_i)(f e_i),$$

and therefore

$$s = \left(1 - f^2\right) s + \sum_{i=1}^m (f f_i)(f e_i).$$

It remains to note that the functions $1 - f^2$, $f f_1$, $\ldots$, $f f_m$ belong to μ_z. ►

The lemma just proved above has a compact reformulation in terms of exact sequences. Recall that a sequence of A-modules

$$\cdots \to P_{i-1} \xrightarrow{\alpha_{i-1}} P_i \xrightarrow{\alpha_i} P_{i+1} \to \cdots$$

is said to be *exact at the term* P_i if $\operatorname{Ker} \alpha_i = \Im \alpha_{i-1}$. The sequence is called *exact* if it is exact at every term.

12.9 Corollary. *For any vector bundle* π, *the sequence*

$$0 \to \mu_z \Gamma(\pi) \to \Gamma(\pi) \to \pi_z \to 0,$$

where the first arrow is the inclusion, while the second assigns to every section its value at point $z \in M$, *is exact. Hence* $\Gamma(\pi)/\mu_z \Gamma(\pi) \cong \pi_z$. ►

To every element $h \in |A|$ we can assign the ideal $\mu_h = \operatorname{Ker} h \subset A$. The above result justifies the following definition: The *fiber* P_h *of an* A-*module* P over a point $h \in |A|$ is the quotient module $P/\mu_h P$. The *value* p_h of an element $p \in P$ at the point h is the image of p under the natural projection $P \to P_h$. For the case in which $A = C^\infty(M)$ and $h = h_z$ for $z \in M$, we can write $P_z \overset{\text{def}}{=} P/\mu_z P$ in the same sense and speak of the value p_z of an element $p \in P$ at the point z.

Exercise. Show that for the module of vector fields $P = D(M)$ over the algebra of smooth functions $A = C^\infty(M)$, we have $D(M)_z = T_z M$, and the value of an element $X \in D(M)$ at the point z is just the vector of the field X at this point (see 9.39). (In other words, the notation X_z in both cases has the same meaning.)

12.10. For the analysis of modules of sections, the following fact is important:

Proposition. *Suppose that the sections $s_1, \ldots, s_l \in \Gamma(\pi)$ have the property that for every point $z \in M$ the vectors $s_1(z), \ldots, s_l(z)$ span the fiber π_z. Then these sections generate the module $\Gamma(\pi)$.*

◀ Let k be the dimension of the bundle π. For an ordered set of integers $I = (i_1, \ldots, i_k)$, $1 \leqslant i_1 < \cdots < i_k \leqslant l$, put

$$U_I = \{z \in M \mid s_{i_1}(z), \ldots, s_{i_k}(z) \in \pi_z \text{ are linearly independent}\}.$$

Evidently, the set U_I is open, and the sections $s_{i_1}|_{U_I}, \ldots, s_{i_k}|_{U_I}$ generate the $C^\infty(U_I)$-module $\Gamma(\pi|_{U_I})$. Moreover, $\bigcup_I U_I = M$. Indeed, for any point $z \in M$, we can choose a basis $s_{i_1}(z), \ldots, s_{i_k}(z)$ among the vectors $s_1(z), \ldots, s_l(z)$ that span the fiber π_z. This means that $z \in U_I$.

For a section $s \in \Gamma(\pi)$, we have

$$s|_{U_I} = \sum_{\alpha=1}^{k} \lambda_{I,\alpha} s_{i_\alpha}|_{U_I}, \qquad \lambda_{I,\alpha} \in C^\infty(U_I).$$

Now let $\mu_I \in C^\infty(M)$ be a function which is strictly positive inside U_I, vanishes outside of this set, and has the property that the functions

$$\nu_{I,i} = \begin{cases} \mu_I \lambda_{I,i} & \text{inside } U_I, \\ 0 & \text{outside } U_I, \end{cases}$$

are smooth on M. Then the function $\mu = \sum_I \mu_I$ is everywhere positive on M and

$$\mu_I s = \sum_i \nu_{I,i} s_i.$$

Therefore,

$$s = \frac{1}{\mu} \sum_I \mu_I s = \sum_{I,i} \frac{\nu_{I,i}}{\mu} s_i. \qquad ▶$$

12.11. Geometrization of modules. With every A-module P over a commutative K-algebra A we can associate a geometrical object

$$|P| = \bigcup_{h \in |A|} P_h \quad (\text{or } |P| = \bigcup_{z \in M} P_z, \text{ if } A = C^\infty(M)),$$

together with a natural projection onto $|A|$:

$$|P| \supset P_h \ni p_h \overset{\pi_P}{\mapsto} h \in |A|.$$

The A-module $P_h = P/\mu_h P$ can be also viewed as a module over A/μ_h, hence, by virtue of the isomorphism $A/\mu_h = K$, as a K-module.

The projection π_P looks very much like a bundle and, as we shall show below, is equivalent to a vector bundle if $A = C^\infty(M)$ and the module P

is projective and finitely generated. We shall refer to such projections as
pseudo-bundles.

In the case $P = D(M)$, every element $X \in D(M)$ corresponds to a
section $s_X \colon z \mapsto X_z \in T_z M = D(M)_z$ of the tangent bundle π_T. This
construction is of general nature and can be used for arbitrary pseudo-
bundles by assigning the map

$$s_p \colon |A| \to |P|, \quad h \mapsto p_h,$$

to an element $p \in P$. This allows us to visualize the elements of an arbitrary
module P as sections of the pseudo-bundles $|P|$ much in the same way as the
elements of an arbitrary algebra A were viewed as functions on its spectrum
$|A|$. One of the main goals of the present chapter is to show that vector
bundles are obtained from projective modules just as smooth manifolds are
obtained from smooth algebras.

Maps $s_p \colon |A| \to |P|$ are referred to as *sections* of the pseudo-bundles
π_P. (There is no other way to distinguish a reasonable class among all
maps $s \colon |A| \to |P|$ such that $\pi_P \circ s = \mathrm{id}_{|A|}$.) The set $\Gamma(P)$ of all sections
of the pseudo-bundles π_P forms an A-module with respect to the natural
operations

$$(s_{p_1} + s_{p_2}) = s_{p_1 + p_2}, \quad p_1, p_2 \in P,$$
$$(as_p) = s_{ap}, \quad a \in A, \quad p \in P.$$

To every A-module P we thus assign the A-module $\Gamma(P)$ of sections of
the pseudo-bundles π_P. Our aim can now be stated more precisely: We
want to show that for $A = C^\infty(M)$ and every projective finitely generated
A-module P, the pseudo-bundles π_P is a vector bundle and the two modules
P and $\Gamma(P)$ are naturally isomorphic.

Exercise. Prove that the assignment $P \mapsto \Gamma(P)$ is a functor in the
category of A-modules.

If P is a $C^\infty(M)$-module and its element $p \in P$ belongs to the intersection $\bigcap_{z \in M} \mu_z P$, then the value of p at every point $z \in M$ is zero. Such
elements can be called *invisible*, or *unobservable*. Indeed, by the principle
of observability, the class $p \bmod \mu_z P$ should be viewed as a certain component of the inner structure of the point $z \in M$, and the fact that p belongs
to all subspaces $\mu_z P$ means that this component is unobservable.

A $C^\infty(M)$-module P is said to be *geometrical*, if $\bigcap_{z \in M} \mu_z P = 0$, i.e., if
all elements of P are observable.

Exercise. Prove that P is geometrical if and only if the two modules P
and $\Gamma(P)$ are isomorphic.

The algebraic paraphrase of the above discussion is as follows. The map

$$\Gamma = \Gamma_P \colon P \to P \Big/ \bigcap_{z \in M} \mu_z P = \Gamma(P)$$

kills all unobservable elements of P. Therefore, the quotient module $\Gamma(P)$ defined in this way can be called the *geometrization* of P. The assignment $P \mapsto \Gamma(P)$ defines a functor from the category $\operatorname{Mod} C^\infty(M)$ of all $C^\infty(M)$-modules to the category $\operatorname{GMod} C^\infty(M)$ of geometrical $C^\infty(M)$-modules. In some situations, it is sufficient to use the smaller category $\operatorname{GMod} C^\infty(M)$ instead of the bigger category $\operatorname{Mod} C^\infty(M)$.

Exercise. Show that the subcategory $\operatorname{GMod} C^\infty(M) \subset \operatorname{Mod} C^\infty(M)$ is stable under the operations $\otimes$ and Hom: if P and Q are both geometrical $C^\infty(M)$-modules, then the modules $P \otimes Q$ and $\operatorname{Hom}_{C^\infty(M)}(P, Q)$ are also geometrical.

The behavior of the fiber P_h when the point $h \in |A|$ varies provides important information about the module P. For example, one can speak of the *support of a module*,

$$\operatorname{supp} P = \overline{\{h \in |A| \mid P_h \neq 0\}} \subset |A|,$$

where the bar means closure in the Zariski topology.

Exercises.   1. The tangent space $T_z M$ at a point $z \in M$ can be considered as a $C^\infty(M)$-module with multiplication defined by the rule $(f, \xi) \mapsto f(z)\xi$, $f \in C^\infty(M)$, $\xi \in T_z M$. Show that the support of this module consists of one point, namely z.

   2. Prove that the support of the $C^\infty(M)$-module $D(M, N)$ (module of vector fields along a submanifold $N \subset M$, see Section 9.46) coincides with N.

The geometrization of A-modules helps to visualize and thus better understand various algebraic constructions. For example, the structure of an A-module homomorphism $f \colon P \to Q$ is displayed through the family of its values at different points of the spectrum of $|A|$. By the *value* of F at the point $h \in |A|$ we understand the map of quotient modules $F_h \colon P_h \to Q_h$, well defined because $F(\mu_h P) \subset \mu_h Q$. In the geometrical situation, when $A = C^\infty(M)$, we can use the notation F_z, P_z, Q_z instead of F_{h_z}, P_{h_z}, Q_{h_z}.

12.12. Topology in $|P|$. The set $|P|$ can be turned into a topological space by using an appropriate generalization of the ideas used in Chapter 9 to prove that the cotangent manifold T^*M is the $\mathbb{R}$-spectrum of the symbol algebra $\mathcal{S}_*$. In the situation under study, a natural candidate for the role of such an algebra is the symmetric algebra $\mathcal{S}(P^*)$ of the module $P^* = \operatorname{Hom}_A(P, A)$. By definition,

$$\mathcal{S}(P^*) = \bigoplus_{k \geqslant 0} \mathcal{S}_k(P^*),$$

where $\mathcal{S}_k(P^*)$ is the kth symmetric power of the module P^*. An element $f \in P^* = \mathcal{S}_1(P^*)$ can be viewed as a function on $|P|$ by setting

$$f(p_h) \overset{\text{def}}{=} f(p) \mod \mu_h \in A/\mu_h = K, \quad h \in |A|, \ p \in P.$$

Since f is a homomorphism, the value $f(p_h)$ does not depend on the choices made. For a generic element $f_1 \otimes \ldots \otimes f_k \in (P^*)^{\otimes k}$, where $(P^*)^{\otimes k}$ denotes the kth tensor power of the A-module P^*, we put

$$(f_1 \otimes \cdots \otimes f_k)(p_h) \stackrel{\text{def}}{=} f_1(p_h) \cdots f_k(p_h) \in K.$$

From this formula, elements of $(P^*)^{\otimes k}$ can be understood as symmetric functions on $|P|$. For an element $\omega \in (P^*)^{\otimes k}$ of the form

$$\omega = f_1 \otimes \cdots \otimes f_i \otimes \ldots \otimes f_j \otimes \cdots \otimes f_k - f_1 \otimes \cdots \otimes f_j \otimes \cdots \otimes f_i \otimes \cdots \otimes f_k$$

the corresponding function ω is identically zero. The quotient algebra of the complete tensor algebra

$$(P^*)^{\otimes} = \sum_{k \geqslant 0} (P^*)^{\otimes k}$$

over the ideal generated by such elements is the symmetric algebra $\mathcal{S}(P^*)$.

Below, we discuss this idea in detail for modules over the algebra of smooth functions $C^\infty(M)$ and show that for any vector bundle π, the two spaces $|\Gamma(\pi)|$ and E_π coincide.

Denote by $\mathcal{F}(|P|)$ the K-algebra of functions on $|P|$ that correspond to elements of the algebra $\mathcal{S}(P^*)$. With the help of this algebra, we can turn the set $|P|$ into a topological space with the Zariski topology, in which the basic closed sets are zero sets of functions belonging to $\mathcal{F}(|P|)$.

Exercise. Show that the maps $\pi_P \colon |P| \to |A|$ and $s_p \colon |A| \to |P|$, $p \in P$, are continuous in this topology.

Using the Zariski topology in $|P|$, one can widen the class of sections of the pseudo-bundles π_P. Namely, a *continuous section* of π_P is a continuous map $s \colon |A| \to |P|$ such that $\pi_P \circ s = \mathrm{id}_{|A|}$. The set of all continuous sections of π_P will be denoted by $\Gamma_0(P)$.

Exercise. Show that the structure of a K-linear space in each fiber $P_h \subset |P|$ induces the structure of an A-module in $\Gamma_0(P)$.

12.13. The functor of sections. The assignment $\pi \mapsto \Gamma(\pi)$ that associates the module of sections $\Gamma(\pi)$ to a given vector bundle π can be made into a *functor* from the category of A-modules to itself as follows. Let $\alpha \in \mathrm{Mor}(\pi, \eta)$. Put

$$\Gamma(\alpha)(s) = \alpha \circ s \quad \text{for any } s \in \Gamma(\pi).$$

Then $\Gamma(\alpha) \colon \Gamma(\pi) \to \Gamma(\eta)$ is a $C^\infty(M)$-module homomorphism, and the assignment $\alpha \mapsto \Gamma(\alpha)$ has the property of *functionality*:

(i) $\Gamma(\mathrm{id}_\pi) = \mathrm{id}_{\Gamma(\pi)}$ for any π,

(ii) $\Gamma(\alpha \circ \beta) = \Gamma(\alpha) \circ \Gamma(\beta)$ for any pair of morphisms $\zeta \xrightarrow{\beta} \pi \xrightarrow{\alpha} \eta$.

Exercise. Let $P = \Gamma(\pi)$, $Q = \Gamma(\eta)$, $\alpha \in \mathrm{Mor}(\pi, \eta)$, and $F = \Gamma(\alpha)$. Prove that the map $F_z \colon P_z \to Q_z$ (see Section 12.11) is canonically identified with $\alpha_z \colon \pi_z \to \eta_z$ under the identifications $P_z = \pi_z$, $Q_z = \eta_z$, described in Lemma 12.8.

The study of the functor Γ that relates the geometry of vector bundles with the algebra of rings and modules is the main point of the present chapter. This functor allows one to express the geometrical properties of vector bundles and operations with them in algebraic language. Here is a simple example.

Proposition. *A vector bundle π is trivial if and only if the module $\Gamma(\pi)$ is free.*

◀ Indeed, choose a trivializing diffeomorphism $\varphi \colon E_\pi \to M \times V$ and a basis $v_1, \ldots, v_n$ of the linear space V. Let $e_i(x) = \varphi^{-1}(x, v_i)$. Then the set $\{e_1, \ldots, e_m\}$ is a free basis of the module $\Gamma(\pi)$.

Conversely, suppose that the module $\Gamma(\pi)$ is free with a basis $e_1, \ldots, e_m$; then we can define a diffeomorphism $\varphi \colon E_\pi \to M \times \mathbb{R}^n$ by setting

$$\varphi\left(\sum_{i=1}^{m} \lambda_i e_i(x)\right) = (x; \lambda_1, \ldots, \lambda_m). \quad ▶$$

Remark. For any free $C^\infty(M)$-module P of *finite type*, i.e., with a finite set of generators, there exists a vector bundle whose module of sections is isomorphic to P. Indeed, a free $C^\infty(M)$-module of rank m is isomorphic to $\Gamma(\pi)$, where π is the product bundle $M \times \mathbb{R}^m \to M$.

12.14. Projective modules. It is natural to suppose that section modules of vector bundles must possess certain specific properties originating from the fact that all fibers of a given bundle are the same. These properties serve as a formalization of our doctrine that the inner structures of all points are identical (see Section 11.1). We shall see that an adequate description can be given by using the notion of projectivity.

A module P over a commutative ring A is said to be *projective* if it has the following property: for any epimorphism of A-modules $\varphi \colon Q \to R$ and any homomorphism $\psi \colon P \to R$ there is a homomorphism $\chi \colon P \to Q$ such that $\varphi \circ \chi = \psi$, i.e., the diagram

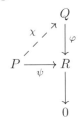

commutes. The homomorphism χ is called the *lift* of ψ along φ. Let us give several equivalent definitions of projectivity.

12.15 Proposition. *The following properties of an A-module P are equivalent*:

(a) *P is projective*;

(b) *any epimorphism $\varphi\colon Q \to P$ of an arbitrary A-module onto P splits, i.e., there exists a homomorphism $\chi\colon P \to Q$ such that $\varphi \circ \chi = \mathrm{id}_P$*;

(c) *P is isomorphic to a direct summand of a free A-module*;

(d) *the functor $\mathrm{Hom}_A(P, \cdot)\colon Q \;\mapsto\; \mathrm{Hom}_A(P, Q)$ on the category of A-modules is exact, i.e., it preserves the class of exact sequences*.

◀ (a) $\Longrightarrow$ (b). It suffices to set $R = P$ and $\psi = \mathrm{id}_P$ in the definition of projectivity.

(b) $\Longrightarrow$ (c). Let us consider a free module Q with free basis $\{e_p\}_{p \in P}$ equipotent to the set P and construct an epimorphism $\varphi\colon Q \to P$ by putting $\varphi(e_p) = p$ for every $p \in P$. By virtue of (b), there is a homomorphism $\chi \in \mathrm{Hom}(P, Q)$ such that $\chi \circ \varphi = \mathrm{id}_P$. Then $P \cong \Im \chi$ and $Q = \Im \chi \oplus \mathrm{Ker}\,\varphi$. Indeed, any element $a \in Q$ can be written as $\chi(\varphi(a)) + (a - \chi(\varphi(a)))$; here the first summand belongs to $\Im \chi$ and the second to $\mathrm{Ker}\,\varphi$. On the other hand, if $a \in \Im \chi \cap \mathrm{Ker}\,\varphi$, then $a = \chi(p)$, $p \in P$, and $0 = \varphi(a) = \varphi(\chi(p)) = p$. Therefore, $a = 0$.

(c) $\Longrightarrow$ (d). Note that for a free module R, the functor $\mathrm{Hom}_A(R, \cdot)$ is exact. This follows from the fact that a homomorphism of a free module R into another module is uniquely determined by its values on the basis elements, and these values can be arbitrary. Now suppose that $R = P \oplus Q$ and

$$S = \{\cdots \to S_k \overset{\varphi_k}{\to} S_{k+1} \to \cdots\}$$

is an exact sequence of A-modules. Then the sequence $\mathrm{Hom}_A(R, S)$,

$$\cdots \to \mathrm{Hom}_A(R, S_k) \to \mathrm{Hom}_A(R, S_{k+1}) \to \cdots,$$

is exact, too. This sequence decomposes into the direct sum of the sequence $\mathrm{Hom}_A(P, S)$ of the form

$$\cdots \to \mathrm{Hom}_A(P, S_k) \to \mathrm{Hom}_A(P, S_{k+1}) \to \cdots,$$

and the sequence $\mathrm{Hom}_A(Q, S)$ of the form

$$\cdots \to \mathrm{Hom}_A(Q, S_k) \to \mathrm{Hom}_A(Q, S_{k+1}) \to \cdots.$$

In other words, every term of the sequence $\mathrm{Hom}_A(R, S)$ is the direct sum of the corresponding terms of two sequences $\mathrm{Hom}_A(P, S)$ and $\mathrm{Hom}_A(Q, S)$, and every homomorphism of the sequence $\mathrm{Hom}_A(R, S)$ is the direct sum of the corresponding homomorphisms in $\mathrm{Hom}_A(P, S)$ and $\mathrm{Hom}_A(Q, S)$. It remains to apply the following simple observation: The direct sum of two sequences of modules is exact if and only if both summands are exact.

Finally, to prove the implication (d) $\implies$ (a) it suffices to apply the property (d) to the exact sequence

$$0 \to \operatorname{Ker}\varphi \to Q \xrightarrow{\varphi} R \to 0. \quad \blacktriangleright$$

Exercise. Suppose that $P \subset R$, where P is projective and R is free. Is it true that there exists a submodule $Q \subset R$ such that $R = P \oplus Q$?

12.16. Examples of projective modules. I. Over a field $\mathbb{F}$, all $\mathbb{F}$-modules are projective, because they are all free.

II. Over the ring of integers $\mathbb{Z}$, all projective modules are free (although in this case not all modules are free).

III. The simplest example of a module that is projective, but not free, is provided by the group $\mathbb{Z}$, considered as a module over the ring $\mathbb{Z} \oplus \mathbb{Z}$ with multiplication $(a,b) \cdot x = ax$.

IV. Modules of sections of vector bundles are projective. This will be proved later, in Theorem 12.32.

Exercise. Describe the projective modules over the ring of residues $\mathbb{Z}/m\mathbb{Z}$ and over the matrix ring.

12.17. Subbundles. We say that a vector bundle $\eta\colon E_\eta \to M$ is a *subbundle* of a vector bundle $\pi\colon E_\pi \to M$ (denoted by $\eta \subset \pi$) if

(i) the total space E_η is a submanifold of E_π;

(ii) the projection η is the restriction of π to E_η;

(iii) for any point $x \in M$ the fiber η_x is a linear subspace of the fiber π_x.

Examples. I. The zero subbundle: its total space coincides with the image of the zero section.

II. The tangent bundle of the 2-sphere does not contain one-dimensional subbundles. To prove this fact, suppose that such a subbundle ξ exists. Then for a smooth oriented closed curve $\Gamma \subset S^2$, we can define an integer invariant $\nu(\Gamma)$ equal to the number of half-turns made by the tangent vector to the curve with respect to the fibers of ξ. The number $\nu(\Gamma)$ does not change under smooth deformations of the curve Γ, and it changes its sign when the orientation of the curve changes. If Γ^+ is a small positively oriented circle and Γ^- the same circle with the negative orientation, then it is evident that $\nu(\Gamma^+) = 2$ and $\nu(\Gamma^-) = -2$. But on the sphere the curve Γ^+ can be smoothly deformed into Γ^-. This contradiction proves our assertion.

12.18. The local structure of subbundles. Suppose that at every point $z \in M$ a linear subspace η_z of the fiber π_z is given. This will define a subbundle of π if the following two properties hold:

(i) the set $\bigcup_{z\in M} \eta_z$ must be a submanifold of E_π;

(ii) the local triviality property must hold for the family $\{\eta_z\}$.

In the case of a trivial bundle π, these requirements can be stated in the form of the following simple lemma.

12.19 Lemma. *Let* $\pi\colon M \times \mathbb{R}^n \to M$ *be the projection on the first factor, and let a* k*-dimensional linear subspace* $\eta_z \subset \pi_z \cong \mathbb{R}^n$ *be given at every point* $z \in M$. *Denote by* $\widetilde{\eta}\colon M \to G_{n,k}$ *the map* $\widetilde{\eta}(z) = \eta_z$, *where the plane* η_z *is regarded as a point in the Grassmann manifold* $G_{n,k}$. *Then the following two conditions are equivalent:*

(a) *The family* $\{\eta_z\}_{z\in M}$ *defines a subbundle* $\eta \subset \pi$.

(b) *The map* $\widetilde{\eta}$ *is smooth.*

◀ (a) $\Longrightarrow$ (b). Pick a point $a \in M$ and a basis $e_1, \ldots, e_k$ of the space η_a. By Lemma 12.8, there exist sections $s_1, \ldots, s_k \in \Gamma(\eta)$ such that $s_i(a) = e_i$. Since the sections s_i are smooth, there is a neighborhood U of a such that the vectors $s_1(z), \ldots, s_k(z)$ are linearly independent for all $z \in U$. Therefore, these vectors form a basis of the space η_z. In the neighborhood U, the map $\widetilde{\eta}$ can be represented as the composition of two maps

$$U \ni z \mapsto \big(s_1(z), \ldots, s_k(z)\big) \mapsto \mathcal{L}\big(s_1(z), \ldots, s_k(z)\big) \in G_{n,k},$$

where $\mathcal{L}\big(s_1(z), \ldots, s_k(z)\big)$ stands for the linear span of the vectors $s_1(z), \ldots, s_k(z)$. Therefore, the map $\widetilde{\eta}$ is smooth.

(b) $\Longrightarrow$ (a). Suppose that the plane $\eta_z \in G_{n,k}$ smoothly depends on the point z. Using the standard coordinate system in $G_{n,k}$ (see 11.11, VI), we can choose a basis $s_1(z), \ldots, s_k(z)$ of η_z that smoothly depends on the point z, if z belongs to a certain neighborhood U of a point $a \in M$.

Define a map $\varphi\colon U \times \mathbb{R}^k \to U \times \mathbb{R}^n$ by

$$\varphi(z, \lambda_1, \ldots \lambda_k) = \left(z, \sum_{i=1}^{k} \lambda_i s_i(z)\right).$$

The linear independence of the vectors $s_1(z), \ldots, s_k(z)$ implies that φ is of maximal rank. By the implicit function theorem (see Theorem 6.23 and Remark 6.24), the image $\Im \varphi$ is a submanifold of the space $U \times \mathbb{R}^n$. According to the construction of the map φ, this means that $\{\eta_z\}_{z\in M}$ defines a subbundle of π. ▶

The lemma describes the structure of subbundles of a trivial vector bundle and thus the local structure of subbundles of any vector bundle. We shall now apply the lemma to investigate the conditions under which the kernel and the image of a bundle morphism $\varphi \in \mathrm{Mor}(\pi, \eta)$ are subbundles in π and η, respectively. By the *kernel* (respectively, *image*) we understand the set $\bigcup_{z\in M} \mathrm{Ker}\,\varphi_z \subset E_\pi$, together with the restriction to this set of the projection π (respectively, the set $\bigcup_{z\in M} \Im \varphi_z \subset E_\eta$, together with the restriction of η).

12.20 Proposition. *For a vector bundle morphism* $\varphi\colon \pi \to \eta$ *over a manifold* M, *the following conditions are equivalent:*

(a) $\dim \operatorname{Ker} \varphi_x$ *does not depend on* x.

(b) $\dim \Im \varphi_x$ *does not depend on* x.

(c) $\operatorname{Ker} \varphi$ *is a subbundle of* π.

(d) $\Im \varphi$ *is a subbundle of* η.

◀ The implications (c) $\Longrightarrow$ (a) and (d) $\Longrightarrow$ (b) are evident. The equivalence (a) $\Longleftrightarrow$ (b) follows from the fact that the sum $\dim \operatorname{Ker} \varphi_x + \dim \Im \varphi_x$ is equal to the dimension of the fiber of π and is thus constant.

Let us prove that (b) implies (d). Since assertion (d) is local, it suffices to prove it in a neighborhood of an arbitrary point of the base space. Let $U \subset M$ be a neighborhood of the given point such that the bundles $\pi\big|_U$ and $\eta\big|_U$ are trivial. Therefore, we can assume that we deal with a morphism φ_U of trivial bundles acting from $\pi_U \colon U \times V \to U$ to $\eta_U \colon U \times W \to U$. To this morphism there corresponds a smooth map $\widetilde{\varphi}_U \colon U \to \operatorname{Hom}(V, W)$ sending each point x to the operator $\widetilde{\varphi}(x)$ whose value on the vector $v \in V$ is equal to the W-component of the element $\varphi(x, v) \in M \times W$. By assumption, the rank of $\widetilde{\varphi}_x$ does not depend on the point x; denote it by r. Suppose that at a given point $a \in U$ the vectors $v_1, \ldots, v_r$ have the property that their images under φ_a are linearly independent. Then, by continuity, there is a neighborhood of a where the vectors $\varphi_x(v_1), \ldots, \varphi_x(v_r)$ form a basis of $\Im \varphi_x$ that smoothly depends on x. Now by Lemma 12.19, $\Im \varphi$ is a subbundle of η.

The previous argument can also be applied to prove the implication (a) $\Longrightarrow$ (c). If a family of operators $\varphi_x \in \operatorname{Hom}(V, W)$ smoothly depends on x and has constant rank r, then $\Im \varphi_x$, regarded as a point of the Grassmannian $G_{W,r}$, is a smooth function of x (by virtue of Lemma 12.19). Note that $\operatorname{Ker} \varphi_x = \operatorname{Ann} \Im \varphi_x^*$ (we recall that $\operatorname{Ann} \Im \varphi_x^*$ denotes the set of mutual zeros of all linear functionals from $\Im \varphi_x^*$). The smooth dependence of φ_x^* on x follows from the fact that the components of this operator in appropriate bases are equal to the components of φ_x. Hence $\Im \varphi_x^*$ is a smooth function of x. It remains to note that the map

$$\operatorname{Ann} \colon G_{V^*,r} \to G_{V, \dim V - r}$$

sending every subspace to its annihilator is smooth.   ▶

12.21. Direct sum of vector bundles. In the case of vector bundles, the construction of the direct sum (Section 11.14) must agree with the linear structure in the fibers. We say that a vector bundle π is the *direct sum* of two subbundles η and ζ (notation: $\pi = \eta \oplus \zeta$) if its fiber over every point $x \in M$ is the direct sum of two subspaces: $\pi_x = \eta_x \oplus \zeta_x$.

If two vector bundles η and ζ over one manifold M are given, we can construct a vector bundle π that decomposes into a direct sum of two subbundles isomorphic to η and ζ. Such a vector bundle π is defined uniquely up to isomorphism; it is called the *outer direct sum* or *Whitney sum* of the bundles η and ζ and is also denoted by $\eta \oplus \zeta$. As in the case of arbitrary locally trivial bundles, the total space of $\eta \oplus \zeta$ can be defined as

$$E_{\eta \oplus \zeta} = \{(y, z) \in E_\eta \times E_\zeta \mid \eta(y) = \zeta(z)\},$$

and the projection is the map sending a point (y, z) to $\eta(y)$. The fiber of the Whitney sum over a point $x \in M$ is in a natural bijection with the space $\eta_x \oplus \zeta_x$; this bijection endows $(\eta \oplus \zeta)_x$ with the structure of a vector space.

12.22. Examples of direct sums. I. Let M be a submanifold of a Euclidean space E. Then the trivial bundle $E \times M \to M$ is a direct sum of two subbundles: the tangent subbundle $\pi_T : TM \to M$ and the normal subbundle $\nu : NM \to M$. The fiber ν_z of the normal subbundle over a point $z \in M$ is by definition the orthogonal complement of the tangent space $T_z M$ in $T_z E$, the latter being identified with E. It is interesting to note that, for example, for the sphere $S^2 \subset \mathbb{R}^3$, the normal bundle ν is trivial, but the tangent bundle π_T is not (since it does not have nonvanishing sections; see Example 11.11, III). We see that the direct sum of a trivial and a nontrivial bundle can be trivial, a fact seems quite unexpected at first glance.

II. The Whitney sum of the Möbius band with the trivial one-dimensional (unit) bundle is nontrivial. Indeed, the total space of this sum is the product $[0, 1] \times \mathbb{R}^2 \subset \mathbb{R}^3$ with points $(0, y, z)$ and $(1, -y, z)$ identified for any $y, z \in \mathbb{R}$. This manifold is nonorientable and therefore is not diffeomorphic to $S^1 \times \mathbb{R}^2$.

12.23 Proposition. *If* $\pi = \eta \oplus \zeta$, *then* $\Gamma(\pi) = \Gamma(\eta) \oplus \Gamma(\zeta)$ (*direct sum of submodules*).

◀ The assertion follows from (11.6) by virtue of the natural identifications $\Gamma(\eta) \times \Gamma(\varphi)$ and $\Gamma(\eta) \oplus \Gamma(\varphi)$. ▶

The following proposition is important because of its relationship with the property of projectivity.

12.24 Proposition. *Every subbundle of a vector bundle has a direct complement.*

◀ The proof is based on a standard technical trick: the introduction of a scalar product. A scalar product on a vector bundle π is by definition a scalar product in every fiber $x \in M$, smoothly depending on x. The smoothness requirement can be stated as follows: the scalar product of any two smooth sections is a smooth function.

Remark. After studying the construction of the tensor product of vector bundles (Section 12.35), the reader will see that the scalar product on a vector bundle π is nothing but a smooth function on the manifold $E_{\pi \otimes \pi}$ whose restriction to every fiber is linear and positive definite.

Example. A scalar product on the tangent bundle is the same thing as a Riemannian metric on the given manifold.

12.25 Lemma. *On any vector bundle there exists a scalar product.*

◁ To prove the lemma, note that, for a trivial bundle, the problem has a trivial solution: it suffices to supply every fiber with one and the same scalar product. Now let $\{U_i\}_{i \in I}$ be a covering of the manifold M with open sets, each of which trivializes π, and let g_i be a scalar product on the vector bundle $\pi\big|_{U_i}$. Choose a partition of unity $\{e_i\}_{i \in I}$, subjected to the covering $\{U_i\}_{i \in I}$ (see 4.18), and for $y_1, y_2 \in \pi_x$ set

$$g(y_1, y_2) = \sum e_i(x) g_i(y_1, y_2),$$

where the summation ranges over all indices i for which $x \in U_i$. All the necessary properties of the function g can be verified in a straightforward way.   ▷

Let us continue the proof of Proposition 12.24. Let π be a bundle over M and $\eta \subset \pi$. Choose a scalar product g on π. Then in the fiber over any point $x \in M$, we can consider the subspace $\eta_x^\perp$, the orthogonal complement to η in π with respect to the scalar product g. We only have to check that $\eta_x^\perp$ smoothly depends on x. Since this is a local property, we can assume that the bundle π is trivial, $\pi \colon M \times V \to M$. Then we have smooth maps $\widetilde{\eta} \colon M \to G_{V,k}$ (see 12.19) and $\widetilde{g} \colon M \to (V \otimes V)^*$ (scalar product). Denote by $N \subset (V \otimes V)^*$ the set of all symmetric positive definite bilinear forms. The map $G_{V,k} \times N \to G_{V,n-k}$ sending a pair (L, φ) to the orthogonal complement of L with respect to the scalar product φ is smooth. The map $x \mapsto \eta_x^\perp$ is the composition

$$M \to M \times M \to G_{V,k} \times N \to G_{V,n-k},$$

where the first arrow is the diagonal map $x \mapsto (x, x)$, the second arrow is the direct product of the maps $\widetilde{\eta}$ and $\widetilde{g}$, and the third arrow is the map introduced above. The resulting map is smooth, and by Lemma 12.19, the proposition is proved.   ▶

12.26. Pullback vector bundles. The geometrical construction of the pullback vector bundles does not differ from the same construction in the general case (see 11.16).

Example. Let $\pi \colon E \to M$ be an arbitrary bundle and suppose that $\lambda = \pi_{T^*} \colon T^*M \to M$ is the cotangent bundle of its base space. Then the sections of the pullback bundle $\pi^*(\lambda)$ are called *horizontal 1-forms* on the total space E (see Section 14.4). Pay attention to the fact that $\Gamma(\pi^*(\lambda))$ is embedded into the module of 1-forms on the manifold E.

The following assertion is one of the key facts in the theory of vector bundles.

12.27 Theorem. *Every vector bundle with connected base can be obtained as the pullback of the tautological bundle over an appropriate Grassmann manifold.*

◀ Let $\eta \colon E_\eta \to M$, $\dim M = n$, $\dim E_\eta = n + k$. Denote by

$$o_z \colon \eta_z \to T_z(\eta_z) \subset T_z(E_\eta)$$

the canonical identification of the vector space η_z with its tangent space at zero. By Whitney's theorem, there exists an immersion $\phi\colon E_\eta \to \mathbb{R}^{2(n+k)}$ of the manifold E_η into a Euclidean space. This means that all differentials

$$d_y\phi\colon T_y(E_\eta) \to T_{\phi(y)}\mathbb{R}^{2(n+k)}, \qquad y \in E_\eta,$$

are injective.

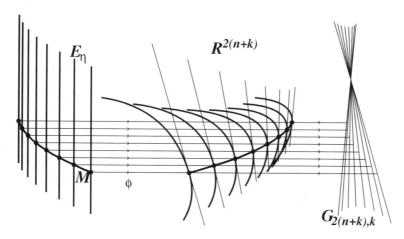

Figure 12.1. The Gauss map.

The *Gauss map* $g\colon M \to G_{2(n+k),k}$ assigns to every point $z \in M$ the k-dimensional subspace of the tangent space $T_O(\mathbb{R}^{2(n+k)}) \cong \mathbb{R}^{2(n+k)}$ which is the image of the subspace $T_z(\eta_z) \subset T_z(E_\eta)$ under the map

$$d_z(r_{\phi(z)} \circ \phi)\colon T_z(E_\eta) \to T_O\left(\mathbb{R}^{2(n+k)}\right)$$

(see Figure 12.1). Here z is understood as a point of the manifold E_η, the zero element in the fiber η_z, and $r_a\colon v \mapsto v - a$ (for $a, v \in \mathbb{R}^{2(n+k)}$) is the translation of the space $\mathbb{R}^{2(n+k)}$ by the vector $-a$. The map g is covered by the morphism of vector bundles

$$\gamma\colon \eta \to \Theta_{2(n+k),k}, \qquad \gamma_z = d_z(r_{\phi(z)} \circ \phi) \circ o_z,$$

where $\Theta_{2(n+k),k}$ denote the tautological bundle (see p. 160). For every point $z \in M$ the map γ_z is an isomorphism of the fiber η_z onto the fiber at $g(z) \in G_{2(n+k),k}$. Now Theorem 11.19 shows that the bundles η and $g^*(\Theta_{2(n+k),k})$ are isomorphic. ▶

12.28 Corollary. *The $C^\infty(M)$-module $\Gamma(\pi)$, $\pi\colon E_\pi \to M$, has a system of generators consisting of no more than N elements, where $N = N(n, k)$ is a natural number that depends only on the dimensions of the base and the fiber (n and k respectively) of the bundle π.*

◀ Note that the sections $s_{I,i}$ of the tautological bundle $\Theta_{m,l}$ described in Example V of Section 11.12 satisfy the assumptions of Proposition

12.10. Therefore, the sections $\widehat{f}(s_{I,i})$ of any pullback bundle $f^*(\Theta_{m,l})$ (see Section 11.16) also satisfy these assumptions and thus generate the module $\Gamma(f^*(\Theta_{m,l}))$. Theorem 12.27 implies that any k-dimensional vector bundle over a manifold of dimension n is the pullback of the tautological bundle $\Theta_{2(n+k),k}$ via the Gauss map. By Proposition 12.10, the sections

$$\widehat{g}(s_{I,i}), \text{ where } 1 \leqslant i \leqslant k, \ I = \{i_1, \ldots, i_k\} \subset \{1, \ldots, 2(n+k)\},$$

generate the module $\Gamma(\pi)$. The number of these sections is

$$k\binom{2(n+k)}{k} = N(n,k). \quad \blacktriangleright$$

We have completed all preparations necessary to state and prove the two main theorems of this chapter (Theorems 12.29 and 12.32), which give an exhaustive description of the section modules of vector bundles.

12.29 Theorem. *For any pair π, η of vector bundles over a manifold M, the section functor Γ determines a one-to-one correspondence*

$$\mathrm{Mor}(\pi, \eta) \cong \mathrm{Hom}_{C^\infty(M)}(\Gamma(\pi), \Gamma(\eta)).$$

◀ We must prove that for any $C^\infty(M)$-homomorphism of modules $F \colon \Gamma(\pi) \to \Gamma(\eta)$ there exists a unique bundle morphism $\varphi \colon \pi \to \eta$ such that $\Gamma(\varphi) = F$.

First we prove the uniqueness. Suppose that

$$\varphi, \psi \in \mathrm{Mor}(\pi, \eta) \text{ and } \Gamma(\varphi) = \Gamma(\psi).$$

This means that $\varphi, \psi \colon E_\pi \to E_\eta$ and $\varphi \circ s = \psi \circ s$ for any section $s \in \Gamma(\pi)$, i.e.,

$$\varphi(s(x)) = \psi(s(x)) \text{ for all } s \in \Gamma(\pi) \text{ and all } x \in M.$$

According to Lemma 12.8(a), every point of E_π can be represented as $s(x)$, which implies that $\varphi = \psi$.

Now suppose that a homomorphism $F \colon \Gamma(\pi) \to \Gamma(\eta)$ is given, and we must define the corresponding map $\varphi \colon E_\pi \to E_\eta$, i.e., we must define its value $\varphi(y) \in E_\eta$ at an arbitrary point $y \in E_\pi$. The point y belongs to a certain fiber: $y \in \pi_x$. By Lemma 12.8(a), we can choose a section $s \in \Gamma(\pi)$ such that $s(x) = y$ and put

$$\varphi(y) = F(s)(x).$$

We must check (a) that φ is well defined, (b) that φ is a bundle morphism, and (c) the equality $\Gamma(\varphi) = F$.

(a) Let s_1 and s_2 be two sections of π such that $s_1(x) = s_2(x)$. Lemma 12.8(b) shows that $s_1 - s_2 \in \mu_x \Gamma(\pi)$. Therefore,

$$F(s_1 - s_2) \in \mu_x \Gamma(\eta) \quad \text{and} \quad F(s_1)(x) = F(s_2)(x).$$

(b) The only thing worth verifying here is the smoothness of the map $\varphi \colon E_\pi \to E_\eta$. It is sufficient to prove that the map $\varphi_U = \varphi\big|_{\pi^{-1}(U)}$ is smooth for an arbitrary open set $U \subset M$, over which both π and η are trivial. Note that

$$\varphi_U(t(x)) = F_U(t)(x), \qquad (12.1)$$

where $t \in \Gamma(\pi|_U)$ and $F_U \colon \Gamma(\pi|_U) \to \Gamma(\eta|_U)$ is the localization of the homomorphism F on the subset U (that is, over the multiplicative system of functions that do not vanish at the points of U; see Section 11.6). Choosing some bases of the free $C^\infty(U)$-modules $\Gamma(\pi|_U)$ and $\Gamma(\eta|_U)$, we can write the homomorphism F_U as a matrix over the ring $C^\infty(U)$. By virtue of equation (12.1), the same matrix gives the coordinate representation of the morphism φ_U (Lemma 12.5). Since the elements of the matrix belong to $C^\infty(U)$, the map φ_U is smooth.

(c) For any $s \in \Gamma_\pi$, we have

$$(\Gamma(\varphi)(s))(x) = (\varphi \circ s)(x) = \varphi(s(x)) = F(s)(x),$$

i.e., $\Gamma(\varphi)(s) = F(s)$. ▶

12.30 Lemma. *If $\varphi \colon \zeta \to \pi$ is a vector bundle morphism and φ_z is an isomorphism of vector spaces $\zeta_z \cong \pi_z$ for any point $z \in M$, then φ is a bundle isomorphism.*

◀ The only thing we must check is that the inverse map $\varphi^{-1} \colon E_\pi \to E_\zeta$ is smooth. By the inverse function theorem (Section 6.21), it suffices to show that the differential $d_y\varphi$ at any point $y \in E_\zeta$ is an isomorphism of the corresponding tangent spaces. Since the dimensions of the manifolds E_ζ and E_π are equal, the dimensions of the tangent spaces $T_y E_\zeta$ and $T_{\varphi(y)} E_\pi$ are equal too. Therefore, for the differential $d_y\varphi$ to be an isomorphism, it is sufficient that it be injective. To check the latter, note that we must have $d_{\varphi(y)}\pi \circ d_y\varphi = d_y\zeta$, because $\pi \circ \varphi = \zeta$. Now suppose that $d_y\varphi(v) = 0$. Then $d_y\zeta(v) = 0$, which means that $v \in T_y(\zeta_z)$, where $z = \zeta(y)$. By assumption, $\varphi|_{\zeta_z} = \varphi_z$ is an isomorphism between the fibers ζ_z and π_z. This implies $v = 0$. ▶

12.31 Corollary. *If $F \in \mathrm{Hom}_{C^\infty(M)}(\Gamma(\zeta), \Gamma(\pi))$ and for any point $x \in M$ the induced map*

$$F_x \colon \Gamma(\zeta)/\mu_x\Gamma(\zeta) \to \Gamma(\pi)/\mu_x\Gamma(\pi)$$

is an isomorphism of vector spaces, then F is a module isomorphism.

◀ This fact is easily reduced to the lemma just proved. Indeed, we have $F = \Gamma(\varphi)$ for an appropriate morphism $\varphi \in \mathrm{Mor}(\zeta, \pi)$ and $F_x = \varphi_x$ at every point $x \in M$. ▶

12.32 Theorem. *Suppose M is a connected smooth manifold and let P be a $C^\infty(M)$-module. Then P is isomorphic to the module of sections $\Gamma(\pi)$ of a smooth vector bundle π over M if and only if P is finitely generated and projective.*

◀ (i) Recall, first of all, that *for any vector bundle π over M, the module $\Gamma(\pi)$ is finitely generated* (Corollary 12.28).

(ii) Let us prove that *the module $\Gamma(\pi)$ is projective.*

◁ According to (i), there exists a finite system of sections $s_1, \dots, s_N$ that generate the module $\Gamma(\xi)$. Let Q be a free $C^\infty(M)$-module of rank N with generators $e_1, \dots, e_N$ and let $F \colon Q \to \Gamma(\xi)$ be the homomorphism given by $F(e_i) = s_i$ for $i = 1, \dots, N$. By construction, F is epimorphic. Note that $Q = \Gamma(\eta)$ for a certain trivial vector bundle η; hence by Theorem 12.29 there is a bundle morphism $\varphi \in \mathrm{Mor}(\pi, \eta)$ such that $F = \Gamma(\varphi)$.

Lemma 12.8 implies that at every point $z \in M$ the map φ_z is surjective, since F is an epimorphism. Therefore, we can apply Proposition 12.20 and infer that $\mathrm{Ker}\, \varphi$ is a subbundle in η. By Proposition 12.24, there is a direct decomposition $\eta = \mathrm{Ker}\, \varphi \oplus \zeta$ for an appropriate subbundle ζ of π. Proposition 12.23 implies that

$$\Gamma(\eta) = \Gamma(\mathrm{Ker}\, \varphi) \oplus \Gamma(\zeta);$$

i.e., the module $\Gamma(\zeta)$ is a direct summand of a free module. By Proposition 12.15, $\Gamma(\zeta)$ is projective.

Now let us check that the map φ restricted to the total space of the subbundle ζ gives an isomorphism between ζ and π. Indeed, $\varphi_z \colon \zeta_z \to \pi_z$ is a linear isomorphism for any point $z \in M$, and we can use Lemma 12.30. The isomorphism of bundles $\zeta \cong \pi$ implies the isomorphism of modules $\Gamma(\zeta) \cong \Gamma(\pi)$ and thus assertion (ii).   ▷

(iii) We shall now prove that *any projective module of finite type is isomorphic to the module of sections of a smooth vector bundle.*

◁ Suppose that P is a projective $C^\infty(M)$-module with a finite number of generators. Then (see Proposition 12.15 and the remark in Section 12) we can write $\Gamma(\eta) = P' \oplus Q$, where η is the trivial bundle over M, P' and Q are submodules of $\Gamma(\eta)$, and $P' \cong P$. Since we are considering P only up to isomorphism, in the sequel by abuse of notation we suppress the prime and write P instead of P'.

Let $P_z = \{p(z) \mid p \in P\}$. This is an $\mathbb{R}$-linear subspace in η_z. The subspace Q_z is defined similarly. We claim that $\eta_z = P_z \oplus Q_z$.

Indeed, let $y \in \eta_z$. Choose a section $s \in \Gamma(\eta)$ such that $s(z) = y$ and represent it as $p + q$, where $p \in P$, $q \in Q$. Then $y = p(z) + q(z) \in P_z + Q_z$. On the other hand, suppose that $y \in P_z \cap Q_z$, i.e., $y = p(z) = q(z)$, where $p \in P$, $q \in Q$. Then $(p - q)(z) = 0$, hence by Lemma 12.8(b) we have

$$p - q = \sum_i f_i s_i = \sum_i f_i p_i + \sum_i f_i q_i$$

for a certain choice of $f_i \in \mu_z$, $p_i \in P$, $q_i \in Q$. Since $P \cap Q = 0$, the last equation implies $p = \sum_i f_i p_i$. Therefore, $p(z) = 0$, i.e., $y = 0$, and thus $\eta_z = P_z \oplus Q_z$.

We must verify that the union of all subspaces P_z constitutes a subbundle in η and that P coincides with the module of its (smooth) sections. We shall show first that $\dim P_z$ does not depend on z. Let $\dim P_z = r$ for some point $z \in M$ and let $p_1, \ldots, p_r \in P$ be a set of sections whose values at z span the subspace P_z. The continuity of sections implies the linear independence of the vectors $p_1(y), \ldots, p_r(y)$ for all points y in a neighborhood U of the point z. Therefore, $\dim P_y \geqslant \dim P_z$.

A similar argument for the submodule Q shows that $\dim Q_y \geqslant \dim Q_z$ in a neighborhood of z. Since the sum $\dim P_y + \dim Q_y$ is constant, we see that $\dim P_y$ is a locally constant function of y. Since M is connected, it is a global constant.

Having in mind Lemma 12.5, we shall now prove that the subspace P_z, viewed as a point of $G_{V,r}$, where V is the fiber of η, smoothly depends on z (here we can assume that η is a trivial bundle, because the problem under consideration is local). Indeed, let $p_1(a), \ldots, p_r(a)$ be a basis of the space P_a at a certain point $a \in M$. Then the vectors $p_1(z), \ldots, p_r(z)$ form a basis of the linear space P_z for all points z belonging to some neighborhood of a. We see that the family P_z locally has a basis that smoothly depends on z and thus represents a smooth family of points in the Grassmannian $G_{V,r}$.

Denote the bundle with fibers P_z by π. By construction, $P \subset \Gamma(\pi)$. Let us prove the reverse inclusion. If $s \in \Gamma(\pi) \subset \Gamma(\eta)$, then there are elements $p \in P$ and $q \in Q$ such that $s = p + q$. Since $P_z \cap Q_z = 0$, the equation $p(z) + q(z) = s(z)$ implies that $q(z) = 0$ for any $z \in M$. Hence $q = 0$ and $s = p \in P$.   ▷

We have shown that $P = \Gamma(\pi)$. This completes the proof of the theorem. ▶

Remark. Our proof shows that the module $\Gamma(\xi)$ is projective also in the case of a disconnected base manifold. On the other hand, any projective module over $C^\infty(M)$ is evidently reduced to the direct sum of modules $\Gamma(\pi_\alpha)$, where π_α is a vector bundle over a connected component M_α of the manifold M. The dimension of the bundle π_α may vary between the connected components.

12.33. Equivalence of the two categories. Theorems 12.29 and 12.32, taken together, establish the equivalence of the category VB_M of vector bundles over the manifold M and the category $\mathrm{Mod}_{\mathrm{pf}}\, C^\infty(M)$ of projective finite-type modules over the algebra $C^\infty(M)$. This result is in full parallel with the result of Section 7.19 about the equivalence between the category of smooth manifolds and the category of smooth $\mathbb{R}$-algebras. It can be used in either direction, i.e., by applying algebra to geometry and vice versa.

Here is a simple example:

Proposition. *For any vector bundle π there is a vector bundle η such that the direct sum $\pi \oplus \eta$ is a trivial bundle.*

This fact, which plays a fundamental role in K-theory is surprising from the geometrical viewpoint, can be reduced to the mere definition of projective modules by the application of Theorem 12.32.

Below (in Section 12.38) we shall give an example of an algebraic statement (*the tensor square of a one-dimensional projective module over* $C^\infty(M)$ *is isomorphic to* $C^\infty(M)$), which becomes evident after a geometrical trick (introduction of a scalar product on the corresponding vector bundle).

In the next sections, we discuss two operations on vector bundles, namely the construction of tensor products and pullback bundles, and the corresponding operations on projective modules.

12.34 Proposition. *The tensor product of two projective modules over a commutative ring is a projective module.*

◀ The well-known property (whose proof we leave to the reader as an algebraic exercise) of Hom$-\otimes$-*associativity* reads

$$\mathrm{Hom}_A(P \otimes_A Q, R) \cong \mathrm{Hom}_A(P, \mathrm{Hom}_A(Q, R))$$

for any three A-modules P, Q, R. In other words, this means that the functor $\mathrm{Hom}_A(P \otimes Q, \cdot)$ is the composition of functors $\mathrm{Hom}_A(P, \cdot)$ and $\mathrm{Hom}_A(Q, \cdot)$. It remains to use the equivalence of (a) and (d) from Proposition 12.15.   ▶

12.35. Tensor product of vector bundles. Let η and ζ be vector bundles over a manifold M. The fiber of the new bundle $\pi = \eta \otimes \zeta$ (*the tensor product of* η *and* ζ) over a point $z \in M$ is, by definition, the linear space $\pi_z = \eta_z \otimes \zeta_z$. The structure of smooth manifold on the total space $E_\pi = \bigcup_{z \in M} \pi_z$ is introduced in the following way.

Every point of the base manifold M has a neighborhood U over which both bundles η and ζ are trivial. The sets $\pi^{-1}(U)$ form a covering of the space E_π. Let V be the "outer" fiber of the bundle η, let W be the "outer" fiber of the bundle ζ, while $\varphi \colon \eta^{-1}(U) \to U \times V$ and $\psi \colon \zeta^{-1}(U) \to U \times W$ are the corresponding trivializing diffeomorphisms. The chart

$$\chi \colon \pi^{-1}(U) \to U \times (V \otimes W)$$

is constructed as follows. Let $u \in \pi^{-1}(U)$. Then $u \in \eta_z \otimes \zeta_z$ for some point $z \in M$. The maps φ and ψ, restricted to the fiber over z, give isomorphisms $\eta_z \to V$ and $\zeta_z \to W$ and hence an isomorphism $\eta_z \otimes \zeta_z \to V \otimes W$. Denote the image of an element u under this isomorphism by v and put $\chi(u) = (z, v)$.

Exercise. Prove that these charts form a smooth atlas on E_π.

The bundle $\eta \otimes \zeta$ is thus defined. Note that it satisfies the axiom of local triviality by construction.

For brevity, we denote by $\eta^{\otimes k}$ the kth tensor power of a bundle η.

Example. If π_T and π_{T^*} are the tangent and cotangent bundles of a manifold, then $\pi_T^{\otimes k} \otimes \pi_{T^*}^{\otimes l}$ is the bundle of tensors of type (k, l) (k times contravariant and l times covariant).

12.36 Proposition. *Let P and Q be projective modules over a commutative ring A. Then the module $\mathrm{Hom}_A(P,Q)$ is also projective. If both P and Q are finitely generated, then the module $\mathrm{Hom}_A(P,Q)$ is also of finite type.*

◀ If R and S are (finitely generated) free modules, then $\mathrm{Hom}_A(R,S)$ is also (finitely generated and) free. Indeed, let $\{r_i\}$ and $\{s_j\}$ be the free generators of R and S, respectively. Then the A-homomorphisms $h_{i,j} \in \mathrm{Hom}_A(R,S)$, defined by the rule $h_{i,j}(r_i) = \delta_{i,j}s_j$, where $\delta_{i,j}$ is the Kronecker delta, constitute a basis of $\mathrm{Hom}_A(R,S)$.

Now suppose that R and S are free A-modules that contain the modules P and Q, respectively, as direct summands. Let

$$\alpha_P \colon P \hookrightarrow R \quad \text{and} \quad \beta_P \colon R \to P, \quad \alpha_P \circ \beta_P = \mathrm{id}_P,$$

(and similarly with α_Q and β_Q for Q) be the injections and projections that realize the corresponding decompositions of R and S into direct sums. Then the A-homomorphism

$$\mathrm{Hom}_A(P,Q) \ni h \mapsto \alpha_Q \circ h \circ \beta_P \in \mathrm{Hom}_A(R,S)$$

is the injection of $\mathrm{Hom}_A(P,Q)$ into the free module $\mathrm{Hom}_A(R,S)$, whose image, together with the kernel of the projection

$$\mathrm{Hom}_A(R,S) \ni H \mapsto \beta_Q \circ H \circ \alpha_P \in \mathrm{Hom}_A(P,Q),$$

turns the module $\mathrm{Hom}_A(P,Q)$ into a direct summand in the free A-module $\mathrm{Hom}_A(R,S)$. ▶

There is a natural map

$$\iota \colon P^* \otimes_A Q \to \mathrm{Hom}_A(P,Q), \quad \text{where } P^* = \mathrm{Hom}_A(P,A), \qquad (12.2)$$

defined by the formula $\iota(p^* \otimes q)(p) = p^*(p)q$. If the modules P and Q are projective and finitely generated, then ι is an isomorphism. This fact has an evident proof for finitely generated free modules and can be generalized to arbitrary projective modules of finite type by an argument similar to the proof of Proposition 12.36.

12.37. To every vector bundle ζ, we can associate the dual bundle ζ^*, whose fiber ζ_z^* over a point $z \in M$ is the vector space dual to ζ_z. The precise construction of a smooth atlas on the set $E_{\zeta^*} = \bigcup_{z \in M} \zeta_z^*$ repeats the corresponding construction for the case of the cotangent bundle from the tangent bundle given in Section 9.24.

Example. $\pi_T^* = \pi_{T^*}$.

For any vector bundle ζ, there is a natural pairing

$$\Gamma(\zeta) \times \Gamma(\zeta^*) \to C^\infty(M), \quad (s, s^*)(z) \overset{\mathrm{def}}{=} (s(z), s^*(z)),$$

where $s \in \Gamma(\zeta)$, $s^* \in \Gamma(\zeta^*)$, $z \in M$.

Exercise. Using local triviality, verify that $(s, s^*)(z)$ is a smooth function for any smooth sections s and s^*.

For the bundle $\zeta = \pi_{TM}$, this pairing, by the equality $(\pi_{TM})^* = \pi_{T^*M}$, turns into the pairing

$$\Gamma(\pi_{TM}) \times \Gamma(\pi_{T^*M}) \to C^\infty(M).$$

A metric on ζ allows us to identify ζ_z and ζ_z^* for any point $z \in M$. Lemma 12.30 implies that vector bundles ζ and ζ^* are isomorphic.

Given a pair of vector bundles η and ζ, one can define the bundle $\mathrm{Hom}(\eta, \zeta)$ with fiber $\mathrm{Hom}_{\mathbb{R}}(\eta_z, \zeta_z)$ over every point $z \in M$ in the same way. The construction follows that of the tensor product of two vector bundles (see Section 12.35).

Another possibility for defining the bundle $\mathrm{Hom}(\eta, \zeta)$ is to explicitly reduce it to the constructions of the tensor product and the dual bundle, using the natural isomorphism of vector spaces $\mathrm{Hom}_{\mathbb{R}}(\eta_z, \zeta_z) = \eta_z^* \otimes \zeta_z$.

12.38. Example. The tensor square of the Möbius band, viewed as a one-dimensional vector bundle over the circle (Example 12.6, III), is $\mathbb{I}_{S^1}$, the trivial one-dimensional bundle. This result can either be understood directly from the construction of the Möbius band and the definition of the tensor product (we recommend to the reader to carry it out) or be deduced from a more general fact:

Proposition. *The tensor square of any one-dimensional bundle over an arbitrary manifold is isomorphic to the unit bundle.*

To prove the latter, we use the above-mentioned isomorphisms

$$\zeta \otimes \zeta \cong \zeta^* \otimes \zeta \cong \mathrm{Hom}(\zeta, \zeta).$$

If $\dim \zeta = 1$, then the bundle $\mathrm{Hom}(\zeta, \zeta)$ is trivial: the trivializing diffeomorphism $\varphi \colon M \times \mathbb{R} \to E_{\mathrm{Hom}(\zeta,\zeta)}$ can be defined by the formula $\varphi(a, \lambda)y = \lambda y$ for all $a \in M$, $\lambda \in \mathbb{R}$, $y \in \pi^{-1}(a)$.

Note, finally, that tensor multiplication introduces a group structure into the set $V^1(M)$ of equivalence classes of one-dimensional vector bundles over the manifold M. The order of any nonunital element of this group is 2. Thus, $V^1(S^1) = \mathbb{Z}_2$, the two elements being represented by the trivial bundle and the Möbius band.

In the next theorem, we establish the relationship between the functors Γ, $\otimes$, and Hom.

12.39 Theorem. *The functor Γ preserves tensor products and homomorphisms, i.e.,*

$$\Gamma(\pi \otimes \eta) \cong \Gamma(\pi) \otimes \Gamma(\eta),$$
$$\Gamma(\mathrm{Hom}(\pi, \eta)) \cong \mathrm{Hom}_{C^\infty(M)}(\Gamma(\pi), \Gamma(\eta)),$$

for any two vector bundles π and η over the manifold M (here and below, tensor products of $C^\infty(M)$-modules are computed over the ring $C^\infty(M)$).

◀ The natural isomorphisms

$$\operatorname{Hom}(\pi,\eta) \cong \pi^* \otimes \eta, \quad \operatorname{Hom}_{C^\infty(M)}(\Gamma(\pi),\Gamma(\eta)) \cong \Gamma(\pi^*) \otimes \Gamma(\eta)$$

(see Section 12.36) together with the natural identifications $(\pi^*)^* = \pi$, $\Gamma((\pi)^*)^* = \Gamma(\pi)$ show that it suffices to prove the assertion only for tensor products.

We shall construct a map from $\Gamma(\pi) \otimes \Gamma(\eta)$ to $\Gamma(\pi \otimes \eta)$ and verify that it is an isomorphism.

Let $s \in \Gamma(\pi)$, $t \in \Gamma(\eta)$. Define the section $s \otimes t \in \Gamma(\pi \otimes \eta)$ by the formula $(s \otimes t)(x) = s(x) \otimes t(x)$. From the construction of the bundle $\pi \otimes \eta$, we see that $s \otimes t$ is a smooth section of $\pi \otimes \eta$. The assignment of $s \otimes t$ to the pair (s,t) is homomorphic with respect to both arguments and thus defines a $C^\infty(M)$-module homomorphism $\iota \colon \Gamma(\pi) \otimes \Gamma(\eta) \to \Gamma(\pi \otimes \eta)$. Let us show that the value of this homomorphism at a point $x \in M$,

$$\iota_x \colon \Gamma(\pi) \otimes \Gamma(\eta)/\mu_x\big(\Gamma(\pi) \otimes \Gamma(\eta)\big) \to \pi_x \otimes \eta_x,$$

is an isomorphism of vector spaces. Note that

$$\iota_x \left[\sum s_i \otimes t_i \right] = \sum s_i(x) \otimes t_i(x),$$

where the square brackets denote the equivalence class of an element in the quotient space.

(i) Surjectivity of ι_x. An arbitrary element of the space $\pi_x \otimes \eta_x$ has the form $\sum y_i \otimes z_i$ with $y_i \in \pi_x$, $z_i \in \eta_x$. By Lemma 12.8(a), there are sections $s_i \in \Gamma(\pi)$, $t_i \in \Gamma(\eta)$ that take values y_i and z_i, respectively, at the point x. Then

$$\iota_x[\sum s_i \otimes t_i] = \sum y_i \otimes z_i.$$

(ii) Injectivity of ι_x. We must show that if $s_i \in \Gamma(\pi)$, $t_i \in \Gamma(\eta)$, and $\sum s_i(x) \otimes t_i(x) = 0$ in the space $\pi_x \otimes \eta_x$, then there exist sections $p_i \in \Gamma(\pi)$, $q_i \in \Gamma(\eta)$ and functions $f_i \in \mu_x$ such that

$$\sum s_i \otimes t_i = \sum f_i p_i \otimes q_i$$

(equality in $\Gamma(\pi) \otimes \Gamma(\eta)$).

The following lemma clarifies the structure of zero elements in the tensor product of vector spaces.

12.40 Lemma. *Let V and W be vector spaces over a certain field. Suppose that $v_i \in V$, $w_i \in W$ are nonzero vectors and*

$$\sum_{i=1}^m v_i \otimes w_i = 0 \quad in \ V \otimes W.$$

Then there exist a natural number k, $1 \leqslant k \leqslant m$, and elements a_{ij}, $1 \leqslant i \leqslant k$, $k < j \leqslant m$, of the ground field such that after an appropriate *renumeration*

$$\{1, \ldots, m\} \to \{1, \ldots, m\},$$

the same one for both $\{v_i\}$ and $\{w_i\}$, the following relations hold:

$$v_j = \sum_{i=1}^{k} a_{ij} v_i, \quad j = k+1, \ldots, m;$$

$$w_i = - \sum_{j=k+1}^{m} a_{ij} w_j, \quad i = 1, \ldots, k.$$

◁ Indeed, if the elements $v_1, \ldots, v_m$ are linearly independent, then the equality $\sum v_i \otimes w_i = 0$ implies that all vectors w_i are zero. If not, choose from $v_1, \ldots, v_m$ a maximal linearly independent subset. Let it be $v_1, \ldots, v_k$. Expand v_j for $j = k+1, \ldots, m$ in terms of this basis: $v_j = \sum_{i=1}^{k} a_{ij} v_i$. Then

$$\sum_{i=1}^{m} v_i \otimes w_i = \sum_{i=1}^{k} v_i \otimes w_i + \sum_{j=k+1}^{m} \left(\sum_{i=1}^{k} a_{ij} v_i \right) \otimes w_j$$

$$= \sum_{i=1}^{k} v_i \otimes \left(w_i + \sum_{j=k+1}^{m} a_{ij} w_j \right),$$

whence $w_i = - \sum_{j=k+1}^{m} a_{ij} w_i$. The lemma is proved.   ▷

Applying the lemma in the current situation, we obtain

$$s_j = \sum_{i=1}^{k} a_{ij} s_i + s'_j, \quad j = k+1, \ldots, m; \quad s'_j \in \mu_x \Gamma(\pi);$$

$$t_i = - \sum_{j=k+1}^{m} a_{ij} t_i + t'_i, \quad i = 1, \ldots, k; \quad t'_i \in \mu_x \Gamma(\eta).$$

Therefore,

$$\sum_{i=1}^{m} s_i \otimes t_i = \sum_{i=1}^{k} s_i \otimes \left(-\sum_{j=k+1}^{m} a_{ij} t_j + t'_i \right) + \sum_{j=k+1}^{m} \left(\sum_{i=1}^{k} a_{ij} s_i + s'_j \right) \otimes t_j$$

$$= \sum_{i=1}^{k} s_i \otimes t'_i + \sum_{j=k+1}^{m} s'_j \otimes t_j \in \mu_x \big(\Gamma(\pi) \otimes \Gamma(\eta) \big),$$

as desired.

We see that the morphism ι is an isomorphism for every point $x \in M$. Using the corollary of Lemma 12.30, we want to conclude that ι is a module isomorphism. For this corollary to apply, both modules must be modules of sections of smooth vector bundles. In our case, only the module $\Gamma(\pi) \otimes \Gamma(\eta)$ is to be checked in this respect. But by Theorem 12.32 and Proposition 12.34, it is projective. Hence, by Theorem 12.32, this module is isomorphic to a certain module of sections, and thus the corollary can be used. This completes the proof of the theorem. ▶

12.41. Change of rings. To translate the construction of the induced bundle into algebraic language, we must understand what relations between modules over different rings arise when a homomorphism from one ring to another is given.

Let $\varphi\colon A \to B$ be a ring homomorphism, and P a module over A. The homomorphism φ allows us to view B as an A-module with multiplication of any element $b \in B$ by $a \in A$ defined as $ab \overset{\text{def}}{=} \varphi(a)b$ and hence define an A-module $\varphi_*(P) = B \otimes_A P$. Setting $b_1 \cdot (b_2 \otimes p) = b_1 \cdot b_2 \otimes p$, we convert $\varphi_*(P)$ into a B-module. The assignment $P \mapsto \varphi_*(P)$ extends to a functor from $\operatorname{Mod} A$ to $\operatorname{Mod} B$, called the *change of rings functor* denoted by Φ_*.

Proposition. *The change of rings functor preserves projectivity.*

◀ We shall show that the projectivity of an A-module P implies the projectivity of the B-module $\varphi_*(P)$, using property (d) from Proposition 12.15. For an arbitrary B-module Q, there is an isomorphism of Abelian groups

$$\operatorname{Hom}_A(P \otimes_A B, Q) \cong \operatorname{Hom}_A(P, \operatorname{Hom}_A(B, Q)). \qquad (12.3)$$

More exactly, the elements

$$\gamma \in \operatorname{Hom}_A(P \otimes_A B, Q) \quad \text{and} \quad \delta \in \operatorname{Hom}_A(P, \operatorname{Hom}_A(B, Q))$$

that correspond to each other under this isomorphism are related by the equations

$$\gamma(p \otimes b) = \delta(p)(b), \quad p \in P, \ b \in B.$$

In particular, if $\gamma \in \operatorname{Hom}_B(P \otimes_A B, Q)$, then for any $p \in P$, $b_1, b_2 \in B$ we have $\gamma(p \otimes b_1 b_2) = b_1 \gamma(p \otimes b_2)$, and so

$$\delta(p)(b_1 b_2) = b_1 \cdot \delta(p)(b_2),$$

i.e., $\delta(p) \in \operatorname{Hom}_B(B, Q) \cong Q$. The converse argument is also valid. Therefore, the isomorphism (12.3) induces an isomorphism

$$\operatorname{Hom}_B(P \otimes_A B, Q) \cong \operatorname{Hom}_A(P, Q).$$

This isomorphism is natural with respect to Q, i.e., it extends to an isomorphism of functors on the category of B-modules with values in the category of Abelian groups,

$$\operatorname{Hom}_B(P \otimes_A B, \cdot) \cong \operatorname{Hom}_A(P, \cdot),$$

which, by Proposition 12.15, implies that the B-module $P \otimes_A B$ is projective. ▶

12.42. Algebraic formulation of pullback bundles. We now establish the algebraic meaning of the pullback procedure of vector bundles. Let

$\varphi\colon N \to M$ be a smooth map of manifolds, $\Phi = \varphi^*\colon C^\infty(M) \to C^\infty(N)$, the corresponding homomorphism of function rings, and let Φ_* be the change of rings functor. According to Proposition 12.41, the functor Φ_* preserves projectivity; besides, it preserves the finite type property. Therefore, the functor Φ_* can be restricted to the subcategory of finitely generated projective modules:

$$\Phi_*\colon \mathrm{Mod}_{\mathrm{pf}}\, C^\infty(M) \to \mathrm{Mod}_{\mathrm{pf}}\, C^\infty(N).$$

12.43 Theorem. *For any smooth map of manifolds $\varphi\colon N \to M$ and any vector bundle π over M, there is an isomorphism of $C^\infty(N)$-modules*

$$\Gamma(\varphi^*(\pi)) \cong \Phi_*(\Gamma(\pi)).$$

This isomorphism can be chosen to be natural with respect to π, so that the functors $\Gamma \circ \varphi^*$ and $\Phi_* \circ \Gamma$ are isomorphic, and the functor diagram

$$
\begin{array}{ccc}
\mathrm{VB}_M & \xrightarrow{\ \ \varphi^*\ \ } & \mathrm{VB}_N \\
{\scriptstyle\Gamma}\big\downarrow & & \big\downarrow{\scriptstyle\Gamma} \\
\mathrm{Mod}_{\mathrm{pf}}\, C^\infty(M) & \xrightarrow[\ \Phi_*\ ]{} & \mathrm{Mod}_{\mathrm{pf}}\, C^\infty(N)
\end{array}
$$

is commutative.

◄ Below we will refer to the lifting of sections $\hat\varphi$ defined in Section 11.18. Let $A = C^\infty(M)$, $B = C^\infty(N)$. The map

$$B \times \Gamma(\pi) \to \Gamma(\varphi^*(\pi)),$$

which sends the pair (f, s) to the section $f \cdot \hat\varphi(s)$, is homomorphic over A with respect to either argument (here the B-module $\Gamma(\varphi^*(\pi))$ is viewed as an A-module with multiplication introduced via the ring homomorphism Φ). Therefore, this map defines an A-homomorphism

$$\nu\colon B \otimes_A \Gamma(\pi) \to \Gamma(\varphi^*(\pi)).$$

Note that in fact, ν is a homomorphism not only over A, but also over B. Indeed, for $f, g \in B$ and $s \in \Gamma(\pi)$ we have

$$\nu(fg \otimes s) = fg\hat\varphi(s) = f\nu(g \otimes s).$$

Let us prove that ν is an isomorphism. The module $B \otimes_A \Gamma(\pi)$ is finitely generated and projective; hence by Theorem 12.32 it is isomorphic to the module of sections of a bundle over N. Using Lemma 12.30, we can consider the value of the homomorphism ν at a point $w \in N$:

$$\nu_w\colon B \otimes_A \Gamma(\pi)/\mu_w \otimes_A \Gamma(\pi) \to (\varphi^*(\pi))_w \cong \pi_{\varphi(w)}.$$

Using the identification of $(\varphi^*(\pi))_w$ with $\pi_{\varphi(w)}$, we obtain

$$\nu_w([g \otimes s]) = g(w)s(\varphi(w)).$$

The map ν_w is epimorphic, because for any $z \in \pi_{\varphi(w)}$ we can, by Lemma 12.8(a), find a section s such that $s(\varphi(w)) = z$ and hence $\nu_w([1 \otimes s]) = z$.

Now let us check that ν_w is monomorphic. Let $\sum_i g_i \otimes s_i$ be an element of $B \otimes_A \Gamma(\pi)$ such that

$$\sum_i g_i(w)s_i(\varphi(w)) = 0.$$

Set $g_i(w) = \beta_i \in \mathbb{R}$ and put $\overline{g}_i = g_i - \beta_i$. The previous equation can be rewritten as $\sum_i \beta_i s_i \in \mu_{\varphi(w)}\Gamma(\pi)$, i.e.,

$$\sum_i \beta_i s_i = \sum_j f_j t_j, \text{ where } f_j \in \mu_{\varphi(w)},\ t_i \in \Gamma(\pi).$$

By the definition of the A-module structure in B,

$$g \otimes ft = \varphi^*(f)g \otimes t \quad \text{for all} \quad g \in B,\ f \in A,\ t \in \Gamma(\pi).$$

Therefore, the following transformations are valid:

$$\sum_i g_i \otimes s_i = \sum_i \overline{g}_i \otimes s_i + \sum_i \beta_i \otimes s_i = \sum_i \overline{g}_i \otimes s_i + 1 \otimes \sum_i \beta_i s_i$$

$$= \sum_i \overline{g}_i \otimes s_i + 1 \otimes \sum_j f_j t_j$$

$$= \sum_i \overline{g}_i \otimes s_i + \sum_j \varphi^*(f_j) \otimes t_j \in \mu_w \otimes \Gamma(\pi)$$

(the last inclusion holds because $\varphi^*(f_j) \in \varphi^*(\mu_{\varphi(w)}) \subset \mu_w$).

We see that ν_w is an isomorphism at any point $w \in N$; hence ν is an isomorphism of B-modules. Its naturality with respect to π is evident. The theorem is proved. ▶

Exercise. Show that the $C^\infty(N)$-module $D_\varphi(M)$ consisting of vector fields along the map $\varphi\colon N \to M$ (see Section 9.47) is naturally isomorphic to the module $\Gamma(\varphi^*(\pi_T))$.

What is the algebraic meaning of the lifting of sections, i.e., what is the map $\Gamma(\pi) \to B \otimes_A \Gamma(\pi)$ it is described by? It is easy to see from the definitions that this is the map that takes each element $s \in \Gamma(\pi)$ to $1 \otimes s$.

12.44. Pseudo-bundles and geometrical modules. Let M be a smooth manifold and $\mathcal{F} = C^\infty(M)$. Theorem 12.32 gives a geometrical meaning to the notion of finitely generated projective $\mathcal{F}$-module. We want to find out, in the spirit of Sections 12.11–12.12, to what extent arbitrary modules over $\mathcal{F}$ possess a geometrical interpretation.

In Section 12.11, to an arbitrary module P over an arbitrary commutative K-algebra A we assigned a pseudo-bundle π_P. For the function algebra $\mathcal{F}(|P|)$ on the total space of the bundle, we took the symmetric algebra $S(P^*)$ of the module P. In the case of the algebra of smooth functions $A = C^\infty(M)$, it is natural to take the smooth envelope $\overline{\mathcal{F}(|P|)} = \overline{S(P^*)}$ instead of just $S(P^*)$, as we did in Section 10.12 for the cotangent bundle and the algebra of symbols.

12.45 Exercises.   1. Show that the maps

$$\pi_P \colon |P| \to |A|, \quad s_p \colon |A| \to |P| \quad (p \in P),$$

defined in Section 12.11, are <u>continuous</u> in the Zariski topology defined by the function algebra $\overline{\mathcal{S}(P^*)}$.

2. Let $P = D(C^\infty(\mathbf{K}))$ be the module of vector fields on the cross (see Exercises 7.14, 9.35, 9.45, 10.10). Describe $|P|$.

From now on, by continuous sections of a pseudo-bundles π_P, we shall understand sections, continuous in the Zariski <u>topology</u>, <u>corresponding</u> to the smooth envelope of the symmetric algebra $\overline{\mathcal{F}(|P|)} = \overline{\mathcal{S}(P^*)}$. The set of all such sections forms a module over the ring $A = C^\infty(M)$, which we denote by $\Gamma_c(\pi)$. The assignment $p \mapsto s_p$ defines a $C^\infty(M)$-module homomorphism

$$\mathbb{S} \colon P \to \Gamma_c(\pi).$$

By Theorem 12.32, for a projective finitely generated module P, this homomorphism is a monomorphism, and its image coincides with the submodule of smooth sections in $\Gamma_c(\pi)$.

We pass to examples of geometrical and nongeometrical modules over algebras of smooth functions.

12.46. Examples. *A. Geometrical nonprojective modules.*

I. A smooth map of manifolds $\varphi \colon M \to N$ gives rise to a homomorphism of the corresponding smooth function rings $\varphi^* \colon B \to A$ and thereby turns A into a B-module. An easy check shows that this module is always geometrical. However, it is projective only in exceptional cases (for instance, if φ is a diffeomorphism). The simplest example of a geometrical nonprojective module of this kind is obtained if M is the manifold consisting of one point.

Exercise. Describe all smooth maps φ for which the B-module A is projective.

II. The ideal μ_a of any point $a \in M$, viewed as a $C^\infty(M)$-module, is obviously geometrical. It turns out that this module is projective if and only if $\dim M = 1$.

Indeed, the value of the module μ_a at a point $b \in M$ is the quotient space $\mu_a/\mu_a\mu_b$. Its dimension is

$$\dim \mu_a/\mu_a\mu_b = \begin{cases} \dim M, & \text{if } b = a, \\ 1, & \text{if } b \neq a. \end{cases}$$

The first equality follows from the fact that μ_a/μ_a^2 is the cotangent space of the manifold M at the point a (see Section 9.27). The second one is a consequence of Lemma 2.11.

The fiber dimension of the vector bundle corresponding to μ_a is thus constant in the case $\dim M = 1$ and nonconstant in the case $\dim M > 1$.

There exist two nondiffeomorphic connected one-dimensional manifolds: the line $\mathbb{R}^1$ and the circle S^1. What vector bundles correspond to μ_a in each case? The answer, at first sight unexpected, is that for the line it is the trivial one-dimensional bundle $\mathbb{I}_{\mathbb{R}^1}$, while for the circle it is the Möbius band bundle described in Example 12.6, I. Here is the proof.

◀ Let $\mathcal{F} = C^\infty(\mathbb{R}^1)$ and let $a \in \mathbb{R}^1$ be an arbitrary point. Hadamard's Lemma 2.10 implies that the map $\mathcal{F} \to \mu_a$ that sends each function f to the product $(x - a)f$ establishes the module isomorphism $\mathcal{F} \to \mu_a$. Therefore, $\mu_a \cong \mathcal{F} \cong \Gamma(\mathbb{I}_{\mathbb{R}^1})$.

This argument does not apply to $\mathcal{F} = C^\infty(S^1)$ and $a \in S^1$, because in this case μ_a is not a principal ideal: there is no smooth function on the circle that vanishes only at one point and has a nonzero derivative at this point. The isomorphism between μ_a and the module of sections of the nontrivial vector bundle π, whose total space E_π is the Möbius band, can be defined as follows. We know that the tensor square of π is isomorphic to $\mathbb{I}_{\mathbb{R}^1}$ (see Example 12.38); hence there is a well-defined multiplication $\Gamma(\pi) \times \Gamma(\pi) \to \mathcal{F}$. Fix a section $f_0 \in \Gamma(\pi)$ with a single simple zero at the point a (i.e., $f_0(a) = 0$, $f_0'(a) \neq 0$, $f_0(b) \neq 0$ for $b \neq a$). Then the map that sends every section $f \in \Gamma(\pi)$ to the product $f_0 f \in \mathcal{F}$ establishes the required isomorphism $\Gamma(\pi) \to \mu_a$.   ▶

B. *Nongeometrical modules.*

III. The $C^\infty(M)$-module of lth order jets $J_z^l M = C^\infty(M)/\mu_z^{l+1}$ (see Section 9.64) is not geometrical if $l \geq 1$. This is due to the facts that:

(i) $\mu_{z'} \cdot J_z^l M = J_z^l M$, if $z' \neq z$;

(ii) $\mu_z \cdot J_z^l M = \mu_z/\mu_z^{l+1}$.

Leaving the proof to the reader, we deduce that

$$\bigcap_{z' \in M} \mu_{z'} \cdot J_z^l M = \mu_z/\mu_z^{l+1}.$$

The last module is composed of all "invisible" elements of $J_z^l M$ (see 12.11) and is nontrivial for $l \geq 1$.

However, $C^\infty(M)$-modules $T_z M = D(M)/\mu_z D(M)$ (see Lemma 10.7) and $T_z^* M = \Lambda^1(M)/\mu_z \Lambda^1(M)$ are geometrical. Indeed, if P is one of them, then $\mu_{z'} \cdot P = P$ for $z' \neq z$ and $\mu_z \cdot P = 0$ (prove that) and therefore $\bigcap_{z' \in M} \mu_{z'} \cdot P = 0$.

IV. Let $A = C^\infty(\mathbb{R})$ and let $I \subset A$ be the ideal that consists of all functions with compact support. The reader can prove, by way of exercise, that the quotient module $P = A/I$ has the property $P = \bigcap_{x \in \mathbb{R}} \mu_x P$, so that the corresponding map $\mathbb{S}$ (see Section 12.45) is identically zero. The module P in this example consists entirely of invisible (unobservable) elements.

12.47. Vector bundles as pseudo-bundles. We are now in a position to keep the promise given previously and explain the relationship between the algebraic treatment of a vector bundle as a module and the treatment of a pseudo-bundle as an embedding of smooth algebras.

Proposition. *The algebra of functions on the total space of a vector bundle π is isomorphic to the smooth envelope of the complete symmetric algebra of the module of sections $\Gamma(\pi)$.*

◄ Since the modules of sections of a given bundle π and its conjugate π^* are isomorphic (see the remark in Section 12.38), it suffices to construct an isomorphism of the algebra $C^\infty(E_\pi)$ with the smooth envelope of the symmetric algebra of $\Gamma(\pi^*)$. Such an isomorphism can be built in a natural way. Indeed, every section $s \in \Gamma(\pi^*) = S^1(\Gamma(\pi^*))$ defines a function on E_π that is linear on the fibers. Elements of $S^2(\Gamma(\pi^*))$ correspond to functions on E_π that are quadratic on the fibers; elements of $S^3(\Gamma(\pi^*))$ correspond to functions on E_π that are cubic on the fibers; etc. Such functions are obviously smooth. The whole symmetric algebra $S(\Gamma(\pi^*))$ can be considered as the algebra of all functions on E_π polynomial on every fiber of π. The construction of the smooth envelope (Section 3.36) extends the set of polynomial functions to the set of all smooth functions.   ►

12.48. In this book, we deal with smooth manifolds, smooth functions, smooth vector fields, smooth sections of vector bundles, etc. What can we say about similar objects that are not infinitely smooth, but for instance of class C^m? These notions of the standard calculus can be treated in the algebraic framework, using different functional algebras and using the change of rings procedure.

Exercises.   1. Let P be the module of smooth (of class C^∞) sections of a vector bundle π_P. For an arbitrary $m \geqslant 0$, give an algebraic definition of the module of sections of this bundle belonging to the class C^m (e.g., to C^0, i.e., continuous sections).

2. Give an algebraic definition of vector fields (differential operators) of class C^m on a smooth manifold M.

13

Localization

13.1. Localization of differential operators. Locality is the most important property of differential operators. It is precisely this property that explains the fundamental role that the differential calculus plays in the natural sciences, especially in physics. Indeed, the experimental information on the basis of which various theories are constructed is, by its very nature, local is space, in time and in many other respects. Therefore, the mathematics that processes this information must be localizable. Ignoring this principle leads to substantial difficulties in the construction of new theories, as we can observe, for instance, in contemporary quantum field theory.

The importance of using a localizable mathematical apparatus in classical physics does not directly jump to the eye mainly because the standard, historically arisen coordinate description of differential operators

$$\Delta = \sum a_\sigma(x) \frac{\partial^{|\sigma|}}{\partial x_\sigma}$$

is automatically localized, i.e., can be applied to any function independently of the domain of the variable in which the function is defined. Therefore, in this "classical" context, the question of localization is not a problem.

In the general framework of the differential calculus on commutative algebras, the localization property of differential operators means, in particular, that it is possible to localize them to any domain "observable" by the given commutative algebra, for example on open (in the Zariski topology) subsets of its spectrum. In this chapter, we first present the necessary general facts having to do with the localization of algebras and modules

© Springer Nature Switzerland AG 2020

J. Nestruev, *Smooth Manifolds and Observables*, Graduate Texts
in Mathematics 220, https://doi.org/10.1007/978-3-030-45650-4_13

(see Sections 11.6–11.8) and conclude the chapter with the derivation of a general formula on the localization of differential operators.

13.2. Universality property of localization. We shall say that, in an A-module P, division by an element $s \in A$ *is possible* if the $\Bbbk$-linear operator $p \mapsto sp$, $p \in P$, is invertible. In that case, $s^{-1}p$ denotes the element of the module P such that $s(s^{-1}p) = p$. If division by the elements s_1 and s_2 is possible in P, then obviously it is possible by s_1s_2. Therefore, all the divisors $s \in A$ of the module P constitute a multiplicative set $S_{\mathrm{div}}(P) \subset A$.

Lemma. *Let $S \subset S_{\mathrm{div}}(P)$ be a multiplicative set. Then P carries the structure of an $S^{-1}A$-module.*

◀ Let us define an $S^{-1}A$-module structure in P by the formula

$$(a/s)p \overset{\text{def}}{=} s^{-1}(ap), \ a \in A, \ s \in S, \ p,$$

about which, however, we must check that it is well defined. Suppose that $a/s = a_1/s_1$, i.e., $s'(as_1 - a_1s) = 0$ for some $s' \in S$. We must verify that $s^{-1}(ap) = s_1^{-1}(a_1p)$. But this follows from the sequence of equalities

$$s^{-1}(ap) = s_1^{-1}(a_1p) \iff (ss_1)[s^{-1}(ap)] = (ss_1)[s_1^{-1}(a_1p)] \iff$$
$$\iff (s_1a - sa_a)p = 0 \iff s'(as_1 - a_1s)p = 0.$$

The fact that P with this structure is an $S^{-1}A$-module is obvious.   ▶

Proposition. *Suppose P and Q are A-modules, and division by elements of the multiplicative set S in Q are possible. Then the correspondence*

$$\mathrm{Hom}_A(P,Q) \ni H \mapsto S^{-1}H \in \mathrm{Hom}_{S^{-1}A}(S^{-1}P, Q),$$

where $(S^{-1}H)(p/s) \overset{\text{def}}{=} s^{-1}H(p)$ is an $S^{-1}A$-module isomorphism.

◀ First of all, we must prove that the map $S^{-1}H$ is well defined. This, however, can be proved exactly like the similar statement in the previous lemma. Further, the verification of the fact that $S^{-1}H$ is a homomorphism of $S^{-1}A$ is obvious and we omit it. Finally, noting that $H = S^{-1}H \circ \iota_{S,P}$, where $\iota_{S,P} \colon P \to S^{-1}P$ is the canonical homomorphism $(p \mapsto (p,1))$, we see that the correspondence $H \to S^{-1}H$ is bijective.   ▶

The homomorphism ι_S takes elements $s \in S$ to invertible elements of the algebra $S^{-1}A$ and this property uniquely characterizes the pair $(\iota_S, S^{-1}A)$. Namely, we have:

Corollary. *Let $H \colon A \to B$ be a homomorphism of commutative algebras with unit (over a field $\Bbbk$) such that all elements $H(s)$, $s \in S$, are invertible. Then H can be uniquely presented in the form $H = H_S \circ \iota_S$, where $S \colon S^{-1}A \to B$ is a $\Bbbk$-algebra homomorphism.*

◀ From the homomorphism H, the algebra B acquires an A-module structure and, by the previous lemma, the structure of an $S^{-1}A$-module as well. Therefore, the homomorphism $S^{-1}H \colon S^{-1}A \to B$ of $S^{-1}A$-modules is well defined. It is easy to check that it is also an algebra homomorphism, so it only remains to put $H_S = S^{-1}H$. ▶

Exercise. Let $S \subset A$ be a multiplicative subset. A homomorphism of commutative $\Bbbk$-algebras $\iota \colon A \to C$ will be called *S-invertible*, if all the elements $\iota(s)$, $s \in S$, are invertible. An S-invertible homomorphism ι is called *universal*, if for any S-invertible homomorphism $H \colon A \to B$ there exists a unique $\Bbbk$-algebra homomorphism $H_1 \colon C \to B$ for which we have $H = h_1 \circ \iota$. Show that there exists an isomorphism $\varphi \colon S^{-1}A \to C$ such that $\iota = \varphi \circ \iota_S$. In other words, any homomorphism (of the type specified above) is equivalent to the homomorphism ι_S.

13.3. Geometrical portrait of localization. In our previous exposition, we were able to appreciate the usefulness of the fact that tangent vectors and the other differential operators that we considered can be localized. There we understood localization as the passage from the algebra $C^\infty(M)$ to the algebra $C^\infty(U)$, as they are *naively* described in Sections 3.23 and 3.24. The algebraic formalization of that "experimental material", which is well known in algebraic geometry, was presented in Section 11.6. It will now be appropriate to illustrate this procedure geometrically.

Let $S \subset A$ be a multiplicative subset. With it, we associate the subset

$$U_S \overset{\text{def}}{=} \{h \in |A| \mid h(s) \neq 0, \forall s \in S\} = \bigcap_{s \in S} U_s \subset |A|.$$

Proposition. *The map* $|\iota| \colon |S^{-1}A| \to |A|$ *is an embedding and*
$$\Im(|\iota|) = U_S.$$

If S is finitely generated, then $|S^{-1}A|$ is an open (in the Zariski topology) subset of $|A|$.

◀ If $\bar{h} \in |S^{-1}A|$, $h = |\iota|(\bar{h})$ and $s \in S$, then
$$1_{\Bbbk} = \bar{h}(1_{S^{-1}A}) = \bar{h}(s/1 \cdot 1/s) = \bar{h}(s/1)\bar{h}(1/s).$$

Therefore,
$$\bar{h}(s/1) = h(s) \neq 0, \, \forall s \in S, \quad \bar{h}(1/s) = h(s)^{-1}, \quad \bar{h}(a/s) = h(a)h(s)^{-1}.$$

The last one of these equalities shows that h uniquely determines $\bar{h}$, i.e., $|\iota|$ is an embedding. Besides, we see that

$$\Im(|\iota|) \subset \{h \in \mathrm{Spec}_{\Bbbk}\, A \mid h(s) \neq 0, \, \forall s \in S\}.$$

Conversely, if $h(s) \neq 0$, $\forall s \in S$, we define a certain element $\bar{h} \in |S^{-1}A|$ by setting $\bar{h}(a/s) = h(a)h(s)^{-1}$. The reader will easily verify that this element is well defined, and the existence of such an $\bar{h}$ proves the converse inclusion. Finally, noting that $U_{ab} = U_a \cap U_b$, we see that

$$\Im(|\iota|) = \bigcap_{s \in S} U_s = \bigcap_{i=1}^{m} U_{s_i}$$

if the elements $s_1, \ldots, s_m$ generate S.   ▶

The equality $s(a_1 s_2 - a_2 s_1) = 0$ that describes the equivalence relation in the definition of formal fractions a/s also has a clear geometrical interpretation. As is easily seen, it is equivalent to the equality $f_{a_1}/f_{s_1} = f_{a_2}/f_{s_2}$ for functions defined on U_S (see Section 8.4).

It is intuitively clear that U_S does not define S, nor $S^{-1}A$. One of the reasons for this is illustrated by the following example.

Example. Let M be the closed interval $[-1, 1]$, while $U = \{-1 < x < 1\}$ is the corresponding open one and $A = C^\infty(M)$. Then, as we have seen, $U = U_S$ if

$$S = \{f \in C^\infty(M) \mid f(x) \neq 0, \forall x \in U\}.$$

On the other hand, $U = U_{S_1}$, where S_1 is generated (multiplicatively) by the function

$$\varphi = (x^2 - 1)|_M.$$

Here the algebras $S^{-1}A$ and $S_1^{-1}A$ are not isomorphic. Indeed, suppose that $g/\varphi^k \in S_1^{-1}A$, $g \in A$. Then $\varphi^k/1 \cdot g/\varphi^k = g|_U$ is a bounded function on U, since $g \in C^\infty(M)$. However, there does not exist any function $\psi \in C^\infty(M)$ such that $\psi^l/1 \cdot \mu$ becomes a bounded function on U for an any value of the degree l independent of the element $\mu \in S^{-1}A$. For example, this will be the case for $\mu = 1/s$, $0 < s \in S$, if $1/s$ tends to infinity as $x \to 1$ faster than any degree of $|\psi^k|$.

Exercise.   1. Give an example of a multiplicative set $S \subset C^\infty(M)$ for which $U_S = \varnothing$.

2. Is there such a multiplicative set in the polynomial algebra $\Bbbk[x_1, \ldots, x_n]$?

13.4. U_S and S_V. Subsets of the spectrum of an algebra A are a natural source of useful multiplicative systems. Namely, to an arbitrary nonempty subset $V \subset |A|$, we can associate the multiplicative system

$$S_V \overset{\text{def}}{=} \{s \in A \mid h(s) \neq 0, \forall h \in V\}.$$

In other words, S_V is generated by elements $a \in A$ such that the function f_a on $|A|$ (see Section 8.4) does not vanish at points of the subset V. If V is empty, then S_V is defined as the set of those elements of the algebra A which are not divisors of zero. In all these cases, S_V is nonempty because it always contains 1_A.

Conversely, to any multiplicative system S in A, we can associate the subset

$$U_S \overset{\text{def}}{=} \{h \in |A| \mid h(s) \neq 0, \forall s \in S\}.$$

We have already dealt with these two constructions previously in the case when the subset V is open. Further, we shall see that they are useful not only in this case.

Iterating these constructions, we obtain a chain of the form

$$V \Longrightarrow S_V \Longrightarrow U_{S_V} \Longrightarrow \ldots, \qquad S \Longrightarrow U_S \Longrightarrow S_{U_S} \Longrightarrow \ldots \qquad (13.1)$$

Obviously, $V \subset U_{S_V}$ and $S \subset S_{U_S}$. The examples presented below show that, generally speaking, U_{S_V} and S_{U_S} are wider than V and S, respectively. Here the chains (13.1) already stabilize at the third step.

Proposition. *Let* $V' = U_{S_V}$ *and* $S' = S_{U_S}$. *Then* $S_{V'} = S_V$ *and* $U_{S'} = U_S$.

◀ Note that $S_{V_1} \subset S_{V_2}$ if $V_1 \supset V_2$. But since $V' \supset V$, it follows that $S_{V'} \subset S_V$. On the other hand, if $\hat{S} = S_V$, then $U_{\hat{S}} = U_{S_V} = V'$ and $\hat{S}' = S_{U_{\hat{S}}} = S_{V'}$. Therefore, $S_V = \hat{S} \subset \hat{S}' = S_{V'}$ and so $S_{V'} = S_V$.

The equality $U_{S'} = U_S$ is proved exactly in the same way. ▶

13.5. Examples. In the examples in this series, we have assumed that $A = \Bbbk[x_1, \ldots, x_n]$ and so $|A| = \Bbbk^n$, while in examples I-V, it is also assumed that $V = \Bbbk^n \setminus \{0\}$.

I. According to its definition, S_V consists, in this case, of polynomials whose restriction to V is everywhere nonzero. Let f be such a polynomial and $L \subset V$ be a straight line. Then $f|_L$ is a polynomial on that line. It the field $\Bbbk$ is algebraically closed, e.g., $\Bbbk = \mathbb{C}$, then this polynomial has roots, provided its degree is positive. Therefore, $f|_L$ is a nonzero constant. But since this is true for all lines not passing through $(0, \ldots, 0)$, it follows that f is also a nonzero constant. Thus, in the case under consideration, S_V consists of nonzero constants and so U_{S_V} equals $\Bbbk^n$, i.e., is wider, than V.

II. S_V becomes much richer if the field $\Bbbk$ is not algebraically closed. For example, if $\Bbbk = \mathbb{R}$, then S_V contains, besides nonzero constants, positive definite quadratic forms in the variables $x_1, \ldots, x_n$. Since each such form vanishes at the point $(0, \ldots, 0)$, it follows that U_{S_V} does not contain that point and so coincides with V. Thus, in this case, $U_{S_V} = V$.

III. Let $f(t) \in \Bbbk[t]$ be a polynomial of degree l without roots in the field $\Bbbk$, and let $\varphi(x) = \varphi(x_1 \ldots, x_{n-1}) \in \Bbbk[x_1 \ldots, x_{n-1}]$ be a homogeneous polynomial of degree m that vanishes only at the point $(0, \ldots, 0) \in \Bbbk^{n-1}$. Then the homogeneous polynomial

$$g(x_1, \ldots, x_n) = x_n^{lm} f(\varphi(x_1/x_n, \ldots, x_{n-1}/x_n))$$

vanishes only at the point $\{0\} \in \Bbbk^n$. Indeed, this is obvious if $x_n \neq 0$, while $g(x_1, \ldots, x_{n-1}, 0) = \varphi(x_1, \ldots, x_{n-1})^l$. Besides, the above construction yields, by induction, polynomials f with the required properties, beginning with the polynomial $x_1 \in \Bbbk[x_1]$. Therefore, for the same reason as in the previous example, we have $U_{S_V} = V$. Thus, a nontrivial widening of the set V at the second step occurs only when the field $\Bbbk$ is algebraically closed.

IV. Since A is an algebra without zero divisors, $S = A \setminus 0_A$ is a multiplicative system, while $S^{-1}A$ is its fraction field. Therefore, $|S^{-1}A| = \varnothing$ and so $U_S = \varnothing$ and $S_{U_S} = S$.

V. Suppose the field $\Bbbk$ is not algebraically closed and the polynomial $g(x_1, \ldots, x_n)$ vanishes only at the point $\{0\} \in \Bbbk^n$ (see Example III above). If $S = \{g^k\}_{k \geq 0}$, then $U_S = \Bbbk^n \setminus \{0\}$ and so S_{U_S} consists of all the polynomials that do not vanish at the points of $\Bbbk^n \setminus \{0\}$. Thus, S_{U_S} is wider than S.

VI. If $V = \Bbbk^n$, then S_V consists of polynomials that do not vanish anywhere. If the field $\Bbbk$ is algebraically closed , then S_V consists of nonzero constants only. For fields that are not algebraically closed, this is no longer so, since there exist polynomials of positive degree without zeros in $\Bbbk^n$. For instance, such is the polynomial $f = g(\psi(x))$, where f is the polynomial from Example III, while the polynomial $\psi \in \Bbbk[x_1, \ldots, x_n]$ is arbitrary.

13.6. Germs. As we have already seen (Proposition 11.7), the localization of the algebra A with respect to a multiplicative set S_U is the analog of the passage from the algebra $C^\infty(M)$ to the algebra $C^\infty(U)$ under the condition that the subset $U \subset |A|$ is open. Now if, on the other hand, the set $U \subset M$ is closed, then the result of such a localization is the *algebra of germs of smooth functions along the subset U*. This will be shown below. Therefore, it is appropriate to regard the algebra of germs of an arbitrary algebra A as the analog of the algebra $S_U^{-1}A$ for some closed subset $U \subset |A|$. It should be noted that in the case of smooth manifolds, this construction is of the most interest when U is a smooth submanifold.

So, let $N \subset M$ be a closed subset. (Here we have replaced U by N because the letter U is systematically used to denote open sets.) Let us call two functions f and g from $C^\infty(M)$ *equivalent*, if they coincide in some neighborhood U of the set N and write $f \sim_N g$. By a *germ of a function $f \in C^\infty(M)$ along N* we mean its equivalence class w.r.t. $\sim_N$. Arithmetical operations with functions can obviously be carried over to germs. Thus, we have defined the *algebra of germs of functions* along a closed subset $N \subset M$; we denote this algebra by $C^\infty_{\mathrm{germ}}(N)$.

Proposition.   (a) *The algebras $C^\infty_{\mathrm{germ}}(N)$ and $S_N^{-1}C^\infty(M)$ are canonically isomorphic.*

  (b) *Let a subset $V \subset |A|$ be closed and let $\iota \colon A \to S_V^{-1}A$ be the canonical isomorphism. Then $\Im|\iota| = V$.*

◄ (a) Let U be an open set containing N. The equality $f|_U = g|_U$ is equivalent to $\iota(f) = \iota(g)$, where $\iota \colon C^\infty(M) \to S_N^{-1}C^\infty(M)$ is the canonical homomorphism. Indeed, the last relation means that $\varphi \cdot (f - g) = 0$ for some function $\varphi \in S_N$. Since, by definition, φ does not vanish anywhere on N, we have

$$N \subset U_\varphi = \{z \in M \mid \varphi(z) \neq 0\} \quad \text{and} \quad f|_{U_\varphi} = g|_{U_\varphi} \, .$$

The last equality is obvious since $\varphi(z) \neq 0$ and $\varphi(z)(f(z) - g(z)) = 0$ imply $f(z) = g(z)$.

Finally, we must establish that the algebra $S^{-1}C^\infty(M)$ consists only of the germs of smooth functions along N. This is equivalent to ι being an epimorphism. Let us prove this. Let $f/\varphi \in S^{-1}C^\infty(M)$. Then to the pair (f, φ) representing this formal fraction, we can associate the function $\psi \in C^\infty(U_\varphi)$, $\psi(z) = f(z)/\varphi(z)$. If the function $f' \in C^\infty(M)$ coincides with ψ on some open set $U \supset N$, then $f/\varphi = f'/1_A = \iota(f')$, i.e., we have $g(f - \varphi f') \equiv 0$ for some function $g \in S_N$. But since the function $f - \varphi f'$ is identically zero on U, we can choose, in the role of g, any function which is strictly positive on U and is zero outside of U. The existence of functions f' and g with the required properties was established previously.

(b) The fact that $V \subset \Im|\iota|$ immediately follows from the proposition in Section 13.3, so that it only remains to show that $(|A| \setminus V) \cap \Im|\iota| = \varnothing$. But the set $(|A| \setminus V)$ is open and so it is the union of some open sets from the base. Let us show that if U_a is one of them, then $U_a \cap (\Im|\iota|) = \varnothing$.

Note that $V \subset V_a \overset{\text{def}}{=} |A| \setminus U_a$. If $h_0 \in U_a$, then $\lambda = h_0(a) \neq 0$. Therefore, $a_0 = a - \lambda 1_A$, implies $h(a_0) = -\lambda \neq 0$ provided $h \in V_a$, so that we have $a_0 \in S_{V_a} \subset S_V$. On the other hand, $h_0(a_0) = 0$ and the proposition from Section 13.3 implies $h_0 \notin \Im|A|$. ▶

13.7. Example IV from Section 11.6 shows that quite different multiplicative systems can lead to the same localization of a given algebra A. So, it is of interest to find out when this can occur. In its turn, this question is a particular case of a more general and less natural question: How are two localizations of the same algebra related? In what follows, we shall begin to study this question, as well as related ones.

We will need to consider formal fractions corresponding to different multiplicative systems. The fact that such a fraction belongs to a given localization of the algebra A under consideration will depend on whether or not the denominator of the fraction belongs to that system.

The multiplicative system itself will be denoted by the capital letter S with additional indications, e.g., S_α or $\bar{S}$, while its elements will be denoted by a lower case s with the same indicator and, possibly, an additional index, e.g., $s_{\alpha,i}$ or $\bar{s}_i$. Finally, $\iota_S \colon A \to S^{-1}A$ will denote the canonical homomorphism .

13.8. Localizations of a localization. Let $V \subset U \subset M$ be open subsets of a manifold M. Then the algebra $C^\infty(V)$ is a localization of both the algebra $C^\infty(M)$ and the algebra $C^\infty(U)$ (Proposition 11.7). Now since $C^\infty(U)$ is a localization of the algebra $C^\infty(M)$, it follows that $C^\infty(V)$ is a "localization of a localization". This coincidence of iterated localizations is not only true for multiplicative sets of the form S_U on smooth manifolds, and in what follows we will show that it takes place for arbitrary multiplicative sets.

Let S and S_+ be multiplicative subsets in A and $S^{-1}A$, respectively. Further, let us consider the set $\bar{S}$ of elements $a \in A$ such that the pair (a, s) for some s is a formal fraction belonging to S_+. Since the pair (s, s) represents the unit $1_{S^{-1}A} \in S_+$, we have $S \subset \bar{S}$. The set $\bar{S}$ is multiplicative. Indeed, if $a_i/s_i \in S_+$, $i = 1, \ldots, m$, then $(a_1 \ldots a_m)/(s_1 \ldots s_m) \neq 0$, which is equivalent to $sa_1 \ldots a_m \neq 0$, for all $s \in S$. This shows that $\bar{S}$ is closed under multiplication and does not contain the zero element.

Proposition. "The localization of localization" $S_+^{-1}(S^{-1}A)$ *is naturally isomorphic to the algebra* $\bar{S}^{-1}A$.

◀ In what follows, we use the notation a_+/s_+ for formal fractions that belong to S_+. Thereby we indicate that $a_+ \in \bar{S}$. Let the pair $(a/s, a_+/s_+)$ represent the formal fraction $\mathbf{f}$ from $S_+^{-1}(S^{-1}A)$. Then the map η

$$S_+^{-1}(S^{-1}A) \ni \mathbf{f} \mapsto as_+/a_+s \in \bar{S}$$

is the required isomorphism. To establish this, we must first show that the map η is well defined.

Let $(a'/s', a'_+/s'_+)$ be another pair representing the fraction $\mathbf{f}$. This means that

$$\frac{\hat{a}_+}{\hat{s}_+} \cdot \left(\frac{aa'_+}{ss'_+} - \frac{a'a_+}{s's_+} \right) = 0 \quad \Longleftrightarrow \quad s_0\hat{a}_+(aa'_+s's_+ - a'a_+ss'_+) = 0,$$

for some $s_0 \in S$. We must now show that the formal fractions as_+/a_+s and $a's'_+/a'_+s'$ are equal, i.e.,

$$\bar{s}(aa'_+s's_+ - a'a_+ss'_+) = 0$$

for some $\bar{s} \in \bar{S}$. But for that it suffices to put $\bar{s} = s_0\hat{a}_+ \in \bar{S}$.

An arbitrary element of the algebra $\bar{S}^{-1}A$ can be expressed in the form $a/a_+s = as_+/a_+ss_+$. It is the image under the map η of the formal fraction represented by the pair $(a/ss_+, a_+/s_+)$. Hence, η is surjective. Finally, if $\eta(\mathbf{f}) = 0$, then $(a'_+s''_+) \cdot (as_+) = 0$ for some element $a'_+s''_+ \in \bar{S}$. In other words, $\hat{s}(aa_+ = 0)$ for $\hat{s} = s_+s''_+ \in S$. This implies

$$(a_+/s_+) \cdot (a/s) = aa_+/ss_+ = 0$$

and so $\mathbf{f} = 0$. Hence, the map η is injective. ▶

For further applications, it is convenient to denote the multiplicative set $\bar{S}$ by $S^{-1}(S_+)$.

Corollary. (a) *Let a multiplicative set* S_1 *in* A *be such that the set* $\iota_S(S_1) \stackrel{\text{def}}{=} \{\iota_S(s_1) \mid s_1 \in S_1\}$ *is multiplicative in* $S^{-1}A$. *Then the set* S_1 *is contained in* $S^{-1}(S_+)$.

(b) *The set* $\iota_S(S^{-1}(S_+))$ *is multiplicative in* $S^{-1}A$.

(c) *The localization of the algebra* $S^{-1}A$ *via the multiplicative set* $\iota_S(S^{-1}(S_+))$ *coincides with the localization of the algebra* A *via the set* $S^{-1}(S_+)$ *and hence coincides with* $S_+^{-1}(S^{-1}A)$.

◀ The first two statements and the first part of the third one are immediate consequences of the construction of the set $S^{-1}(S_+)$, while the remaining part is the assertion of the proposition from Section 13.8.   ▶

13.9. Compatibility of multiplicative sets. Assertion (c) of the previous corollary shows that any localization of the algebra $S^{-1}A$ coincides with its localization via a multiplicative set of the form $\iota_S(S')$, where S' is some multiplicative set in the algebra A. In this connection, it makes sense to characterize those multiplicative sets in A which are taken to multiplicative sets of the algebra $S^{-1}A$ by the homomorphism ι_S.

Let us call two multiplcative sets S_1 and S_2 of the algebra A *compatible*, if $s_1 s_2 \neq 0$ for all $(s_1, s_2) \in S_1 \times S_2$. If the algebra A is geometrical, then the inequality $s_1 s_2 \neq 0$ is equivalent to the existence of an element $h \in |A|$ such that $0 \neq h(s_1 s_2) = h(s_1)h(s_2)$. In its turn, this is equivalent to $U_{S_1} \cap U_{S_2} \neq \varnothing$.

Since, in the algebra of subsets of a given set, the intersection operation plays the role of multiplication, one can use the geometrical interpretation of the compatibility relation given above in order to naturally introduce the product of compatible multiplicative sets by setting

$$S_1 S_2 \overset{\text{def}}{=} \{s_1 s_2 \mid (s_1, s_2) \in S_1 \times S_2\}.$$

Note that $S_1 S_2 = S_2 S_1$ and $S_1, S_2 \subset S_1 S_2$. Besides, if the multiplcative sets $S_1 S_2$ and S_3 are compatible, then

$$s_1 s_2 s_3 \neq 0 \text{ for all } (s_1, s_2, s_3) \in S_1 \times S_2 \times S_3.$$

This obviously implies the compatibility of the sets S_i and S_j, as well as that of the sets S_i and $S_j S_k$, where i, j, k are different indices taking values in $\{1, 2, 3\}$. In particular, we have associativity, $(S_1 S_2)S_3 = S_1(S_2 S_3)$, for the product of compatible multiplicative sets.

Proposition. (a) *The set $\iota_{S_1}(S_2)$ is multiplicative in the algebra $S_1^{-1}A$ if and only if the multiplcative sets S_1 and S_2 are compatible in the algebra A.*
   (b) *If two multiplcative sets S_1 and S_2 are compatible, then*

$$\left(\iota_{S_1}(S_2)\right)^{-1}(S_1^{-1}A) = (S_1 S_2)^{-1}A = \left(\iota_{S_2}(S_1)\right)^{-1}(S_2^{-1}A).$$

◀ (a) The multiplicativity of the set $\iota_{S_1}(S_2)$ means that $0 \neq \iota_{S_1}(s_2) = s_2/1$ for all $s_2 \in S_2$. In its turn, the inequality $s_2/1 \neq 0$ means that $s_1 s_2 \neq 0$ for all $s_1 \in S_1$.

(b) In view of the commutativity of the "product" $S_1 S_2$, it suffices to prove only the first of these two equations. To do that, we only need to note that the set $S^{-1}(S_+)$ for $S = S_1$ and $S_+ = \iota_{S_1}(S_2)$ coincides with $S_1 S_2$ and use the proposition from Section 13.8.   ▶

Exercise.   1. Show that the multiplicative sets S_1 and S_2 are compatible whenever $U_{S_1} \cap U_{S_2} \neq \varnothing$.

2. Give an example of compatible multiplicative sets S_1 and S_2 for which $U_{S_1} \cap U_{S_2} = \varnothing$.

3. Show that the compatibility relation is not transitive in general.

13.10. Noncompatible multiplicative sets. The multiplicative sets S_1 and S_2 are called *noncompatible*, if there exist elements $s_i \in S_i$, $i = 1, 2$, such that $s_1 s_2 = 0$. The condition $U_{S_1} \cap U_{S_2} = \varnothing$ is geometrically necessary for the noncompatibility of S_1 and S_2 (see Exercises 1 and 2 from the previous section).

For noncompatible multiplicative sets, there is an analog of the operation of union for nonintersecting sets. Namely, to each pair $s_1 \in S_1$, $s_2 \in S_2$ such that $s_1 s_2 = 0$, let us associate the element $s_1 + s_2$. The set consisting of all products of elements of that form and of the unit element 1_A will be denoted by $S_1 \sqcup S_2$. Let us show that it is multiplicative, i.e., that an arbitrary product of the form $w = (s_{1,1} + s_{2,1}) \cdot \ldots \cdot (s_{1,m} + s_{2,m})$, where $s_{i,j} \in S_i$ and $s_{1,j} s_{2,j} = 0$, is nonzero. Let $s = s_{1,1} \cdot \ldots \cdot s_{1,m}$. Then

$$sw = (s_{1,1}^2 + s_{1,1}s_{2,1}) \cdot \ldots \cdot (s_{1,m}^2 + s_{1,m}s_{2,m}) = s_{1,1}^2 \cdot \ldots \cdot s_{1,m}^2 \in S_1 \, .$$

Therefore, $sw \neq 0$ and so $w \neq 0$.

Examples. I. Let U_1, U_2 be open subsets of a manifold M such that $U_1 \cap U_2 = \varnothing$, and let $S_i = S_{U_i}$, $i = 1, 2$. Then $S_1 \sqcup S_2$ consists of all functions $f_1 + f_2$, $f_i \in C^\infty(M)$, such that f_1 does not vanish anywhere on U_1, is identically zero on U_2, and similarly for f_2. So, $U_{S_1 \sqcup S_2} = U_1 \cup U_2$.

II. Let $\mathbf{K}$ be the coordinate cross on the plane (see Exercise 1 in Section 2.11) and let $\bar{x} = x|_{\mathbf{K}}, \bar{y} = y|_{\mathbf{K}}$. Let also $S_1 = \{\bar{x}^k\}_{k \geq 0}$ and $S_2 = \{\bar{y}^k\}_{k \geq 0}$. Then U_{S_1} and U_{S_2} are the coordinate axes with the point $(0,0)$ deleted, while $S_1 \sqcup S_2$ consists of the unit $1_{\mathbf{K}}$ and all possible functions of the form $\bar{x}^k + \bar{y}^l$, $k \geq 1$, $l \geq 1$. Here $U_{S_1 \sqcup S_2} = \mathbf{K} \setminus \{(0,0)\}$.

13.11. Natural homomorphisms of localizations. The algebra homomorphism $H \colon S_1^{-1}A \to S_2^{-1}A$ is called *natural* if $\iota_{S_2} = H \circ \iota_{S_1}$.

Proposition. (a) *A necessary and sufficient condition for the existence of the natural homomorphism $H \colon S_1^{-1}A \to S_2^{-1}A$ is the invertibility of all elements $\iota_{S_2}(s_1)$, $s_1 \in S_1$.*

(b) *The natural homomorphism is unique (if it exists).*

(c) *The element $\iota_{S_2}(s_1)$ is invertible if and only if it divides some element $s_2 \in S_2$, i.e., if there exists an element $a \in A$ such that $s_1 a = s_2$.*

(d) *If the natural homomorphism H exists, then the multiplcative sets S_1 and S_2 are compatible.*

◀ (a)–(c) Further it will be convenient to use the notation $a/1_i \overset{\text{def}}{=} \iota_{S_i}(a)$ which stresses that $a/1_i$ is a formal fraction in $S_i^{-1}A$. The elements $\iota_{S_1}(s_1)$ are obviously invertible, and therefore, their images under the homomorphism H, i.e., the elements $\iota_{S_2}(s_1)$, are also invertible. Therefore,

$$H(a/s_1) = H((a/1_1) \cdot (1/s_1)) = H(\iota_{S_1}(a))H(\iota_{S_1}(s_1)^{-1})$$
$$= \iota_{S_2}(a)[\iota_{S_2}(s_1)]^{-1},$$

which proves the uniqueness of the homomorphism H.

Further, the invertibility of the element $\iota_{S_2}(s_1) = s_1/1_2$ means that for some $a' \in A$ and $s_2', s_2'' \in S_2$, we have

$$(s_1/1_2) \cdot (a'/s_2') = 1/1_2 \iff s_2''(s_1 a' - s_2') = 0 \implies s_1 a = s_2,$$

where $a = a' s_2''$, $s_2 = s_2' s_2''$. Conversely, the relation $s_1 a = s_2$ implies $(s_1/1_2) \cdot (a/s_2) = 1/1_2$, i.e., $[\iota_{S_2}(s_1)]^{-1} = a/s_2$. This proves assertion (c).

Finally, assuming that the elements $\iota_{S_2}(s_1)$ are invertible, we define the homomorphism H by setting

$$H(b/s_1) \stackrel{\text{def}}{=} \iota_{S_2}(b) \cdot (a/s_2), \quad \text{where} \quad s_1 a = s_2.$$

We leave to the reader the routine verification of the fact that h is well defined.

(d) The fact that $s_1 s_2' \neq 0$ for all $s_1 \in S_1$, $s_2' \in S_2$ follows from the relation $s_1 s_2' a = ss_2 s_2' \neq 0$, where $s_1 a = s_2$. ▶

Example. If $S_1 \subset S_2$, then, in view of the fact that any element $\iota_{S_2}(s_1)$ is (obviously) invertible, the natural homomorphism $H \colon S_1^{-1}A \to S_2^{-1}A$ exists. In particular, the homomorphism $\iota_{S_+} \colon S^{-1}A \to S_+^{-1}(S^{-1}A)$, taking the "localization" to the "localization of localization", is identified with a similar homomorphism for $S_1 = S$ and $S_2 = \bar{S}$ (see Section 13.8). In particular, if $U_1 \supset U_2$ are open sets of a manifold M, then the restriction homomorphism

$$C^\infty(U_1) \to C^\infty(U_1), \ f \mapsto f|_{U_2},$$

is natural. If $U_1 \not\supset U_2$, then there does not exist any natural homomorphism from $C^\infty(U_1)$ to $C^\infty(U_2)$.

The fact that the natural homomorphism $H \colon S_1^{-1}A \to S_2^{-1}A$ is a more delicate version of the restriction of functions follows from the relation $V_{S_1} \supset V_{S_2}$. Indeed, in this case any element $s_1 \in S_1$ divides some element $s_2 \in S_2$, so that $h(s_1) \neq 0$, if $h(s_2) \neq 0$. Moreover, as can be easily seen, $f_{a/s_1}|_{V_{S_2}} = f_{H(a/s_1)}$.

13.12 Exercise. Prove that the composition of natural homomorphisms is also a natural homomorphism.

13.13. Equivalence of multiplicative sets. Two multiplicative sets S_1 and S_2 are said to be *equivalent* if the localizations $S_1^{-1}A$ and $S_2^{-1}A$ of the algebra A coincide, i.e., are naturally isomorphic.

Corollary. (a) *Two multiplicative sets S_1 and S_2 are equivalent if and only if any element $s_1 \in S_1$ divides some element $s_2 \in S_2$ and conversely.*

(b) *The homomorphism $\iota_S \colon A \to S^{-1}A$ is a isomorphism if and only S consists of invertible elements.*

(c) *Two localizations $S_1^{-1}A$ and $S_2^{-1}A$ are naturally isomorphic, if $S_{ij} \overset{\text{def}}{=} \iota_{S_i}(S_j)$, $i \neq j$, is a multiplicative set in $S_i^{-1}A$ such that the canonical homomorphism $\iota_{S_{ij}} \colon S_i^{-1}A \to S_{ij}^{-1}(S_i^{-1}A)$ is an isomorphism.*

(d) *If two multiplicative sets S_1 and S_2 are equivalent, then the sets S_i and $S_1 S_2$, $i = 1, 2$, are also equivalent.*

◄ Assertions (a)–(c) directly follow from the proposition in the previous section. To prove (c), one must additionally have in mind assertion (b).

(d) It suffices to prove that any element $s_1 s_2 \in S_1 S_2$ divides, say, some element from S_1, since the converse is obvious. But s_2 divides some element $s_1' \in S_1$, hence $s_2 a = s_1'$ and so $s_1 s_2 a = s_1 s_1' \in S_1$. ►

Examples. I. Let U_1 and U_2 be diffeomorphic open domains in a manifold M and let $S_i = S_{U_1}$, $i = 1, 2$. Then the localizations $S_i^{-1}C^\infty(M) = C^\infty(U_i)$ of the algebra $C^\infty(M)$ are naturally isomorphic only when $U_1 = U_2$, although as abstract algebras they are always isomorphic.

II. Let $A = C^\infty(\mathbf{K})$ and $S_1 = \{\bar{x}^k\}_{k \geq 0}$, $S_2 = \{\bar{y}^k\}_{k \geq 0}$ be given in the notation of Example II from Section 13.10. Then S_1 and S_2 not equivalent, although the algebras $S_1^{-1}A$ and $S_2^{-1}A$ are isomorphic.

III. Let $s \in A$ be a nonnilpotent element and let l, k be nonnegative integers. Then the multiplcative sets $\{s^{lm}\}_{m \geq 0}$ and $\{s^{km}\}_{m \geq 0}$ are equivalent if $lk > 0$, and nonequivalent, if $k + l = 0$, while the element s is noninvertible.

13.14. Maximal multiplicative sets. Note that any pair of equivalent multiplicative sets is contained in some equivalent multiplicative set (assertion (d) of the proposition from the previous section). In connection with this, one could suppose that there exists a *maximal multiplicative set*, containing all those equivalent to it. This is indeed so, moreover, the construction of maximal sets is very simple.

Proposition. *Let S be some multiplicative set in an algebra A and let $\hat{S}$ be the sets of divisors of all elements of S. Then the set $\hat{S}$ is multiplicative and contains any multiplicative set equivalent to S.*

◄ Let $a, a' \in \hat{S}$. Then $ab = s$, $a'b' = s'$ for some $b, b' \in A$, $s, s' \in S$. Since $(aa')(bb') = ss' \in S$, it follows that $\hat{S}$ is closed w.r.t. multiplication and does not contain zero. Obviously, $S \subset \hat{S}$ and, in particular, $1_A \in \hat{S}$. Thus, the set $\hat{S}$ is multiplicative.

If a multiplicative set S_1 is equivalent to S and $s_1 \in S_1$, then $s_1 b_1 = s''$ for some $b_1 \in A$, $s'' \in S$. Hence, $s_1 \in \hat{S}$. ►

It is important to note that there is a bijection between localizations of the algebra A and maximal multiplicative sets.

Examples. The following multiplicative sets are maximal.

I. The set S_V for any $V \subset |A|$.

II. The set of polynomials of the form $\lambda \varphi^k$, $\varphi \in \Bbbk[x_1, \dots, x_n]$ under the condition that $0 \neq \lambda \in \Bbbk$ and the polynomial φ is irreducible.

III. The set $A \setminus \mathcal{I}$, where $\mathcal{I}$ is an ideal.

13.15. Geometricity and localization. The geometricity property of algebras or modules is not preserved in general under localization, and the reasons for that should be understood. First of all, let us note that the canonical homomorphism $\iota = \iota_P : P \to S^{-1}P$ generates a linear map of $\Bbbk$-vector spaces

$$\iota_{P,h} : P_h \to (S^{-1}P)_{\bar{h}}, \ p_h \mapsto (\iota(p))_{\bar{h}}, \quad \bar{h} \in |S^{-1}A|, \ h = |\iota|(\bar{h}).$$

Lemma.   1. $\operatorname{Ker} \bar{h} = S^{-1}(\operatorname{Ker} h)$.

2. $\iota_{P,h}$ *is an isomorphism.*

◄ The relation $\bar{h}(a/s) = h(s)^{-1}h(a)$ immediately implies the first assertion. To prove the second one, we first note that the map $\iota_{P,h}$ is surjective because

$$(p/s)_{\bar{h}} = (1/s)_{\bar{h}}(p/1)_{\bar{h}} = h(s)^{-1}\iota(p)_{\bar{h}}.$$

Further, if $\iota_{P,h}(p_h) = 0$, then

$$\iota(p) \in \operatorname{Ker} \bar{h} \cdot S^{-1}P \Leftrightarrow \iota(p) = \sum_i (a_i/s_i) \cdot (p_i'/s_i'),$$

where $a_i \in \operatorname{Ker} h$, $s_i, s_i' \in S$ and $p_i' \in P$. Bringing this sum to a common denominator, we obtain

$$\sum (a_i/1) \cdot (p_i/s) = \left(\sum a_i p_i \right) / s'', \quad \text{where} \quad s'' \in S.$$

Thus $\iota(p) = p''/s''$, where $p'' = \sum a_i p_i \in \operatorname{Ker} h \cdot P$, or, equivalently, we can write $\tilde{s}p'' = \tilde{s}s''p$ for some $\tilde{s} \in S$. Therefore,

$$h(\tilde{s}s'')p_h = (\tilde{s}s''p)_h = (\tilde{s}p'')_h = h(\tilde{s})p_h'' = 0$$

But since we have $h(\tilde{s}s'') \neq 0$, it follows from Proposition 13.3 that $p_h = 0$, i.e., the map $\iota_{P,h}$ is injective.   ►

Corollary. *The family of vector spaces* $(S^{-1}P)_{\bar{h}}, \bar{h} \in |S^{-1}A|$ *associated with the $S^{-1}A$-module $S^{-1}P$ can be canonically identified with the family* $\{P_h\}, h \in \Im|\iota|$.

Exercise. Prove that the $S^{-1}A$-module $S^{-1}P$ is projective, provided the A-module P is projective.

13.16. Let $U \subset M$ be an open subset of a manifold M, let $A = C^{\infty}(M)$ and $P = \Gamma(\pi)$, where π is a vector bundle over M.

Proposition. *The $S_U^{-1}A$-module $S_U^{-1}P$ and the $C^{\infty}(U)$-module $\Gamma(\pi|_U)$ are naturally isomorphic.*

◀ Recall that $C^\infty(U) = S_U^{-1}A$, where the equality is understood as the natural identification (Proposition 11.7). Here the fiber $\pi^{-1}(z)$ is identified with $P_z \overset{\text{def}}{=} P_{h_z}$, while the element $p \in P$, with the section

$$\sigma_p\colon M \ni z \mapsto p_z \overset{\text{def}}{=} p_{h_z} \in \pi^{-1}(z)$$

and this section is smooth. Besides, there are natural identifications $(S_U^{-1}P)_z \leftrightarrow P_z \leftrightarrow \pi^{-1}(z)$, if $z \in U$ (obtained in Section 13.15). This shows that

(1) the set of maps $U \ni z \mapsto (S_U^{-1}P)_z$ is identical to the family of fibers of the bundle $\pi|_U$;

(2) the section $\sigma_p|_U$ coincides with the section $U \ni z \mapsto \iota_P(p)_z$.

But since any element of the module $S_U^{-1}P$ is of the form p/f, $f \in S_U$, it follows that the corresponding section of the vector bundle $\pi|_U$ is $1/f \cdot \sigma_p|_U$. This establishes an isomorphism between the $S_U^{-1}A$-module $S_U^{-1}P$ and the $C^\infty(U)$-module $\Gamma(\pi|_U)$. ▶

13.17 Proposition. *Let P be a geometrical A-module and let $S \subset A$ be a multiplicative subset. Then*

(a) $p/s \in \mathrm{Inv}(S^{-1}P) \Leftrightarrow \iota(p) \in \mathrm{Inv}(S^{-1}P)$, *here Inv is defined in Section 8.4;*

(b) $\iota(p) \in \mathrm{Inv}(S^{-1}P) \Leftrightarrow \Im(|\iota|) \cap U_p = \varnothing$. *Here $\iota(p) \neq 0$ only when we additionally have $U_p \cap U_s \neq \varnothing$, $\forall s \in S$;*

(c) *if the given algebra A is geometrical, $0 \neq a \in A$ and $S = S_{U_a}$, then the algebra $S^{-1}A$ is also geometrical.*

◀ (a) Since $(p/s)_{\bar h} = h(s)^{-1}\iota(p)_{\bar h} = h(s)^{-1}\iota_{P,h}(p_h)$ and $h(s)^{-1} \neq 0$, the assertion follows from the fact that $\iota_{P,h}$ is an isomorphism (see the lemma in Section 13.15) for all $h \in \Im(|\iota|)$.

(b) First note that

$$\{\iota(p) \in \mathrm{Inv}(S^{-1}P)\} \Leftrightarrow \{0 = \iota(p)_{\bar h} = \iota_{P,h}(p_h), \forall \bar h \in |S^{-1}A|\},$$

and since $\iota_{P,h}$ is an isomorphism, the last one of these two assertions is equivalent to $p_h = 0$, for any $h \in \Im(|\iota|)$. But this means precisely that we have $\Im(|\iota|) \cap U_p = \varnothing$.

Further, $\iota(p) = 0$ means that $sp = 0$ for some $s \in S$. Therefore, $\iota(p) \neq 0$ if and only if $sp \neq 0$ for all $s \in S$. But the module P is geometrical, so this is equivalent to $(sp)_h = h(s)p_h \neq 0$ for some $h = h_s \in |A|$. Therefore, $h_s \in U_p \cap U_s$, i.e., $U_p \cap U_s \neq \varnothing$.

(c) Since the algebra A is geometrical, we have $U_a \neq \varnothing$ provided $a \neq 0$. Also note that $a \in S_{U_a}$ and $U_s \supset U_a$ for all $s \in S_{U_a}$. Hence, $\cap_{s \in S_{U_a}} U_s = U_a$ and Proposition 13.3 implies $\Im(|\iota|) = U_a$. If $\iota(a') \in \mathrm{Inv}(A_{U_a})$, then item (b) implies $U_a \cap U_{a'} = \varnothing$. On the other hand, since $a \in S_{U_a}$, we again see, by the same proposition, that the intersection $U_{a'} \cap U_a$ must be nonempty if $a' \neq 0$. Therefore, $a' = 0$. ▶

Corollary. *Suppose the field $\Bbbk$ is not algebraically closed and A is some $\Bbbk$-algebra. Suppose also that $a_1, \ldots, a_m \in A$, $a_i \neq 0$, for all i, and $U = U_{a_1} \cup \cdots \cup U_{a_m}$. Then the algebra $S_U^{-1} A$ is geometrical.*

◀ In this case $U = U_a$ for $a = \varphi(a_1, \ldots, a_m)$, where $\varphi(x_1, \ldots, x_m)$ is a polynomial that vanishes only at the point $(0, \ldots, 0) \in \Bbbk^m$ (see Example III in Section 13.5). ▶

Example. Let M be a manifold and let $\varnothing \neq V \subset M$ be a closed subset. The set of functions $S = \{f \in C^\infty(M) \mid U_f \supset V\}$, obviously, is multiplicative and $\Im|_{\iota_S}| = \cap_{f \in S} U_f = V$. Then $\varnothing \neq \iota_S(g) \in \mathrm{Inv}(S^{-1}C^\infty(M))$ if $g|_V = 0$, but $g|_U \neq 0$ for any open set $U \supset V$—this follows from assertions (a) and (b) of the previous proposition. In particular, this shows that the algebra $S^{-1}C^\infty(M)$ is not geometrical and does not contain any nilpotent elements.

Exercise.   1. Let $A = \Bbbk[x_1, \ldots, x_n]$, where $\Bbbk$ is a field of zero characteristic. Show that the algebra $S^{-1}A$ is geometrical if the set S is finitely generated. Is this assertion true if $\Bbbk$ has finite characteristics?

2. In the polynomial algebra from the previous exercise, find a multiplicative system S such that the algebra $S^{-1}A$ is not geometrical.

3. Let the algebras A and $S^{-1}A$, as well as the A-module P, be geometrical. Will the $S^{-1}A$-module $S^{-1}P$ necessarily be geometrical?

13.18. Localization of differential operators. Let $\Delta \in \mathrm{Diff}_k(P, Q)$. Suppose that there exists a localization $\Delta_S : S^{-1}P \to S^{-1}Q$ of the operator Δ corresponding to the localization $S^{-1}A$ of the algebra A, i.e., an operator such that

$$
\begin{array}{ccc}
P & \xrightarrow{\Delta} & Q \\
{\scriptstyle S^{-1}}\downarrow & & \downarrow{\scriptstyle S^{-1}} \\
S^{-1}P & \xrightarrow{S^{-1}\Delta} & S^{-1}Q
\end{array}
$$

commutes. Let us put $\tilde{s} = \iota(s)$, $s \in S$. Applying formula (9.24) to Δ_S for

$$a_1 = \tilde{s}^{-1}, \quad a_2 = \cdots = a_{k+1} = \tilde{s},$$

we obtain

$$
0 = [(\delta_{\tilde{s}})^k \delta_{\tilde{s}^{-1}}](\Delta_S) = \sum_{0 \leq i \leq k} (-1)^i \binom{k}{i} \tilde{s}^i \circ \Delta_S \circ \tilde{s}^{k-i-1} +
$$

$$
+ \sum_{1 \leq j \leq k+1} (-1)^j \binom{k}{j-1} \tilde{s}^{j-2} \circ \Delta_S \circ \tilde{s}^{k-j+1},
$$

or, equivalently,

$$\sum_{0 \le i \le k+1} (-1)^{i-1} \binom{k+1}{i} \tilde{s}^{i-1} \circ \Delta_S \circ \tilde{s}^{k-i} = 0 \,.$$

The last equality may be rewritten in the form

$$\tilde{s}^k \circ \Delta_S \circ \tilde{s}^{-1} = \sum_{0 \le j \le k} (-1)^{k-j} \binom{k+1}{j} \tilde{s}^{j-1} \circ \Delta_S \circ \tilde{s}^{k-j} \,.$$

Dividing it by $\tilde{s}^k$ and applying the obtained result to $p \in P$, we obtain

$$\Delta_S \left(\frac{p}{s} \right) = \sum_{i=0}^{k} (-1)^i \binom{k+1}{i+1} \frac{\Delta(s^i p)}{s^{i+1}} = \frac{L_k(s, \Delta)(p)}{s^{k+1}}, \qquad (13.2)$$

where

$$L_k(s, \Delta) = \sum_{i=0}^{k} (-1)^i \binom{k+1}{i+1} s^{k-i} \circ \Delta \circ s^i. \qquad (13.3)$$

This heuristic consideration brings us to the following definition.

13.19. Definition. The *localization of an operator* $\Delta \in \mathrm{Diff}_k(P, Q)$ $\Delta : P \to Q$ *over the multiplicative set* $S \subset A$ *is defined as the operator given by formula* (13.2).

However, we must prove that this notion is well defined, i.e., does not depend on the choice of the pair (p, s) representing the formal fraction p/s. To do that, we shall need certain properties of the operator $L_k(s, \Delta)$.

13.20 Lemma. *The following relations hold for the operators* $L_k(s, \Delta)$, $\Delta \in \mathrm{Hom}_{\Bbbk}(P, Q)$:

(1) $L_k(s, a \circ \Delta) = a \circ L_k(s, \Delta)$, $a \in A$;

(2) $L_k(s, \Delta \circ a) = L_k(s, \Delta) \circ a$, $a \in A$;

(3) $L_k(s, \Delta) = s^k \Delta - L_{k-1}(s, \delta_s(\Delta))$;

(4) $L_k(s, \Delta) \circ s = s^{k+1} \Delta$;

(5) $L_k(ss_1, \Delta) \circ s_1 = s_1^{k+1} L_k(s, \Delta)$;

(6) $L_k(s, \Delta) - s \circ L_{k-1}(s, \Delta) = (-1)^k \delta_s^k(\Delta)$.

◄ The first two relations are obvious, the third can be proved by a simple direct calculation that we leave to the reader. In its turn, the fourth relation follows from the third by induction over k. Indeed, assuming that (4) holds for $k - 1$, for any $\Bbbk$-linear map, and, in particular, for $\delta_s(\Delta)$, we have

$$\begin{aligned} L_k(s, \Delta) \circ s &= s^k \Delta \circ s - L_{k-1}(s, \delta_s(\Delta)) \circ s \\ &= s^k \circ \Delta \circ s - s^k \delta_s(\Delta) = s^{k+1} \Delta. \end{aligned}$$

Relation (5) is also proved by induction over k. The base of induction is obvious, while the induction step from $k-1$ to k is carried out in the following way. First using relation (3), then the induction hypothesis, and finally relations (1)–(2), we obtain

$$L_k(ss_1, \Delta) \circ s_1 = s^k s_1^k \circ \Delta \circ s_1 - L_{k-1}(ss_1, \delta_{ss_1}(\Delta)) \circ s_1$$
$$= s^k s_1^k \circ \Delta \circ s_1 - s_1^k \circ L_{k-1}(s, \delta_{ss_1}(\Delta))$$
$$= s^k s_1^k \circ \Delta \circ s_1 - s_1^k \circ L_{k-1}(s, s_1 \circ \delta_s(\Delta) + \delta_{s_1}(\Delta) \circ s)$$
$$= s^k s_1^k \circ \Delta \circ s_1 - s_1^{k+1} \circ L_{k-1}(s, \delta_s(\Delta)) - s_1^k L_{k-1}(s, \delta_{s_1}(\Delta)) \circ s.$$

Further, taking into account $L_{k-1}(s, \delta_{s_1}(\Delta)) \circ s = s^k \delta_{s_1}(\Delta)$ and using relation (4), we can rewrite the last of these equalities as

$$s_1^{k+1}(s^k \Delta - L_{k-1}(s, \delta_s(\Delta))) \overset{(3)}{=} s_1^{k+1} L_k(s, \Delta).$$

Finally, relation (6) follows from the fact that

$$L_k(s, \Delta) - s \circ L_{k-1}(s, \Delta) = \sum_{i=0}^{k} (-1)^i \binom{k}{i} s^{k-i} \circ \Delta \circ s^i$$

if we take into account the relation

$$\sum_{i=0}^{k} (-1)^i \binom{k}{i} s^{k-i} \circ \Delta \circ s^i = (-1)^k \delta_s^k(\Delta)$$

(see (9.24)).

13.21 Proposition. *Let* $\Delta \in \mathrm{Diff}_k(P, Q)$. *Then*

(a) *Definition 13.19 well defines* Δ_S, *which is a differential operator of order* $\leq k$ *over the algebra* $S^{-1}A$;

(b) $\iota_Q \circ \Delta = \Delta_S \circ \iota_P$;

(c) *if* $\nabla \in \mathrm{Diff}_l(R, Q)$, *then* $(\Delta \circ \nabla)_S = \Delta_S \circ \nabla_S$;

(d) *if* $A = C^\infty(M), P = \Gamma(\pi_1), Q = \Gamma(\pi_2)$ *and the set* $U \subset M$ *is open, then* Δ_{S_U} *coincides with the restriction* $\Delta_U : \Gamma(\pi_1|_U) \to \Gamma(\pi_2|_U)$ *of the operator* Δ *to* U *in the usual sense of the word.*

◀ (a). Relation (5) of Lemma 13.20 implies that the pairs

$$(L_k(ss_1, \Delta)(s_1 p), (ss_1)^{k+1}) \text{ and } (L_k(s, \Delta)(p), s^{k+1})$$

are equivalent. This proves that the right-hand side of relation (13.2) does not change when the pair (p, s) representing the formal fraction p/s in the left-hand side is replaced by the proportional pair $(s_1 p, s_1 s)$. On the other hand, the pairs (p_1, s_1) and (p_2, s_2) representing the same formal fraction are proportional to the pair

$$(ss_1 p_2, ss_1 s_2) = (ss_2 p_1, ss_1 s_2),$$

where the element $s \in S$ satisfies $s(s_2p_1 - ss_1p_2) = 0$. This proves that the right-hand side of relation (13.2) does not depend on the choice of the pair representing the formal fraction p/s.

To complete the proof of the fact that 13.19 well defines Δ_S, we must also prove that this definition does not depend on k, i.e., that

$$\frac{L_k(s, \Delta)(p)}{s^{k+1}} = \frac{L_l(s, \Delta)(p)}{s^{l+1}}$$

if the operator operator Δ is of order $l < k$. This, obviously, needs to be done only in the case $l = k - 1$, but in that case, it directly follows from item (6) of Lemma 13.20.

Now let us establish that Δ_S is a differential operator of order $\leq k$ over the algebra $S^{-1}A$. In the subsequent formulas and computations needed for this, it will be convenient to use the more compact notation $\tilde{a}$ instead of $\iota(a)$ for $a \in A$. For the proof, we shall use induction on the order k of the localized operator. The expression (13.2) when $k = 0$ has the following form: $\Delta_S(p/s) = \Delta(p)/s$. Since in this case Δ is a homomorphism of A-modules, it follows that

$$\Delta_S \left(\frac{a}{s'} \cdot \frac{p}{s} \right) = \Delta_S \left(\frac{ap}{s's} \right) = \frac{\Delta(ap)}{s's} = \frac{a\Delta(p)}{s's}$$
$$= \frac{a}{s'} \cdot \frac{\Delta(p)}{s} = \frac{a}{s'} \cdot \Delta_S \left(\frac{p}{s} \right).$$

This shows that Δ_S is a homomorphism of $S^{-1}A$-modules, i.e., a differential operators of zero order.

To carry out the induction step, we shall use the relation

$$\delta_{a/s}(X) = \delta_{a \cdot 1/s}(X) = \delta_{\tilde{a}}(X) \circ s^{-1} + \tilde{a} \circ \delta_{s^{-1}}(X), \qquad (13.4)$$

where $a \in A$, $s \in S$ and X is a $\Bbbk$-linear map of $S^{-1}A$-modules. Using this formula, we obtain

$$0 = \delta_{1_{S^{-1}A}}(X) = \delta_{s/s}(X) = \delta_{\tilde{s}}(X) \circ s^{-1} + \tilde{s} \circ \delta_{s^{-1}}(X),$$

and therefore,

$$\delta_{s^{-1}}(X) = -s^{-1} \circ \delta_{\tilde{s}}(X) \circ s^{-1}. \qquad (13.5)$$

The formulas (13.4) and (13.5) reduce the computation of $\delta_{a/s}$ to that of $\delta_{\tilde{b}}$, $b = a, s$, which is obvious enough. Namely, using relations (1) and (2) of Lemma 13.20 and the fact that the fraction $L_k(s, \nabla)/s^{k+1}$ does not depend on the order of the operator ∇ provided it is not greater than k, we obtain

$$\delta_{\tilde{b}}(\Delta_S) \left(\frac{p}{s} \right) = \frac{L_k(s, \delta_b(\Delta))(p)}{s^{k+1}} = \delta_b(\Delta)_S \left(\frac{p}{s} \right), \ b \in A.$$

This implies $\delta_{\tilde{b}}(\Delta_S) = \delta_b(\Delta)_S$. Together with formulas (13.4) and (13.5) for $\square = \Delta_S$, this yields the required formula

$$\delta_{a/s}(\Delta_S) = \delta_a(\Delta)_S \circ s^{-1} - \tilde{a}s^{-1} \circ \delta_s(\Delta)_S \circ s^{-1}. \qquad (13.6)$$

Indeed, the differential operators $\delta_a(\Delta)$ and $\delta_s(\Delta)$ are of order $\leq k - 1$. Therefore, by the induction hypothesis, their localization also has the same order and so, in the right-hand side of formula (13.6), an operator of order $\leq k - 1$ over the algebra $S^{-1}A$ appears. Thus, the operator $\delta_{a/s}(\Delta_S)$ is also of order $\leq k - 1$ for all $p/s \in S^{-1}A$, which implies that the operator Δ_S is of order $\leq k$.

(b) This is obvious.

(c) Since

$$\Delta_S\left(\nabla_S\left(\frac{p}{s}\right)\right) = \Delta_S\left(\frac{L_l(s,\nabla)(p)}{s^{l+1}}\right) = \frac{L_k(s^{l+1},\Delta)(L_l(s,\nabla)(p))}{s^{(k+1)(l+1)}},$$

it suffices to prove

$$s^{kl}L_{k+s}(s,\Delta \circ \nabla) = L_k(s^{l+1},\Delta) \circ (L_l(s,\nabla)$$

Now division by $\iota(s)$ is possible in $S^{-1}A$, and so it suffices to prove that

$$s^{kl}L_{k+s}(s,\Delta \circ \nabla) \circ s = L_k(s^{l+1},\Delta) \circ (L_l(s,\nabla) \circ s. \qquad (13.7)$$

Item (4) in Lemma 13.20 allows us to bring the right-hand side of this hypothetical equality to the form

$$L_k(s^{l+1},\Delta) \circ (L_l(s,\nabla) \circ s) = L_k(s^{l+1},\Delta) \circ s^{l+1} \circ \nabla = s^{(k+1)(l+1)}\Delta \circ \nabla.$$

Similarly, we have $L_{k+s}(s,\Delta \circ \nabla) \circ s = s^{k+l+1}\Delta \circ \nabla$, which proves (13.7).

(d). Note that the relation $L_k(1,\Delta) = \Delta$ is actually relation (4) from Lemma 13.20 for $s = 1$. Hence,

$$\Delta_S\left(\frac{p}{1}\right) = \frac{L_k(1,\Delta)(p)}{1} = \frac{\Delta(p)}{1}. \qquad (13.8)$$

In the case under consideration, we have $S = S_U$ and we can identify $\Gamma(\pi|_U)$ and $S^{-1}R$ if $R = \Gamma(\pi)$ (see 13.16). Here $p/1 = p|_U$, $\Delta(p)/1 = \Delta(p)|_U$. Therefore, 13.8 may be rewritten in the form $\Delta_{S_U}(p|_U) = \Delta(p)|_U$. But since $\Delta(p)|_U = \Delta|_U(p|_U)$, we have $\Delta_{S_U}(p|_U) = \Delta|_U(p|_U)$. From this we can conclude that $\Delta_{S_U} = \Delta|_U$. Indeed, Δ_{S_U} is a differential operator over the algebra $S^{-1}C^\infty(M) = C^\infty(U)$ and so it is a differential operator in the usual sense of the word (see Section 14.22). Therefore, this operator and hence the operator $\Delta|_U$ are local (Proposition 14.22). But since in some neighborhood of an arbitrary point $z \in U$ any element of the module $\Gamma(\pi_1|_U)$ may be represented in the form $p|U$, it follows that the operators coincide in this neighborhood. But the point z was arbitrary, so they coincide everywhere. ▶

Remark. If the element $s \in A$ is invertible, then formula (13.2) can be rewritten in the form

$$\Delta\left(s^{-1}p\right) = \sum_{i=0}^{k}(-1)^i\binom{k+1}{i+1}s^{-(i+1)}\Delta(s^i p) = s^{-(k+1)}L_k(s,\Delta)(p), \quad (13.9)$$

describes the result of applying the operator Δ to the "nonformal" fraction $p/s = s^{-1}p$. The reader can once again appreciate the advantage of the algebraic approach by trying to find this formula on the basis of the classical definition of differential operators.

13.22. Localization is well defined. Definition 13.19 describes the localization of differential operators using the description of the localized algebra itself in terms of multiplicative sets S. Therefore, we must show that the result does not depend on that description.

Proposition. *Let the multiplcative sets S_1 and S_2 be equivalent and let $\eta\colon S_1^{-1}A \to S_1^{-1}A$ be the natural isomorphism. Then $\eta \circ \Delta_{S_1} = \Delta_{S_2} \circ \eta$.*

◀ Since the composition of natural isomorphisms is also natural, see Exercise 13.12, it suffices to prove this only in the case when S_2 is a maximal multiplicative set equivalent to S_1. In this case, $S_1 \subset S_2$ and any element of $S_2^{-1}A$ may be represented by a formal fraction whose denominator belongs to S_1. Therefore, the expressions for $\Delta_{S_1}(p/s_1)$ and $\Delta_{S_2}(p/s_1)$ in the defining formula 13.2 are exactly identical except that in the first one the formal fraction is understood as an element of $S_1^{-1}A$, while in the second one, as an element of $S_2^{-1}A$. But in that case, the required assertion becomes obvious, ▶

14
Differential 1-forms and Jets

14.1. Duality: vectors and covectors. In Chapter 9, considering vector fields of various types, we came to a generalization that covers them all, namely to the notion of P-valued vector field or derivation of the main algebra A with values in an arbitrary A-module P (see Section 9.50). As we have seen, the set of all such vector fields constitutes an A-module, denoted by $D(P)$. In this way, we obtain the functor D (see 9.51).

On the other hand, geometry, mechanics, and physics naturally require the introduction of notions dual to tangent vectors and to vector fields. In elementary differential geometry, this is the notion of cotangent vector, i.e., the differential of a function at the given point (Sections 9.21 and 9.22). We have previously noted the duality between velocity and momentum in mechanics.

We also noted previously that tangent vectors and covectors at a point z of a manifold M are dual to each other in the sense of elementary linear algebra. Recall that tangent covectors are linear functionals on the tangent space $T_z M$, while tangent vectors are linear functionals on the cotangent space $T_z^* M$. This is equivalent to saying that a pairing

$$T_z M \oplus T_z^* M \to \mathbb{R}$$

is defined, i.e., a bilinear form $\langle \xi, \eta \rangle \in \mathbb{R}$, where $\xi \in T_z M, \eta \in T_z^* M$. Similarly, for the quantity dual to the notion of the vector field, it is natural to take the notion of *covector field*, which is a family of tangent covectors $\omega = \{\omega_z\}_{z \in M}, \omega_z \in T_z^* M$ *smoothly* depending on z. One of the possible definitions of smoothness here consists in that, for any vector field

© Springer Nature Switzerland AG 2020
J. Nestruev, *Smooth Manifolds and Observables*, Graduate Texts
in Mathematics 220, https://doi.org/10.1007/978-3-030-45650-4_14

$X \in D(M)$, the function

$$M \ni z \mapsto \langle X_z, \omega_z \rangle \in \mathbb{R}$$

is smooth. If this function is denoted by $\omega(X)$, then, by definition, we have $\omega(X)(z) = \langle X_z, \omega_z \rangle$. It is easy to verify that $\omega(fX) = f\omega(X)$ for $f \in C^\infty(M)$. This implies that the covector field ω can also be defined as a $C^\infty(M)$–linear map

$$\omega \colon D(M) \to C^\infty(M), \quad X \mapsto \omega(X),$$

i.e., as an element of the $C^\infty(M)$–module

$$D^*(M) \overset{\text{def}}{=} \operatorname{Hom}_{C^\infty(M)}(D(M), C^\infty(M))$$

Moreover, we have the isomorphism

$$D(M) = \operatorname{Hom}_{C^\infty(M)}(D^*(M)), C^\infty(M)), \tag{14.1}$$

establishing the duality of vector and covector fields in the sense of linear algebra. However, the duality understood in this sense no longer holds for vector fields of other types, as will be shown below, where we shall be using the simplified notation $P^* = \operatorname{Hom}_A(P, A)$ for any P.

14.2 Lemma. *Let $N \subset M$ be a nowhere dense closed subset of a manifold M. If P is a geometrical $C^\infty(M)$-module and $\operatorname{supp} P \subset N$ (see 12.11), then $P^* = 0$.*

◀ Let a function $f \in C^\infty(M)$ satisfy $f|_N = 0$. Then $fp = 0$, for all $p \in P$. Indeed, $(fp)_z = f(z)p_z \in P_z$ (see Section 12.11) and so $(fp)_z = 0$, provided $z \in N$. On the other hand, $p_z = 0$ if $z \notin N$. Thus, $(fp)_z = 0$, for all $z \in M$. So, since the module P is geometrical, this implies $fp = 0$.

Now let $\gamma \in P^*$. If $f|_N = 0$, then $f\gamma(p) = \gamma(fp) = 0$. Besides, since the set $M \setminus N$ is open, it follows that the function f may be chosen so that $f(z) = 1$ if $z \in M \setminus N$. Therefore, $\gamma(p)(z) = 0$, i.e., $\gamma(p)|_{M \setminus N} = 0$. Now since $M \setminus N$ is everywhere dense, this implies that $\gamma(p) = 0$, for all $p \in P$, i.e., we have $\gamma = 0$. ▶

Corollary. *Let $N \subset M$ be a nowhere dense closed subset of a manifold M. Then $D(M, N)^* = 0$ and so $(D(M, N)^*)^* = 0$ (see 9.46).*

◀ It suffices to note that $\operatorname{supp} D(M, N) = N$. ▶

Exercise.   1. State and prove the general algebraic version of Lemma 14.2.

2. Suppose the smooth map $\varphi \colon M \to N$ is not a diffeomorphism. Can it happen that $(D(\varphi)^*)^* = D(\varphi)$ (see 9.47), if we regard $D(\varphi)$ as a
   I. $C^\infty(M)$-module?
   II. $C^\infty(N)$-module?

14.3. What instead of duality? By historical inertia, the duality (14.1) has remained in the physics literature, where vectors and covectors are often called *contravariant* and *covariant* vectors, respectively. On the contrary, mathematicians prefer to call covector fields differential forms of first order or simply 1-forms.

In view of the importance of the duality that we are now discussing, the following natural question arises: Does this notion have a replacement that would work for all kinds of vector fields? Thanks to the fact that we are living at a time when mathematicians are familiar with the notions of functor and *representing object of a functor*, this question does have a positive answer, and we now begin to prepare to explain this.

14.4. Differential 1-forms. Thus, we begin with the commonly accepted notion of differential 1-form. Fix a manifold M. The module of sections $\Gamma(\pi_{T^*})$ of the bundle $\pi_{T^*} = \pi_{T^*M}$ is called the *module of differential 1-forms* of the manifold M and denoted by $\Lambda^1(M)$. The elements of this module, i.e., smooth sections of the bundle π_{T^*}, are referred to as *differential 1-forms* on the manifold M.

According to Section 9.25, any function $f \in C^\infty(M)$ gives rise to a section of π_{T^*} defined by

$$s_{df} \colon M \to T^*M, \quad s_{df}(z) = d_z(f).$$

Sections of this kind are called *differentials* of smooth functions. The differential of a function f, viewed as a purely algebraic object, will be denoted by df. The same thing, viewed geometrically, as a map from M to T^*M, will be denoted by s_{df} (the graph of the function of df).

Exercise 9.22 implies that the map

$$d \colon C^\infty(M) \to \Lambda^1(M), \quad f \mapsto df,$$

is a derivation called the *universal derivation* of the algebra $C^\infty(M)$ with values in the $C^\infty(M)$-module $\Lambda^1(M)$. The origin of this term will be clarified later.

Now let (U, x) be a chart on M and let $\left(\pi_{T^*}^{-1}(U), T^*x\right)$ be the corresponding chart on T^*M. In Section 9.24, we denoted by T^*x the system of coordinate functions $\{x_i, p_j\}$, where the value of x_i at $(z, \theta) \in T^*U$ is the ith coordinate of the point z, while p_j is the jth component in the expansion of the covector θ in the basis dx_i. Within this chart, every smooth section s has the coordinate representation

$$x_i = x_i, \quad i = 1, \ldots, n,$$
$$p_j = p_j(x), \quad j = 1, \ldots, n,$$

where $p_j(x) \in C^\infty(U)$. For simplcity, the section $s_{dx_i} \in \Gamma(T^*U) = \Lambda^1(U)$ will be denoted by dx_i. It follows that the sections dx_i, $i = 1, \ldots, n$, form a basis of the free $C^\infty(U)$-module $\Lambda^1(U)$. In particular, the restriction of the differential form $\omega \in \Lambda^1(M)$ to U belongs to $\Lambda^1(U)$ and can, therefore, be written as

$$\omega = \sum_i p_i(x)dx_i.$$

For the differential of a function df, we can write

$$df = \sum \frac{\partial f}{\partial x_i} dx_i$$

(see the exercise in Section 9.25).

Vector fields on M can be viewed as sections of the tangent bundle (see Section 9.40), and the pairing

$$\Gamma(\pi_{TM}) \times \Gamma(\pi_{T^*M}) \to C^\infty(M)$$

becomes

$$D(M) \times \Lambda^1(M) \to C^\infty(M). \tag{14.2}$$

Returning to local coordinates, we recall that for any point $z \in U \subset M$ the basis $d_z x_1, \ldots, d_z x_n$ of the linear space $T_z^* M$ is by definition dual to the basis

$$\frac{\partial}{\partial x_1}\bigg|_z, \ldots, \frac{\partial}{\partial x_n}\bigg|_z$$

of the space $T_z M$. If X is a vector field and ω is a 1-form represented in special local coordinates by

$$X = \sum_i \alpha_i(x)\frac{\partial}{\partial x_i}, \quad \omega = \sum_i p_i(x)dx_i,$$

then the result of the pairing (X, ω) restricted to U is the function

$$\langle X, \omega \rangle\big|_U = \sum_i \alpha_i p_i(x) \in C^\infty(U).$$

In particular, for $\omega = df$ we obtain

$$\langle X, df \rangle\big|_U = \sum_i \alpha_i \frac{\partial f}{\partial x_i} = X(f)\big|_U.$$

14.5. The universal 1-form. Recall that any 1-form $\lambda \in \Lambda^1(M)$, just as the differential of a function $df \in \Lambda^1(M)$, can be understood as a section of the cotangent bundle $\pi \colon T^*M \to M$:

$$s_\lambda \colon M \to T^*M, \ M \ni x \mapsto (x, \lambda|_x) \in T^*M.$$

It determines the dual maps

$$s_\lambda^* \colon C^\infty(T^*M) \to C^\infty(M) \text{ and } s_\lambda^* \colon \Lambda^1(T^*M) \to \Lambda^1(M).$$

Definition. A form $\rho \in \Lambda^1(T^*M)$ is said to be a *universal 1-form*, if for every 1-form $\lambda \in \Lambda^1(M)$ we have

$$s_\lambda^*(\rho) = \lambda.$$

Proposition. *Such a form ρ exists and is unique.*

◀ An arbitrary point of the manifold T^*M can be written as $(x, df|_x)$, where $x \in M$, and $f \in C^\infty(M)$ is some function. Let us put:

$$\rho|_{(x,df|_x)} \stackrel{\text{def}}{=} d(\pi^*(f))|_{(x,df|_x)},$$

Let $\lambda \in \Lambda^1(M)$ and let $f \in C^\infty(M)$ be a function such that $\lambda|_x = df|_x$. Then

$$s_\lambda^*(\rho)|_{(x,df|_x)} = s_\lambda^*(d(\pi^*(f))|_{(x,df|_x)}) = d(s_\lambda^*(\pi^*(f))|_x) = df|_x = \lambda|_x.$$

The uniqueness of ρ is a consequence of the fact that if, for some form $\rho' \in \Lambda^1(T^*M)$ and any form $\lambda \in \Lambda^1(M)$, we have $s_\lambda^*(\rho') = 0$, then $\rho' = 0$. ▶

Exercise. Show that in the system of special coordinates (see Section 9.24), $\rho = \sum_{i=1}^n p_i dx_i$.

14.6. Universal derivation. The definition of the module of differential 1-forms $\Lambda^1(M)$ given above was descriptive. Here we intend to give a conceptual definition of the same notion.

Let $\mathfrak{M}$ be a certain category of modules over a $\Bbbk$-algebra A, and $\Lambda \in \mathfrak{M}$. A derivation $\delta\colon A \to \Lambda$ is called *universal* in the category $\mathfrak{M}$ if for any module P from $\mathfrak{M}$ the correspondence

$$\mathrm{Hom}_A(\Lambda, P) \ni h \mapsto h \circ \delta \in D(P)$$

is an isomorphism of the A-modules $\mathrm{Hom}_A(\Lambda, P)$ and $D(P)$.

Proposition. *The universal derivation, if it exists, is unique up to isomorphism; i.e., if (δ', Λ') is another universal derivation, then there exists an isomorphism of A-modules $\gamma\colon \Lambda \to \Lambda'$ such that $\delta' = \gamma \circ \delta$.*

◀ The universality of the derivation δ implies that there exists a homomorphism $\gamma\colon \Lambda \to \Lambda'$ such that $\delta' = \gamma \circ \delta$. Similarly, $\delta = \gamma' \circ \delta'$ for an appropriate $\gamma' \in \mathrm{Hom}_A(\Lambda', \Lambda)$. Therefore, $\delta = \gamma' \circ \gamma \circ \delta$ and $\Im\delta \subset \Lambda_0$, where

$$\Lambda_0 = \{\omega \in \Lambda \mid \gamma'(\gamma(\omega)) = \omega\} \subset \Lambda.$$

Let $\alpha\colon \Lambda \to \Lambda/\Lambda_0$ be the natural projection. Then $0 = \alpha \circ \delta \in D(\Lambda/\Lambda_0)$, hence $\alpha = 0$ by the universality of δ. Since α is surjective, it follows that $\Lambda = \Lambda_0$ and $\gamma' \circ \gamma = \mathrm{id}_\Lambda$.

Symmetrically, $\gamma \circ \gamma' = \mathrm{id}_{\Lambda'}$; hence γ and γ' are mutually inverse. ▶

14.7 Theorem. *The pair $(d, \Lambda^1(C^\infty(M)))$ is the universal derivation in the category of geometrical $C^\infty(M)$-modules.*

◄ Let us prove that the natural homomorphism

$$\eta_P \colon \mathrm{Hom}_{C^\infty(M)}\left(\Lambda^1(M), P\right) \to D(P), \quad h \mapsto h \circ d,$$

is an isomorphism if the module P is geometrical. To this end, we shall construct the inverse homomorphism

$$\nu_P \colon D(P) \to \mathrm{Hom}_{C^\infty(M)}\left(\Lambda^1(M), P\right), \quad X \mapsto h_X.$$

We use the fact that any 1-form $\omega \in \Lambda^1(M)$ can be written as $\omega = \sum_i f_i dg_i$ (this will be independently proved below; see Corollary 14.17). Put

$$h_X(\omega) \overset{\mathrm{def}}{=} \sum_i f_i X(g_i)$$

and let us check that h_X is well-defined.

Let $z \in M$ and $f = \sum_i c_i g_i$, where $c_i = f_i(z) \in \mathbb{R}$. Then we can write $\omega_z = \sum_i c_i d_z g_i = d_z f$. Moreover,

$$h_X(\omega)(z) = \sum_i f_i(z) X_z(g_i) = X_z\left(\sum_i c_i g_i\right) = X_z(f).$$

Here X_z denotes the composition $C^\infty(M) \overset{X}{\to} P \to P_z$ (see the end of Section 12.11), therefore, the value of $h_X(\omega)$ at an arbitrary point $z \in M$ is well-defined, i.e., it does not depend on the choice of the representation $\omega = \sum_i f_i dg_i$. Since P is geometrical, this implies that $h_X(\omega)$ is well-defined.

In the case $\omega = dg$, we have by definition $h_X(dg) = X(g)$, which means that $X = h_X \circ d \Leftrightarrow \eta_P \circ \nu_P = \mathrm{id}_{D(P)}$. If $X = h \circ d$, then

$$h_X(\omega) = \sum_i f_i h(dg_i) = h\left(\sum_i f_i dg_i\right) = h(\omega),$$

and so $h_X = h \Leftrightarrow \nu_P \circ \eta_P = \mathrm{id}_{D(P)}$. ►

The theorem yields the pairing

$$\Lambda^1(M) \times D(P) \to P, \quad (\omega, X) \mapsto h_X(\omega),$$

whose result can be written as $\omega(X) = h_X(\omega)$.

Exercise. Show that for $P = C^\infty(M)$ this pairing coincides with the one defined in Section 14.4.

14.8. Conceptual definition of differential forms. The theorem proved in the previous section suggests a conceptual approach to the theory of differential forms over an arbitrary algebra A. Namely, *differential 1-forms* are defined as elements of the A-module Λ, the target of the universal derivation which we now denote $d_{\mathfrak{M}} \colon A \to \Lambda$. We stress that this module depends on the choice of the category of A-modules $\mathfrak{M}$ (see Section 14.6). This module is called the *representing object* for the functor D in this category and is denoted by $\Lambda_{\mathfrak{M}}$.

Exercise. Indicate a category of $C^\infty(M)$-modules in which the functor D is not representable, i.e., does not determine a universal derivation.

14.9 Theorem. *In the category of all A-modules over an arbitrary commutative K-algebra A, the functor D is representable. We denote the corresponding universal derivation in that category by*

$$d_{\mathrm{alg}} \colon A \to \Lambda_{\mathrm{alg}}(A).$$

◄ The proof that follows is a typical category theory argument. Consider the free A-module $\widetilde{\Lambda}$ generated by the symbols $\widetilde{d}a$ for all $a \in A$. Let $\widetilde{\Lambda}_0$ be the submodule spanning all relations of the form

$$\widetilde{d}(ka) - k\widetilde{d}a, \quad \widetilde{d}(a+b) - \widetilde{d}a - \widetilde{d}b, \quad \widetilde{d}(ab) - a\widetilde{d}b - b\widetilde{d}a, \quad k \in K,\ a,b \in A.$$

Let us put

$$\Lambda_{\mathrm{alg}}(A) = \widetilde{\Lambda}/\widetilde{\Lambda}_0 \quad \text{and} \quad d_{\mathrm{alg}}a = \widetilde{d}a \mod \widetilde{\Lambda}_0, \quad a \in A.$$

Since $\widetilde{d}(ab) - a\widetilde{d}b - b\widetilde{d}a \in \widetilde{\Lambda}_0$, it follows that $d_{\mathrm{alg}}(ab) - ad_{\mathrm{alg}}b - bd_{\mathrm{alg}}a = 0$, and so d_{alg} is a derivation. Further, since the A-module $\widetilde{\Lambda}$ is free, any map of its generators $\psi \colon \widetilde{d}a \mapsto q \in Q$ can be uniquely extended to an A-homomorphism $\psi \colon \widetilde{\Lambda} \to Q$. In particular, this allows to assign to any $X \in D(P)$ the A-homomorphism $\tilde{h}_X \colon \widetilde{\Lambda} \to P$ determined by the map $\widetilde{d}a \mapsto X(a)$ of the generators of the module $\widetilde{\Lambda}$. It is obvious that $\tilde{h}_X$ annihilates the generators of the A-module $\widetilde{\Lambda}_0$. Therefore, $\tilde{h}_X$ can be represented as the composition

$$\widetilde{\Lambda} \xrightarrow{\ \mathrm{pr}\ } \frac{\widetilde{\Lambda}}{\widetilde{\Lambda}_0} = \Lambda_{\mathrm{alg}}(A) \xrightarrow{\ h_X\ } P,$$

where pr is the natural projection, while h_X is an A-homomorphism. Here $X(a) = \tilde{h}_X(\widetilde{d}a) = h_X(\mathrm{pr}(\widetilde{d}a)) = h_X(d_{\mathrm{alg}}a)$, i.e., $X = h_X \circ d_{\mathrm{alg}}$. ►

14.10. Is this any good? Now the following natural question arises: Does the previous algebraic construction of 1-forms coincide with what is well known and approved in differential geometry? In other words, does $\Lambda_{\mathrm{alg}}(A)$ coincide with $\Lambda^1(M)$ if $A = C^\infty(M)$? More precisely, since the differential $d \colon C^\infty(M) \to \Lambda^1(M)$ is a derivation, the universality of the pair $(\Lambda_{\mathrm{alg}}(A), d_{\mathrm{alg}})$ implies that there exists a unique homomorphism, namely $h_d \colon \Lambda_{\mathrm{alg}}(A) \to \Lambda^1(M)$ such that $d = h_d \circ d_{\mathrm{alg}}$. In these terms, our question can be stated as follows: Is h_d an isomorphism? The answer turns out to be negative, as the next statement implies.

Proposition. *Let $M = \mathbb{R}$. Then $d_{\mathrm{alg}}e^x - e^x d_{\mathrm{alg}}x \neq 0$, i. e., $d_{\mathrm{alg}}e^x - e^x d_{\mathrm{alg}}x$ is an example of an invisible element.*

◄ First let us prove that x and e^x are algebraically independent, i.e., the function

$$f_n(x)e^{nx} + \cdots + f_1(x)e^x + f_0(x),$$

where the $f_i(x)$ are polynomials over $\mathbb{R}$, is nonzero if at least one of these polynomials may be nonzero. But this easily follows from the fact that e^{nx} increases more rapidly than any polynomial or than e^{kx}, $k < n$, for $x \to \infty$.

Further, let us consider the localization of the main algebra $A = C^\infty(\mathbb{R})$ w.r.t. the multiplicative set consisting of the polynomials $P(x, e^x)$. Then the algebra A_1 defined in this way contains the field F of rational functions in the variables x and $y = e^x$, and we consider the derivation $\delta = \partial/\partial y$ of this field. If, moreover, $\mathcal{I}$ is a maximal ideal of the algebra A_1, then the field $\bar{F} = A_1/\mathcal{I}$ contains the isomorphic image of the field F, which we identify with F itself. It is not too hard to prove that any derivation on a subfield can be extended to a derivation of the whole field for a field of characteristic zero. In particular, we can extend the derivation δ to the derivation $\bar{\delta}$ of the field $\bar{F}$. Then

$$\bar{\delta}(\bar{x}) = \overline{\delta(x)} = 0, \quad \bar{\delta}(\bar{y}) = \overline{\delta(y)} = 1,$$

where $\bar{z} = z \mod \mathcal{I} \in A_1/\mathcal{I} = \bar{F}$ for $z \in A_1$.

Finally, consider the composition

$$A \to A_1 \to \bar{F} = A_1/\mathcal{I} \xrightarrow{\delta} \bar{F},$$

which we denote by ∂. Obviously, it is a derivation of the algebra A with values in the A-module $\bar{F}$ and, besides, we have $\partial(x) = 0$ and $\partial(e^x) = 1$.

But $\partial = h \circ d_{\mathrm{alg}}$, in view of the universality of the derivation d_{alg}, where $h\colon \Lambda_{\mathrm{alg}}(A) \to \bar{F}$, is an A-homomorphism. Therefore, the assumption that $d_{\mathrm{alg}}(e^x) = e^x d_{\mathrm{alg}}(x)$ implies the equality $\partial(e^x) = e^x \partial(x)$, which is a contradiction, since $\partial(e^x) = 1$, whereas $\partial(x) = 0$. ▶

Exercise. Let the functions $f, g \in C^\infty(M)$ be algebraically independent and $g(x) = \phi(f(x))$, where $\phi \in C^\infty(\mathbb{R})$. Prove that $d_{\mathrm{alg}}(g) \neq \phi'(f) d_{\mathrm{alg}}(f)$.

14.11. Geometrical differential forms. The fact that $\Lambda_{\mathrm{alg}}(A)$ and $\Lambda^1(M)$ do not coincide for the algebra $C^\infty(M)$ might lead us to suspect that the "conceptual" approach to differential forms is inadequate. However, considerations of observability here will help us out once again. Namely, let $\Lambda_{\mathrm{geom}}(A) = \Gamma(\Lambda_{\mathrm{alg}}(A))$ and $d_{\mathrm{geom}} = \Gamma_{\Lambda_{\mathrm{alg}}(A)} \circ d_{\mathrm{alg}}$ (see Section 12.11).

Theorem. *The pair $(d_{\mathrm{geom}}, \Lambda_{\mathrm{geom}}(A))$ is the universal derivation of the algebra A in the category of geometrical A-modules. In particular, if $A = C^\infty(M)$, then $\Lambda^1(M) \cong \Lambda_{\mathrm{geom}}(A)$.*

◀ Let the A-module P be geometrical and $X \in D(P)$. Then $X = h_X^{\mathrm{alg}} \circ d_{\mathrm{alg}}$ in view of the universality of d_{alg}, where $h_X^{\mathrm{alg}}\colon \Lambda_{\mathrm{alg}}(A) \to P$ is the corresponding A-homomorphism. In its turn, since the module P is geometrical, it follows that h_X^{alg} can be represented as the composition

$$\Lambda_{\mathrm{alg}}(A) \xrightarrow{\Gamma_{\Lambda_{\mathrm{alg}}(A)}} \Lambda_{\mathrm{geom}}(A) \xrightarrow{h_X^{\mathrm{geom}}} P,$$

where h_X^{geom} is an A-homomorphism (see Proposition in Section 14.6 and the discussion in Section 12.11), which implies that $X = h_X^{\text{geom}} \circ d_{\text{geom}}$.

Further, as can be easily seen, the uniqueness of the homomorphism h_X^{geom} follows from the fact that the A-module $\Lambda_{\text{geom}}(A)$ is generated by elements of the form $d_{\text{geom}}a$, $a \in A$. But this is indeed so, because the elements $d_{\text{alg}}a$, $a \in A$, generate $\Lambda_{\text{alg}}(A)$, while $\Gamma_{\Lambda_{\text{alg}}(A)}$ is surjective.   ▶

Corollary. *Let* $f(x) \in C^\infty(\mathbb{R})$. *Then*

$$d_{\text{alg}}(f(x)) - f'(x)d_{\text{alg}}x \in \text{Inv}(\Lambda_{\text{alg}}^1(A)).$$

◀ Indeed, $\Gamma(d_{\text{alg}}(f(x)) - f'(x)d_{\text{alg}}x) = d_{\text{geom}}(f(x)) - f'(x)d_{\text{geom}}x$ and d_{geom} can be identified with the ordinary differential d in view of the uniqueness of the universal derivation.   ▶

14.12. Universal derivation in other categories. Universal derivations may be constructed in other categories of A-modules. To see this, we will consider some examples of such categories. Let $\mathcal{N}$ be a complete subcategory of the category $\text{Mod}\,A$ of all A-modules, i.e., a category such that $\text{Mor}_{\mathcal{N}}(P, Q)$, $P, Q \in \mathcal{N} \subset \text{Mod}\,A$ coincides the set of morphisms $\text{Mor}_{\text{Mod}\,A}(P, Q)$ in $\text{Mod}\,A$. The simplest such category is the following one.

Let $A = C^\infty(M)$ and $z \in M$. Consider the category $\text{GMod}_z\,C^\infty(M)$ whose objects are the geometrical $C^\infty(M)$-modules whose support is the single point z. In other words, they are modules P such that $\mu_z P = 0$, or, equivalently, $P = P_z$. Then

$$d_z \colon C^\infty(M) \to T^*M, \quad f \mapsto d_z f,$$

is the universal derivation (or *differential*) in this category. Indeed, if $X \in D(P)$, then $X(1) = 0$ and $X(\mu_z^2) \subset \mu_z P = 0$. Therefore, the value $X(f)$ is uniquely determined by the coset of the function $(f - f(z)) \in \mu_z$ by mod μ_z^2. But, as we have seen (see Section 9.27), this coset can be identified with $d_z f$.

Exercise. Let $h \in |A|$ and $\text{GMod}_h\,A$ be the category of A-modules whose support is the point h. Prove that if $\Bbbk$ is a field, then

$$d_h \colon A \to T_h^* A, \quad d_h a = (a - h(a)1_A) \mod \mu_h^2,$$

is the universal derivation in the category $\text{GMod}_h\,A$.

Let W be a closed subset in the Zariski topology of $|A|$ and let $\text{Mod}_W\,A$ and $\text{GMod}_W\,A$ be the categories whose objects are all, or, respectively, all the geometrical A-modules with support in W. Both of them may be considered as generalizations of $\text{GMod}_h\,A$. Let

$$\mathcal{I}_W = \{a \in A \mid h(a) = 0, \forall h \in W\}.$$

Obviously, $\mathcal{I}_W$ is an ideal of the algebras A and $C^\infty(N) = C^\infty(M)/\mathcal{I}_N$ if N is a submanifold of the manifold M.

Lemma. *Let* $\Bbbk$ *be a field, let* A *be a* $\Bbbk$-*algebra, let* P *and* Q *be* A-*modules and let the subset* $W \subset |A|$ *be closed. Then*

(i) *if $W = \mathrm{supp}(P)$, then $\mathcal{I}_W \cdot P \subset \mathrm{Inv}(P)$, in particular, $\mathcal{I}_W \cdot P = 0$ if the module P is geometrical;*

(ii) $\mathrm{supp}\left(Q/(\mathcal{I}_W \cdot Q)\right) \subset W$;

(iii) *if the module P is geometrical and $W = \mathrm{supp}\, P$, then any homomorphism $\varphi \colon Q \to P$ can be uniquely represented as at the composition of homomorphisms*

$$Q \xrightarrow{\mathrm{pr}} Q/(\mathcal{I}_W \cdot Q + \mathrm{Inv}(Q)) \xrightarrow{\varphi'} P,$$

where pr *is the natural projection.*

◀ (i) Let $a \in \mathcal{I}_W$, $p \in P$ and $h \in |A|$. Let us show that $(ap)_h = h(a)p_h = 0$. Indeed, $p_h = 0$ if $h \in |A| \setminus W$, and $h(a) = 0$ if $h \in W$.

(ii) Let $h \in |A| \setminus W$ and $q \in Q$. Since $|A| \setminus W$ is open, it follows that $h \in U_a \subset |A| \setminus W$ for some $a \in A$. The latter means that $h(a) \neq 0$ and $a \in \mathcal{I}_W$. Therefore, $(a'q)_h = q_h$, where $a' = 1/h(a) \cdot a \in \mathcal{I}_W$, and so $Q = \mathcal{I}_W \cdot Q + \mu_h \cdot Q$. It remains to note that

$$\left(Q/(\mathcal{I}_W \cdot Q)\right)_h = Q/(\mathcal{I}_W \cdot Q + \mu_h \cdot Q).$$

(iii) It suffices to prove that $(\mathcal{I}_W \cdot Q + \mathrm{Inv}(Q)) \subset \mathrm{Ker}\,\varphi$. Since we have $\varphi(\mathcal{I}_W \cdot Q) \subset \mathcal{I}_W \cdot P$, it follows that $\mathcal{I}_W \cdot Q \subset \mathrm{Ker}\,\varphi$ in view of (i). Besides, $\varphi(\mathrm{Inv}(Q)) \subset \mathrm{Inv}(P) = 0$, since the module P is geometrical. ▶

Proposition. *The composition*

$$A \xrightarrow{d_{\mathrm{geom}}} \Lambda^1_{\mathrm{geom}}(A) \xrightarrow{\mathrm{pr}} \Lambda^1_{\mathrm{geom}}(A)/\left(\mathcal{I}_W \cdot \Lambda^1_{\mathrm{geom}}(A)\right),$$

denoted by d^W_{geom}, where pr *is the factorization map, is the universal derivation in the category* $\mathrm{GMod}_W A$.

◀ Let the A-module P be geometrical, $\mathrm{supp}\, P \subset W$ and $X \in D(P)$. Then, in view of the universality of the differential d_{geom}, we have $X = h_X \circ d_{\mathrm{geom}}$, where $h_X \colon \Lambda^1_{\mathrm{geom}}(A) \to P$ is a homomorphism. In its turn, assertion (iii) of the previous lemma shows that h_X can be represented as the composition

$$\Lambda^1_{\mathrm{geom}}(A) \xrightarrow{\mathrm{pr}} \Lambda^1_{\mathrm{geom}}(A)/\left(\mathcal{I}_W \cdot \Lambda^1_{\mathrm{geom}}(A)\right), \xrightarrow{h^W_X} P$$

for the natural choice of h^W_X, and so $X = h^W_X \circ d^W_{\mathrm{geom}}$. Moreover, the uniqueness of this representation follows from the fact that the quotient module $\Lambda^1_{\mathrm{geom}}(A)/\left(\mathcal{I}_W \cdot \Lambda^1_{\mathrm{geom}}(A)\right)$ is an object of the category $\mathrm{GMod}_W A$, as statement (ii) of the previous lemma asserts. ▶

Similar facts hold for other functors of the differential calculus. For the functor Diff_l, they are discussed below.

14.13. Behavior of differential forms under smooth maps of manifolds. The universality property is the cause of the fact that 1-forms, just like functions, are naturally transformed by smooth maps. Let $F \colon M \to N$

be a smooth map of manifolds. Note that the composition

$$C^\infty(N) \xrightarrow{F^*} C^\infty(M) \xrightarrow{d} \Lambda^1(M)$$

is a derivation of the algebra $C^\infty(N)$ with values in the $C^\infty(N)$-module $\Lambda^1(M)$. We recall that the $C^\infty(N)$-module structure in $\Lambda^1(M)$ is defined by

$$(f, \omega) \mapsto F^*(f)\omega, \quad f \in C^\infty(N), \ \omega \in \Lambda^1(M).$$

Exercise. Prove that $\Lambda^1(M)$ is a geometrical $C^\infty(N)$-module.

By Theorem 14.7, there exists a $C^\infty(N)$-homomorphism

$$h_{d \circ F^*} : \Lambda^1(N) \to \Lambda^1(M)$$

such that $h_{d \circ F^*} \circ d = d \circ F^*$. For the sake of brevity, we shall write F^* instead of $h_{d \circ F^*}$. We thus have an $\mathbb{R}$-linear map

$$F^* : \Lambda^1(N) \to \Lambda^1(M),$$

with the following properties:

(i) $F^* \circ d = d \circ F^*$;

(ii) $F^*(f\omega) = F^*(f)F^*(\omega)$, if $f \in C^\infty(N)$, $\omega \in \Lambda^1(N)$.

In view of (i) and (ii), we see that

$$F^*(\omega) = \sum_i F^*(f_i)dF^*(g_i), \quad \text{if} \quad \omega = \sum_i f_i dg_i.$$

Exercise. Show that

1. $(F \circ G)^* = G^* \circ F^*$, where $L \xrightarrow{G} M \xrightarrow{F} N$;

2. $(F^*)^{-1} = (F^{-1})^*$;

3. $F^*(\omega)_z(\xi) = \omega_{F(z)}(d_z F(\xi))$, or, equivalently,

$$F^*(\omega)_z = (d_z F)^*(\omega_{F(z)}), \quad \text{where } z \in M, \ \xi \in T_z^*(M) \text{ and } \omega \in \Lambda^1(N).$$

14.14. The jet algebras $\mathcal{J}^l(M)$. We return to the case $A = C^\infty(M)$ and for every natural l define the $C^\infty(M)$-module $\mathcal{J}^l(M)$ as the module of sections of the vector bundle $\pi_{\mathcal{J}^l} : J^l M \to M$ (see Example IX in Section 11.11). The elements of this module are referred to as *l-jets* on the manifold M.

According to Section 9.65, every function $f \in C^\infty(M)$ gives rise to the section

$$s_{j_l(f)} : M \to J^l M, \quad z \mapsto [f]_z^l.$$

Sections of this kind are called *l-jets* of smooth functions.

Note that the multiplication in the algebra $C^\infty(M)$ induces an algebra structure in each fiber of the bundle $\pi_{\mathcal{J}^l}$. In fact, let $z \in M$, $f, g \in C^\infty(M)$, and $h \in \mu_z^{l+1}$. Then $fg = f(g + h) \bmod \mu_z^{l+1}$. Therefore, the formula

$$[f]_z^l \cdot [g]_z^l \overset{\text{def}}{=} [fg]_z^l$$

gives a well-defined product in the fiber $J_z^l M$. This multiplication induces a $C^\infty(M)$-algebra structure in the module $\mathcal{J}^l(M) = \Gamma(\pi_{J^l})$.

Using this multiplication, we can give a more transparent coordinate expression of jets. Let $\delta x_i = j_l(x_i) - x_i j_l(1)$ and

$$\delta^\sigma = \delta x_{i_1} \cdots \delta x_{i_k} \text{ if } \sigma = (i_1, \dots, i_k), \ 0 < |\sigma| \leqslant l.$$

We also put $\delta x^\varnothing = j_l(1)$, and then the sections δ^σ, $|\sigma| \leqslant l$, form a basis of the vector bundle π_{J^l} over U. This follows from the fact that the l-jets of the polynomials $(x - z)^\sigma$, $|\sigma| \leqslant l$, form a basis of $J_z^l M$. We see that a jet of order l on the manifold M in the special coordinate system corresponding to a local chart (U, x) can be written as $\sum_{|\sigma| \leqslant l} \alpha_\sigma \delta x^\sigma$, where $\alpha_\sigma \in A$.

14.15. Jet algebras $\mathcal{J}^l(M)$ as representing objects. With respect to differential operators in the algebra $C^\infty(M)$, the jet algebras $\mathcal{J}^l(M)$ play a role similar to the role of the module of differential forms $\Lambda^1(M)$ with respect to derivations. This fact can be proved by an argument very close to the one that we used in Section 14.4 (also see Theorem 14.19 below).

If we forget the geometrical meaning of the module $\mathcal{J}^l(M)$ as the module of sections of the bundle $J^l M$ and view its elements from a purely algebraic standpoint, we can still denote the jet of a function $f \in C^\infty(M)$ by $j_l(f)$; otherwise, we denote it by $s_{j_l(f)}$.

14.16 Theorem. *There is a finite set of functions $f_1, \dots, f_m \in C^\infty(M)$ whose l-jets $j_l(f_1), \dots, j_l(f_m)$ generate the $C^\infty(M)$-module $\mathcal{J}^l(M)$.*

◀ If $M = \mathbb{R}^k$, then as such we can take the set of all monomials x^σ with $|\sigma| \leqslant l$, where $x = (x_1, \dots, x_k)$ are the usual coordinates in $\mathbb{R}^k$. In the general case, choose an appropriate $k \in \mathbb{N}$ and consider an immersion $F: M \to \mathbb{R}^k$ (Whitney's immersion theorem). Then the family $F^*(x^\sigma)$, $|\sigma| \leqslant l$, will have the required property. Indeed, by Corollary 12.28, it suffices to show that for every point $z \in M$ the l-jets $[F^*(x^\sigma)]_z^l$ generate the vector space $J_z^l M$. Since F is an immersion, the differential $d_z F$ is injective; hence $(d_z F)^*$ is surjective. Therefore, among the differentials $d_z F^*(x_i)$, $i = 1, \dots, k$, there are $n = \dim M$ linearly independent ones, say $d_z F^*(x_1), \dots, d_z F^*(x_n)$. The functions $F^*(x_1), \dots, F^*(x_n)$ form a local system of coordinates near the point z. As Corollary 2.9 to the generalized Hadamard lemma shows, the monomials of degree $\leqslant l$ in these variables generate the space $J_z^l M$. ▶

14.17 Corollary. *There is a finite set of functions $f_1, \dots, f_m$ from $C^\infty(M)$ whose differentials $df_1, \dots, df_m$ generate the $C^\infty(M)$-module $\Lambda^1(M)$.*

◀ Let $f_1, \ldots, f_m$ be the functions whose jets generate $\mathcal{J}^1(M)$. Then their differentials generate $\Lambda^1(M)$. Indeed, the canonical direct decomposition $J_z^1 M = \mathbb{R} \oplus T_z^* M$ (see the proof of Corollary 9.27) shows that the bundle π_{J^1} is the direct sum of the two bundles $\mathbb{I}_M \colon M \times \mathbb{R} \to M$ and π_{T^*}. Passing to the modules of sections and using Proposition 12.23, we infer that

$$\mathcal{J}^1(M) = C^\infty(M) \oplus \Lambda^1(M).$$

Therefore, the images of the elements that generate $\mathcal{J}^1(M)$ under the projection $\mathcal{J}^1(M) \to \Lambda^1(M)$ generate $\Lambda^1(M)$. It remains to notice that the image of $j_1(f)$ is df.   ▶

14.18 Proposition. *The $\mathbb{R}$-homomorphism $j_l \colon C^\infty(M) \to \mathcal{J}^l(M)$, given by $f \mapsto j_l(f)$ is a differential operator of order l, i.e., it satisfies Definition 9.57.*

◀ We must prove that $\big(\delta_{g_0} \circ \cdots \circ \delta_{g_l}\big)(j_l) = 0$ for any functions $g_0, \ldots, g_l$. Let $\theta \in \mathcal{J}^l(M)$ and

$$\Delta = \theta \cdot j_l \colon C^\infty(M) \to \mathcal{J}^l(M) \to M, \quad \Delta(f) = (\theta \circ j_l)(f).$$

Then

$$\delta_g(\Delta)(f) = \theta \cdot j_l(gf) - g\theta \cdot j_l(f) = \theta \cdot j_l(g) \cdot j_l(f) - \theta \cdot g j_l(f)$$
$$= \big(j_l(g) - g j_l(1)\big) \cdot \theta \cdot j_l(f) = \delta_l(g) \cdot \theta \cdot j_l(f) = \delta_l(g) \cdot \Delta(f),$$

where $\delta_l(g) = j_l(g) - g j_l(1)$. Therefore,

$$\big(\big(\delta_{g_0} \circ \cdots \circ \delta_{g_l}\big)(j_l)\big)(f) = \delta_l(g_0) \cdots \delta_l(g_l) \cdot j_l(f).$$

The required fact follows from the equality $\delta_l(g_0) \cdots \delta_l(g_l) = 0$. To prove the latter, note that the image of the element $\delta_l(g)$ under the natural projection

$$\mathcal{J}^l(M) \to \mathcal{J}^0(M) = C^\infty(M), \quad j_l(f) \mapsto j_0(f),$$

is equal to $\delta_0(g) = 0$. This means that the value of the section $\delta_l(g)$ at any point $z \in M$ belongs to $\mu_z J_z^l M = \mu_z/\mu_z^{l+1}$. Hence

$$\delta_l(g_0) \cdots \delta_l(g_l) \in \mu_z^{l+1} J_z^l M = \mu_z^{l+1}/\mu_z^{l+1} = 0.$$

The element $\delta_l(g_0) \cdots \delta_l(g_l)$ is thus the zero section of the vector bundle π_{J^l}, i.e., the zero element of the module $\mathcal{J}^l(M)$.   ▶

The operator $j_l \colon C^\infty(M) \to \mathcal{J}^l(M)$ is referred to as the *universal differential operator of order $\leqslant l$* on the manifold M. The origin of the word "universal" will become clear in a little while.

If P is a geometrical $C^\infty(M)$-module, then there is a natural pairing

$$\mathrm{Diff}_l P \times \mathcal{J}^l(M) \to P.$$

Indeed, suppose that $\Delta \in \mathrm{Diff}_l P$, $\Theta \in \mathcal{J}^l(M)$, $z \in M$, and let $f \in C^\infty(M)$ be a smooth function such that $\Theta(z) = [f]_z^l$. Put

$$(\Delta, \Theta)(z) = \Delta(f)(z) \in P_z.$$

By virtue of Corollary 9.64, the value $(\Delta, \Theta)(z)$ does not depend on the choice of f.

Exercise. As in the proof of Theorem 14.7, show that the totality of all values $\Delta(f)(z) \in P_z$ uniquely determines the element $(\Delta, \Theta) \in P$.

To an arbitrary operator $\Delta \in \mathrm{Diff}_l\, P$, we can assign the homomorphism of $C^\infty(M)$-modules $h_\Delta \colon \mathcal{J}^l(M) \to P$ by putting $h_\Delta(\Theta) = (\Delta, \Theta)$. It follows from the definition of the pairing that we have $(\Delta, j_l(f)) = \Delta(f)$. Therefore, $j_l \circ h_\Delta = \Delta$. On the other hand, if $h \colon \mathcal{J}^l(M) \to P$ is an arbitrary $C^\infty(M)$-homomorphism, then the composition

$$\Delta_h = h \circ j_l \colon C^\infty(M) \to P,$$

according to 9.67, 9.59, is a differential operator of order $\leqslant l$ that satisfies $h_{\Delta_h} = h$. We thus arrive at the following important result:

14.19 Theorem. *For any geometrical $C^\infty(M)$-module P, the assignment*

$$\mathrm{Hom}_{C^\infty(M)}(\mathcal{J}^l(M), P) \ni h \mapsto h \circ j_l \in \mathrm{Diff}_l\, P$$

defines a natural isomorphism of $C^\infty(M)$-modules

$$\mathrm{Hom}_{C^\infty(M)}(\mathcal{J}^l(M), P) \cong \mathrm{Diff}_l\, P.$$

In other words, the functor Diff_l defined on the category of geometrical $C^\infty(M)$-modules is representable, with representing object $\mathcal{J}^l(M)$. ▶

This theorem explains why the differential operator

$$j_l \colon C^\infty(M) \to \mathcal{J}^l(M)$$

is called universal. As in the case of differential forms, the assumption that P is geometrical is essential: without it, Theorem 14.19 is not valid.

The significance of Theorem 14.19 is also due to the fact that it shows how to correctly introduce the notion of the jet in the differential calculus over any commutative K-algebra A. To do this, one must, first of all, choose the corresponding category of A-modules, say $\mathfrak{M}$ (see Section 14.6) and then define the module of l-jets $\mathcal{J}^l_{\mathfrak{M}}(A)$ as the range of values of the universal differential operator $j_l^{\mathfrak{M}} \colon A \to \mathcal{J}^l_{\mathfrak{M}}(A)$. The universality of the operator $j_l^{\mathfrak{M}}$ means that for any module P in the category $\mathfrak{M}$, the correspondence

$$\mathrm{Hom}_A(\mathcal{J}^l_{\mathfrak{M}}(A), P) \ni h \mapsto h \circ j_l \in \mathrm{Diff}_l\, P$$

determines a natural A-module isomorphism $\mathrm{Hom}(\mathcal{J}^l_{\mathfrak{M}}(A), P) \cong \mathrm{Diff}_l\, P$.

Exercise. Describe the module of l-jets $\mathcal{J}^l_{\mathfrak{M}}(A)$ and the module of 1-forms $\Lambda^1_{\mathfrak{M}}(A)$, where A is the algebra of smooth functions on the cross $\mathbf{K}$, while $\mathfrak{M}$ is the category of geometrical modules over this algebra.

14.20. Jet bundles. Suppose that P is a projective $C^\infty(M)$-module, say, the module of sections of a vector bundle $\pi_P \colon E \to M$, and $\mu_z \in C^\infty(M)$ is the maximal ideal corresponding to the point $z \in M$. Note that $\mu_z^{l+1} P$ is a submodule of P, and let $J_z^l P \stackrel{\text{def}}{=} P/\mu_z^{l+1}P$ be the quotient module. The image of the element $p \in P$ under the natural projection will be denoted by $[p]_z^l \in J_z^l P$.

The vector space $J_z^l P$ is a module over the algebra $J_z^l M$ with respect to the multiplication

$$[f]_z^l[p]_z^l \stackrel{\text{def}}{=} [fp]_z^l, \quad f \in C^\infty(M),\ p \in P.$$

Exercise. Prove that this multiplication is well defined.

Put $J^l P \stackrel{\text{def}}{=} \bigcup_{z \in M} J_z^l P$. Our nearest aim is to equip the set $J^l P$ with the structure of a smooth manifold in such a way that the natural projection

$$\pi_{J^l P} \colon J^l P \to M, \quad J_z^l P \mapsto z \in M,$$

will define a vector bundle structure over M on the smooth manifold $J^l P$. This vector bundle will be called the bundle of *jets of order l* (or *l-jets*) of the bundle π_P.

On the total space E of the bundle π_P, there is an adapted atlas (see Section 12.3). Its charts are of the form $(\pi^{-1}(U), x, u)$, where (U, x) is a chart of the corresponding atlas on M and $u = (u^1, \ldots, u^m)$, $m = \dim \pi_P$, are the fiber coordinates. Then, according to Proposition 12.13, the localization $P_U = \Gamma(\pi_P|_U)$ is a free $C^\infty(U)$-module. Let $e_1, \ldots, e_m$ be its basis. The restriction of an element $q \in P$ to U can be written as

$$q|_U = \sum_{i=1}^m f^i e_i, \quad \text{where } f^i \in C^\infty(U).$$

In other words, in the adapted coordinates, any section of the vector bundle is represented by a vector function $(f^1, \ldots, f^m)$ in the variables $(x_1, \ldots, x_n)$. This implies that $[q]_z^l$ is uniquely determined by the m-uple $([f^1]_z^l, \ldots, [f^m]_z^l)$, and therefore, the collection of numbers

$$\left(x_1, \ldots, x_n, u^1, \ldots, u^m, \ldots, p_\sigma^i, \ldots\right), \quad |\sigma| \leqslant l, \quad p_\sigma^j = \frac{\partial^{|\sigma|} f^j}{\partial x^\sigma}\bigg|_z,$$

uniquely determines the point $[q]_z^l \in \pi_{J^l P}^{-1}(U)$, where $\left(u^j, \ldots, p_\sigma^i, \ldots\right)$ are the special local coordinates of the l-jet of the function f^j in $J^l M$ (see Section 11.11, IX). The functions

$$x_i,\ u^j,\ p_\sigma^j, \quad 1 \leqslant i \leqslant n,\ 1 \leqslant j \leqslant m,\ 0 < |\sigma| \leqslant l,$$

form a coordinate system in the domain $\pi_{J^l P}^{-1}(U) \subset J^l P$. Charts of this type will be referred to as *special charts* on $J^l P$.

Exercise. Show that the special charts on $J^l P$ that correspond to compatible charts on M are compatible as well. In other words, these charts form an atlas, thus defining the structure of a smooth manifold in $J^l P$.

The projection

$$\pi_{J^l P} \colon J^l P \to M, \quad J^l_z P \mapsto z \in M,$$

is evidently smooth. The special charts described above are direct products of the form $U \times \mathbb{R}^{mN}$, where N is the number of all derivatives of order $\leqslant l$ (see Example IX in Section 11.11). Therefore, $\pi_{J^l P}$ is a locally trivial bundle. Moreover, the trivializing diffeomorphisms $\pi^{-1}_{J^l P}(U) \to U \times \mathbb{R}^{mN}$ are linear on each fiber, so that $\pi_{J^l P}$ is a vector bundle over M.

14.21. Let $A = C^\infty(M)$, $z \in M$, and let P and Q be $C^\infty(M)$-modules corresponding to the vector bundles π_P and π_Q over M, respectively. We have the following generalization of Corollary 9.61.

14.22 Proposition. *Suppose that p_1, $p_2 \in P$, $\Delta \in \mathrm{Diff}_l(P,Q)$, and $p_1 - p_2 \in \mu_z^{l+1} P$. Then $\Delta(p_1)(z) = \Delta(p_2)(z)$. In particular, if the elements p_1, $p_2 \in P$ coincide in a neighborhood $U \ni z$, then $\Delta(p_1)(z) = \Delta(p_2)(z)$ for any differential operator $\Delta \in \mathrm{Diff}(P,Q)$. In other words, the differential operators that act on sections of vector bundles are local.*

◀ Indeed, if $p_1 - p_2 \in \mu_z^{l+1} P$, then $\Delta(p_1 - p_2) \in \mu_z Q$ by Proposition 9. Now, if the two sections p_1 and $p_2 \in P$ coincide in a neighborhood of $U \ni z$, then $p_1 - p_2 \in \mu_z^{l+1} P$ for any l.   ▶

This proposition allows us to well define the localization

$$\Delta\big|_U \colon P\big|_U \to Q\big|_U, \quad \Delta\big|_U(\bar{p})(z) = \Delta(p)(z), \quad \bar{p} \in P\big|_U, \ p \in P, \ z \in U,$$

for any differential operator $\Delta \in \mathrm{Diff}(P,Q)$ and any open set $U \subset M$, where p is an arbitrary element of the module P coinciding with $\bar{p}$ in a certain neighborhood of the point z. According to this definition, $\Delta_U(p\big|_U)(z) = \Delta(p)\big|_U(z)$ if $p \in P$. The operator Δ is uniquely determined by its restrictions to the charts of an arbitrary atlas of the manifold M.

Now fix a system of local coordinates $x_1, \dots, x_n$ in a neighborhood $U \subset M$ so that both vector bundles $\pi_P\big|_U$ and $\pi_Q\big|_U$ are trivial. Let $e_1, \dots, e_m$ and $\varepsilon_1, \dots, \varepsilon_k$ be bases of the modules $P\big|_U$ and $Q\big|_U$, respectively. Then the restriction of the elements $p \in P$ and $q \in Q$ to U can be represented as

$$p\big|_U = \sum_{i=1}^m f^i e_i, \quad q\big|_U = \sum_{r=1}^k g^r \varepsilon_r, \quad \text{where } f^i, \ g^r \in C^\infty(U).$$

Fixing the bases $e_1, \dots, e_m$ and $\varepsilon_1, \dots, \varepsilon_r$, we can define the following $C^\infty(U)$-linear maps:

$$\alpha_i \colon C^\infty(U) \to P\big|_U, \quad f \mapsto f e_i, \quad 1 \leqslant i \leqslant m,$$

$$\beta_j \colon Q\big|_U \to C^\infty(U), \quad \sum_{r=1}^k g^r \varepsilon_r \mapsto g^j, \quad 1 \leqslant j \leqslant r.$$

The composition $\Delta_{i,j} \overset{\text{def}}{=} \beta_j \circ \Delta \circ \alpha_i \colon C^\infty(U) \to C^\infty(U)$ is, according to Sections 9.67 and 9.59, a differential operator of order $\leqslant l$. For scalar differential operators, we have already proved that the algebraic definition 9.57 coincides with the conventional one. Now we see that the action of the operator $\Delta|_U$ on $p|_U$, i.e., on a vector function $(f^1, \ldots, f^m)$, is given by

$$\begin{pmatrix} \Delta_{1,1} & \cdots & \Delta_{1,m} \\ \vdots & \ddots & \vdots \\ \Delta_{k,1} & \cdots & \Delta_{k,m} \end{pmatrix} \begin{pmatrix} f_1 \\ \vdots \\ f_m \end{pmatrix} = \begin{pmatrix} \Delta_{1,1}(f_1) + \ldots + \Delta_{1,m}(f_m) \\ \vdots \\ \Delta_{k,1}(f_1) + \ldots + \Delta_{k,m}(f_m) \end{pmatrix}.$$

It follows that the standard notion of matrix differential operator is a particular case of the general algebraic definition 9.67, since for the scalar differential operators, like $\Delta_{i,j}$, this fact has already been established (see Section 9.62). Matrix differential operators are the coordinate description of differential operators (in the sense of Definition 9.67) that send the sections of one vector bundle to the sections of the other over the algebra $A = C^\infty(M)$.

14.23. Jet modules. The module of smooth sections of the vector bundle $\pi_{J^l P}$ is called the *module of l-jets of the bundle* π_P and is denoted by $\mathcal{J}^l(P)$. The elements of this module are called *geometrical l-jets of the module P* or simply *l-jets*. It is worth noting that the $C^\infty(M)$-module $\mathcal{J}^l(P)$ is also a $\mathcal{J}^l(M)$-module with respect to the following multiplication,

$$(\theta \cdot \Theta)(z) \overset{\text{def}}{=} \theta(z)\Theta(z), \quad \theta \in \mathcal{J}^l(M), \ \Theta \in \mathcal{J}^l(P),$$

where the multiplication $J^l_z M \times J^l_z P \to J^l_z P$ that appears on the right-hand side was defined in Section 14.20.

As in the scalar case, any element of the module P gives rise to a section $j_l(p)$ of the bundle $\pi_{J^l P}$ defined by $j_l(p)(z) = [p]^l_z$. Suppose that in the adapted coordinates p is represented by the vector function $(f^1, \ldots, f^m)$. Then in the corresponding special coordinates on $J^l P$, the section $j_l(p)$ takes the form of the vector function

$$\left(f^1, \ldots, f^m, \ldots, \frac{\partial^{|\sigma|} f^i}{\partial x^\sigma}, \ldots \right), \quad |\sigma| \leqslant l.$$

The coordinate expression of $j_l(p)$ shows, first, that this section is smooth, i.e., that $j_l(p) \in \mathcal{J}^l(P)$, and, second, that the $\mathbb{R}$-linear map

$$j_l \colon P \to \mathcal{J}^l(P), \quad p \mapsto j_l(p),$$

is a differential operator of order $\leqslant l$.

Exercise. Give a coordinate-free proof of these facts.

The operator j_l is referred to as the *universal differential operator* of order l in the bundle π_P.

14.24 Proposition. *Let P be a projective $C^\infty(M)$-module. There exists a finite set of elements $p_1, \ldots, p_m \in P$ such that the corresponding l-jets $j_l(p_1), \ldots, j_l(p_m)$ generate the $C^\infty(M)$-module $\mathcal{J}^l(P)$.*

◀ Let $\bar{p}_1, \ldots, \bar{p}_k \in P$ be a finite system of generators of the module P (see Corollary 12.28). Let $f_1, \ldots, f_s$ be a finite set of functions whose l-jets generate $\mathcal{J}^l(M)$ (Proposition 14.16). Then the l-jets

$$j_l(f_i \bar{p}_j), \quad 1 \leqslant i \leqslant s, \ 1 \leqslant j \leqslant k,$$

generate $\mathcal{J}^l(P)$. To establish this fact, it is sufficient, by Proposition 12.10, to show that the l-jets $[f_i \bar{p}_j]_z^l$ generate the fiber $J_z^l P$ for any $z \in M$. We know that any element of this fiber has the form $[p]_z^l$ for a certain $p \in P$. Let

$$p = \sum_j g_j \bar{p}_j, \quad g_j \in C^\infty(M), \quad \text{and} \quad [g_j]_z^l = \sum_i \lambda_{ji} [f_j]_z^l, \ \lambda_{ji} \in \mathbb{R}.$$

Then

$$[p]_z^l = \sum_j [g_j \bar{p}_j]_z^l = \sum_j [g_j]_z^l [\bar{p}_j]_z^l = \sum_{i,j} \lambda_{ji} [f_i]_z^l [\bar{p}_j]_z^l = \sum_{i,j} \lambda_{ji} [f_i \bar{p}_j]_z^l. \quad ▶$$

Suppose that Q is a geometrical $C^\infty(M)$-module. Following the approach used in Section 14.18, we define the pairing

$$(\cdot, \cdot) \colon \mathrm{Diff}_l(P, Q) \times \mathcal{J}^l(P) \to Q.$$

For a point $z \in M$ and a jet $\Theta \in \mathcal{J}^l(M)$, we can choose an element $p \in P$ such that $\Theta(z) = [p]_z^l$. For an arbitrary differential operator $\Delta \in \mathrm{Diff}_l(P, Q)$, let us put $(\Delta, \Theta)(z) = \Delta(p)(z) \in Q_z$. By virtue of Proposition 14.22, the value $(\Delta, \Theta)(z)$ does not depend on the choice of the representative in the class $[p]_z^l$.

If $\Theta = \sum_i h_i j_l(p_i)$ (see Proposition 14.24), then for p we can take the element $\sum_i \lambda_i p_i$, where $\lambda_i = h_i(z) \in \mathbb{R}$. Therefore,

$$\Delta(p)(z) = \sum_i \lambda_i \Delta(p_i)(z) = \left[\sum_i h_i \Delta(p_i) \right](z).$$

Since the module Q is geometrical, it follows that

$$(\Delta, \Theta) = \sum_i h_i \Delta(p_i) \in Q.$$

This proves the existence of the pairing.

Proceeding as at the end of Section 14.18, we assign to any operator $\Delta \in \mathrm{Diff}_l(P, Q)$ the homomorphism of $C^\infty(M)$-modules

$$h_\Delta \colon \mathcal{J}^l(P) \to Q, \quad h_\Delta(\Theta) = (\Delta, \Theta).$$

It follows from the definition of the pairing that $(\Delta, j_l(p)) = \Delta(p)$. Therefore, $h_\Delta \circ j_l = \Delta$. On the other hand, if $h \colon \mathcal{J}^l(P) \to Q$ is an arbitrary

$C^\infty(M)$-homomorphism, then the composition

$$\Delta_h = h \circ j_l \colon P \to Q$$

is, according to Sections 9.59 and 9.67, a differential operator of order $\leqslant l$. Also, evidently, $h_{\Delta_h} = h$. We have thus established the following important fact.

14.25 Theorem. *Let a projective $C^\infty(M)$-module P be given. For any geometrical $C^\infty(M)$-module Q, the correspondence*

$$\operatorname{Hom}_{C^\infty(M)}(\mathcal{J}^l(P), Q) \ni h \mapsto h \circ j_l \in \operatorname{Diff}_l(P, Q)$$

defines a natural isomorphism of $C^\infty(M)$-modules

$$\operatorname{Hom}_{C^\infty(M)}(\mathcal{J}^l(P), Q) \cong \operatorname{Diff}_l(P, Q).$$

In other words, the functor $Q \mapsto \operatorname{Diff}_l(P, Q)$ is representable in the category of geometrical $C^\infty(M)$-modules, and the $C^\infty(M)$-module $\mathcal{J}^l(P)$ is its representing object. ▶

The last theorem makes it possible to change our point of view and define the module $\mathcal{J}^l(P)$ (together with the operator $j_l \colon P \to \mathcal{J}^l(P)$) as the representing object of the functor $Q \mapsto \operatorname{Diff}_l(P, Q)$ in the category of geometrical $C^\infty(M)$-modules. This approach is conceptual and thereby immediately extends to arbitrary algebras and categories of modules. Of course, the question of existence must be answered separately in each particular case.

Exercise. Prove that the A-module $\operatorname{Diff}_l(P, Q)$ is geometric, provided that the module Q is geometrical.

15

Functors of the Differential Calculus and their Representations

15.1. In this chapter, we shall describe the main functors of the differential calculus in commutative algebras, the corresponding objects and the natural transformations that represent them, as well as the main rules that they satisfy. In the previous chapter, we used, for the same goal, standard constructions coming from differential geometry, whereas now we change our approach to the categorial algebraic one. The latter turns out to be much more flexible and universal and works perfectly in situations where the classical geometrical approach leads to difficulties, sometimes unsurmountable. Moreover, thanks to this approach, many generalizations are carried out automatically.

One of the most important applications of this algebraic approach to the differential calculus is the contemporary theory of nonlinear differential equations, in which the analogs of the standard differential geometrical structures are studied on infinite-dimensional manifolds of special type. These are the infinite extensions of differential equations and, in particular, the spaces of jets of infinite order, their differential coverings, and other inverse limits of projective systems of smooth manifolds. Manifolds of this class do not possess a norm, nor a sufficiently strong topology or any other structure traditionally used in the construction of the differential of a function and other elements of the differential calculus on finite-dimensional manifolds. Classical tradition itself lies in the belief that a necessary prerequisite to the construction of the basic elements of the differential calculus in any context is the notion of limit.

Another important application, where the approach to the differential calculus developed in this book appears to have no valid alternative, initially arose in theoretical physics. Its different parts are specified by the prefix "super", for instance, super-manifolds, super-symmetries, and so on. It eventually turned out that this "super-mathematics", which initially

© Springer Nature Switzerland AG 2020
J. Nestruev, *Smooth Manifolds and Observables*, Graduate Texts
in Mathematics 220, https://doi.org/10.1007/978-3-030-45650-4_15

seemed enigmatic and mysterious, actually enters in a natural way in the much wider context of the differential calculus over *graded* commutative algebras as a very particular case. This important generalization is the subject matter of the last chapter of this book.

As we have already seen, such basic structures of the differential calculus as vector fields, differential 1-forms, jets, and even the entire Hamiltonian formalism, may be constructed in any commutative algebra. Moreover, in that context, geometrical portraits of these algebras arise on their spectra. Actually, all this occurs for all the structures of the differential calculus, in particular—all the differential geometrical structures. Thus, commutative algebras become objects of differential geometry. This possibility is inherent in the structure of the calculus of *functors of the differential calculus* and the present chapter is an introduction to that calculus. Informally, it might be called the "logic" of the differential calculus.

In what follows, we replace the long expression "functor of the differential calculus" by the term *diffunctor*. This is a new term and is introduced here for the first time. Since it is impossible to give a general definition of this notion within the framework of the present book, in what follows we will limit ourselves to the description of several main examples of diffunctors and show how to work with them. In a certain sense, these diffunctors generate the *algebra of diffunctors*. Further, talking of diffunctors in general, we simply stress that some fact or some construction holds for all diffunctors, and not only those that are explicitly described in this chapter.

Below we will be considering modules over a certain commutative $\Bbbk$-algebra A, where, for simplicity, we assume that $\Bbbk$ is a field. In the cases where this must be specified, we denote the operator of multiplication of elements of a module P by an element $a \in A$ by a_P. The fact that $\mathcal{O}$ is an object of the category $\mathcal{K}$ is expressed, by abuse of notation, as $\mathcal{O} \in \mathrm{Ob}\,\mathcal{K}$.

Remark. Let P, Q be A-modules and $\varphi \colon P \to Q$ be a $\Bbbk$-linear map; then $\varphi \circ a_P$ and $a_Q \circ \varphi$, $a \in A$, will also be maps of P to Q:

$$(\varphi \circ a_P)(p) = \varphi(ap), \quad (a_Q \circ \varphi)(p) = a\varphi(p), \quad p \in P.$$

In what follows, we will often omit the lower indices indicating the modules on which the operators of multiplication by elements of the algebra A work, especially in the cases where these modules have cumbersome notations such as $\mathrm{Diff}_l(\mathrm{Diff}_k(C^\infty(M)))$. It seems to us that the expression

$$a \circ \varphi, \quad \varphi \in \mathrm{Diff}_l(\mathrm{Diff}_k(C^\infty(M))),$$

looks simpler and—most important—is easier to understand than

$$a_{\mathrm{Diff}_k(C^\infty(M))} \circ \varphi, \quad \varphi \in \mathrm{Diff}_l(\mathrm{Diff}_k(C^\infty(M))).$$

We hope that this will not confuse the reader, since what specific modules are understood is always clear from the context.

15.2. Representing objects and homomorphisms. In this section, we recall the notion of representing the object and the general categorial scheme in accordance with which to a natural transformation of functors one assigns a morphism of their representing objects. Here we assume, since this is the only case needed in what follows, that everything takes place in the category $\text{Mod}\,A$ and, in particular, that representing objects are A-modules, while morphisms are homomorphisms of A-modules.

First recall that a *representing object* for a covariant functor Φ from some category of A-modules $\mathfrak{M}$ to itself is an A-module $\mathcal{O}_\Phi$ such that for any $P \in \text{Ob}\,\mathfrak{M}$ there is an isomorphism $\mathbf{i}_\Phi = \mathbf{i}_{\Phi,P} \colon \Phi(P) \to \text{Hom}_A(\mathcal{O}_\Phi, P)$. Here we assume that for any homomorphism $h \colon P \to Q$, we have the commutative diagram

$$
\begin{array}{ccc}
\Phi(P) & \xrightarrow{\ \Phi(h)\ } & \Phi(Q) \\[2pt]
{\scriptstyle \mathbf{i}_{\Phi,P}}\Big\downarrow & & \Big\downarrow {\scriptstyle \mathbf{i}_{\Phi,Q}} \\[2pt]
\text{Hom}_A(\mathcal{O}_\Phi, P) & \xrightarrow[\text{Hom}_A(\mathcal{O}_\Phi,h)]{} & \text{Hom}_A(\mathcal{O}_\Phi, Q)\,,
\end{array}
\tag{15.1}
$$

where $\text{Hom}_A(\mathcal{O}_\Phi, h)(\varphi) = h \circ \varphi$, $\varphi \in \text{Hom}_A(\mathcal{O}_\Phi, P)$. In other words, the above means that the system of isomorphisms $\mathbf{i}_\Phi$ allows to identify the functor Φ with the functor

$$
\text{Hom}_A(\mathcal{O}_\Phi, \cdot) \colon P \mapsto \text{Hom}_A(\mathcal{O}_\Phi, P).
$$

Let Φ_i be a certain functor and $\mathcal{O}_{\Phi_i}$ be its representing A-module, $i = 1, 2$. Assume also that F is a *natural transformation of the functor* Φ_1 to the functor Φ_2. This means that for any arbitrary homomorphism of A-modules $h \colon Q \to R$ we have the two following commutative diagrams

$$
\begin{array}{ccc}
\Phi_1(Q) & \xrightarrow{\ \Phi_1(h)\ } & \Phi_1(R) \\[2pt]
{\scriptstyle F(Q)}\Big\downarrow & & \Big\downarrow {\scriptstyle F(R)} \\[2pt]
\Phi_2(Q) & \xrightarrow[\ \Phi_2(h)\ ]{} & \Phi_2(R)
\end{array}
\qquad
\begin{array}{ccc}
\Phi_1(\mathcal{O}_{\Phi_1}) & \xrightarrow{\ \Phi_1(h)\ } & \Phi_1(P) \\[2pt]
{\scriptstyle F(\mathcal{O}_{\Phi_1})}\Big\downarrow & & \Big\downarrow {\scriptstyle F(P)} \\[2pt]
\Phi_2(\mathcal{O}_{\Phi_1}) & \xrightarrow[\ \Phi_2(h)\ ]{} & \Phi_2(P)\,.
\end{array}
\tag{15.2}
$$

From now on we simplify the notation $\mathcal{O}_{\Phi_1}$ to $\mathcal{O}_1$ and similarly for Φ_2. The diagram on the right is a particular case of the one on the left for $Q = \mathcal{O}_1$, $R = P$. From the latter, using the isomorphisms $\mathbf{i}_{\Phi_1,\mathcal{O}_1}$ and $\mathbf{i}_{\Phi_1,P}$, we obtain the diagram

$$
\begin{array}{ccc}
\text{Hom}_A(\mathcal{O}_1, \mathcal{O}_1) & \xrightarrow{\ \text{Hom}_A(\mathcal{O}_1,h)\ } & \text{Hom}_A(\mathcal{O}_1, P) \\[2pt]
{\scriptstyle \mathbf{i}_F(\mathcal{O}_1)}\Big\downarrow & & \Big\downarrow {\scriptstyle \mathbf{i}_F(P)} \\[2pt]
\text{Hom}_A(\mathcal{O}_2, \mathcal{O}_1) & \xrightarrow[\ \text{Hom}_A(\mathcal{O}_2,h)\ ]{} & \text{Hom}_A(\mathcal{O}_2, P)\,,
\end{array}
\tag{15.3}
$$

where we have used the notation $\mathbf{i}_F(Q) = \mathbf{i}_{\Phi_2,Q} \circ F(Q) \circ \mathbf{i}_{\Phi_1,Q}^{-1}$.

Note further that $\text{Hom}_A(\mathcal{O}_1, h)(\text{id}_{\mathcal{O}_1}) = h$. Hence, the commutativity of diagram (15.3) for the identity homomorphism $\text{id}_{\mathcal{O}_1} \in \text{Hom}_A(\mathcal{O}_1, \mathcal{O}_1)$

means that
$$\mathbf{i}_F(P)(h) = h \circ H_F, \quad \text{where} \quad H_F \overset{\text{def}}{=} \mathbf{i}_F(\mathcal{O}_1)(\mathrm{id}_{\mathcal{O}_1}) \in \mathrm{Hom}_A(\mathcal{O}_2, \mathcal{O}_1).$$

Thus, we obtain

Proposition. *If two functors Φ_i, $i = 1, 2$, are given each of which is identified with the functor $\mathrm{Hom}_A(\mathcal{O}_i, \cdot)$, then any natural transformation $F \colon \Phi_1 \to \Phi_2$ is described by the homomorphism of the representing objects $H_F \colon \mathcal{O}_2 \to \mathcal{O}_1$:*
$$\Phi_1(P) \cong \mathrm{Hom}_A(\mathcal{O}_1, P) \ni h \mapsto h \circ H_F \in \mathrm{Hom}_A(\mathcal{O}_2, P) \cong \Phi_2(P).$$

The homomorphism H_F is called *representing* for the transformation F.

Exercise. Show that any homomorphism $H \colon \mathcal{O}_2 \to \mathcal{O}_1$ determines a natural transformation of the functor Φ_1 to the functor Φ_2.

15.3. Exact sequence of functors. In what follows, we will meet *sequences of transformations of diffunctors*
$$\cdots \xrightarrow{F_{i-2}} \Phi_{i-1} \xrightarrow{F_{i-1}} \Phi_i \xrightarrow{F_i} \Phi_{i+1} \xrightarrow{F_{i+1}} \cdots \tag{15.4}$$

Such a sequence is called *exact* (respectively, is said to be a *complex*), if for any A-module P the following sequence of A-homomorphisms is exact (resp., is a complex):
$$\cdots \xrightarrow{F_{i-2}(P)} \Phi_{i-1}(P) \xrightarrow{F_{i-1}(P)} \Phi_i(P) \xrightarrow{F_i(P)} \Phi_{i+1}(P) \xrightarrow{F_{i+1}(P)} \cdots; \tag{15.5}$$

recall that this means that for all indices i we have the condition $\mathrm{Im}\, F_{i-1}(P) = \mathrm{Ker}\, F_i(P)$ (resp., $\mathrm{Im}\, F_{i-1}(P) \subset \mathrm{Ker}\, F_i(P)$).

Let $\mathcal{O}_i$ be an A-module representing the functor Φ_i, and let
$$\cdots \xleftarrow{H_{i-2}} \mathcal{O}_{i-1} \xleftarrow{H_{i-1}} \mathcal{O}_i \xleftarrow{H_i} \mathcal{O}_{i+1} \xleftarrow{H_{i+1}} \cdots, \quad H_i \overset{\text{def}}{=} H_{F_i}, \tag{15.6}$$

be the corresponding (15.4) sequence of representing homomorphisms.

Proposition. *The sequence of functors* (15.4) *is exact (resp., is a complex) if and only if such is the sequence of representing modules and homomorphisms* (15.6).

◀ Identify the functors Φ_i and $\mathrm{Hom}_A(\mathcal{O}_i, \cdot)$. In that case the sequence (15.5) acquires the form
$$\cdots \xrightarrow{F_{i-2,P}} \mathrm{Hom}_A(\mathcal{O}_{i-1}, P) \xrightarrow{F_{i-1,P}} \mathrm{Hom}_A(\mathcal{O}_i, P) \xrightarrow{F_{i,P}}$$
$$\xrightarrow{F_{i,P}} \mathrm{Hom}_A(\mathcal{O}_{i+1}, P) \xrightarrow{F_{i+1,P}} \cdots, \tag{15.7}$$

where $F_{i,P}(\varphi) = \varphi \circ H_i$, $\varphi \in \mathrm{Hom}_A(\mathcal{O}_i, P)$.

First let us show that the exactness of (15.4) implies the exactness of (15.6). Let $P = \mathcal{O}_{i-1}$ and $\varphi = \mathrm{id}_{\mathcal{O}_{i-1}}$. Then, $F_{i-1,P}(\varphi) = H_{i-1}$. But $F_{i,P} \circ F_{i-1,P} = 0$, so that $H_{i-1} \circ H_i = F_{i,P}(H_{i-1}) = 0$, therefore,

$\operatorname{Im} H_i \subset \operatorname{Ker} H_{i-1}$. Further, let $P = \operatorname{coker} H_i$ and $\varphi \colon \mathcal{O}_i \to P$ be the natural projection. Then $F_{i,P}(\varphi) = \varphi \circ H_i = 0$, i.e., $\varphi \in \operatorname{Ker} F_{i,P}$, and, in view of the exactness of (15.7), we have $\varphi = F_{i-1,P}(\psi) = \psi \circ H_{i-1}$ for some $\psi \in \operatorname{Hom}_A(\mathcal{O}_{i-1}, P)$. But

$$\operatorname{Im} H_i = \operatorname{Ker} \varphi = \operatorname{Ker}(\psi \circ H_{i-1}) \supset \operatorname{Ker} H_{i-1},$$

i.e., $\operatorname{Ker} H_{i-1} \subset \operatorname{Im} H_i$. Together with the previous inclusion, this shows that $\operatorname{Ker} H_{i-1} = \operatorname{Im} H_i$ and so, the sequence (15.6) is exact at the term $\mathcal{O}_i$.

How let us show that the exactness of (15.6) implies the exactness of (15.4). For that it suffices to verify the exactness of the sequence (15.7) for any module P. If $\varphi \in \operatorname{Ker} F_{i,P}$, then $\varphi \circ H_i = 0$, which is equivalent to $\operatorname{Im} H_i \subset \operatorname{Ker} \varphi$. But $\operatorname{Im} H_i = \operatorname{Ker} H_{i-1}$ in view of the exactness of the sequence (15.6), and so we obtain $\operatorname{Ker} H_{i-1} \subset \operatorname{Ker} \varphi$. Finally, the last inclusion allows to represent φ in the form $\varphi = \psi \circ H_{i-1}$ for some element $\psi \in \operatorname{Hom}_A(H_{i-1}, P)$. But since $\psi \circ H_{i-1} = F_{i-1,P}(\psi)$, we see that $\operatorname{Ker} F_{i,P} \subset \operatorname{Im} F_{i-1,P}$.

Conversely, if $\varphi = F_{i-1,P}(\psi) = \psi \circ H_{i-1}$, then $F_{i,P}(\varphi) = \psi \circ H_{i-1} \circ H_i = 0$, so that $\operatorname{Ker} F_{i,P} \supset \operatorname{Im} F_{i-1,P}$. Therefore, $\operatorname{Ker} F_{i,P} = \operatorname{Im} F_{i-1,P}$.

Finally, note that the proof of the fact that the sequences (15.4) and (15.6) are actually complexes is part of the proof of their simultaneous exactness given above. ▶

If a sequence of functors

$$0 \to \Phi_0 \xrightarrow{F_0} \Phi_1 \xrightarrow{F_1} \Phi_2 \tag{15.8}$$

is exact, then it is natural to call the functor Φ_0 the *kernel of the transformation of functors* F_1. Similarly, one defines the *image of a transformation of functors*. One of the useful and simple applications of Proposition 15.3 is the proof of the existence of the representing object for the kernel functor.

Corollary. *Let the functor Φ_0 be the kernel of the natural transformations of functors $F \colon \Phi_1 \to \Phi_2$ and let $\mathcal{O}_i$ be the representing object for the functor Φ_i, $i = 1, 2$. Then $\mathcal{O}_0 \overset{\text{def}}{=} \operatorname{coker} H_F$ is the representing object for the functor Φ_0.*

◀ If $H_0 \colon \mathcal{O}_1 \to \mathcal{O}_0 = \operatorname{coker} H_F$ is the natural projection, then the sequence

$$0 \leftarrow \mathcal{O}_0 \xleftarrow{H_0} \mathcal{O}_1 \xleftarrow{H_F} \mathcal{O}_2 \tag{15.9}$$

is exact and Proposition 15.3 shows that the functor $\operatorname{Hom}_A(\mathcal{O}_0, \cdot)$ is the kernel of the natural transformation of functors

$$\operatorname{Hom}_A(\mathcal{O}_1, \cdot) \to \operatorname{Hom}_A(\mathcal{O}_2, \cdot),$$

generated by the homomorphism H_F, which can be identified with the kernel of the transformation F. ▶

15.4. Composition of diffunctors and their representations. The description of representing objects for the composition of diffunctors is based on the following general algebraic fact, which will be constantly used in this chapter. For commutative algebras, it can be stated as follows:

Let A and B be commutative algebras over a common field $\Bbbk$ and let $P \in \operatorname{Mod} A$, $R \in \operatorname{Mod} B$. Let also Q be an (A, B)-bimodule, i.e., a $\Bbbk$-vector space possessing the structure of a left A-module and of a right B-module, and let these structures be compatible,

$$(aq)b = a(qb), \ a \in A, \ b \in B, \ q \in Q.$$

Then $P \otimes_A Q$ and $\operatorname{Hom}_B(Q, R)$ are obviously also (A, B)-bimodules. In this situation, we have an isomorphism of (A, B)-bimodules

$$\iota \colon \ \operatorname{Hom}_B(P \otimes_A Q, R) \cong \operatorname{Hom}_A(P, \operatorname{Hom}_B(Q, R)), \tag{15.10}$$

specified by the relation

$$\big((\iota(\varphi))(p)\big)(q) = \varphi(p \otimes_A q), \ p \in P, \ q \in Q, \ \varphi \in \operatorname{Hom}_B(P \otimes_A Q, R). \tag{15.11}$$

Below the isomorphism ι will be used in the situation when $A = B$, and we will speak of A-bimodules rather than (A, A)-bimodules. Forgetting the *right* A-module structure of Q, we obtain its *left* A-module, denoted $Q_<$, similarly, its *right* A-module, denoted $Q_>$. Depending on the situation, the indices ">" and "<" may be lower or upper. They are needed to describe more complicated constructions involving A-bimodules. For example, the symbol $\operatorname{Hom}_A^<(Q_>, R)$ denotes the $\Bbbk$-vector space of homomorphisms of the A-module $Q_>$ to the A-module R, whose A-module structure is generated by the left multiplication in the bimodule Q. The symbols $\operatorname{Hom}_A^<(R, Q_>)$, $_<Q_> \otimes R$, etc., are understood similarly.

15.5. Differentially closed categories. An example of how one may prove the existence of representing objects for diffunctors over any commutative algebra is the construction of the representing object for the diffunctor D, i.e., for the module of differential 1–forms presented in Section 14.4. Actually, this is a very general construction. However, it is insufficiently constructive to be used for concrete calculations, which is nicely illustrated by the example of *algebraic* differential 1-forms for the algebra $C^\infty(M)$ (see Section 14.10). The reason for this is that the category of all $C^\infty(M)$-modules is too wide, and we should limit ourselves only to "observable" modules, i.e., geometrical modules. The necessity of restricting the category of modules in one way or another also arises in many other situations, so that we must face the problem of the actual existence of objects that represent diffunctors in the given category of modules over the given algebra. The following definition allows us to adequately formalize the setting of that problem.

15.6. Definition. A category of A-modules $\mathfrak{M}$ is called *differentially closed* if

(a) the result of applying any diffunctor to an object of the category $\mathfrak{M}$ is again an object of this category;

(b) the natural transformation of diffunctors applied to objects of the category $\mathfrak{M}$ yield morphisms in this category;

(c) objects representing diffunctors in the category $\mathfrak{M}$ are objects of this same category.

Now the question that concerns us can be stated as follows:

> Is the category of A-modules under consideration differentially closed or not?

A simple but conceptually important fact is that the notion of differential closedness is in natural agreement with the classical mechanism of observability. Formally this means the following.

Proposition. *The category of geometrical modules over a geometrical algebra is differentially closed.*

◄ The fact that the module $\mathcal{O}$ is the representing object for some covariant diffunctor Φ means that we have a natural isomorphism $\Phi(P) \cong \mathrm{Hom}_A(\mathcal{O}, P)$. If the module P is geometrical, then the canonical isomorphism

$$\mathrm{Hom}_A(\mathcal{O}, P) \to \mathrm{Hom}_A(\Gamma(\mathcal{O}), P)$$

(see the proposition in Section 8.6) shows that $\Gamma(\mathcal{O})$ represents the diffunctor Φ in the category GMod A.

Thus, everything is reduced to the proof of the existence of representing objects for the diffunctors in the category of all A-modules. But such a proof presupposes the description of the whole algebra of diffunctors, and this, as we already pointed out at the beginning of this chapter, is beyond the scope of this book. For the diffunctors that will be discussed further in this chapter and their various combinations, the question of existence will be solved as we go along. Hence, from the formal point of view, the above proposition can be regarded as proved only for those diffunctors.   ►

Note that the general proof of existence is of simple algebraic character, in the spirit of how it was carried out for differential 1-forms (see Section 14.6).

15.7. Sufficient condition for differential closedness. For our aims, a condition sufficient for differential closedness is given in the next proposition.

Proposition. *A category of A-module $\mathfrak{M}$ is differentially closed if*

(i) *the algebra A belongs to $\mathfrak{M}$;*

(ii) *the module representing the diffunctor $\mathrm{Diff}_k^<$ in the category $\mathfrak{M}$ is an object of the same category;*

(iii) $\mathfrak{M}$ *is closed with respect to the operations* $\otimes_A$ *and* Hom_A;

(iv) *submodules of any module belonging to the category* $\mathfrak{M}$ *also belong to* $\mathfrak{M}$.

◀ We will carry out the proof only for the diffunctors studied later in this chapter for the same reason as for the proof of the proposition from Section 15.6. The construction of representing objects for diffunctors is a step-by-step procedure, at each step of which the representing object of the subsequent, more complicated diffunctor, is constructed on the basis of the previous ones. This requires the application of the operations $\otimes_A$ and Hom_A and the consideration of certain submodules of modules that arise in the process. The fact that this can be done for the diffunctors studied below can be seen directly from their description. The starting point of the process is the construction of the module of jets as representing objects for the diffunctors $\mathrm{Diff}_k^<$, $k \geq 0$. Here we must pay attention to the fact that $\mathrm{Diff}_0^<$ is a functor equivalent to the identity, whose representing object is the algebra A (see Section 15.8). Thus, the algebra A itself must be an object of the category $\mathfrak{M}$. ▶

15.8. The diffunctors $\mathrm{Diff}_k^<$ **and** $D_{(k)}$. Let P be an A-module and let k be some nonnegative integer. Then the correspondence $Q \mapsto \mathrm{Diff}_k^<(P, Q)$, where $\mathrm{Diff}_k^<(P, Q)$ is defined in Section 9.67, is a covariant functor, denoted $\mathrm{Diff}_{k,P}^<$, from the category $\mathrm{Mod}\, A$ to itself. Here to the homomorphism of A-modules $F\colon Q \to R$ corresponds the homomorphism

$$\mathrm{Diff}_{k,P}^<(Q) = \mathrm{Diff}_k^<(P,Q) \ni \Delta \mapsto F \circ \Delta \in \mathrm{Diff}_k^<(P,R) = \mathrm{Diff}_{k,P}^<(R).$$

This is one of the main diffunctors.

Recall (see Section 9.57) that the diffunctor $\mathrm{Diff}_{k,A}^<$ is denoted simply by $\mathrm{Diff}_k^<$ and we write $\mathrm{Diff}_k^<(P)$ instead of $\mathrm{Diff}_k^<(A,P)$. Its important characteristic is that it splits naturally. Namely, to each element $p \in P$ we, associate the homomorphism of A-modules $p_A\colon A \to P$, $p_A(a) = ap$. The correspondence $p \mapsto p_A$ establishes an isomorphism of A-modules

$$P \to \mathrm{Hom}_A(A, P) = \mathrm{Diff}_0^<(P) \tag{15.12}$$

or, in other words, the equivalence of the functor $\mathrm{Diff}_0^< = \mathrm{Hom}_A(A, \cdot)$ to the identity. Besides, the homomorphism[1]

$$Д_{k,P}^<\colon \mathrm{Diff}_k^<(P) \to \mathrm{Diff}_0^<(P), \qquad \Delta \mapsto \Delta(1_A), \tag{15.13}$$

that determines the splitting

$$\mathrm{Diff}_k(P) = \mathrm{Ker}(Д_{k,P}^<) \oplus \mathrm{Diff}_0^<(P) \tag{15.14}$$

---

[1] The cyrillic letter Д looks like Δ and reads as D.

is defined. The correspondence $P \mapsto \operatorname{Ker}(\text{Д}^{<}_{k,P})$ is a new diffunctor, which will be denoted by $D_{(k)}$. Thus,

$$D_{(k)}(P) = \{\Delta \in \operatorname{Diff}^{<}_{k}(P) \mid \Delta(1_A) = 0\}. \qquad (15.15)$$

In this connection, let us note that $D_{(1)} = D$.

In this new notation, and having in mind the isomorphism (15.12), we can write the splitting (15.14) as

$$\operatorname{Diff}^{<}_{k}(P) = D_{(k)}(P) \oplus P. \qquad (15.16)$$

The above may also be summarized as an exact split sequence of diffunctors

$$0 \to D_{(k)} \longrightarrow \operatorname{Diff}^{<}_{k} \xrightarrow{\text{Д}^{<}_{k}} \operatorname{id}_{\operatorname{Mod} A} \to 0. \qquad (15.17)$$

Applying it to an A-module P, we obtain the exact sequence

$$0 \to D_{(k)}(P) \longrightarrow \operatorname{Diff}^{<}_{k}(P) \xrightarrow{\text{Д}^{<}_{k,P}} P \to 0. \qquad (15.18)$$

For $k = 1$, this is the *Spencer Diff-sequence of 1-st order*.

Exercise. Show that the map

$$\operatorname{Diff}^{>}_{k}(Q) \to D_{(k)}, \quad \Delta \mapsto \Delta - \Delta(1_A),$$

is a differential operator of order $\leq k - 1$.

An important property of the diffunctors $\operatorname{Diff}^{<}_{k}$ and $D_{(k)}$, as well as of all the others, is their compatibility with the mechanism of observability.

Proposition. *If the module Q is geometrical, then the module $\operatorname{Diff}^{<}_{k}(P,Q)$ is also geometrical, i.e., the functor $\operatorname{Diff}^{<}_{k,P}$ takes objects of $\operatorname{GMod} A$ to objects of $\operatorname{GMod} A$.*

◀ The proof is a consequence of the obvious inclusion

$$\operatorname{Ker} h \cdot \operatorname{Diff}^{<}_{k}(P,Q) \subset \operatorname{Diff}^{<}_{k}(P, \operatorname{Ker} h \cdot Q), \quad h \in |A|,$$

which immediately implies that

$$\operatorname{Inv}(\operatorname{Diff}^{<}_{k}(P,Q)) \subset \operatorname{Diff}^{<}_{k}(P, \operatorname{Inv} \cdot Q) = 0. \quad \blacktriangleright$$

If $l \leq k$, then, since $\operatorname{Diff}^{<}_{l}(P,Q) \subset \operatorname{Diff}^{<}_{k}(P,Q)$, we obtain a *natural transformation* of the diffunctors $\operatorname{Diff}^{<}_{l,P} \to \operatorname{Diff}^{<}_{k,P}$, which may be called an *embedding*. The left and right multiplication of differential operators by some element $a \in A$ induce the following natural transformations:

$$l_a, r_a \colon \operatorname{Diff}^{<}_{k,P} \to \operatorname{Diff}^{<}_{k,P} \colon$$

$$l_a(\Delta) = a_Q \circ \Delta, r_a(\Delta) = \Delta \circ a_P, \Delta \in \operatorname{Diff}^{<}_{l}(P,Q).$$

We will call them left and right *product of the diffunctor $\operatorname{Diff}^{<}_{k,P}$ by the element $a \in A$*, respectively.

15.9. The diffunctor $\mathrm{Diff}_k^>$. Now we consider the diffunctor

$$\mathrm{Diff}_{k,Q}^> \colon P \mapsto \mathrm{Diff}_k^>(P, Q),$$

which is the right analog of the diffunctor $\mathrm{Diff}_{k,P}^<$. In this situation, to the homomorphism $F \colon P \to R$ corresponds the homomorphism

$$\mathrm{Diff}_{k,Q}^>(R) = \mathrm{Diff}_k^>(R, Q) \ni \Delta \mapsto \Delta \circ F \in \mathrm{Diff}_k^>(P, Q) = \mathrm{Diff}_{k,P}^>(P),$$

which means that $\mathrm{Diff}_{k,P}^>$ is a contravariant functor.

Further, note that the operators of *change of structure*

$$\mathrm{id}_k^< \colon \mathrm{Diff}_k^<(P, Q) \to \mathrm{Diff}_k^>(P, Q),$$
$$\mathrm{id}_k^> \colon \mathrm{Diff}_k^>(P, Q) \to \mathrm{Diff}_k^<(P, Q),$$

which are identity maps of the *set* $\mathrm{Diff}_k(P, Q)$, are differential operators of order $\leq k$. Indeed, $\delta_a(\mathrm{id}_k^<) = (a_Q - a_P)|_{\mathrm{Diff}_k(P,Q)}$ or, equivalently,

$$\delta_a(\mathrm{id}_k^<)(\Delta) = \delta_a(\Delta) \in \mathrm{Diff}_k^<(P, Q), \quad \Delta \in \mathrm{Diff}_k^<(P, Q),$$

and therefore

$$[(\delta_{a_1} \circ \delta_{a_0} \circ \cdots \circ \delta_{a_k})(\mathrm{id}_k^<)](\Delta) = (\delta_{a_1} \circ \delta_{a_0} \circ \cdots \circ \delta_{a_k})(\Delta) = 0.$$

Similarly, one can show that $\mathrm{id}_k^>$ is an operator of order $\leq k$.

Further, consider the homomorphism $Д_k^< = Д_{k,P}^< \colon \mathrm{Diff}_k^< P \to P$ (see (15.14)). Then the operator

$$Д_k^> \overset{\mathrm{def}}{=} Д_k^< \circ \mathrm{id}_k^< \colon \mathrm{Diff}_k^> P \to P,$$

which coincides with $Д_k^<$ as a map of sets, being the composition of operators of order 0 and k, is an operator of order k. The importance of this operator is in its *co-universality*, a property that we will now clarify.

Let $\Delta \in \mathrm{Diff}_k(P, Q)$. The map

$$h^\Delta \colon P \to \mathrm{Diff}_k^> Q, \quad p \mapsto \Delta \circ p, \quad \text{where} \quad (\Delta \circ p)(a) \overset{\mathrm{def}}{=} \Delta(ap),$$

is obviously a homomorphism of A-modules, which makes the following diagram commutative

$$
\begin{array}{ccc}
P & \overset{h^\Delta}{\longrightarrow} & \mathrm{Diff}_k^> Q \\
 & \Delta \searrow & \downarrow Д_k^> \\
 & & Q.
\end{array}
\tag{15.19}
$$

Indeed, $(Д_k^> \circ h^\Delta)(p) = Д_k^>(\Delta \circ p) = (\Delta \circ p)(1_A) = \Delta(p).$

Proposition. *The map*

$$\mathrm{Diff}_k^>(P, Q) \to \mathrm{Hom}_A(P, \mathrm{Diff}_k^> Q), \quad \Delta \mapsto h^\Delta, \tag{15.20}$$

is an isomorphism of A-modules.

◀ The map

$$\mathrm{Hom}_A(P, \mathrm{Diff}_k^> Q) \to \mathrm{Diff}_k^>(P, Q), \quad h \mapsto Д_k^> \circ h,$$

is obviously inverse to the map (15.20). (Check that it is a homomorphism of A-modules.)   ▶

The map (15.20) may be regarded as a map of left A-modules:

$$\mathrm{Diff}_k^<(P, Q) \to \mathrm{Hom}_A^<(P, \mathrm{Diff}_k^> Q). \tag{15.21}$$

Obviously, it is also an isomorphism.

The isomprphism (15.20) means precisely that the module $\mathrm{Diff}_k^> Q$ co-*represents* the diffunctor $\mathrm{Diff}_{k,Q}^>$, while the operator $Д_k^>$ is *co-universal*. Here the prefix "co" stresses the duality with respect to the space of jets (see Section 15.9), which *represents* the diffunctor $\mathrm{Diff}_{k,P}^<$.

15.10. Prolongation of co-representing homomorphisms. The composition operation of differential operators acquires its functorial expression in iterations of the diffunctor $\mathrm{Diff}_k^>$. In their turn, the relations between them are described by *prolongations* of co-representing homomorphisms. The necessity of the operation of prolongation stems from the following commutative diagram:

$$\tag{15.22}$$

where $\Delta \in \mathrm{Diff}_k^>(P, Q)$ and $h_{(l)}^\Delta \overset{\text{def}}{=} h^{\left(h_\Delta \circ Д_l^>\right)}$. The lower left triangle of this diagram is the diagram (15.19) for the operator $h^\Delta \circ Д_l^>$. The homomorphism $h_{(l)}^\Delta$ is called the *l-th prolongation* homomorphism h^Δ.

Similarly, the operator $\Delta_{(l)} \overset{\text{def}}{=} \Delta \circ Д_l^>$ is called the *l-th prolongation* of the operator Δ. Since $\Delta_{(l)}$ is an operator of order $\leq k + l$, we have the next commutative diagram, whose upper triangle is the diagram (15.19) for the operator $\Delta_{(l)}$:

$$\tag{15.23}$$

The diagram

$$\mathrm{Diff}_l^> (\mathrm{Diff}_k^> Q) \xrightarrow{\;C_{l,k}^>\;} \mathrm{Diff}_{k+l}^> Q \qquad (15.24)$$

$$\Big\downarrow \text{Д}_l^> \qquad\qquad \Big\downarrow \text{Д}_{k+l}^>$$

$$\mathrm{Diff}_k^> Q \xrightarrow{\;\text{Д}_k^>\;} Q$$

is a particular case of the previous one for the operator $\Delta = \text{Д}_k^>$. Namely,

$$C_{l,k}^> = C_{l,k}^>(Q) = h_\nabla, \quad \text{where} \quad \nabla = \text{Д}_k^> \circ \text{Д}_l^>: \ \mathrm{Diff}_l^> (\mathrm{Diff}_k^> Q) \to Q.$$

This diagram can be understood as the definition of the homomorphism $C_{l,k}^>$, which is called the *gluing homomorphism* and is the required functorial interpretation of the composition operation of differential operators.

Finally, combining the diagrams (15.22)–(15.24), it is not difficult to construct the following commutative diagram.

$$\mathrm{Diff}_l^> P \xrightarrow{\;\text{Д}_l^>\;} P \qquad (15.25)$$

$$h^{\Delta_{(l)}} \Big\downarrow \qquad \searrow^{\Delta_{(l)}} \qquad \Big\downarrow \Delta$$

$$\mathrm{Diff}_{k+l}^> Q \xrightarrow[\;\text{Д}_{k+l}^>\;]{} Q$$

According to the definition, $h^{\Delta_{(l)}}(\nabla) = \Delta \circ \nabla$. It is obvious from the description that $h^{\Delta_{(l)}}|_{\mathrm{Diff}_m^> P} = h^{\Delta_{(m)}}$.

15.11 Exercise. According to the definition, the module $\mathrm{Diff}^> P$ (see Section 9.68) is filtered by the submodules $\mathrm{Diff}_l^> P$. Let $\Delta \in \mathrm{Diff}_k(P,Q)$. Recall that a homomorphism of a filtered module is defined as a homomorphism that shifts the filtration by a certain constant. Such, in particular, is the A-homomorphism

$$h_*^\Delta: \ \mathrm{Diff}^> P \to \mathrm{Diff}^> Q, \quad h_*^\Delta|_{\mathrm{Diff}_l^> P} = h^{\Delta_{(l)}}.$$

Describe the structure of the filtered $\mathrm{Diff}(A)$-module in the A-module $\mathrm{Diff}\, R$ and show that h_*^Δ is a homomorphism of $\mathrm{Diff}(A)$-modules.

We now pass to modules of jets and related constructions, which are dual to those discussed in the last three sections.

15.12. Modules of jets. The object representing the diffunctor $\mathrm{Diff}_{k,P}^<$ in a category of A-modules $\mathfrak{M}$ is called the *module of jets of k-th order* or *k-jets* in the category $\mathfrak{M}$. This object, denoted $\mathcal{J}^k(P;\mathfrak{M})$, is the range of the differential operator

$$j_k^{\mathfrak{M}}: \ P \to \mathcal{J}^k(P;\mathfrak{M}).$$

universal in $\mathfrak{M}$. This means that for any module $Q \in \mathrm{Ob}\,\mathfrak{M}$, each operator $\Delta \in \mathrm{Diff}_k(P,Q)$ can be uniquely represented in the form $\Delta = h_\Delta \circ j_k^{\mathfrak{M}}$, where

$$h_\Delta = h_\Delta^{\mathfrak{M}} \colon \mathcal{J}^k(P; \mathfrak{M}) \to Q$$

is a homomorphism of A-modules. Recall that representing objects are unique up to equivalence if, of course, they exist.

Let us first construct the module of k-jets in the category of all A-modules Mod A. To this end, we consider the A-bimodule $A \otimes_\Bbbk P$ whose left (right) multiplication is defined as

$$(a, a' \otimes_\Bbbk p) \mapsto aa' \otimes_\Bbbk p \qquad (a' \otimes_\Bbbk p, a) \mapsto a' \otimes_\Bbbk ap,$$

respectively.

To an element $a \in A$ assign the (A, A)-homomorphism

$$\delta^a \colon A \otimes_\Bbbk P \to A \otimes_\Bbbk P, \quad \delta^a(a' \otimes_\Bbbk p) = a' \otimes_\Bbbk ap - aa' \otimes_\Bbbk p.$$

Further, let μ_{k+1} be a submodule of the bimodule $A \otimes_\Bbbk P$ generated by all elements of the form $(\delta^{a_0} \circ \delta^{a_1} \circ \cdots \circ \delta^{a_k})(a \otimes_\Bbbk p)$. Put

$$\mathcal{J}^k(P) \overset{\text{def}}{=} \frac{A \otimes_\Bbbk P}{\mu_{k+1}},$$

$$j_k \colon P \to \mathcal{J}^k_<(P), \quad j_k(p) = 1 \otimes_\Bbbk p \mod \mu_{k+1}.$$

Now we can consider the left module $\mathcal{J}^k_<(P)$, where the lower index "<" means that the bimodule

$$\mathcal{J}^k(P) \overset{\text{def}}{=} (A \otimes_\Bbbk P)/\mu_{k+1}$$

is regarded as a left A-module only. The symbol $\mathcal{J}^k_>(P)$ is understood similarly.

Theorem.   (a) *The map j_k is a differential operator of order $\leq k$.*

(b) *Any operator $\Delta \in \mathrm{Diff}_k(P, Q)$ can be uniquely expressed in the form $\Delta = h_\Delta \circ j_k$, where $h_\Delta \colon \mathcal{J}^k_<(P) \to Q$ is some homomorphism of A-modules.*

(c) *The map $\Delta \mapsto h_\Delta$ is an isomorphism between the A-modules $\mathrm{Diff}^<_k(P, Q)$ and $\mathrm{Hom}_A(\mathcal{J}^k_<(P), Q)$.*

◀ (a) Since $[\delta_a(j_k)](p) = \delta^a(1 \otimes_\Bbbk p) \mod \mu_{k+1}$, we have

$$\big((\delta_{a_0} \circ \delta_{a_1} \circ \cdots \circ \delta_{a_k})(j_k)\big)(p)$$
$$= (\delta^{a_0} \circ \delta^{a_1} \circ \cdots \circ \delta^{a_k})(1 \otimes_\Bbbk p) \mod \mu_{k+1} = 0.$$

(b) Consider the homomorphism of A-modules

$$\tilde{h}_\Delta \colon A \otimes_\Bbbk P \to P, \quad a \otimes_\Bbbk p \mapsto a\Delta(p),$$

where $A \otimes_\Bbbk P$ is understood as a left A-module. It is well-defined, since the map $A \times P \to Q$, $(a, p) \mapsto a\Delta(p)$, is $\Bbbk$-bilinear, while $A \otimes_\Bbbk P$ is the universal object for $\Bbbk$-bilinear forms on $A \times P$. Further

$$\tilde{h}_\Delta(\delta^{a'}(a \otimes_\Bbbk p)) = \tilde{h}_\Delta(a \otimes_\Bbbk a'p - aa' \otimes_\Bbbk p)$$
$$= a\Delta(a'p) - aa'\Delta(p) = a\delta_{a'}(\Delta)(p) = \tilde{h}_{\delta_{a'}(\Delta)}(a \otimes_\Bbbk p).$$
$$(15.26)$$

This shows that $\tilde{h}_\Delta \circ \delta^{a'} = \tilde{h}_{\delta_{a'}(\Delta)}$, and therefore

$$\tilde{h}_\Delta \circ \delta^{a_0} \circ \delta^{a_1} \circ \cdots \circ \delta^{a_k} = \tilde{h}_{(\delta_{a_0} \circ \delta_{a_1} \circ \cdots \circ \delta_{a_k})(\Delta)} = 0.$$

This implies $\tilde{h}_\Delta(\mu_{k+1}) = 0$. Hence, $\tilde{h}_\Delta$ can be represented in the form of the composition $A \otimes_\Bbbk P \to \mathcal{J}^k_<(P) \xrightarrow{h_\Delta} Q$, where the first of these maps is the factorization, while h_Δ is a homomorphism. Since

$$h_\Delta(aj_k(p)) = \tilde{h}_\Delta(a \otimes_\Bbbk p) = a\Delta(p),$$

and elements of the form $aj_k(p)$ additively generate the module $\mathcal{J}^k_<(P)$, this implies the uniqueness of h_Δ. ▶

The property of universality of jets can be viewed conveniently in the following commutative diagram.

$$\begin{array}{ccc} P & \xrightarrow{j_k} & \mathcal{J}^k_<(P) \\ & \Delta \searrow & \downarrow h_\Delta \\ & & Q. \end{array} \qquad (15.27)$$

15.13. Geometrical jets. The next statement is a concrete illustration of the proposition from Section 15.6.

Corollary. *The diffunctor* $\mathrm{Diff}^<_{k,P}$ *can be represented in the category of geometrical A-modules* $\mathrm{GMod}\,A$. *Namely,*

$$\mathcal{J}^{k,\Gamma}_<(P) \stackrel{\mathrm{def}}{=} \Gamma(\mathcal{J}^k_<(P)),$$

where Γ *is the geometrization homomorphism,* $\mathcal{J}^k_<(P)$ *is the module of k-jets of the A-module P in this category. The corresponding universal differential operator is written as*

$$j^\Gamma_k = j^\Gamma_{k,P} \colon P \to \mathcal{J}^{k,\Gamma}_<(P).$$

◀ If $\Delta \in \mathrm{Diff}_k(P,Q)$ and the module Q is geometrical, then the homomorphism $h_\Delta \colon \mathcal{J}^k_<(P) \to Q$ can be represented in the form of a composition as follows $h_\Delta = h^\Gamma_\Delta \circ \Gamma$, where $h^\Gamma_\Delta \colon \Gamma(\mathcal{J}^k_<(P)) \to Q$ is a certain well-defined homomorphism (see the proposition in Section 8.6). ▶

Exercise. Verify that the method used in Section 14.14 to construct the module of geometrical jets for the algebra $A = C^\infty(M)$ is a particular case of the above construction.

15.14. As we saw in Chapter 14, the diffunctor D possesses representing objects in various categories of A-modules. In this connection, it is natural to ask if there exist such objects for the diffunctor $\mathrm{Diff}^<_{k,P}$ not only in the categories $\mathrm{Mod}\,A$ and $\mathrm{GMod}\,A$.

If $\mathfrak{M}$ is some category of A-modules and $P \in \mathrm{Ob}\,\mathfrak{M}$, then the universal differential operator for $\mathrm{Diff}^<_{k,P}$ and the corresponding module of k-jets, if they exist, will be denoted as follows:

$$j_k^\mathfrak{M} = j_{k,P}^\mathfrak{M} \colon P \to \mathcal{J}^k_<(P;\mathfrak{M}).$$

By construction, the module of jets $\mathcal{J}^k_<(P)$ is generated by the elements $j_k(p)$, $p \in P$. But since the geometrization functor Γ is surjective, it follows that the elements $j_k^\Gamma(p) = \Gamma(j_k(p))$, $p \in P$, generate the module $\mathcal{J}^k_<(P;\Gamma)$. This suggests what must be done in the general case.

Namely, let us consider the intersection of all the kernels of the homomorphisms $h_\Delta \colon \mathcal{J}^k_<(P) \to Q$ for all possible operators $\Delta \in \mathrm{Diff}_k(P,Q)$ and all possible modules $Q \in \mathrm{Ob}\,\mathfrak{M}$. This intersection forms a submodule of $\mathcal{J}^k_<(P)$, which we denote by $\mathcal{I}_\mathfrak{M} = \mathcal{I}_{\mathfrak{M},P}$. We then write

$$\mathcal{J}^k_<(P;\mathfrak{M}) \overset{\mathrm{def}}{=} \mathcal{J}^k_<(P)/\mathcal{I}_\mathfrak{M}, \qquad j_{k,P}^\mathfrak{M} \overset{\mathrm{def}}{=} \pi_\mathfrak{M} \circ j_{k,P},$$

where $\pi_\mathfrak{M} \colon \mathcal{J}^k_<(P) \to \mathcal{J}^k_<(P;\mathfrak{M})$ is the factorization map. Since we have $\mathcal{I}_\mathfrak{M} \subset \mathrm{Ker}\,h_\Delta$, it follows that $Q \in \mathrm{Ob}\,\mathfrak{M}$, and so h_Δ can be represented in the form $h_\Delta = h_\Delta^\mathfrak{M} \circ \pi_\mathfrak{M}$, where $h_\Delta^\mathfrak{M} \colon \mathcal{J}^k_<(P;\mathfrak{M}) \to Q$ is some homomorphism. This homomorphism is unique, since for any $p \in P$ we have

$$h_\Delta^\mathfrak{M}(j_{k,P}^\mathfrak{M}(p)) = h_\Delta(j_{k,P}(p)) = \Delta(p),$$

and the elements $j_{k,P}^\mathfrak{M}(p)$ obviously generate the module $\mathcal{J}^k_<(P;\mathfrak{M})$.

15.15 Lemma–Exercise. *Let $\mathfrak{M}$ be a subcategory of some category of A-modules $\mathfrak{N}$ and let $P,Q \in \mathrm{Ob}\,\mathfrak{M}$. Show that there exists a natural surjection*

$$\Pi = \Pi_\mathfrak{M}^\mathfrak{N} \colon \mathcal{J}^k(P;\mathfrak{N})) \to \mathcal{J}^k_<(P;\mathfrak{M}), \tag{15.28}$$

which makes the following diagram commutative

$$\mathcal{J}^k(P;\mathfrak{N})) \tag{15.29}$$

for any $\Delta \in \mathrm{Diff}_k(P,Q)$.

An operator $\Delta \in \mathrm{Diff}_k(P,Q)$, as the lemma shows, may be represented in many different ways as a homomorphism of the form $h_\Delta^\mathfrak{K} \colon \mathcal{J}^k_<(P;\mathfrak{K}) \to Q$ depending on the category $\mathfrak{K}$ containing the modules P and Q.

It is important to note that the module $\mathcal{J}_<^k(P; \mathfrak{M})$ does not necessarily belong to the category $\mathfrak{M}$. As we will see below, this gives us an obstruction to the construction of representing objects for more complicated diffunctors in categories of that type. That is precisely the point in introducing categories of modules in which the representing objects of any diffunctor are again objects of the same category, i.e., *differentially closed* categories. (see Section 15.5).

15.16. Homorphisms representing multiplications. Let $\mathfrak{M}$ be a category of A-modules such that $\mathcal{J}_<^k(P; \mathfrak{M}) \in \mathrm{Ob}\,\mathfrak{M}$ for all $k \in \mathbb{Z}_+$. Recall that $l_a(\Delta) = a_Q \circ \Delta$ if $\Delta \in \mathrm{Diff}_k(P, Q)$. Since $h_\Delta^{\mathfrak{M}}$ is a homomorphism, it follows that

$$h_{l_a(\Delta)}^{\mathfrak{M}} = a_Q \circ h_\Delta^{\mathfrak{M}} = h_\Delta^{\mathfrak{M}} \circ a_{\mathcal{J}_<^k(P;\mathfrak{M})}.$$

This shows that to the transformation l_a of the diffunctor $\mathrm{Diff}_k^<$ corresponds the multiplication homomorphism by a of elements of the module $\mathcal{J}_<^k(P; \mathfrak{M})$.

The previous argument can be presented in a more elegant form, which will be used in what follows. For the transformation r_a, it involves considering the following commutative diagram.

$$
\begin{array}{ccc}
P & \xrightarrow{\;j_k^{\mathfrak{M}}\;} & \mathcal{J}_<^k(P; \mathfrak{M}) \\
& {\scriptstyle\nabla}\searrow & \downarrow{\scriptstyle h_\nabla^{\mathfrak{M}}} \\
& & \mathcal{J}_<^k(P; \mathfrak{M}),
\end{array}
\qquad (15.30)
$$

where $\nabla = j_k^{\mathfrak{M}} \circ a_P$. This diagram expresses the universality of the operator $j_k^{\mathfrak{M}}$ with respect to the operator ∇, see (15.27). Note that the universality property may only be used in the case when $\mathcal{J}_<^k(P; \mathfrak{M}) \in \mathrm{Ob}\,\mathfrak{M}$. This is yet another one of the numerous reasons for which differentially closed categories are important.

From the diagram (15.30), one can deduce several useful consequences. First of all, the diagram allows us to define the operation of *right* multiplication of elements of the module $\mathcal{J}_<^k(P; \mathfrak{M})$ by an element $a \in A$ by putting $\theta a = h_{j_k^{\mathfrak{M}} \circ a_P}^{\mathfrak{M}}(\theta)$. As can be easily seen, $(a' j_k^{\mathfrak{M}}(p))a = a' j_k^{\mathfrak{M}}(ap)$.

Exercise.   1. Show that the right multiplication defined above supplies the vector space $\mathcal{J}_<^k(P; \mathfrak{M})$ with a second A-module structure, which, together with its initial ("left") structure, defines a A-bimodule structure on it.

    2. Using a diagram similar to (15.30), describe the endomorphism of the module $\mathcal{J}_<^k(P; \mathfrak{M})$ corresponding to the transformation l_a of the diffunctor $\mathrm{Diff}_k^<$, that was described above in a different way.

The symbol $\mathcal{J}^k_>(P; \mathfrak{M})$ will be used to denote the module of jets with respect to right multiplication, while the symbol $\mathcal{J}^k(P; \mathfrak{M})$ will describe the corresponding A-bimodule.

Further, we have

$$h^{\mathfrak{M}}_{\Delta \circ a_P}(j^{\mathfrak{M}}_k(p)) = (\Delta \circ a_P)(p) = \Delta(ap),$$

$$(h^{\mathfrak{M}}_\Delta \circ h^{\mathfrak{M}}_{j^{\mathfrak{M}}_k \circ a_P})(j^{\mathfrak{M}}_k(p)) = h^{\mathfrak{M}}_\Delta((j^{\mathfrak{M}}_k \circ a_P)(p)) = h^{\mathfrak{M}}_\Delta(j^{\mathfrak{M}}_k(ap)) = \Delta(ap).$$

This implies $h^{\mathfrak{M}}_{\Delta \circ a_P} = h^{\mathfrak{M}}_\Delta \circ h^{\mathfrak{M}}_{j^{\mathfrak{M}}_k \circ a_P} = h^{\mathfrak{M}}_\Delta \circ a^>$, where $a^>$ is actually the operator of right multiplication by a on $\mathcal{J}^k_<(P; \mathfrak{M})$, since elements of $j^{\mathfrak{M}}_k(p)$, $p \in P$, generate $\mathcal{J}^k_<(P; \mathfrak{M})$. Thus, the endomorphism of the module $\mathcal{J}^k_<(P; \mathfrak{M})$, representing the transformation r_a of the diffunctor $\mathrm{Diff}^<_k$, is $h^{\mathfrak{M}}_{j^{\mathfrak{M}}_k \circ a_P}$, or the operator of right multiplication by a in $\mathcal{J}^k_<(P; \mathfrak{M})$.

15.17. Representation of the diffunctor. $\mathrm{Diff}^>_{k,P}$ Let $P, Q \in \mathrm{Ob}\ \mathfrak{M}$. The description of right multiplication in the A-bimodule $\mathcal{J}^k_<(P; \mathfrak{M})$, proposed above yields the following isomorphism:

$$\mathrm{Diff}^>_k(P, Q) = \mathrm{Hom}^>_A(\mathcal{J}^k_<(P; \mathfrak{M}), Q). \tag{15.31}$$

Recall (see Section 15.4) that the symbol in the right-hand side stands for the $\Bbbk$-vector space of all A-homomorphisms of the A-module $\mathcal{J}^k_<(P; \mathfrak{M})$ to the A-module Q supplied with an A-module structure by means of right multiplication from the A-bimodule $\mathcal{J}^k(P; \mathfrak{M})$. In this sense, we say that the bimodule represents the diffunctor $\mathrm{Diff}^>_{k,P}$.

15.18. Jets of algebras. The modules of jets of the main algebra A, for which we shall also use the simplified notation $\mathcal{J}^k_<(\mathfrak{M}) = \mathcal{J}^k_<(A; \mathfrak{M})$, deserve special attention. One reason for this is that $\mathcal{J}^k(\mathfrak{M}) = \mathcal{J}^k(A; \mathfrak{M})$ is an algebra, while $\mathcal{J}^k(P; \mathfrak{M})$ is a $\mathcal{J}^k(A; \mathfrak{M})$-module. Both these objects are natural transformations of the diffunctor

$$\mathrm{Diff}^<_{k,P} \xrightarrow{\ \mathfrak{p}\ } \mathrm{Diff}^<_{k,P} \circ \mathrm{Diff}^<_k$$

and we have a similar transformation for $P = A$.

The transformation $\mathfrak{p}$ is defined in the following way. With the elements $\Delta \in \mathrm{Diff}_k(P, Q)$ and $p \in P$, we associate the operator

$$\Delta \circ p_A \in \mathrm{Diff}_k Q, \quad (p_A)(a) = ap \quad (\text{see } 15.12)$$

and note that

$$(\mathrm{Diff}^<_{k,P} \circ \mathrm{Diff}^<_k)(Q) = \mathrm{Diff}^<_k(P, \mathrm{Diff}^<_k Q).$$

Then, considering the $\Bbbk$-linear map

$$\mathfrak{p}_Q\colon\ \mathrm{Diff}^<_k(P, Q) \to \mathrm{Diff}^<_k(P, \mathrm{Diff}^<_k Q), \quad \mathfrak{p}_Q(\Delta)(p) \stackrel{\mathrm{def}}{=} \Delta \circ p_A,$$

we define the transformation $\mathfrak{p}$ by setting $\mathfrak{p}(Q) \stackrel{\mathrm{def}}{=} \mathfrak{p}_Q$. The following simple exercise shows that $\mathfrak{p}$ is well-defined.

Exercises.   1. Show that $\mathfrak{p}_Q$ is a homomorphism of A-modules;

2. Prove that $\delta_a(\mathfrak{p}_Q(\Delta)) = \mathfrak{p}_Q(\delta_a(\Delta))$, and consequently, $\mathfrak{p}_Q(\Delta)$ is an operator of order $\leq k$.

We must also describe the representing object for the diffunctor $\mathrm{Diff}_{k,P}^< \circ \mathrm{Diff}_k^<$. If $P, Q \in \mathrm{Ob}\,\mathfrak{M}$, then

$$\mathrm{Diff}_k^<(P, \mathrm{Diff}_k^< Q) \cong \mathrm{Hom}_A(\mathcal{J}_<^k(P; \mathfrak{M}), \mathrm{Diff}_k^< Q)$$

$$\cong \mathrm{Hom}_A(\mathcal{J}_<^k(P; \mathfrak{M}), \mathrm{Hom}_A(\mathcal{J}_<^k(\mathfrak{M}), Q)) \qquad (15.32)$$

$$\cong \mathrm{Hom}_A(\mathcal{J}_<^k(P; \mathfrak{M}) \otimes_A \mathcal{J}_<^k(\mathfrak{M}), Q).$$

This shows that the required representing object is the A-module $\mathcal{J}_<^k(P; \mathfrak{M}) \otimes_A \mathcal{J}_<^k(\mathfrak{M})$.

Now we can describe the above-mentioned algebra structure in $\mathcal{J}^k(\mathfrak{M})$ and in the $\mathcal{J}^k(\mathfrak{M})$-module $\mathcal{J}^k(P; \mathfrak{M})$. Let $\nabla \in \mathrm{Diff}_k^<(P, \mathrm{Diff}_k^< Q)$. Then, in the first of the modules, the homomorphism (15.32) assigns to this operator the homomorphism $h_\nabla^{\mathfrak{M}}$. Denote by F_{φ_∇} and φ_∇ the homomorphisms corresponding to it in the last two modules. Then, by definition of the isomorphisms (15.32), we have

$$\big(\nabla(p)\big)(a) = \big(h_\nabla^{\mathfrak{M}}(j_{k,P}^{\mathfrak{M}}(p))\big)(a) = \big(F_{\varphi_\nabla}(j_{k,P}^{\mathfrak{M}}(p))]\big)(j_k^{\mathfrak{M}}(a)$$

$$= \varphi_\nabla(j_{k,P}^{\mathfrak{M}}(p) \otimes_A j_k^{\mathfrak{M}}(a)).$$

This implies that to the operator $\nabla \in \mathrm{Diff}_k^<(P, \mathrm{Diff}_k^< Q)$ corresponds the homomorphism φ_∇, defined by the relation

$$\varphi_\nabla(j_{k,P}^{\mathfrak{M}}(p) \otimes_A j_k^{\mathfrak{M}}(a)) = \big(\nabla(p)\big)(a). \qquad (15.33)$$

Now if $\nabla = \mathfrak{p}(\Delta)$, then $\nabla(p) = \Delta \circ p$ and from (15.33) we can deduce the relation

$$\Delta(ap) = \varphi_{\mathfrak{p}(\Delta)}(j_{k,P}^{\mathfrak{M}}(p) \otimes_A j_k^{\mathfrak{M}}(a)) \qquad (15.34)$$

For $\Delta = j_{k,P}^{\mathfrak{M}}$, let us put $\varphi_{\mathfrak{p}(\Delta)} = \varphi_{\mathfrak{p},P}$. Hence,

$$\varphi_{\mathfrak{p},P} \colon \mathcal{J}_<^k(P; \mathfrak{M}) \otimes_A \mathcal{J}_<^k(\mathfrak{M}) \to \mathcal{J}_<^k(P; \mathfrak{M}) \qquad (15.35)$$

and in view of (15.34),

$$\varphi_{\mathfrak{p},P}(j_k^{\mathfrak{M}}(p) \otimes_A j_k^{\mathfrak{M}}(a)) = j_{k,P}^{\mathfrak{M}}(ap). \qquad (15.36)$$

The homomorphism $\varphi_{\mathfrak{p},A} \colon \mathcal{J}_<^k(\mathfrak{M}) \otimes_A \mathcal{J}_<^k(\mathfrak{M}) \to \mathcal{J}_<^k(\mathfrak{M})$ defines a multiplication in $\mathcal{J}_<^k(\mathfrak{M})$ that we denote ".". The relation (15.36) also implies

$$j_k^{\mathfrak{M}}(a') \cdot j_k^{\mathfrak{M}}(a) \stackrel{\text{def}}{=} \varphi_{\mathfrak{p},A}(j_k^{\mathfrak{M}}(a') \otimes_A j_k^{\mathfrak{M}}(a)) = j_k^{\mathfrak{M}}(aa')$$

and so

$$(a_1 j_k^{\mathfrak{M}}(a_2)) \cdot (a_3 j_k^{\mathfrak{M}}(a_4)) = a_1 a_3 j_k^{\mathfrak{M}}(a_2 a_4). \qquad (15.37)$$

Associativity and commutativity of this multiplication follow directly from (15.37).

Similarly, $\varphi_{\mathfrak{p},P}$ defines the structure of a $\mathcal{J}^k(\mathfrak{M})$-module in $\mathcal{J}^k(P;\mathfrak{M})$ and (15.36) implies

$$(a_1 j_k^{\mathfrak{M}}(a_2)) \cdot (a_3 j_{k,P}^{\mathfrak{M}}(p)) = a_1 a_3 j_{k,P}^{\mathfrak{M}}(a_2 p). \tag{15.38}$$

Note that the algebra structure in $\mathcal{J}^k(\mathfrak{M})$ and the $\mathcal{J}^k(\mathfrak{M})$-module structure in $\mathcal{J}^k(P;\mathfrak{M})$ could have been introduced directly by means of the sufficiently obvious formulas (15.37) and (15.38). However, one must prove that the formulas are well-defined, and this approach hides the functorial origin of the structures , i.e., their meaning and nature. The latter is important in order to establish the relationship with other natural structures, as well as in more complicated situations, when the direct approach is difficult.

15.19. Right minus left. It is convenient to "specify" the difference between right and left multiplication by an element $a \in A$ in the bimodule $\mathcal{J}^k(P;\mathfrak{M})$ via the element

$$\delta_{k,\mathfrak{M}}^a \overset{\text{def}}{=} j_k^{\mathfrak{M}}(a) - a j_k^{\mathfrak{M}}(1_A) \in \mathcal{J}^k(A;\mathfrak{M}),$$

having in mind that $\delta_{k,\mathfrak{M}}^a \cdot j_k^{\mathfrak{M}}(p) = j_k^{\mathfrak{M}}(ap) - a j_k^{\mathfrak{M}}(1_A)$. It is also convenient to denote

$$\delta_{k,\mathfrak{M}}^{a_1,\ldots,a_m} \overset{\text{def}}{=} \delta_{k,\mathfrak{M}}^{a_1} \cdot \ldots \cdot \delta_{k,\mathfrak{M}}^{a_m}.$$

Obviously, the symbol $\delta_{k,\mathfrak{M}}^{a_1,\ldots,a_m}$ is symmetric with respect to the upper indices.

Lemma–Exercise. *Show that*

1. $\delta_{k,\mathfrak{M}}^{a_0,a_1,\ldots,a_k} = 0$;

2. $\delta_{k,\mathfrak{M}}^{a_1,\ldots,a_m} \cdot j_{k,A}^{\mathfrak{M}}(a) = \delta_{k,\mathfrak{M}}^{a_1,\ldots,a_{m-1},aa_m} - a\delta_{k,\mathfrak{M}}^{a_1,\ldots,a_m}$;

3. $\delta_{k,\mathfrak{M}}^{a_1 a_1',a_2\ldots,a_k} = a_1 \delta_{k,\mathfrak{M}}^{a_1',a_2\ldots,a_k} + a_1' \delta_{k,\mathfrak{M}}^{a_1,a_2\ldots,a_k}$;

4. $\delta_{a_1 a_1',a_2\ldots,a_k}(\Delta) = a_1 \delta_{a_1',\ldots,a_k}(\Delta) + a_1' \delta_{a_1,\ldots,a_k}(\Delta)$ if $\Delta \in \text{Diff}_k(P,Q)$;

5. $\left(\delta_{a_1,\ldots,a_m}(j_{k,P}^{\mathfrak{M}})\right)(p) = \delta_{k,\mathfrak{M}}^{a_1,\ldots,a_m} \cdot j_{k,P}^{\mathfrak{M}}(p)$.

The third assertion of the lemma means that the correspondence

$$(a_1,\ldots,a_k) \mapsto \delta_{k,\mathfrak{M}}^{a_1,\ldots,a_k}$$

is a symmetric derivation of the algebra A w.r.t. each of the k arguments, in other words, a *symmetric k-derivation*. From the first assertion, it follows that the right and left products by the same element of the algebra A coincide on elements of the form $\delta_{k,\mathfrak{M}}^{a_1,\ldots,a_k}$. Hence, the A-bimodule that they generate in $\mathcal{J}^{k,\mathfrak{M}}$ actually turns out to be simply an A-module. Below (see Proposition 16.3) we will show that this submodule coincides with the kernel of the homomorphism

$$\pi_{k,A}^{k-1,\mathfrak{M}} \colon \mathcal{J}^k(A;\mathfrak{M}) \to \mathcal{J}_{\mathfrak{M}}^{k-1}(A), \quad j_{k,A}^{\mathfrak{M}}(a) \mapsto j_{k-1,A}^{\mathfrak{M}}(a).$$

In its turn, in the category $\mathfrak{M} = \mathrm{GMod}\, C^\infty(M)$, this kernel is identified with the $C^\infty(M)$-module of symmetric covariant tensors of order k on the manifold M. Thus, the above construction shows the similarity of the notion of tensor of the type indicated with that of arbitrary commutative algebra.

15.20. The modules of jet of algebras are also important because, as the next proposition shows, the description jets of arbitrary modules reduce to that of jets of algebras. For an exact formulation of this, we will need the symbol ${}_<R_>\otimes_A P$, where R is some (A,A)-bimodule. This symbol denotes the tensor product of the right bimodule R by P, i.e., $R_>\otimes_A P$, the structure of A-module in which is defined by the left multiplication in R. In particular, in the obvious notation, this means that

$$ra \otimes_A p = r \otimes_A ap \quad \text{and} \quad a(r \otimes_A p) = (ar) \otimes_A p.$$

Proposition. *If $\mathfrak{M}$ is a differentially closed category, then*

$$\mathcal{J}_<^k(P;\mathfrak{M}) = {}_<\mathcal{J}_>^k(A;\mathfrak{M}) \otimes_A P, \quad \mathcal{J}_>^k(P;\mathfrak{M}) = \mathcal{J}_>^k(A;\mathfrak{M}) \otimes_A P.$$

◀ Taking $Q = \mathcal{J}_>^k(A;\mathfrak{M})$, let us use the isomorphism (15.10):

$$\mathrm{Hom}_A^<(\mathcal{J}_>^k(A;\mathfrak{M}) \otimes_A P, R) \cong \mathrm{Hom}_A(P, \mathrm{Hom}_A^>({}_<\mathcal{J}^k(A;\mathfrak{M}), R)) \quad (15.39)$$

$$\cong \mathrm{Hom}_A^<(P, \mathrm{Diff}_k^> R). \quad (15.40)$$

It remains to recall that $\mathrm{Hom}_A^<(P, \mathrm{Diff}_k^> R) = \mathrm{Diff}_k(P,R)$ (see the proposition in Section 15.9).   ▶

Here it is important to note that the isomorphism established in this proposition identifies $j_k(p)$ and $j_k(1_A) \otimes_A p$; we will express this fact by the formula

$$j_k(p) = j_k(1_A) \otimes_A^> p, \qquad p \in P, \quad (15.41)$$

where, using the symbol $\otimes_A^>$, we stress that the tensor product is considered in the right structure on the A-module $\mathcal{J}^k(A;\mathfrak{M})$.

15.21. Reduced jets. The splitting (15.16) of the diffunctor $\mathrm{Diff}_k^<$ restricted to the category $\mathfrak{M}$, induces the splitting

$$\mathcal{J}_<^k(A;\mathfrak{M}) = \Lambda_{\mathfrak{M}}^{(k)}(A) \oplus A, \quad (15.42)$$

where $\Lambda_{\mathfrak{M}}^{(k)}(A)$ is the module of *reduced jets*, representing the diffunctor $D_{(k)}$ in the category $\mathfrak{M}$, while the algebra A is interpreted as the representing object of the diffunctor $\mathrm{id}_{\mathrm{Mod}\,A} = \mathrm{Diff}_0$. Since the diffunctor $D_{(k)}$ is the kernel of the transformation Д$_k^<$ (see 15.17), it follows from the Corollary from Section 15.3, that the module $\Lambda_{\mathfrak{M}}^{(k)}(A)$ representing $D_{(k)}$ is isomorphic to the cokernel of the homomorphism that represents the transformation $Д_k^<$. It is easy to see that this homomorphism is the correspondence

$$\mathrm{Diff}_0(A) \cong A \ni a \mapsto aj_k^{\mathfrak{M}}(1_A) \in \mathcal{J}_<^k(A; \mathfrak{M})$$

and so $\Lambda_{\mathfrak{M}}^{(k)}(A) = \mathcal{J}_<^k(A; \mathfrak{M})/A \cdot j_k^{\mathfrak{M}}(1_A)$.

Exercise. Show that

1. The factorization homomorphism

$$\Pi_k^{\mathfrak{M}} \colon \mathcal{J}_<^k(A; \mathfrak{M}) \to \Lambda_{\mathfrak{M}}^{(k)}(A)$$

represents the embedding transformation of the diffunctors

$$\mathrm{Df}_{(k)} \to \mathrm{Diff}_k^<$$

in the category $\mathfrak{M}$.

2. The module $\Lambda_{\mathfrak{M}}^{(k)}(A)$ is isomorphic to the submodule of the module $\mathcal{J}_<^k(A; \mathfrak{M})$ generated by elements of the form $\delta_{k,\mathfrak{M}}^a = j_k^{\mathfrak{M}}(a) - aj_k^{\mathfrak{M}}(1_A)$ and that the homomorphism

$$\mathcal{J}_<^k(A; \mathfrak{M}) \ni a'j_k^{\mathfrak{M}}(a) \mapsto a'j_k^{\mathfrak{M}}(a) - a'aj_k^{\mathfrak{M}}(1_A) \in \Lambda_{\mathfrak{M}}^{(k)}(A)$$

representing the embedding transformation $\mathrm{Df}_{(k)} \to \mathrm{Diff}_k^<$ of the diffunctors $\mathrm{Df}_{(k)}$.

3. The submodule $\Lambda_{\mathfrak{M}}^{(k)}(A) \subset \mathcal{J}_<^k(A; \mathfrak{M})$ is actually a sub-bimodule of the A-bimodule $\mathcal{J}^k(A; \mathfrak{M})$.

Finally, note that $\Lambda_{\mathfrak{M}}^{(1)}(A) \cong \Lambda_{\mathfrak{M}}(A)$, since this module represents the diffunctors $\mathrm{Df}_{(1)}$ and D, respectively, while $D_{(1)} = D$. Hence, the right one of the two commutative diagrams (15.43) below is a particular case of the left one for $k = 1$, so that it is natural to call $d_{(k)}^{\mathfrak{M}} \stackrel{\text{def}}{=} \Pi_k, \circ j_k^{\mathfrak{M}}$ a *derivation of k-th order* in the category $\mathfrak{M}$.

$$\Lambda_{\mathfrak{M}}^{(k)}(A) \xleftarrow{\Pi_k^{\mathfrak{M}}} \mathcal{J}_<^k(A; \mathfrak{M}) \qquad \Lambda_{\mathfrak{M}}^1(A) \xleftarrow{\Pi_1^{\mathfrak{M}}} \mathcal{J}_<^{1,\mathfrak{M}}(A) \qquad (15.43)$$

$$d_{(k)}^{\mathfrak{M}} \nwarrow \quad \uparrow j_k^{\mathfrak{M}} \qquad\qquad d^{\mathfrak{M}} \nwarrow \quad \uparrow j_1^{\mathfrak{M}}$$

$$A, \qquad\qquad\qquad A.$$

15.22. Jets of infinite order. The commutative diagram (15.27) in the case when $\Delta = j_{l,P}^M$

$$P \xrightarrow{j_{k,P}^{\mathfrak{M}}} \mathcal{J}_<^k(P; \mathfrak{M}) \qquad (15.44)$$

$$\nabla \searrow \quad \downarrow h_\nabla^{\mathfrak{M}}$$

$$\mathcal{J}_<^l(P; \mathfrak{M}),$$

where $\nabla = j_{l,P}^{\mathfrak{M}}$, allows one to immediately find the homomorphism $\pi_{k,P}^{l,\mathfrak{M}}$ representing the embedding of the diffunctors $\text{Diff}_{l,P}^{<} \subset \text{Diff}_{k,P}^{<}$. Namely, $\pi_{k,P}^{l,\mathfrak{M}} = h_{j_{l,P}^{\mathfrak{M}}}^{\mathfrak{M}}$ and, in particular, $\pi_{k,P}^{l,\mathfrak{M}}(j_{k,P}^{\mathfrak{M}}(p)) = j_l^{\mathfrak{M}}(p)$. The last equality implies the surjectivity of the homomorphism $\pi_{k,P}^{l,\mathfrak{M}}$. From the conceptual point of view, this is a consequence of the fact that $\pi_{k,P}^{l,\mathfrak{M}}$ represents an embedding, i.e., an injective transformation of diffunctors. (Check this!) Obviously, we have the relation $\pi_{l,P}^{m,\mathfrak{M}} \circ \pi_{k,P}^{l,\mathfrak{M}} = \pi_{k,P}^{m,\mathfrak{M}}$. Besides, the equality

$$\pi_{k,P}^{l,\mathfrak{M}}(j_{k,P}^{\mathfrak{M}}(ap)) = j_{l,P}^{\mathfrak{M}}(ap)$$

implies that the projection $\pi_{k,P}^{l,\mathfrak{M}}$ commutes with the right multiplication in $\mathcal{J}^{k,\mathfrak{M}}(P)$ and in $\mathcal{J}^{l,\mathfrak{M}}(P)$, and hence is a homomorphism of these A-bimodules.

Thus, we obtain a "tower" of jets:

$$P = \mathcal{J}^{0,\mathfrak{M}}(P) \overset{\pi_{1,P}^{0,\mathfrak{M}}}{\longleftarrow} \cdots \overset{\pi_{k,P}^{k-1,\mathfrak{M}}}{\longleftarrow} \mathcal{J}^{k,\mathfrak{M}}(P) \overset{\pi_{k+1,P}^{k,\mathfrak{M}}}{\longleftarrow} \mathcal{J}^{k+1,\mathfrak{M}}(P) \overset{\pi_{k+2,P}^{k+1,\mathfrak{M}}}{\longleftarrow} \cdots \tag{15.45}$$

The inverse limit of this chain of homomorphisms is the A-bimodule of *infinite* jets that we denote $\mathcal{J}(P;\mathfrak{M})$. Recall that its elements, by definition, are sequences of the form

$$\theta = \{\theta_k\}_{k\geq 0}, \quad \text{where} \quad \theta_k \in \mathcal{J}^{k,\mathfrak{M}}(P) \quad \pi_{k,k-1}^{\mathfrak{M}}(\theta_k) = \theta_{k-1}.$$

The left and right product of such a sequence by an element $a \in A$ is defined in the obvious way:

$$a^{<}\theta \overset{\text{def}}{=} \{a\theta_k\}_{k\geq 0}, \quad a^{>}\theta \overset{\text{def}}{=} j_P^{\mathfrak{M}}(a)\theta = \{j_{k,P}^{\mathfrak{M}}(a)\theta_k\}_{k\geq 0},$$

where $j_Q^{\mathfrak{M}}(q) \overset{\text{def}}{=} \{j_{k,Q}^{\mathfrak{M}}(q)\}_{k\geq 0}$, $q \in Q$, denotes the infinite jet of the element q. Similarly, the operation of multiplication

$$\theta\theta' \overset{\text{def}}{=} \{\theta_k\theta_k'\}_{k\geq 0}$$

transforms $\mathcal{J}^{\mathfrak{M}}(A)$ into a $\Bbbk$-algebra, while the bimodule $\mathcal{J}(P;\mathfrak{M})$ becomes a $\mathcal{J}^k(A;\mathfrak{M})$-module .

The maps

$$\pi_{\infty,P}^{k,\mathfrak{M}} : \mathcal{J}(P;\mathfrak{M}) \to \mathcal{J}^k(P;\mathfrak{M}), \quad \theta \mapsto \theta_k ,$$

are, obviously, homomorphisms of A-bimodules and, respectively, of algebras if $P = A$. Their kernels form a filtration

$$\mathcal{J}(P;\mathfrak{M}) \supset \text{Ker}(\pi_{\infty,P}^{0,\mathfrak{M}}) \supset \cdots \supset \text{Ker}(\pi_{\infty,P}^{k,\mathfrak{M}}) \supset \cdots \tag{15.46}$$

of the bimodule $\mathcal{J}(P;\mathfrak{M})$ and, respectively, of the algebra $\mathcal{J}(A;\mathfrak{M})$ if $P = A$.

Exercise. Establish the isomorphism

$$\mathrm{Ker}(\pi_{k,P}^{l,\mathfrak{M}}) = \frac{\mathrm{Ker}(\pi_{\infty,P}^{l,\mathfrak{M}})}{\mathrm{Ker}(\pi_{\infty,P}^{k,\mathfrak{M}})}. \tag{15.47}$$

Remark. The modules and their corresponding spaces of infinite jets play a fundamental role in the contemporary geometrical theory of nonlinear differential equations. By using them, one can construct *diffieties*—objects that are analogs of varieties for differential equations.

15.23. Naturality of jets. Another specific property of jet algebras is their *naturality*. This means that any homomorphism of a commutative algebra $\varphi\colon A \to B$ under a certain condition (formulated below) can be extended to the following commutative diagram:

$$
\begin{array}{ccc}
\mathcal{J}_<^{k,\mathfrak{M}}(A) & \xrightarrow{\;\mathcal{J}^k(\varphi)\;} & \mathcal{J}_<^{k,\mathcal{K}}(B) \\
{\scriptstyle j_k^{\mathfrak{M}}}\Big\uparrow & & \Big\uparrow{\scriptstyle j_k^{\mathcal{K}}} \\
A & \xrightarrow{\quad\varphi\quad} & B
\end{array}
\tag{15.48}
$$

The reason for this is that the given algebra homomorphism allows to consider any B-module Q as also being an A-module. Namely, the operation of multiplication of elements of the module Q by an element $a \in A$, denoted $\cdot_\varphi$, is defined by the rule

$$a \cdot_\varphi q \stackrel{\mathrm{def}}{=} \varphi(a)q, \quad q \in Q. \tag{15.49}$$

The A-module defined in this way is denoted Q_φ. In particular, the algebra B itself may be regarded as an A-module, and φ, as a homomorphism of A-modules. Moreover, any differential operator connecting two B-modules is, obviously, at the same time, a differential operator between A-modules. Hence, the composition $j_k^{\mathcal{K}} \circ \varphi$ is a differential operator of order $\leq k$ if $\mathcal{J}_>^{k,\mathcal{K}}(B)$ is regarded as an A-module. If, additionally, this A-module belongs to the category $\mathfrak{M}$, then $j_k^{\mathcal{K}} \circ \varphi = h_{j_k^{\mathcal{K}} \circ \varphi}^{\mathfrak{M}} \circ j_k^{\mathfrak{M}}$ in view of the universality of the operator $j_k^{\mathfrak{M}}$. Thus, we arrive at the commutative diagram (15.48) by setting $\mathcal{J}^k(\varphi) \stackrel{\mathrm{def}}{=} h_{j_k^{\mathcal{K}} \circ \varphi}^{\mathfrak{M}}$.

The diagram (15.48) may be reduced to the diagram

$$
\begin{array}{ccc}
\Lambda_{\mathfrak{M}}^{(k)}(A) & \xrightarrow{\;\Lambda^{(k)}(\varphi)\;} & \Lambda_{\mathcal{K}}^{(k)}(B) \\
{\scriptstyle d_{(k)}^{\mathfrak{M}}}\Big\uparrow & & \Big\uparrow{\scriptstyle d_{(k)}^{\mathcal{K}}} \\
A & \xrightarrow{\quad\varphi\quad} & B,
\end{array}
\tag{15.50}
$$

provided we additionally assume that $\Lambda_{\mathcal{K}}^{(k)}(A)$, regarded as an A-module, also belongs to $\mathfrak{M}$. Indeed, the commutativity of the diagram (15.48) implies

$$\left(\mathcal{J}^k(\varphi)\right)(j_k^{\mathfrak{M}}(1_A)) = j_k^{\mathcal{K}}(1_B).$$

Hence, the map $\mathcal{J}^k(\varphi)$ takes $\operatorname{Ker}\Pi_k^{\mathfrak{M}} = A \cdot j_k^{\mathfrak{M}}(1_A)$ to $\operatorname{Ker}\Pi_k^{\mathcal{K}} = B \cdot j_k^{\mathcal{K}}(1_B)$ (see (15.43)). Therefore, taking the quotients of the upper terms of the diagram (15.48) by $\operatorname{Ker}\Pi_k^{\mathfrak{M}}$ and $\operatorname{Ker}\Pi_k^{\mathcal{K}}$, respectively, we obtain the diagram (15.50).

The diagrams (15.48) and (15.50) express the *property of naturality* of the operators j_k and $d_{(k)}$ (see Section 15.28 below). In particular, we now see that the naturality property of the classical de Rham differential specified by the equality $F^* \circ d = d \circ F^*$ (see Section 14.13) is a particular case of the general algebraic situation described above, when $F\colon M \to N$ is a smooth map and

$$\varphi = F^*\colon A = C^\infty(N) \to C^\infty(M) = B,$$
$$\mathfrak{M} = \operatorname{GMod} C^\infty(N), \ \mathcal{K} = \operatorname{GMod} C^\infty(M).$$

15.24. φ-connected categories. The assumption made above about $\mathcal{J}_<^{k,\mathcal{K}}(B)$ as an A-module appears rather artificial and should be included in a more meaningful context. Namely, the category of B-modules $\mathcal{K}$ will be called *φ-connected* with the category of A-modules $\mathfrak{M}$ if any B-module from the category $\mathcal{K}$ belongs to the category $\mathfrak{M}$ as an A-module.

Proposition. *If a differentially closed category $\mathcal{K}$ is φ-connected with the category $\mathfrak{M}$, then the diagram (15.48) is commutative. In particular, this is so if $\mathcal{K}$ and $\mathfrak{M}$ are categories of geometrical modules over the algebras A and B, respectively.*

◄ By assumption, $\mathcal{J}_<^{k,\mathcal{K}}(B)$ as an A-module belongs to the category $\mathfrak{M}$. Hence, the previous argument proves the first assertion. The second one follows because the proposition in Section 15.29 implies that the categories of geometrical modules over A and B are φ-connected for any algebra homomorphism $\varphi\colon A \to B$. ►

Exercise. Denote by $\operatorname{Mod}_h A$ the category of A-modules with support at one point $h \in |A|$. Under what condition on the algebra homomorphism $\varphi\colon A \to A'$ will the category $\operatorname{Mod}_h A$ be φ-connected with $\operatorname{Mod}_{h'} A'$?

15.25. Naturality of the representing modules. The naturality property, which we have established previously for differential 1-forms and jets, actually holds for modules representing various diffunctors. The reason for this can be explained by the following argument from the category theory.

Let Φ be a diffunctor, let $\varphi\colon A \to B$ be a homomorphism of commutative algebras, $\mathfrak{M}$ and $\mathcal{K}$ be φ-connected differentially closed categories of A- and B-modules, respectively.

Let also $\mathcal{O}_{\mathfrak{M}}$ and $\mathcal{O}_{\mathcal{K}}$ be the modules representing the functor Φ in the categories $\mathfrak{M}$ and $\mathcal{K}$, respectively. Recall (see Section 15.23), that the homomorphism φ allows one to consider any B-module, in particular $\mathcal{O}_{\mathcal{K}}$, as also being an A-module. In this connection, it should be noted that for the B-module $Q \in \mathrm{Ob}\,\mathcal{K}$, the symbol $\Phi(Q)$ becomes ambiguous if we do not indicate to what category the object Q belongs. In order to avoid this ambiguity, we denote by $\Phi^{\varphi}(Q)$ the value of the diffunctor Φ on the A-module Q_{φ} (see Section 15.23). Thus, we obtain a functor Φ^{φ} from the category of B-modules to the category of A-modules. On the other hand, there is another such functor, namely $\Phi_{\varphi} \colon Q \mapsto \Phi(Q)_{\varphi}$. The following *compatibility property* is a distinctive trait of diffunctors:

> The functors Φ_{φ} and Φ^{φ} are related by a certain natural transformation, denoted T_{Φ}^{φ}.

In other words, this means that for any homomorphism of B-modules $H \colon Q \to R$ the following diagram is commutative:

$$\begin{array}{ccc} \Phi(Q)_{\varphi} & \xrightarrow{\ T_{\Phi}^{\varphi}(Q)\ } & \Phi^{\varphi}(Q) \\ {\scriptstyle \Phi_{\varphi}(H)} \downarrow & & \downarrow {\scriptstyle \Phi^{\varphi}(H)} \\ \Phi(R)_{\varphi} & \xrightarrow{\ T_{\Phi}^{\varphi}(R)\ } & \Phi^{\varphi}(R)\,. \end{array} \tag{15.51}$$

From the same considerations as in 15.6, we do not prove this statement in full generality. The reason for which the statement holds is that all diffunctors are constructed in a specific way from the diffunctors Diff_k, for which it does hold, as we have seen (Section 15.23). Hence, the transformation T_{Φ}^{φ} for the diffunctors Φ discussed in this book is described individually each time.

For instance, let φ be the embedding homomorphism of the subalgebra A of the algebra B and $\Phi = D$. Then the homomorphism T_D^{φ} assigns to the derivation $X \in D(Q)$ of the algebra B the derivation $X \circ \varphi \in D(Q_{\varphi})$. This example shows that, generally speaking, the homomorphism $T_{\Phi}^{\varphi}(Q)$ is neither surjective nor injective.

If $Q \in \mathrm{Ob}\,\mathcal{K}$, then we have the identifications

$$\Phi(Q)_{\varphi} = \big(\mathrm{Hom}_B(\mathcal{O}_{\mathcal{K}}, Q) \big)_{\varphi}, \quad \Phi^{\varphi}(Q) = \mathrm{Hom}_A(\mathcal{O}_{\mathfrak{M}}, Q_{\varphi})\,.$$

For these identifications, the homomorphism $T_{\Phi}^{\varphi}(\mathcal{O}_{\mathcal{K}})$ assigns to the identity homomorphism of the B-module $\mathcal{O}_{\mathcal{K}}$ a certain A-homomorphism

$$\varphi_{\mathfrak{M},\mathcal{K}}^{*} \colon \mathcal{O}_{\mathfrak{M}} \to \big(\mathcal{O}_{\mathcal{K}} \big)_{\varphi}\,. \tag{15.52}$$

Such homomorphisms, relating the representing modules of one and the same diffunctor over various algebras, express the *naturality property* of representing modules. Further, we use the following less precise, but more expressive, notation $\varphi_{\mathfrak{M},\mathcal{K}}^{*} \colon \mathcal{O}_{\mathfrak{M}} \to \mathcal{O}_{\mathcal{K}}$, sometimes simplifying it to φ^{*}, if the context specifies the categories $\mathfrak{M}$ and $\mathcal{K}$.

Let $\varphi_1 \colon A_1 \to A_2$ and $\varphi_2 \colon A_2 \to A_3$ be homomorphisms of commutative algebras and let $\mathfrak{M}_i$ be a differentially closed category of A_i-modules,

$i = 1, 2, 3$. It is readily shown that

$$\left(\varphi_2^{\mathfrak{M}_2, \mathfrak{M}_3}\right)^* \circ \left(\varphi_1^{\mathfrak{M}_1, \mathfrak{M}_2}\right)^* = \left(\varphi_2^{\mathfrak{M}_2, \mathfrak{M}_3} \circ \varphi_1^{\mathfrak{M}_1, \mathfrak{M}_2}\right)^*,$$

if the category $\mathfrak{M}_1$ is φ_1-connected with $\mathfrak{M}_2$, while $\mathfrak{M}_2$ is φ_2-connected with $\mathfrak{M}_3$.

In classical differential geometry, it is well known that any smooth map $F \colon M \to N$ is accompanied by a map of covariant tensors on the manifold N to covariant tensors on the manifold M. Such maps are particular cases of the construction described above for

$$\varphi = F^*, \quad \mathfrak{M} = \operatorname{GMod} C^\infty(N), \quad \mathcal{K} = \operatorname{GMod} C^\infty(M).$$

In the next chapter, this is discussed in more detail for differential forms.

15.26 Proposition. *The homomorphism $\varphi^*_{\mathfrak{M}, \mathcal{K}}$ represents the transformation T^φ_Φ, i.e.,*

$$T^\varphi_\Phi(Q) = \operatorname{Hom}_A(\varphi^*_{\mathfrak{M}, \mathcal{K}}, Q) \colon \left(\operatorname{Hom}_B(\mathcal{O}_\mathcal{K}, Q)\right)_\varphi \to \operatorname{Hom}_A(\mathcal{O}_\mathfrak{M}, Q_\varphi).$$

◄ The proof is identical to that of the Proposition from Section 15.2 and so is omitted.   ►

15.27. Compatibility and transformations of diffunctors. Suppose that $\varphi \colon A \to B$, $\mathfrak{M}$, $\mathcal{K}$ are as in Section 15.25, Φ and Ψ are two of the diffunctors considered above and $\mathrm{T} \colon \Phi \to \Psi$ is their natural transformation. Denote by $\mathcal{O}^{\mathfrak{M}}_\Phi$ and $\mathcal{O}^{\mathfrak{M}}_\Psi$ the representing modules of these diffunctors in the category $\mathfrak{M}$, and similarly for $\mathcal{K}$. Then we have the following commutative diagram:

$$
\begin{array}{ccc}
\mathcal{O}^{\mathfrak{M}}_\Phi & \xrightarrow{\;H^{\mathfrak{M}}_{\mathrm{T}}\;} & \mathcal{O}^{\mathfrak{M}}_\Psi \\
{\scriptstyle \varphi^*}\big\downarrow & & \big\downarrow{\scriptstyle \varphi^*} \\
\mathcal{O}^{\mathcal{K}}_\Phi & \xrightarrow{\;H^{\mathcal{K}}_{\mathrm{T}}\;} & \mathcal{O}^{\mathcal{K}}_\Psi,
\end{array}
\tag{15.53}
$$

where $H^{\mathfrak{M}}_{\mathrm{T}}$ (resp., $H^{\mathcal{K}}_{\mathrm{T}}$) is the homomorphism representing according to Section 15.2 the transformation T in the category $\mathfrak{M}$ (resp., in $\mathcal{K}$).

As in Section 15.25, we will omit the proof of this statement in the general case and present it for the concrete situations that will be considered below.

15.28. Natural differential operators. The considerations that led us in Section 15.23 to the natural property of jets actually arise in a much more general context and lead to the notion of *natural differential operator*. Such operators, as we have seen, are d and j_k. Roughly speaking, these are operators that commute with homomorphisms of the commutative algebra, or with smooth maps, when we limit ourselves to the category of smooth manifolds. More precisely, this means the following.

Keeping the notation from the two previous sections, let us consider the natural transformation $\Phi \xrightarrow{\text{T}} \Psi^<(\mathrm{Diff}_k^>)$. It follows from the formula (15.10) that the A-module $\mathcal{J}_<^{k,\mathfrak{M}}(\mathcal{O}_\Psi^{\mathfrak{M}})$ represents the diffunctor $\Psi^<(\mathrm{Diff}_k^>)$ in the category $\mathfrak{M}$.

Let us bring together the data that we need in the following diagram:

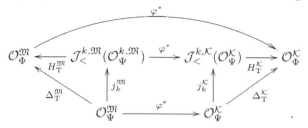

$$(15.54)$$

Here $H_{\mathrm{T}}^{\mathfrak{M}}$ is the homomorphism representing the transformation T in the category $\mathfrak{M}$ according to Section 15.2, and $\Delta_{\mathrm{T}}^{\mathfrak{M}} \overset{\text{def}}{=} H_{\mathrm{T}}^{\mathfrak{M}} \circ j_k^{\mathfrak{M}}$. The symbols $H_{\mathrm{T}}^{\mathcal{K}}$ and $\Delta_{\mathrm{T}}^{\mathcal{K}}$ have a similar meaning, while φ^* denotes the homomorphism of the corresponding representing module in the sense of Section 15.25. The commutativity of the diagram (15.53) obviously implies the commutativity of the diagram (15.54). An important consequence of this is the equality

$$\Delta_{\mathrm{T}}^{\mathcal{K}} \circ \varphi^* = \varphi^* \circ \Delta_{\mathrm{T}}^{\mathfrak{M}}. \tag{15.55}$$

The commutation with smooth maps of the classical de Rham differential $d = d_L \colon C^\infty(L) \to \Lambda^1(L)$, discussed previously, is expressed by the equality

$$d_M \circ F^* = F^* \circ d_N, \quad \text{where} \quad F \colon M \to N, \tag{15.56}$$

which is a very particular case of formula (15.55).

Exercise. Indicate categories $\mathfrak{M}$ and $\mathcal{K}$, together with a natural transformation T between them, for which formula (15.56) becomes equivalent to formula (15.55).

The equality (15.56) is customarily written in the form $d \circ F^* = F^* \circ d$ and is known as the *naturality of the differential*. However, this terminology is not entirely appropriate, since here we are not dealing with a specific operator, but with a functor in the category of smooth manifolds, which relates to the manifold M the concrete operator d_M that "lives" on it. In other words, the term *differential* should be understood as a functor $d \colon M \to d_M$. In the more general algebraic situation, we deal with a "functor", which, to each differentially closed category $\mathfrak{M}$ of modules over a commutative algebra, assigns the operator $d^{\mathfrak{M}}$.

The above motivates the following terminology.

Definition. A *natural differential operator* is, by definition, an assignment, denoted by Δ_{T}, corresponding to some transformation of diffunctors $\Phi \xrightarrow{\text{T}} \Psi^<(\mathrm{Diff}_k^>)$, which to each differential closed category $\mathfrak{M}$ assigns the operator $\Delta_{\mathrm{T}}^{\mathfrak{M}}$ and satisfies (15.55).

We stress that the property of naturality is expressed by formula (15.55) and presupposes that this formula holds in the category of all commutative algebras. We should have in mind that the above construction allows to define various special classes of natural operators by appropriately restricting the category of commutative algebras under consideration. For example, this approach leads to the definition of the natural operators related to Poincaré duality.

15.29. Naturality of geometrization. The category of geometrical modules stands out among other categories due to the fact that it is *natural*. This means that the categories GMod A and GMod B are φ-connected for any homomorphism $\varphi \colon A \to B$, as the next proposition shows

Proposition. *If the B-module Q is geometrical, then the A-module Q_φ is also geometrical.*

◀ The proof is a consequence the following inclusion:

$$\mathrm{Inv}(Q_\varphi) = \bigcap_{h \in |A|} (\mathrm{Ker}\, h \cdot_\varphi Q_\varphi) \subset \bigcap_{\tilde{h} \in |B|} (\mathrm{Ker}(|\varphi|(\tilde{h})) \cdot_\varphi Q_\varphi)$$

$$= \bigcap_{\tilde{h} \in |B|} (\mathrm{Ker}(\tilde{h} \circ \varphi) \cdot_\varphi Q_\varphi) = \bigcap_{\tilde{h} \in |B|} (\varphi^{-1}(\mathrm{Ker}\, \tilde{h}) \cdot_\varphi Q_\varphi)$$

$$= \bigcap_{\tilde{h} \in |B|} (\mathrm{Ker}\, \tilde{h} \cdot Q) = \mathrm{Inv}(Q) = 0.$$

Here we used the fact that $a \cdot_\varphi q = bq$ whenever $a \in \varphi^{-1}(b)$.   ▶

15.30. Representing modules and localization. In view of the importance of the procedure of localization, it is necessary to find out what localization does to representing modules. As in many cases studied previously, it is appropriate here to consider an "example worth imitating", i.e., to consider the case of smooth manifolds. Recall that we have $C^\infty(U) = S_U^{-1} C^\infty(M)$, where the subset $U \subset M$ is open. For instance, $\Lambda^1(U) = S^{-1}\Lambda^1(M)$, as follows from the proposition in Section 13.16. In other words, the localization of the representing module for the functor D in the category of geometrical $C^\infty(M)$-modules represent this functor in the category of geometrical $S_U^{-1} C^\infty(M)$-modules. This and other similar examples lead to the following question: Is it true that the localization of the representing module of some diffunctor will represent it in the localized category? This question must be made more precise.

Let $\mathfrak{M}$ be some category of A-modules and S be some multiplicative set. The *localization $S^{-1}\mathfrak{M}$ of the category $\mathfrak{M}$* has as objects the $S^{-1}A$-modules of the form $S^{-1}P$, while its morphisms are homomorphisms of the form $S^{-1}H$, where H is a homomorphism of A-modules belonging to $\mathfrak{M}$. The lemma from Section 13.2 shows that the categories $\mathfrak{M}$ and $S^{-1}\mathfrak{M}$ are compatible with respect to the canonical homomorphism $\iota \colon A \to S^{-1}A$.

Example. $S^{-1}(\operatorname{Mod} A) = \operatorname{Mod} S^{-1}A$ (see Section 13.2).

Further, let Φ be some diffunctor and $\mathcal{O}_\Phi^{\mathfrak{M}}$ be the module in the category $\mathfrak{M}$ that represents it. Then, in particular, $\Phi(S^{-1}P) = \operatorname{Hom}_A(\mathcal{O}_\Phi^{\mathfrak{M}}, S^{-1}P)$ if $S^{-1}P$ is understood as an A-module. On the other hand, the canonical isomorphism $\operatorname{Hom}_A(\mathcal{O}_\Phi^{\mathfrak{M}}, S^{-1}P) = \operatorname{Hom}_{S^{-1}A}(S^{-1}\mathcal{O}_\Phi^{\mathfrak{M}}, S^{-1}P)$ (see the proposition in Section 13.2) shows that

$$\Phi(S^{-1}P) = \operatorname{Hom}_{S^{-1}A}(S^{-1}\mathcal{O}_\Phi^{\mathfrak{M}}, S^{-1}P),$$

where $S^{-1}P$ is now understood as an $S^{-1}A$-module. This answers the question under discussion.

Proposition. *There is a canonical isomorphism between* $\mathcal{O}_\Phi^{S^{-1}\mathfrak{M}}$ *and* $S^{-1}\mathcal{O}_\Phi^{\mathfrak{M}}$.

Now let $\mathcal{K}$ be some category of $S^{-1}A$-modules, which is ι-connected with the category $\mathfrak{M}$ and let $\mathcal{O}_\Phi^{\mathcal{K}}$ be an $S^{-1}A$-module representing the diffunctor Φ in $\mathcal{K}$. Then the natural homomorphism of A-modules

$$\varphi_{\mathfrak{M},\mathcal{K}}^* \colon \mathcal{O}_\Phi^{\mathfrak{M}} \to \mathcal{O}_\Phi^{\mathcal{K}}$$

splits canonically into the composition $\varphi_{\mathfrak{M},\mathcal{K}}^* = S^{-1}\varphi_{\mathfrak{M},\mathcal{K}}^* \circ \iota_{\mathcal{O}_\Phi^{\mathfrak{M}}}$ (see Section 13.2). This yields the natural homomorphism

$$S^{-1}\varphi_{\mathfrak{M},\mathcal{K}}^* \colon S^{-1}\mathcal{O}_\Phi^{\mathfrak{M}} \to \mathcal{O}_\Phi^{\mathcal{K}},$$

which, in view of the last proposition, can be identified with

$$S^{-1}\varphi_{\mathfrak{M},\mathcal{K}}^* \colon \mathcal{O}_\Phi^{S^{-1}\mathfrak{M}} \to \mathcal{O}_\Phi^{\mathcal{K}}. \tag{15.57}$$

15.31. Geometrization and localization. As we saw in Section 15.29, the categories $\mathfrak{M} = \operatorname{GMod} A$ and $\mathcal{K} = \operatorname{GMod} S^{-1}A$ are ι_S-connected. Hence, in this case, the homomorphism $\iota_\Gamma^{S,*} \stackrel{\text{def}}{=} \iota_{\mathfrak{M},\mathcal{K}}^{S,*}$ is defined, and therefore, so is the homomorphism

$$S^{-1}\left(\iota_\Gamma^{S,*}\right) \colon S^{-1}\mathcal{O}_\Phi^{A,\Gamma} \to \mathcal{O}_\Phi^{S^{-1}A,\Gamma}, \tag{15.58}$$

where $\mathcal{O}_\Phi^{B,\Gamma}$ denotes the B-module representing the diffunctor Φ in the category of geometrical B-modules.

Proposition. *If the localization with respect to the multiplicative set S preserves the property of the A-modules to be geometrical, then $S^{-1}\mathcal{O}_\Phi^{A,\Gamma}$ represents the diffunctor Φ in the category of geometrical S^{-1}-modules.*

◀ Under the assumption of the proposition, the category $\operatorname{GMod} S^{-1}A$ coincides with the category $S^{-1}\operatorname{GMod} A$. Hence, in this case, the homomorphism (15.58) is an isomorphism, as the proposition from the previous section claims.   ▶

Remark. As we saw in Section 13.15, the property of an A-module to be geometrical is not always preserved under localization. When this happens, the algebra $S^{-1}A$ is not geometrical, while the category $S^{-1}\operatorname{GMod} A$ contains nongeometrical $S^{-1}A$-modules. Since the representing modules of any diffunctor in this category are localizations corresponding to the representing module from the category GMod A, this category is differentially closed. This is an example of a differentially closed category intermediate between the category of all modules and the category of geometrical modules. Speaking informally, we can say that in such categories ghost elements can be definitely controlled.

16
Cosymbols, Tensors, and Smoothness

16.1. The order of differential operators yields a natural filtration in any module of differential operators. This filtration, in its turn, induces a filtration in the representing modules. Thus, in the differential calculus, as it is understood in this book, many filtered and, as a consequence, many graded objects appear. The Hamiltonian formalism, which we considered in Chapter 10, is one of the simplest graded constructions of that type, and it shows the importance and necessity of graded objects. One of the main elements of this formalism, the Hamilton-Jacobi equation (the eikonal in geometrical optics), is part of a complicated and almost unstudied system of equations, which describes the behavior of the singularities of solutions of (nonlinear) partial differential equations.

Many natural differential complexes arising in the differential calculus also turn out to be similarly filtered and thereby it becomes possible to use the technique of spectral sequences for the computation of their (co)homology. Independently of this, the corresponding graded complexes carry diverse useful data about the main algebra A itself, the differential operators over A, and so on. An example of this is the Spencer complex, which we shall study in the next chapter.

In the present chapter, we consider the simplest graded objects, symbols and cosymbols, and related diffunctors and constructions.

16.2. Cosymbols and generalizations. Let $\mathfrak{M}$ be a differentially closed category of A-modules and let $P \in \mathrm{Ob}\,\mathfrak{M}$. The restriction of the diffunctor $\mathrm{Diff}_{k,P}$ to $\mathfrak{M}$ will be denoted by $\mathrm{Diff}_{k,P}^{\mathfrak{M}}$. The tower of surjections (15.45)

© Springer Nature Switzerland AG 2020
J. Nestruev, *Smooth Manifolds and Observables*, Graduate Texts
in Mathematics 220, https://doi.org/10.1007/978-3-030-45650-4_16

is a representation of the chain of embedded functors

$$\mathrm{Diff}^{\mathfrak{M}}_{0,P} \to \cdots \to \mathrm{Diff}^{\mathfrak{M}}_{k,P} \to \mathrm{Diff}^{\mathfrak{M}}_{k+1,P} \to \cdots,$$

which it is appropriate to call the *filtration diffunctor* $\mathrm{Diff}^{\mathfrak{M}}_{P} \overset{\mathrm{def}}{=} \mathrm{Diff}^{\mathfrak{M}}(P,\cdot)$; this diffunctor is the direct limit of the tower (15.45). Hence, by duality considerations, it is natural to call the tower the *cofiltration* of the bimodule $\mathcal{J}(P;\mathfrak{M})$. This cofiltration, in its turn, is related to the filtration (15.46).

In this context, the diffunctors $\mathrm{Smbl}^{l}_{k,P}$, which generalize the construction of symbols (see Section 10.1), automatically arise. On the category $\mathfrak{M}$, they are defined as follows:

$$\mathrm{Ob}\,\mathfrak{M} \ni Q \mapsto \mathrm{Smbl}^{l}_{k,P}(Q) \overset{\mathrm{def}}{=} \frac{\mathrm{Diff}^{\mathfrak{M}}_{k,P}(Q)}{\mathrm{Diff}^{\mathfrak{M}}_{l,P}(Q)} = \frac{\mathrm{Diff}^{\mathfrak{M}}_{k}(P,Q)}{\mathrm{Diff}^{\mathfrak{M}}_{l}(P,Q)}.$$

It is natural to understand these diffunctors as the *cokernels* of the embeddings of the diffunctors $\mathrm{Diff}_{l,P} \to \mathrm{Diff}_{k,P}$.

Having in mind Section 10.1, it is appropriate to call the diffunctor $\mathrm{Smbl}_{k,P} \overset{\mathrm{def}}{=} \mathrm{Smbl}^{k-1}_{k,P}$ the *diffunctor of symbols* of order k, and the elements of the module that it represents, denoted $\mathrm{Csml}^{k,P}_{\mathfrak{M}}$, the *cosymbols of order k of the module P* in the category $\mathfrak{M}$, or simply k-cosymbols (of the module P in $\mathfrak{M}$).

16.3. Representation of the diffunctors $\mathrm{Smbl}^{l}_{k,P}$. In order to describe the A-module that represents the diffunctor $\mathrm{Smbl}^{l}_{k,P}$, we shall use the following formula:

$$h_{\Delta}\big(\delta_{a_1,\dots,a_m}(j^{\mathfrak{M}}_{k,P})(p)\big) = \delta_{a_1,\dots,a_m}(\Delta)(p), \tag{16.1}$$

where h_{Δ} was defined in 15.12 and $\Delta \in \mathrm{Diff}^{\mathfrak{M}}_{k}(P,Q)$.

Exercise. Prove formula (16.1) by induction on m, using the fact that $h_{\Delta}(j^{\mathfrak{M}}_{k,P}(p)) = \Delta(p)$.

Proposition. (a) *The kernel of the projection $\pi^{l,\mathfrak{M}}_{k,P}$ is generated by elements of the form $[\delta_{a_0,a_1\dots,a_l}(j^{\mathfrak{M}}_{k,P})](p)$, $a_i \in A$, $p \in P$.*

(b) *The diffunctor $\mathrm{Smbl}^{l}_{k,P}$ in the category $\mathfrak{M}$ is represented by the A-module $\mathrm{Ker}(\pi^{l,\mathfrak{M}}_{k,P})$. In particular, $\mathrm{Csml}^{k,P}_{\mathfrak{M}} = \mathrm{Ker}(\pi^{k-1,\mathfrak{M}}_{k,P})$.*

◀ (a) The operator $j^{\mathfrak{M}}_{l,P}$ may be regarded as an operator of order $\leq k$, and therefore, can be represented as $j^{\mathfrak{M}}_{l,P} = h_{j^{\mathfrak{M}}_{l,P}} \circ j^{\mathfrak{M}}_{k,P}$ by using the universality of $j^{\mathfrak{M}}_{k,P}$. But since

$$h_{j^{\mathfrak{M}}_{l,P}}(j^{\mathfrak{M}}_{k,P}(p)) = j^{\mathfrak{M}}_{l,P}(p) = \pi^{l,\mathfrak{M}}_{k,P}(j^{\mathfrak{M}}_{k,P}(p)), \quad \forall p \in P,$$

it follows that $h^{\mathfrak{M}}_{j^{\mathfrak{M}}_{l,P}} = \pi^{l,\mathfrak{M}}_{k,P}$. Using this relation and formula (16.1) for the operator $\Delta = j^{\mathfrak{M}}_{l,P}$, we find that

$$\pi^{l,\mathfrak{M}}_{k,P}(\delta_{a_0,a_1\ldots,a_l}(j^{\mathfrak{M}}_{k,P})(p)) = \delta_{a_0,a_1\ldots,a_l}(j^{\mathfrak{M}}_{l,P})(p) = 0\,.$$

This shows that elements of the form indicated do lie in $\mathrm{Ker}(\pi^{l,\mathfrak{M}}_{k,P})$.

To verify that the indicated elements generate $\mathrm{Ker}(\pi^{l,\mathfrak{M}}_{k,P})$, consider the submodule that they generate, $I \subset \mathcal{J}^{k,\mathfrak{M}}_<(P)$, and the quotient module $J^l_k = \mathcal{J}^{k,\mathfrak{M}}_<(P)/I$. The differential operator given by $j^k_l \overset{\mathrm{def}}{=} \pi \circ j^{\mathfrak{M}}_{k,P}$, where $\pi\colon \mathcal{J}^{k,\mathfrak{M}}_<(P) \to J^l_k$ is the factorization map

$$\big(\delta_{a_0,a_1\ldots,a_l}(j^k_l)\big)(p) = \big(\delta_{a_0,a_1\ldots,a_l}(j^{\mathfrak{M}}_{k,P})\big)(p) = 0 \mod I,$$

is of order l. Hence, $j^k_l = h \circ j^{\mathfrak{M}}_{l,P}$, where h is a homomorphism. Since the elements $j^{\mathfrak{M}}_{k,P}(p)$ generate the A-module $\mathcal{J}^{k,\mathfrak{M}}_<(P)$, while π is a surjection, it follows that the elements $j^k_l(p) = \pi(j^{\mathfrak{M}}_{k,P}(p))$ generate the A-module J^l_k. This implies that h is surjective.

On the other hand, since $I \subset \mathrm{Ker}(\pi^{l,\mathfrak{M}}_{k,P})$, we have $\pi^{l,\mathfrak{M}}_{k,P} = \Pi \circ \pi$, where $\Pi\colon J^l_k \to \mathcal{J}^l_<(P;\mathfrak{M})$ is a homomorphism. Therefore,

$$j^{\mathfrak{M}}_{l,P} = \pi^{l,\mathfrak{M}}_{k,P} \circ j^{\mathfrak{M}}_{k,P} = \Pi \circ \pi \circ j^{\mathfrak{M}}_{k,P} = \Pi \circ j^k_l = \Pi \circ h \circ j^{\mathfrak{M}}_{l,P}.$$

In particular, $j^{\mathfrak{M}}_{l,P}(p) = (\Pi \circ h)(j^{\mathfrak{M}}_{l,P}(p))$, for all $p \in P$. This implies that $\Pi \circ h = \mathrm{id}_{\mathcal{J}^l_<(P;\mathfrak{M})}$, since the elements $j^{\mathfrak{M}}_{l,P}(p)$ generate $\mathcal{J}^l_<(P;\mathfrak{M})$. Hence, the homomorphism h is injective; together with the above, this shows that h is an isomorphism. Thus, $I = \mathrm{Ker}(\pi^{l,\mathfrak{M}}_{k,P})$.

(b) Let the operator $\Delta \in \mathrm{Diff}^{\mathfrak{M}}_k(P,Q)$ satisfy the equality

$$h_\Delta\big|_{\mathrm{Ker}(\pi^{l,\mathfrak{M}}_{k,P})} = 0.$$

Then $h_\Delta = h^l_\Delta \circ \pi^{l,\mathfrak{M}}_{k,P}$, where $h^l_\Delta\colon \mathcal{J}^l_<(P;\mathfrak{M}) \to Q$ is a homomorphism. This implies that Δ is an operator of order $\le l$. Indeed,

$$\Delta(p) = h_\Delta(j^{\mathfrak{M}}_{k,P}(p)) = h^l_\Delta(\pi^{l,\mathfrak{M}}_{k,P}(j^{\mathfrak{M}}_{k,P}(p))) = h^l_\Delta(j^{\mathfrak{M}}_{l,P}(p))\,.$$

and so $\Delta = h^l_\Delta \circ j^{\mathfrak{M}}_{l,P}$.

Conversely, if the operator $\Delta \in \mathrm{Diff}^{\mathfrak{M}}_l(P,Q)$ is regarded as an operator of order $\le k$, then, obviously, $h_\Delta\big|_{\mathrm{Ker}(\pi^{l,\mathfrak{M}}_{k,P})} = 0$. Therefore, the coset $[\Delta \mod \mathrm{Diff}^{\mathfrak{M}}_l(P,Q)]$ is well defined by the homomorphism $h_\Delta\big|_{\mathrm{Ker}(\pi^{l,\mathfrak{M}}_{k,P})}$.
▶

16.4 Corollary. (a) *The left and right A-modules generated by elements of the form $\delta^{a_0,a_1,\ldots,a_l}_{k,\mathfrak{M}}$ in the A-bimodule $\mathcal{J}^k(A;\mathfrak{M})$ coincide as sets with* $\mathrm{Ker}(\pi^{l,\mathfrak{M}}_{k,A})$.

(b) $\mathrm{Ker}(\pi^{l,\mathfrak{M}}_{k,P}) = \mathrm{Ker}(\pi^{l,\mathfrak{M}}_{k,A})_> \otimes_A P.$

(c) $\text{Csml}^{k,P}_{\mathfrak{M}} = \text{Csml}^{k,A}_{\mathfrak{M}} \otimes_A P$.

◀ Assertions (a) and (b) are direct consequences of assertion (a) in the proposition from Section 16.3. To prove (a), we must additionally take into account the second and fourth relation from the lemma in Section 15.19, for (b), the isomorphism given by the proposition from Section 15.20. Finally, assertion (c) is a particular case of assertion (b), if we take into account the fact that the right and left multiplications in $\text{Csml}^{k,A}_{\mathfrak{M}}$ coincide, and also use assertion b) from the proposition in Section 16.3.   ▶

16.5. The algebra of cosymbols. Having in mind the definition of the module of cosymbols, we can present relation (15.47) for $l = k - 1$ in the form

$$\text{Csml}^{k,P}_{\mathfrak{M}} = \text{Ker}(\pi^{k-1,\mathfrak{M}}_{k,P}) = \frac{\text{Ker}(\pi^{k-1,\mathfrak{M}}_{\infty,P})}{\text{Ker}(\pi^{k,\mathfrak{M}}_{\infty,P})}.$$

Hence, the graded A-module corresponding to the filtration (15.46) is of the form

$$\text{Csml}^{P}_{\mathfrak{M}} = \bigoplus_{k \geq 0} \text{Csml}^{k,P}_{\mathfrak{M}}.$$

Since $\mathcal{J}(P; \mathfrak{M})$ is an A-bimodule, from the formal point of view it follows that $\text{Csml}^{P}_{\mathfrak{M}}$ is also an A-bimodule. But, since the operations of right and left multiplication by elements of the algebra A in $\text{Csml}^{k,P}_{\mathfrak{M}}$ coincide, it follows that $\text{Csml}^{P}_{\mathfrak{M}}$ is simply an A-module.

For $P = A$, the filtration (15.46) yields the filtered algebra $\mathcal{J}(A; \mathfrak{M})$. Hence, $\text{Csml}^{A}_{\mathfrak{M}}$ is a graded algebra, which we shall distinguish by means of the following notation:

$$\text{Csml}_{\mathfrak{M}}(A) \overset{\text{def}}{=} \text{Csml}^{A}_{\mathfrak{M}}, \quad \text{Csml}^{k}_{\mathfrak{M}}(A) \overset{\text{def}}{=} \text{Csml}^{k,A}_{\mathfrak{M}}.$$

Obviously, $\text{Csml}^{P}_{\mathfrak{M}}$ is a graded module over the graded algebra $\text{Csml}_{\mathfrak{M}}(A)$. It follows directly from the corresponding definitions that the result of multiplying the element $\delta^{a_1\ldots,a_k}_{k,\mathfrak{M}} \cdot j^{\mathfrak{M}}_{k,P}(p) \in \text{Csml}^{k,P}_{\mathfrak{M}}$ by the element $\delta^{b_1\ldots,b_l}_{l,\mathfrak{M}} \in \text{Csml}^{l}_{\mathfrak{M}}(A)$ is

$$\delta^{b_1\ldots,b_l,a_1\ldots,a_k}_{k+l,\mathfrak{M}} \cdot j^{\mathfrak{M}}_{k+l,P}(p) \in \text{Csml}^{k+l,P}_{\mathfrak{M}}.$$

(Check this!)

16.6. Symbols and cosymbols. The duality of these two notions is expressed by the fact that they can be paired; here we describe the corresponding construction. Below, $\text{smbl}_{k,P}(\nabla) \in \text{Smbl}_{k,P}$ (see Section 16.2) denotes the *symbol* of the operator $\nabla \in \text{Diff}_k P$, i.e., its coset w.r.t. the module $\text{Diff}_{k-1} P$. Note also that in Chapter 10 the A-module $\text{Smbl}_{k,A}(A)$ was denoted by the $S_k(A)$, for which we shall use the notation $\text{Smbl}_k(A)$.

Recall that $\text{Diff}(P, Q)$ is a left filtered $\text{Diff}(Q, Q)$-module and, accordingly, a right $\text{Diff}(P, P)$-module (see Exercise 15.11). Therefore, passing to the corresponding graded objects, we see that the *A-module of P-symbols*

$$\mathrm{Smbl}_P(Q) \stackrel{\mathrm{def}}{=} \bigoplus_{k \geq 0} \mathrm{Smbl}_{k,P}(Q), \text{ where } \mathrm{Smbl}_{k,P}(Q) = \mathrm{Smbl}_{k,P}^{k-1}(Q),$$

is a left graded module over the graded algebra

$$\mathrm{Smbl}_Q(Q) \stackrel{\mathrm{def}}{=} \bigoplus_{k \geq 0} \mathrm{Smbl}_{k,Q}(Q).$$

in which the operation of module multiplication is defined by the rule

$$\mathrm{smbl}_{l,A}(\Delta) \cdot \mathrm{smbl}_{k,P}(\nabla) = \mathrm{smbl}_{k+l,A}(\Delta \circ \nabla),$$

where $\Delta \in \mathrm{Diff}_l(Q, Q)$, $\nabla \in \mathrm{Diff}_k(P, Q)$. Similarly, $\mathrm{Smbl}_P(Q)$ acquires the structure of a graded $\mathrm{Smbl}_P(P)$-module.

Further, note that with any operator $\Delta \in \mathrm{Diff}_l(P, R)$ we can associate the transformation of diffunctors $\mathfrak{C}_{P,\Delta}^k$,

$$\mathfrak{C}_{P,\Delta}^k(Q): \ \mathrm{Diff}_k(R, Q) \to \mathrm{Diff}_{k+l}(P, Q),$$

$$\mathrm{Diff}_k(R, Q) \ni \nabla \mapsto \nabla \circ \Delta \in \mathrm{Diff}_{k+l}(P, Q). \quad (16.2)$$

If $P, R \in \mathrm{Ob}\,\mathfrak{M}$, then, as can easily be seen,

$$\mathfrak{C}_k^{P,\Delta}: \ \mathcal{J}^{k+l}(P; \mathfrak{M}) \to \mathcal{J}^k(R; \mathfrak{M}), \quad j_{k+l,P}^{\mathfrak{M}}(p) \mapsto j_{k,R}^{\mathfrak{M}}(\Delta(p)). \quad (16.3)$$

is the homomorphism representing this category in $\mathfrak{M}$.

Since the submodule $\mathrm{Diff}_{k-1}(R, Q) \subset \mathrm{Diff}_k(R, Q)$ is taken by the homomorphism $\mathfrak{C}_{P,\Delta}^k(Q)$ to $\mathrm{Diff}_{k+l-1}(P, Q)$, it follows that $\mathfrak{C}_{P,\Delta}^k$ induces a transformation of diffunctors

$$\beth_{P,\Delta}^k: \ \mathrm{Smbl}_{k,R} \to \mathrm{Smbl}_{k+l,P}.$$

Hence, by duality considerations, this transformation is represented in the category $\mathfrak{M}$ by the homomorphism

$$\beth_k^{P,\Delta} \stackrel{\mathrm{def}}{=} \mathfrak{C}_k^{P,\Delta}\big|_{\mathrm{Ker}(\pi_{k+l,P}^{k+l-1,\mathfrak{M}})}: \ \mathrm{Ker}(\pi_{k+l,P}^{k+l-1,\mathfrak{M}}) \to \mathrm{Ker}(\pi_{k,R}^{k-1,\mathfrak{M}}). \quad (16.4)$$

Here, on the basis of the proposition from Section 16.3, we identify $\mathrm{Csml}_{\mathfrak{M}}^{m,P}$ with $\mathrm{Ker}(\pi_{m,P}^{m-1,\mathfrak{M}})$.

Now, having in mind (16.4) and the proposition from Section 16.3, we can calculate the homomorphism $\beth_k^{P,\Delta}$. Namely, taking $n = k + l$ in the notation of Section 9.58, we obtain

$$\mathbf{J}_k^{P,\Delta}(\delta_{a_1,\ldots,a_{k+l}}(j_{k+l,P}^{\mathfrak{M}})(p)) = \mathfrak{C}_k^{P,\Delta}\big(\delta_{a_1,\ldots,a_{k+l}}(j_{k+l,P}^{\mathfrak{M}})(p)\big)$$

$$= \delta_{a_1,\ldots,a_{k+l}}(j_{k,P}^{\mathfrak{M}}\circ\Delta)(p)$$

$$= \sum_{|\varkappa|=k}\big(\delta_{a_\varkappa}(j_{k,P}^{\mathfrak{M}})\circ\delta_{a_{\bar{\varkappa}}}(\Delta)\big)(p)$$

$$= \sum_{|\varkappa|=k}\delta_{a_\varkappa}(j_{k,P}^{\mathfrak{M}})(\delta_{a_{\bar{\varkappa}}}(\Delta)(p))\,.$$

Since $|\bar{\varkappa}| = l$, it follows that $\delta_{a_{\bar{\varkappa}}}(\Delta)$ depends only on the symbol of the operator Δ, i.e., on $\mathfrak{s} = \mathrm{smbl}_l\,\Delta$. Hence, the homomorphism $\mathbf{J}_k^{P,\Delta}$ can be interpreted as the substitution of the symbol $\mathfrak{s}$ in the cosymbols. Having this in mind, we will simplify the notation by putting

$$\mathbf{J}_\mathfrak{s}^m \overset{\text{def}}{=} \mathbf{J}_{m-l}^{P,\Delta}\colon \mathrm{Csml}_{\mathfrak{M}}^{m,P} \to \mathrm{Csml}_{\mathfrak{M}}^{m-l,R}\,,$$

where it is assumed that $\mathbf{J}_\mathfrak{s}^m = 0$ when $m < l$. Accordingly, the formula that describes $\mathbf{J}_\mathfrak{s}^m$ acquires the form

$$\mathbf{J}_\mathfrak{s}^m(\delta_{a_1,\ldots,a_m}(j_{m,P}^{\mathfrak{M}})(p)) = \sum_{|\varkappa|=m-l}\delta_{\mathfrak{M}}^{a_\varkappa}\cdot\big(\mathfrak{s}(a_{\bar{\varkappa}})(p)\big),\qquad(16.5)$$

where $\mathfrak{s}(a_{\bar{\varkappa}}) = \delta_{a_{\bar{\varkappa}}}(\Delta) \in \mathrm{Hom}_A(P,R)$. If $P = A$, then we have $\mathfrak{s}(a_{\bar{\varkappa}}) \in A$, by the canonical isomorphism $\mathrm{Hom}_A(A,R) = R$ and the formula (16.5) for $p = 1_A$ now acquires the form

$$\mathbf{J}_\mathfrak{s}^m(\delta_{\mathfrak{M}}^{a_1,\ldots,a_m}) = \sum_{|\varkappa|=m-l}\delta_{\mathfrak{M}}^{a_\varkappa}\mathfrak{s}(a_{\bar{\varkappa}}) \in \mathrm{Csml}_{\mathfrak{M}}^{m-l}(A)\otimes_A R\,.\qquad(16.6)$$

If $A = C^\infty(M)$ and $\mathfrak{M} = \mathrm{GMod}\,A$, then an operation of this type is the *convolution* operation of a symmetric covariant tensor of order m with a symmetric contravariant tensor of order l.

16.7. Symmetric polyderivatio'n and cosymbols. The *symmetric k-derivations of the algebra A* (with values in the A-module P) are defined as the assignment

$$(a_1,\ldots,a_k) \mapsto \square(a_1,\ldots,a_k) \in P,$$

which is symmetric w.r.t. all its arguments and is a derivation in each one of them. The set of all such derivations obviously forms an A-module, denoted $D_{\mathrm{sym}}^k(P)$. This module can be identified with the following submodule of the A-module

$$D^{\circ k}(P) \overset{\text{def}}{=} D(D(\ldots(P)\ldots))\ (k\ \text{times}).$$

Indeed, if $\nabla \in D^{\circ k}(P)$, then the k-form inductively defined by the rule

$$(a_1, \ldots, a_k) \mapsto \nabla(a_1, \ldots, a_k) \overset{\text{def}}{=} \nabla(a_1)(a_2, \ldots, a_k)$$

is a derivation in each argument. Obviously, the above definition is invertible, in particular, $\nabla \in D^{\circ k}(P)$ can be uniquely recovered from the expression $\nabla(a_1, \ldots, a_k)$. Thus, we can understand an element of the A-module $D^{\circ k}(P)$ as a k-derivation, and this obviously gives us the embedding $D^k_{\text{sym}}(P) \subset D^{\circ k}(P)$.

No less obvious is the fact that the assignments

$$P \mapsto D^k_{\text{sym}}(P) \text{ and } P \mapsto D^{\circ k}(P)$$

define functors, which we accept as members of the family of diffunctors. Here the A-module $(\Lambda^1_{\mathfrak{M}})^{\otimes k}$ represents, as it follows from (15.10), the diffunctor $D^{\circ k}$ in the category $\mathfrak{M}$, while

$$(a_1, \ldots, a_k) \mapsto d^{\mathfrak{M}} a_1 \otimes_A \ldots \otimes_A d^{\mathfrak{M}} a_k$$

is the universal k-derivation. It is easy to see that the representing module for the diffunctor D^k_{sym} is the symmetrization of the module $(\Lambda^1_{\mathfrak{M}})^{\otimes k}$; it is denoted by $S_k(\Lambda^1_{\mathfrak{M}})$. Recall that it is the quotient module of the module $(\Lambda^1_{\mathfrak{M}})^{\otimes k}$ by the submodule generated by all elements of the form

$$\omega_1 \otimes \ldots \otimes \omega_i \otimes \ldots \otimes \omega_j \otimes \ldots \otimes \omega_k - \omega_1 \otimes \ldots \otimes \omega_j \otimes \ldots \otimes \omega_i \otimes \ldots \otimes \omega_k,$$

where $\omega_i \in \Lambda^1_{\mathfrak{M}}$ for all i (see Section 12.12). The image of $\omega_1 \otimes \ldots \otimes \omega_k$ in $S_k(\Lambda^1_{\mathfrak{M}})$ is obviously symmetric w.r.t. the factors ω_i. It is denoted by $\omega_1 \odot \cdots \odot \omega_k$ and is called the *symmetric tensor product* of 1-forms ω_i. In this notation, the assignment

$$(a_1, \ldots, a_k) \mapsto d^{\mathfrak{M}} a_1 \odot_A \cdots \odot_A d^{\mathfrak{M}} a_k \qquad (16.7)$$

is the *universal symmetric k-derivation* in the category $\mathfrak{M}$.

Now note that with the module $\text{Csml}^{k,P}_{\mathfrak{M}}$ we can associate the symmetric k-derivation

$$(a_1, \ldots, a_k) \mapsto \delta_{a_1, \ldots, a_k}(j^{\mathfrak{M}}_{k,P}) = \delta^{a_1, \ldots, a_k}_{k, \mathfrak{M}} \cdot j^{\mathfrak{M}}_{k,P} \in \text{Hom}_A(P, \text{Csml}^{k,P}_{\mathfrak{M}}),$$

(see Section 15.19). Therefore, in view of the universality of the symmetric k-derivation (16.7), the A-homomorphism

$$\text{sc}^{k,P}_{\mathfrak{M}} : S_k(\Lambda^1_{\mathfrak{M}}) \to \text{Hom}_A(P, \text{Csml}^{k,P}_{\mathfrak{M}}),$$

$$d^{\mathfrak{M}} a_1 \odot_A \ldots \odot_A d^{\mathfrak{M}} a_k \mapsto \delta_{a_1, \ldots, a_k}(j^{\mathfrak{M}}_{k,P}).$$

is defined. Below, having in mind the "associativity formula" (15.10), it will be convenient to interpret the homomorphism $\text{sc}^{k,P}_{\mathfrak{M}}$ as follows:

$$\text{sc}^{k,P}_{\mathfrak{M}} : S_k(\Lambda^1_{\mathfrak{M}}) \otimes_A P \to \text{Csml}^{k,P}_{\mathfrak{M}}$$

280   Chapter 16

In particular, if $P = A$, then it is natural to identify the tensor product $S_k(\Lambda_{\mathfrak{M}}^1) \otimes_A P$ with $S_k(\Lambda_{\mathfrak{M}}^1)$, and identify $\mathrm{sc}_{\mathfrak{M}}^{k,A}$ with the homomorphism

$$\mathrm{sc}_{\mathfrak{M}}^k = \mathrm{sc}_{\mathfrak{M}}^{k,A} \colon S_k(\Lambda_{\mathfrak{M}}^1) \to \mathrm{Csml}_{\mathfrak{M}}^k(A), \qquad (16.8)$$

given by $d^{\mathfrak{M}} a_1 \odot_A \cdots \odot_A d^{\mathfrak{M}} a_k \mapsto \delta_{k,\mathfrak{M}}^{a_1,\dots,a_k}$. Since the elements $\delta_{k,\mathfrak{M}}^{a_1,\dots,a_k}$ generate the A-module $\mathrm{Csml}_{\mathfrak{M}}^k(A)$, this homomorphism is surjective. Moreover, using the isomorphism

$$\mathrm{Csml}_{\mathfrak{M}}^{k,P} = \mathrm{Csml}_{\mathfrak{M}}^{k,A} \otimes_A P,$$

we can identify the homomorphism $\mathrm{sc}_{\mathfrak{M}}^{k,P}$ with $\mathrm{sc}_{\mathfrak{M}}^{k,A} \otimes_A \mathrm{id}_P$. This obviously shows the subjectivity of $\mathrm{sc}_{\mathfrak{M}}^{k,P}$.

As can be easily seen, the map

$$\mathrm{sc}_{\mathfrak{M}}^A \colon S_*(\Lambda_{\mathfrak{M}}^1) \to \mathrm{Csml}_{\mathfrak{M}}(A), \quad \mathrm{sc}_{\mathfrak{M}}^A|_{\mathrm{Csml}_{\mathfrak{M}}^{k,A}} = \mathrm{sc}_{\mathfrak{M}}^{k,A},$$

is a surjective algebra homomorphism, while the map

$$\mathrm{sc}_{\mathfrak{M}}^P \colon S_*(\Lambda_{\mathfrak{M}}^1) \otimes_A P \to \mathrm{Csml}_{\mathfrak{M}}^P, \quad \mathrm{sc}_{\mathfrak{M}}^P|_{\mathrm{Csml}_{\mathfrak{M}}^{k,P}} = \mathrm{sc}_{\mathfrak{M}}^{k,P},$$

is a surjective homomorphism of $S_*(\Lambda_{\mathfrak{M}}^1)$-modules, provided we consider the $S_*(\Lambda_{\mathfrak{M}}^1)$-module structure in $\mathrm{Csml}_{\mathfrak{M}}^P$ induced by the homomorphism of the algebras $\mathrm{sc}_{\mathfrak{M}}^A$.

It is useful to note for what follows that if A is an algebra over a field of zero characteristic, then $S_k(\Lambda_{\mathfrak{M}}^1)$ can be embedded in $(\Lambda_{\mathfrak{M}}^1)^{\otimes k}$ by means of the map

$$d^{\mathfrak{M}} a_1 \odot_A \cdots \odot_A d^{\mathfrak{M}} a_k \mapsto \frac{1}{k!} \sum_\sigma d^{\mathfrak{M}} a_{\sigma(1)} \otimes_A \cdots \otimes_A d^{\mathfrak{M}} a_{\sigma(k)}, \qquad (16.9)$$

where the sum is over all the permutations of the indices $1,\dots,k$. In this case, the homomorphism $\mathrm{sc}_{\mathfrak{M}}^k$ looks like the standard symmetrization operation

$$d^{\mathfrak{M}} a_1 \otimes_A \cdots \otimes_A d^{\mathfrak{M}} a_k \mapsto \frac{1}{k!} \sum_\sigma d^{\mathfrak{M}} a_{\sigma(1)} \otimes_A \cdots \otimes_A d^{\mathfrak{M}} a_{\sigma(k)},$$

16.8. Permutations as differential operators. The permutation operators $\mathbf{J}_s^m$, $m \geq 0$, are naturally brought together in the following homomorphism of A-modules

$$\mathbf{J}_s \colon \mathrm{Csml}_{\mathfrak{M}}^P \to \mathrm{Csml}_{\mathfrak{M}}^R, \quad \mathbf{J}_s|_{\mathrm{Csml}_{\mathfrak{M}}^{m,P}} \overset{\mathrm{def}}{=} \mathbf{J}_s^m \qquad (16.10)$$

(see Section 16.6). But $\mathrm{Csml}_{\mathfrak{M}}^P$ and $\mathrm{Csml}_{\mathfrak{M}}^R$ are also $\mathrm{Csml}_{\mathfrak{M}}(A)$-modules, so it is necessary to find out how the operators $\mathbf{J}_s$ interact with this structure.

Proposition. *The permutation operator $\mathbf{J}_s$ is an A-linear differential operator of order l over the algebra $\mathrm{Csml}_{\mathfrak{M}}(A)$.*

◀ The following remark will be useful. Let $\varphi\colon B \to \tilde{B}$ be an epimorphism of commutative algebras over the same ring K, let $\tilde{P}$ and $\tilde{Q}$ be $\tilde{B}$-modules, while $\Delta\colon \tilde{P} \to \tilde{Q}$ is a K-linear map. Assume that Δ is a differential operator of order l over the algebra B, where we regard $\tilde{P}$ and $\tilde{Q}$ as B-modules w.r.t. the homomorphism φ. Then Δ is a differential operator of order l over the algebra $\tilde{B}$. This is because, in the present situation, we have $\delta_{\tilde{b}_1,\ldots,\tilde{b}_m}(\Delta) = \delta_{b_1,\ldots,b_m}(\Delta)$ for any $\tilde{b}_1,\ldots,\tilde{b}_m \in \tilde{B}$, where $b_i \in \varphi^{-1}(\tilde{b}_i)$, $\forall i$.

We will use this fact in the case when $\varphi = \mathrm{sc}_{\mathfrak{M}}^A$ and $\Delta = \beth_{\mathfrak{s}}$. For this, we need to check that $\beth_{\mathfrak{s}}$ is a differential operator of order l over the algebra $S_*(\Lambda_{\mathfrak{M}}^1)$. This is equivalent to this operator being the composition $\beth_{\mathfrak{s}} \circ \mathrm{sc}_{\mathfrak{M}}^P$, since $\mathrm{sc}_{\mathfrak{M}}^P$ is an epimorphism of $S_*(\Lambda_{\mathfrak{M}}^1)$-modules. But the latter fact can be seen from the commutative diagram

$$
\begin{array}{ccc}
S_*(\Lambda_{\mathfrak{M}}^1) \otimes_A P & \xrightarrow{\ i_{\Lambda_{\mathfrak{M}}^1}^l \otimes_A \mathrm{id}_P\ } & S_*(\Lambda_{\mathfrak{M}}^1) \otimes_A S_l(\Lambda_{\mathfrak{M}}^1) \otimes_A P \\
\Big\downarrow{\scriptstyle \mathrm{sc}_{\mathfrak{M}}^P} & & \Big\downarrow{\scriptstyle \mathrm{sc}_{\mathfrak{M}}^A \otimes_A H_{\mathfrak{s}}} \\
\mathrm{Csml}_{\mathfrak{M}}^P & \xrightarrow{\quad \beth_{\mathfrak{s}} \quad} & \mathrm{Csml}_{\mathfrak{M}}^R = \mathrm{Csml}_{\mathfrak{M}}(A) \otimes_A R,
\end{array}
\qquad (16.11)
$$

where $H_{\mathfrak{s}}$ is the compositon

$$
S_l(\Lambda_{\mathfrak{M}}^1) \otimes_A P \xrightarrow{\ \mathrm{sc}_{\mathfrak{M}}^{l,P}\ } \mathrm{Csml}_{\mathfrak{M}}^{l,P} \xrightarrow{\ \beth_{\mathfrak{s}}^l\ } \mathrm{Csml}_{\mathfrak{M}}^{0,R} = R,
$$

while $i_{\Lambda_{\mathfrak{M}}^1}^l$ is the universal symmetric permutation (see Section 16.7). Indeed, $\mathrm{sc}_{\mathfrak{M}}^A \otimes_A H_{\mathfrak{s}}$ is a $S_*(\Lambda_{\mathfrak{M}}^1)$-homomorphism, while the symmetric permutation $i_{\Lambda_{\mathfrak{M}}^1}^l \otimes_A \mathrm{id}_P$ is an A-linear differential operator of order l over the algebra $S_*(\Lambda_{\mathfrak{M}}^1)$. Hence, the composition $\beth_{\mathfrak{s}} \circ \mathrm{sc}_{\mathfrak{M}}^P$, which coincides with

$$
(\mathrm{sc}_{\mathfrak{M}}^A \otimes_A H_{\mathfrak{s}}) \circ (i_{\Lambda_{\mathfrak{M}}^1}^l \otimes_A \mathrm{id}_P),
$$

is also such an operator. ▶

Exercise. Prove the previous proposition by a direct calculation.

16.9. Why smooth manifolds are smooth. The structure of the differential calculus on smooth manifolds has a whole series of simplifying specific traits as compared to commutative algebras of general form. This manifests itself in that differential 1-forms on a manifold M can be defined, as it is done everywhere, as A-linear functions on $D(M)$, i.e., by means of the formula $\Lambda^1(M) = D(M)^*$. This approach considerably simplifies this definition and similar ones, since here it is not necessary to think about functors, representing objects, and the like. But this approach is *conceptually wrong*, because it leads to unsatisfactory results for commutative algebras of general nature. The fact that the definition of 1-forms "accidentally" happened to be adequate is due to the property of the module $\Lambda^1(M)$ of being projective and of finite type. For this reason, if we start from this definition, we have $\Lambda^1(M)^* = (D(M)^*)^* = D(M)$, i.e., we obtain the conceptually correct relation $D(M) = \Lambda^1(M)^*$.

A specific trait of smooth manifolds is, in particular, the fact that its algebra of symbols $S_*(C^\infty(M))$ is isomorphic to the symmetric algebra of the module $D(M)$, whereas any operator $\Delta \in \mathrm{Diff}_k\, C^\infty(M)$ is the sum of compositions of k first-order operators. All these facts are well-known, we have already used them in Chapter 10. But as the next exercise shows, the above-mentioned facts no longer hold on the cross $\mathbf{K}$ (see 2.12).

Exercise. 1. Consider the submodule $\mathrm{Diff}_k^{\mathrm{cmp}}\, C^\infty(\mathbf{K}) \subset \mathrm{Diff}_k\, C^\infty(\mathbf{K})$, consisting of operators that can be represented as the sum of k-fold compositons operators of order one. Describe the quotient module

$$\mathrm{Diff}_k\, C^\infty(\mathbf{K}) / \mathrm{Diff}_k^{\mathrm{cmp}}\, C^\infty(\mathbf{K}).$$

2. Describe the quotient module

$$S_k(D(C^\infty(\mathbf{K}))) / \mathrm{Smbl}_k(C^\infty(\mathbf{K})).$$

It is remarkable that the specific traits of smooth manifolds mentioned above, and many similar ones, are due to one simple circumstance, namely to the fact that $\Lambda^1(M)$ is a projective module of finite type. This will be proved below, but now we shall specify this fact by means of the following definition.

1. A differentially closed category of A-modules $\mathfrak{M}$ is called *smooth* if $\Lambda^1_{\mathfrak{M}}$ is a projective module of finite type.

2. An algebra A is called *smooth* if the category $\mathrm{GMod}\, A$ is smooth. (Note that, the definition given in Chapter 4 is a particular case of this definition; check this!)

Since $D(A) = (\Lambda^1_{\mathfrak{M}})^*$, the module $D(A)$ is also of finite type and is projective provided the category $\mathfrak{M}$ is smooth. In that case,

$$D(A)^* = ((\Lambda^1_{\mathfrak{M}})^*)^* = \Lambda^1_{\mathfrak{M}}.$$

Thus, the existence of a smooth category of A-modules is possible only if the module $D(A)$ is projective and is of finite type. Here the condition $\Lambda^1_{\mathfrak{M}} = D(A)^*$ must be satisfied. .

Example. I. The algebra $C^\infty(M)$ is smooth.

II. The algebra $C^0(M)$ of continuous functions on a smooth manifold M is smooth (!), since the module $\Lambda^1_{\mathrm{GMod}\, C^0(M)}$ is trivial (why?).

III. The category $\mathrm{Mod}\, C^\infty(M)$ is not smooth (see Section 14.10).

Exercise. 1. Show that the algebra of polynomials $\Bbbk[x_1, \dots, x_n]$ over a field of zero characteristic is smooth.

2. Is the algebra $\mathbb{F}_p[x_1, \dots, x_n]$, where $\mathbb{F}_p$ is the field of p elements, smooth?

16.10. The following simple technical result will be needed for the proof of Theorem 16.11, the main general result about smooth categories.

Lemma–Exercise. *Let $F\colon P \to Q$ be an epimorphism of A-modules such that $F^*\colon Q^* \to P^*$ is an isomorphism. If at the same time the modules P, P^* and Q^* are projective and of finite type, then F is an isomorphism.*

◀ For the proof, use the properties of projective modules which are given in Section 12.15   ▶

16.11 Theorem. *Let A be a commutative algebra over a field of zero characteristic and let $\mathfrak{M}$ be a smooth category of A-modules. Then*

$$\mathrm{Csml}_{\mathfrak{M}}^k(A) = S_k(\Lambda_{\mathfrak{M}}^1)$$

and any operator $\Delta \in \mathrm{Diff}_k(A)$ can be represented in the form

$$\Delta = \sum_\alpha \Delta_1^\alpha \circ \cdots \circ \Delta_k^\alpha, \quad \Delta_i \in \mathrm{Diff}_1 A.$$

In particular, the A-module $\mathrm{Csml}_{\mathfrak{M}}^k(A)$ is projective and of finite type.

◀ First, since the module $\Lambda_{\mathfrak{M}}^1$ is projective and we have $(\Lambda_{\mathfrak{M}}^1)^* = D(A)$, it follows that

$$((\Lambda_{\mathfrak{M}}^1)^{\otimes k})^* = ((\Lambda_{\mathfrak{M}}^1)^*)^{\otimes k} = D(A)^{\otimes k} \tag{16.12}$$

(see the exercise in Section 12.15). Therefore, in particular, the elements of the form $X_1 \otimes \ldots \otimes X_k \in D(A)^{\otimes k}$, understood as A-linear functions on $(\Lambda_{\mathfrak{M}}^1)^{\otimes k}$, generate the A-module $((\Lambda_{\mathfrak{M}}^1)^{\otimes k})^*$.

Further, the module $S_k(\Lambda_{\mathfrak{M}}^1)$, being the image of the symmetrization homomorphism, is a direct summand of the projective module $(\Lambda_{\mathfrak{M}}^1)^{\otimes k}$. Hence, it is also projective, while the homomorphism $S_k(\Lambda_{\mathfrak{M}}^1) \to (\Lambda_{\mathfrak{M}}^1)^{\otimes k}$ dual to the embedding $(\Lambda_{\mathfrak{M}}^1)^{\otimes k})^* \to S_k(\Lambda_{\mathfrak{M}}^1)^*$ is surjective. This implies that the A-module $S_k(\Lambda_{\mathfrak{M}}^1)^*$ is generated by the elements

$$X_1 \diamond \cdots \diamond X_k \overset{\mathrm{def}}{=} (X_1 \otimes \ldots \otimes X_k)|_{S_k(\Lambda_{\mathfrak{M}}^1)}, \quad X_i \in D(A).$$

In this situation, we have (see 16.8)

$$(X_1 \diamond \cdots \diamond X_k)(d^{\mathfrak{M}}a_1 \odot \cdots \odot d^{\mathfrak{M}}a_k)$$
$$= (X_1 \otimes \ldots \otimes X_k)(d^{\mathfrak{M}}a_1 \odot \cdots \odot d^{\mathfrak{M}}a_k)$$
$$= \sum_\sigma \delta_{a_1}(X_{\sigma(1)}) \ldots \delta_{a_k}(X_{\sigma(k)}). \tag{16.13}$$

This shows that

$$S_k(\Lambda^1_{\mathfrak{M}})^* = S_k((\Lambda^1_{\mathfrak{M}})^*).$$ (16.14)

On the other hand, we have

$$\mathrm{Csmbl}^k_{\mathfrak{M}}(A)^* = \mathrm{Smbl}_{k,A}(A) = \mathrm{Smbl}_k(A),$$

due to the proposition from Section 16.3. Therefore, having in mind (16.14), we see that the homomorphism sc_k dual to the homomorphism $\mathrm{sc}^k_{\mathfrak{M}}$ can be interpreted as the map

$$\mathrm{sc}_k\colon \mathrm{Smbl}_k(A) \to S_k((\Lambda^1_{\mathfrak{M}})^*) = S_k(D(A)).$$

Finally, by definition, the value of the element $\theta = \mathrm{sc}_k(\mathrm{smbl}_k(\Delta))$, as an A-linear functional on $S_k(\Lambda^1_{\mathfrak{M}})$, is given by the formula

$$\theta(d^{\mathfrak{M}} a_1 \odot \cdots \odot d^{\mathfrak{M}} a_k) = \delta_{a_1, \ldots, a_k}(\Delta).$$ (16.15)

Now if $\Delta = X_1 \circ \cdots \circ X_k$, then

$$\delta_{a_1, \ldots, a_k}(X_1 \circ \cdots \circ X_k) = \sum_\sigma \delta_{a_1}(X_{\sigma(1)}) \cdot \ldots \cdot \delta_{a_k}(X_{\sigma(k)})$$ (16.16)

Comparing this value with (16.13), we conclude that the homomorphism sc_k is surjective. But, being dual to a surjective homomorphism (namely, $\mathrm{sc}^k_{\mathfrak{M}}$), it is also injective. Therefore, sc_k is an isomorphism and we have $\mathrm{Smbl}_k(A) = S_k(D(A))$. Since the module $D(A)$ is projective, we see, from the same considerations as above, that the module $S_k(D(A))$ is also projective.

As the result, we have come to the situation described in the lemma-exercise from Section 16.10 for $P = S_k(\Lambda^1_{\mathfrak{M}})$, $Q = \mathrm{Csmbl}^k_{\mathfrak{M}}(A)$ and $F = \mathrm{sc}^k_{\mathfrak{M}}$. Since the modules appearing in the previous arguments are obviously of finite type, this proves the first assertion of the theorem.

Now let $\Delta \in \mathrm{Diff}_k A$, $\quad k > 0$. Then $\mathrm{sc}_k(\mathrm{smbl}_k \Delta)$ can be presented as

$$\mathrm{sc}_k(\mathrm{smbl}_k \Delta) = \sum_j X^j_1 \odot \ldots \odot X^j_k, \quad X^j_i \in \mathrm{Diff}_1 A.$$

Then, as we see from (16.15) and (16.16),

$$\mathrm{sc}_k(\mathrm{smbl}_k(\Delta)) = \mathrm{sc}_k(\mathrm{smbl}_k(\sum_j X^j_1 \circ \ldots \circ X^j_k))$$

Since sc_k is an isomorphism, we have $\mathrm{smbl}_k(\Delta - \sum_j X^j_1 \circ \ldots \circ X^j_k) = 0$ and so the operator $\Delta - \sum_j X^j_1 \circ \ldots \circ X^j_k$ is of order $\leq k - 1$. Continuing this process, we come to the required result. ▶

Corollary. *If $\mathfrak{M}$ is a smooth category, then the algebra of cosymbols $\mathrm{Csmbl}_{\mathfrak{M}}(A)$ (see Section 16.5) is isomorphic to $S_*(\Lambda^1_{\mathfrak{M}})$, i.e., to the symmetric algebra of the A-module $\Lambda^1_{\mathfrak{M}}$.*

◀ This is now obvious. ▶

16.12. Projectivity of representing modules. It is remarkable that modules representing diffunctors in a smooth category $\mathfrak{M}$ are all projective modules of finite type. This is a consequence of the fact that they are obtained from $\Lambda^1_{\mathfrak{M}}$ via standard linear algebra operations, which preserve the property of finite projectivity. First of all, these are the operations of tensor product (Proposition 12.34), $\mathrm{Hom}_A(\cdot,\cdot)$ and $P \mapsto P^*$ (see Proposition 12.36 and the remark that follows). Another example is the symmetric tensor power, using which we were able to prove above that the module $\mathrm{Csml}^k_{\mathfrak{M}}(A)$ representing the diffunctor Smbl_k is finite projective (Theorem 16.11). The property of finite projectivity of the representing modules of those diffunctors that we study below will be established along the way in each case. The following proposition is the first of this series.

Proposition. *Let $\mathfrak{M}$ be a smooth category of A-modules and P be a projective A-module of finite type. Then the A-modules $\mathcal{J}^k(P;\mathfrak{M})$ and $\mathrm{Csml}^{k,P}_{\mathfrak{M}}$ are also projective of finite type.*

◄ First, note the chain of isomorphisms

$$\mathrm{Ker}\,\pi^{k-1,\mathfrak{M}}_{k,P} = \mathrm{Csml}^{k,P}_{\mathfrak{M}} = \mathrm{Csml}^k_{\mathfrak{M}}(A) \otimes_A P = S_k(\Lambda^1_{\mathfrak{M}}) \otimes_A P,$$

extracted from the proposition in Section 16.3, Corollary 16.4, and Theorem 16.11. It follows from the same theorem that the module $S_k(\Lambda^1_{\mathfrak{M}}) \otimes_A P$ is of finite type and is projective, being the tensor product of such modules. Hence, such also is the module $\mathrm{Ker}\,\pi^{k-1,\mathfrak{M}}_{k,P}$.

Assume that, further, the module $\mathcal{J}^{k-1}(\mathfrak{M})$ is projective and of finite type. Since the homomorphism $\pi^{k-1,\mathfrak{M}}_{k,P}$ is surjective, we can find a homomorphism $\alpha\colon \mathcal{J}^{k-1}(P;\mathfrak{M}) \to \mathcal{J}^k(P;\mathfrak{M})$ for which we have $\pi^{k-1,\mathfrak{M}}_{k,P} \circ \alpha = \mathrm{id}_{\mathcal{J}^{k-1}(P;\mathfrak{M})}$. This implies

$$\mathcal{J}^k(P;\mathfrak{M}) \cong \mathrm{Im}\,\alpha \oplus \mathrm{Ker}\,\pi^{k-1,\mathfrak{M}}_{k,P} \cong \mathcal{J}^{k-1}(P;\mathfrak{M}) \oplus \mathrm{Ker}\,\pi^{k-1,\mathfrak{M}}_{k,P}.$$

Hence, the A-module $\mathcal{J}^k(P;\mathfrak{M})$ is also projective and of finite type. This yields a proof by induction of the required assertion, in which the base of induction is the finiteness and projectivity of the module $\mathcal{J}^0(P;\mathfrak{M}) = P$. ►

16.13. Tensors in smooth categories. In classical differential geometry, covariant tensors of order k on a manifold M are usually defined as $C^\infty(M)$-polylinear functions $\rho(X_1,\ldots,X_k) \in C^\infty(M)$ of vector arguments $X_i \in D(M)$. This can be restated by saying that ρ is a homomorphism of $C^\infty(M)$-modules

$$\rho\colon D(M)^{\otimes k} \to C^\infty(M),\ X_1 \otimes_A \cdots \otimes_A X_k \mapsto \rho(X_1,\ldots,X_k),$$

i.e., $\rho \in (D(M)^{\otimes k})^*$. In particular, if $k = 1$, this is the standard definition of a first order differential form on the manifold M: $\Lambda^1(M) \overset{\mathrm{def}}{=} D(M)^*$ (*cf.* Section 14.4). Since $D(M)$ is a projective module of finite type, we have

$$(D(M)^{\otimes k})^* = (D(M)^*)^{\otimes k} = \Lambda^1(M)^{\otimes k}. \qquad (16.17)$$

But the module $(\Lambda^1(M))^{\otimes k}$, obviously, represents the diffunctor $D^{\circ k}$ in the category $\mathrm{GMod}\, C^\infty(M)$ of geometrical $C^\infty(M)$-modules. Hence, it will be conceptually correct to change our approach by defining covariant tensors as elements of the module representing the diffunctor $D^{\circ k}$ in some category or other. Then, as we have already seen, we will have the equality

$$D^{\circ k}(A) = D(A)^{\otimes k} = ((\Lambda^1_{\mathfrak{M}})^{\otimes k})^*.$$

If $\mathfrak{M} = \mathrm{GMod}\, C^\infty(M)$, then (16.17) follows from this last equality precisely because the module $\Lambda^1_{\mathfrak{M}}$ is finite projective, i.e., because the category $\mathfrak{M}$ is smooth.

Thus, it would be conceptually incorrect to define covariant tensors over an arbitrary commutative algebra by generalizing relation (16.17), i.e., as elements of the module $(D(A)^{\otimes k})^*$. However, this turns out to be "accidentally" correct, provided the algebra A (or the category $\mathfrak{M}$) is smooth.

The above considerations remain in force not only for covariant tensors of special type but also, e.g., for symmetric or skew-symmetric tensors (differential forms). They can be applied without any modifications to all other *covariant quantities* in standard differential geometry. These quantities, in the framework of our approach, are modules representing various diffunctors. Here an interesting new aspect is that these modules are determined by the choice of the differential closed category, so that, generally speaking, several different versions of the same "covariant quantity" may arise. An example of this is "algebraic" and "geometrical" differential forms of the first order (see Sections 14.8–14.11).

In conclusion, speaking informally, let us say that in smooth categories "covariant quantities" may be defined in the same way as in standard differential geometry, but one must have in mind that such definitions are "conceptually incorrect".

16.14. The "tangent bundle" of a commutative algebra. In Chapter 10, we learned that, informally, a natural instrument for observing the cotangent manifold T^*M is the algebra of symbols $\mathrm{Smbl}(C^\infty(M))$. Obvious duality considerations suggest that the algebra of cosymbols must be an instrument of the same type for the tangent manifold TM. And this is indeed so.

Proposition. (a) *Let A be a smooth algebra. Consider the algebra consisting of all pairs (h, φ), $h \in |A|$, $\varphi \in \Lambda^1_\Gamma(A)^*_h$; then this algebra is geometrical; it will be denoted by $\mathrm{Csml}_\Gamma(A)$.*

(b) *The $\mathbb{R}$-spectrum $|\mathrm{Csml}_\Gamma(C^\infty(M))|$ of the algebra $\mathrm{Csml}_\Gamma(C^\infty(M))$ can be naturally identified with TM.*

◀ Since the algebra A is smooth, we have $\mathrm{Csml}_\Gamma(A) = S_*(\Lambda^1_\Gamma)$. Therefore, the first part of assertion (a) is a direct consequence of the proposition in 16.12, while the second one is a particular case of the assertion of the exercise that follows that proposition.

If $A = C^\infty(M)$, then $\Lambda^1_\Gamma(C^\infty(M)) = \Lambda^1(M)$ (Theorem 14.11), and therefore,

$$\Lambda^1_\Gamma(C^\infty(M))^*_{h_z} = \Lambda^1(M)^*_{h_z} = D(M)_{h_z} = T_z M, \ z \in M.$$

This shows that assertion (b) is a particular case of assertion (a).   ▶

Exercise. The manifold TM may be interpreted either as the $\mathbb{R}$-spectrum of the algebra of cosymbols $\mathrm{Csml}_\Gamma(C^\infty(M))$, or as the $\mathbb{R}$-spectrum of the algebra $C^\infty(TM)$. Accordingly, two Zariski topologies coexist on TM. Do they coincide?

16.15. Projection map in fiber bundles. So far we have only constructed the "total space of the tangent bundle" of the algebra A, and now we will "fiber it". This means that we must indicate the analog of the homomorphism

$$\pi_T^*\colon C^\infty(M) \to C^\infty(TM),$$

having in mind that $\pi_T = |\pi_T^*|$. The answer here is very simple: this analog is the algebra homomorphism $\top_{\mathfrak{M}}\colon A \to \mathrm{Csml}_{\mathfrak{M}}(A)$ that identifies A with $\mathrm{Csml}^0_{\mathfrak{M}}(A) \subset \mathrm{Csml}_{\mathfrak{M}}(A)$.

Exercise. Show that the projection $\pi_T\colon TM \to M$ can be naturally identifed with

$$|\top_{\mathfrak{M}}|\colon |\,\mathrm{Csml}_\Gamma(C^\infty(M))| \to |C^\infty(M)|,$$

if $A = C^\infty(M)$ and $\mathfrak{M} = \mathrm{GMod}\, C^\infty(M)$.

16.16. How many tangent bundles does a commutative algebra have? In view of the fact that the algebra of cosymbols depends on the choice of the differentially closed category $\mathfrak{M}$, the following question arises: How does the $\Bbbk$-spectrum of the algebra $\mathrm{Csml}_{\mathfrak{M}}(A)$ depend on the choice of $\mathfrak{M}$, i.e., how many "tangent bundles" can the given algebra A have?

Proposition. *The $\Bbbk$-spectrum of the algebra of cosymbols $\mathrm{Csml}(A)$ in the category of all A-modules coincides with $|\,\mathrm{Csml}_\Gamma(A)|$.*

◀ Recall that the geometrization homomorphism $\Gamma\colon P \to \Gamma(P)$ induces the isomorphism $\mathrm{Hom}_A(P,Q) = \mathrm{Hom}_A(\Gamma(P),Q)$, provided Q is a geometrical A-module. For this reason, $\mathcal{O}^\Gamma_\Phi = \Gamma(\mathcal{O}_\Phi)$, where the module $\mathcal{O}_\Phi$ (resp., $\mathcal{O}^\Gamma_\Phi$) represents the diffunctor Φ in the category of all (resp., geometrical) A-modules. In particular, $\mathrm{Csml}_\Gamma(A) = \Gamma(\mathrm{Csml}(A))$ and for any geometrical A-module Q, we have

$$\begin{aligned}
\mathrm{Hom}_A(\mathrm{Csml}(A), Q) &= \mathrm{Hom}_A(\Gamma(\mathrm{Csml}(A)), Q) \\
&= \mathrm{Hom}_A(\mathrm{Csml}_\Gamma(A), Q).
\end{aligned} \qquad (16.18)$$

Further, with any $h \in |\,\mathrm{Csml}(A)|$, we can associate the A-module $\Bbbk_h$, which as a $\Bbbk$-vector space coincides with $\Bbbk$, while multiplication by an element $a \in A$ is determined by the rule $a \cdot \lambda \overset{\mathrm{def}}{=} h(a)\lambda$, $\lambda \in \Bbbk$. The module $\Bbbk_h$, obviously, is a geometrical A-module and so by (16.18) we have

$$\mathrm{Hom}_A(\mathrm{Csml}(A), \Bbbk_h) = \mathrm{Hom}_A(\mathrm{Csml}_\Gamma(A), \Bbbk_h).$$

This isomorphism, as is easy to see, is compatible with the multiplicative structures of the algebras $\mathrm{Csml}(A)$ and $\mathrm{Csml}_\Gamma(A)$, and therefore, allows to identify their $\Bbbk$-spectra.   ▶

The proposition that we have just proved, together with the next exercise, essentially closes the question about the number of "tangent bundles".

Exercise. Prove that $|\,\mathrm{Csml}_{\mathfrak{M}}(A)| = |\,\mathrm{Csml}_\Gamma(A)|$ if GMod A is a subcategory of the category $\mathfrak{M}$.

16.17. Structures on tangent bundles. The fact that the tangent manifold TM is the $\mathbb{R}$-spectrum of the algebra of cosymbols determines the existence of specific additional structures on it, some of which will be described below. The most important ones of them were discovered "experimentally" in the process of developing the *Lagrangian approach* in classical mechanics and related questions of geometry. Initially, the existence of these structural elements of Lagrangian mechanics appeared in the form of certain combinations of partial derivatives, which kept *stably* arising during concrete computations with functions describing mechanical systems. Later, these "experimental data" were understood, and these structural elements were described in terms of local coordinates. However, the mystery of their origins remained unsolved. One of the reasons for this is that it is wrong to "observe" the tangent manifold TM by means of the algebra $C^\infty(TM)$, i.e., to look at it simply as it were just any other smooth manifold. More precisely, in that case, the family of "observation devices" constituting this algebra does not distinguish the positions of the mechanical system from its velocities.

In what follows, having in mind that the manifold TM must be observed by means of the algebra $\mathrm{Csml}_\Gamma(C^\infty(M))$ (see Section 16.14), we shall indicate certain simple structures, which may be directly discovered by using the algebra $\mathrm{Csml}_{\mathfrak{M}}(A)$ on the "tangent bundle" of the algebra A.

16.18. Sections of the tangent bundle. Recall that to any vector field $X \in D(M)$ can be associated a certain natural section $s_X : M \to TM$ of the tangent bundle $\pi_T : TM \to M$ (see Section 9.40,). This gives us a geometrical portrait of the field X, although the opinion that this portrait should actually be the definition of the notion of field is rather widespread. The fact that such a definition is unsatisfactory and becomes clear when one asks what is the "section of the tangent bundle" of an arbitrary commutative algebra.

To answer that question, we need an analog of the homomorphism s_X^*, i.e., the algebra homomorphism by means of which one can "observe" the tangent bundle of the algebra A, and the algebra A itself. In other words, we must associate with any "vector field" $X \in D(A)$ a homomorphism

$$\Theta_X^{\mathfrak{M}} \colon \operatorname{Csml}_{\mathfrak{M}}(A) \to A,$$

which we define by the formula

$$\Theta_X^{\mathfrak{M}}|_{\operatorname{Csml}_{\mathfrak{M}}^k(A)} \overset{\text{def}}{=} \frac{1}{k!} \mathbf{J}_{\mathfrak{s}_k}, \quad \text{where} \quad \mathfrak{s}_k = \operatorname{smbl}_k X^k, \tag{16.19}$$

assuming that A is an algebra over a ring of zero characteristic (see Section 16.6).

Proposition. $\Theta_X^{\mathfrak{M}}$ is an algebra homomorphism.

◀ It is only the multiplicativity property of $\Theta_X^{\mathfrak{M}}$ that merits a proof. To obtain it, note that

$$\mathbf{J}_{\mathfrak{s}_k}(\delta^{a_1,\dots,a_k}) = \mathfrak{s}_k(\delta^{a_1,\dots,a_k}) = \delta_{a_1,\dots,a_k}(X^k)$$
$$= \sum_{\sigma} X(a_{\sigma(1)}) \cdots X(a_{\sigma(k)}) = k! X(a_1) \cdots X(a_k) \tag{16.20}$$

where the sum is over all permutations of the indices $1, \dots, k$. Hence, we have

$$\Theta_X^{\mathfrak{M}}(a\delta^{a_1,\dots,a_k}) \overset{\text{def}}{=} a X(a_1) \cdots X(a_k), \quad a, a_1, \dots, a_k \in A, \tag{16.21}$$

and this implies multiplicativity, as required. ▶

Corollary. *The analog of the section s_X for an arbitrary commutative algebra with zero characteristic is the map defined by* (16.21)

$$|\Theta_X^{\mathfrak{M}}| \colon |A| \to |\operatorname{Csml}_{\mathfrak{M}}(A)|.$$

Exercise.   1. Express in coordinates the map $|\Theta_X^{\mathfrak{M}}|$ for the algebra $A = C^\infty(M)$ and the category $\mathfrak{M} = \operatorname{GMod} C^\infty(M)$, and verify that it coincides with the expression of the section s_X.

2. Show that the map $|\Theta_X^{\mathfrak{M}}|$ for the categories $\mathfrak{M} = \operatorname{Mod} A$ and $\operatorname{GMod} A$ are the same.

3. Why can't the homomorphism $\Theta_X^{\mathfrak{M}}$ for an algebra A with nonzero characteristic be defined by formula 16.21?

16.19. The "universal vector field" of an algebra A. The analog of the universal vector field on a manifold M (see Section 9.48) can easily be constructed for an arbitrary commutative algebra. Let

$$I_A^{\mathfrak{M}} \colon \Lambda^1(A)^{\mathfrak{M}} \to \mathrm{Csml}_{\mathfrak{M}}^1(A)$$

be the natural identification of A-modules and

$$Z_A^{\mathfrak{M}} \overset{\mathrm{def}}{=} I_A^{\mathfrak{M}} \circ d_1^{\mathfrak{M}} \colon A \to \mathrm{Csml}_{\mathfrak{M}}(A).$$

Then for any $X \in D(A)$, we have the formula

$$X = \Theta_X^{\mathfrak{M}} \circ Z_A^{\mathfrak{M}}. \tag{16.22}$$

Indeed, $(\Theta_X^{\mathfrak{M}} \circ Z_A^{\mathfrak{M}})(a) = \Theta_X^{\mathfrak{M}}(I_A^{\mathfrak{M}}(d^{\mathfrak{M}}(a))) = \Theta_X^{\mathfrak{M}}(\delta_{\mathfrak{M}}^a) = X(a)$ (see (16.21)). Here it is useful to note that the right-hand side of this formula does not depend on the choice of the category $\mathfrak{M}$. (Why?)

Exercise. Check that a particular case of (16.22) is $X = s_X^* \circ Z, X \in D(M)$ (see (9.19)).

16.20. The Legendre transformation. Historically, the first system of classical mechanics, proposed by Lagrange, is based on variational principles; its study lies outside the framework of this book, although variational principles are also an inherent part of the differential calculus on commutative algebras. The corresponding mathematical construction, known as the *Lagrangian formalism*, is dual to the Hamiltonian formalism (studied in Chapter 10), which historically appeared much later. These two approaches are related via the *Legendre transformation*, which we describe below.

If M is the configuration space of the mechanical system S under consideration, then the arena on which the Lagrangian formalism acts is the tangent bundle TM, sometimes called the *state space* of this system. Let $\{x_i, v_i\}_{1 \leq i \leq n}$ be the special system of local coordinates on TM (see Section 9.19). Here the quantities $\{x_i\}$ are interpreted as coordinates of position or configuration of the system S, while the quantities $\{v_i\}$ are the components of its velocity. The system itself is described by a certain function $L \in C^\infty(TM)$, which is called its *Lagrangian*.

The *Legendre transformation* corresponding to the Lagrangian L is defined as a map $\mathfrak{L} \colon TM \to T^*M$. In coordinates, it is described by the formulas

$$x_i = x_i, \; p_i = L_{v_i}, \; i = 1, \ldots, n,$$

where $\{x_i, p_i\}_{1 \leq i \leq n}$ are the special coordinates in T^*M (see Section 9.24). It is easy to check that the map $\mathfrak{L}$ is defined invariantly, i.e., it does not depend on the choice of local coordinates. Its more appropriate coordinate-free definition turns out to be very simple if, for the previously explained reasons, we replace the "wrong algebra of observables" $C^\infty(TM)$ by the "correct" one.

16.21. So, let $L \in \mathrm{Csml}_\Gamma(A)$. Our aim is to relate L with a certain algebra homomorphism

$$\mathcal{L}_L \colon \mathrm{Smbl}(A) \to \mathrm{Csml}_\Gamma(A)$$

so that the homomorphism $\mathfrak{L}^* \colon C^\infty(T^*M) \to C^\infty(TM)$, restricted to $\mathrm{Smbl}(C^\infty(M))$, would be its particular case. Here, however, we must take into account the variational origins of the Legendre transformation, which presupposes the smoothness of the algebra A. In that case, the algebra $\mathrm{Smbl}(A)$ is generated by its first-order component $\mathrm{Smbl}_1(A) = D(A)$, as Theorem 16.11 asserts. Hence, it suffices to define $\mathcal{L}_L$ only on $\mathrm{Smbl}_1(A)$. But this can be done quite obviously:

$$\mathcal{L}_L(\mathrm{smbl}_1 X) \stackrel{\mathrm{def}}{=} \beth^1_X(L), \quad X \in D(A), \qquad (16.23)$$

Now since this same theorem asserts that any operator $\Delta \in \mathrm{Diff}_k A$ can be represented in the form

$$\Delta = a + \sum_{l \le k,\, j} X^j_1 \circ \cdots \circ X^j_l, \quad a \in A, \quad X^j_i \in D(A),$$

we have

$$\mathcal{L}_L(\mathrm{smbl}_k \Delta) = \sum_\alpha \beth^1_{X^\alpha_1}(L) \cdots \beth^1_{X^\alpha_k}(L), \quad k > 0, \qquad (16.24)$$

$$\mathcal{L}_L(a) \stackrel{\mathrm{def}}{=} a, \quad a \in A = \mathrm{smbl}_0 A, \quad k = 0. \qquad (16.25)$$

Note that (16.23) means that the homomorphism of $\Bbbk$-algebras $\mathcal{L}_L$ is actually a homomorphism of A-algebras. Geometrically, this is equivalent to the claim that $|\mathcal{L}_L|$ is a morphism from the "tangent bundle" of the algebra A to its cotangent bundle.

Exercise. Let $A = C^\infty(M)$. Then, in the special coordinates on TM,

$$\{(x_i, v_i)\}_{1 \le i \le n},\ v_i = d^\Gamma x_i,$$

where d^Γ is the universal differential in the geometrical category, the Lagrangian L may be regarded as a polynomial in the variables $v_1, \ldots, v_n$ with coefficients in $C^\infty(M)$. Show that in this case

$$\mathcal{L}_L(\mathrm{smbl}(\partial/\partial x_i)) = L_{v_i}.$$

This exercise demonstrates that the classical Legendre transformation for Lagrangians polynomial in velocity is a particular case of Definition 16.23.

Example. In coordinates, a typical Lagrangian mechanical system has the form

$$L = \sum_{i=1}^n \frac{m v_i^2}{2} - U(x),$$

where $U(x)$ is its potential energy. In this case, $\mathcal{L}_L(p_i) = mv_i$ is a well-known formula of elementary mechanics; it relates velocity and momentum, while the map corresponding to it, $\mathfrak{L} = |\mathcal{L}_L|\colon TM \to T^*M$, is an isomorphism of vector bundles.

16.22. "Anti"-Legendre transformation. Dually, with each function $H \in C^\infty(T^*M)$, one can associate the transformation

$$\mathfrak{H}\colon T^*M \to TM, \quad (x_i, p_i) \mapsto (x_i, v_i = H_{p_i}(x, p)). \qquad (16.26)$$

It is easy to check that it does not depend on the choice of local coordinates, and so must have some hidden meaning. Just as in the case of the Legendre transformation, this meaning can be clarified by passing to the "correct" algebras. Therefore we shall try to find the algebra homomorphism $\mathcal{H}_H\colon \mathrm{Csml}_\Gamma(A) \to \mathrm{Smbl}(A)$ described by formulas (16.26) in the case when $A = C^\infty(M)$.

Let $\theta \in \mathrm{Csml}_\Gamma^1(A)$. Then $\theta = I^{\mathfrak{m}}(\omega)$, $\omega \in \Lambda_\Gamma^1(A)$ (see Section 16.19). On the other hand,

$$X_H|_{\mathrm{Smbl}_0(A)}\colon \mathrm{Smbl}_0(A) = A \to \mathrm{Smbl}(A)$$

is a $\mathrm{Smbl}(A)$-valued derivation of the algebra A. Therefore, if we assume that the algebras A and $\mathrm{Smbl}(A)$ are geometrical, we can write

$$X_H = h_H \circ d_\Gamma^1,$$

where $h_H\colon \Lambda_\Gamma^1(A) \to \mathrm{Smbl}(A)$ is a homomorphism of A-modules. Further, let us put

$$\mathcal{H}_H(\theta) = h_H(\omega). \qquad (16.27)$$

Finally, if the algebra A is smooth, then, as we saw (look at the corollary in Section 16.11), the algebra $\mathrm{Csml}_\Gamma(A)$ is isomorphic to the symmetric algebra of the A-module $\mathrm{Csml}_\Gamma^1(A) = \Lambda^1(A)$. Hence, the multiplicative extension of the map (16.27) to the whole algebra $\mathrm{Csml}_\Gamma(A)$ is the required algebra homomorphism $\mathcal{H}_H$.

Returning to the algebra $A = C^\infty(M)$ and using formula (16.27), we see that

$$\mathcal{H}_H(p_i) = h_H(dx_i) = X_H(x_i),$$

i.e., the homomorphism $\mathfrak{H}^*\colon C^\infty(TM) \to C^\infty(T^*M)$ restricted to the subalgebra of functions is polynomial in the p_i's and coincides with $\mathcal{H}_H$.

16.23. It is interesting that the transformation inverse to the Legendre transformation, if it is defined, is the anti-Legendre transformation. This is illustrated in the next exercise.

Exercise. Let $L = \sum_{i=1}^n mv_i^2 - U(x)$. Find a Hamiltonian $H(p,q)$ such that $\mathcal{L}_L^{-1} = \mathcal{H}_H$.

17
Spencer Complexes and Differential Forms

17.1. In the previous chapter, we saw that modules representing various diffunctors are in a certain sense "made out of" differential forms. For this reason, this object deserves special attention, and so in the present chapter we consider those diffunctors which are necessary for the *conceptual* definition of differential forms and the study of their properties. Let us stress the importance of the "conceptual definition", which at first may appear unnecessarily complicated. It is precisely because of this definition that we will avoid various difficulties that otherwise arise in attempts to carry over to arbitrary commutative algebras many well-known differential-geometrical constructions, and will also discover many new results, including applications to real physics and mechanics.

Another goal of this chapter is to demonstrate to the reader how to apply the typical manipulation techniques to concrete diffunctors and their representing objects. To this end, in the first half of the chapter, we consider the simpler diffunctors, related to Spencer complexes and to second-order differential forms, so as to make the technique and the logic eventually used in the general case more understandable. It should especially be noted that the theory of differential forms over commutative algebras of general nature is intimately connected with Spencer complexes, which are not generally known to the mathematical community even in the context of smooth manifolds. And this is so despite their fundamental role in partial differential equations, especially nonlinear ones.

17.2. Gluing transformations. The more complicated diffunctors that we shall need below arise as *natural kernels* in the transformations of other

© Springer Nature Switzerland AG 2020
J. Nestruev, *Smooth Manifolds and Observables*, Graduate Texts
in Mathematics 220, https://doi.org/10.1007/978-3-030-45650-4_17

diffunctors. If $F\colon \Phi_1 \to \Phi_2$ is a transformation of functors, then the kernel of the map $F(P)\colon \Phi_1(P) \to \Phi_2(P)$ depends, in general, on specific traits of the module P. A natural kernel is a kernel that can be described independently of these specific traits. For example, the specific kernel of the homomorphism $Д^<_{k,P}\colon \mathrm{Diff}^<_k(P) \to P$ is $D_{(k)}(P)$ (see Section 15.8). Certain basic diffunctors are defined as the kernels of *gluing transformations*, which will be described below.

Let $V^\diamond$ be an A-bimodule. We shall denote the left A-module structure in it by replacing $\diamond$ by $<$, and the right one, by replacing $\diamond$ by $>$. Consider the biimodule $\mathrm{Diff}^\diamond_k(V^>)$ and recall (see Section 9.57) that the module structures in it were defined as follows:

the left one: $(a \circ \Delta)(b) = a^> \Delta(b) = \Delta(b)a, \quad a, b \in A, \ \Delta(b) \in V^\diamond.$

the right one: $(\Delta \circ a)(b) = \Delta(ab).$

Note that in our case the bimodule $\mathrm{Diff}^\diamond_k(V^>)$ may now be supplied with a third A-module structure. Namely, we can put

$$(a \bullet \Delta)(b) \overset{\text{def}}{=} a^< \Delta(b) = a\Delta(b), \quad a, b \in A, \ \Delta(b) \in V^\diamond.$$

Let us denote the A-module defined in this way by $\mathrm{Diff}^\bullet_k(V^>)$. Obviously, this third A-module structure commutes with the left and right structures of the A-bimodule $\mathrm{Diff}^\diamond_k(V^>)$. Thereby, it becomes a *trimodule* that we will still denote by $\mathrm{Diff}_k(V^>)$. The A-modules corresponding to it are

$$\mathrm{Diff}^\bullet_k(V^>), \quad \mathrm{Diff}^<_k(V^>) \quad \text{and} \quad \mathrm{Diff}^>_k(V^>).$$

Now let us consider the case $V^> = \mathrm{Diff}^>_l(Q)$, where Q is an arbitrary A-module.

A *gluing transformation* of $\Bbbk$-vector spaces

$$\mathrm{Diff}_k(\mathrm{Diff}^>_l(Q)) \to \mathrm{Diff}_{k+l}(Q)$$

is defined by the rule

$$\Delta \mapsto Д^>_l \circ \Delta, \quad \Delta\colon A \to \mathrm{Diff}^>_l(Q). \tag{17.1}$$

It is easy to see that this is a homomorphism of A-modules

$$\begin{aligned} C^<_{k,l}(Q)\colon \ & \mathrm{Diff}^\bullet_k(\mathrm{Diff}^>_l(Q)) \longrightarrow \mathrm{Diff}^<_{k+l}(Q), \\ C^>_{k,l}(Q)\colon \ & \mathrm{Diff}^>_k(\mathrm{Diff}^>_l(Q)) \longrightarrow \mathrm{Diff}^>_{k+l}(Q). \end{aligned} \tag{17.2}$$

The symbols $C^<_{k,l}(Q)$ and $C^>_{k,l}(Q)$ indicate what specific modules are involved in the homomorphism, while the corresponding homomorphisms of vector spaces are the same (17.1).

In what follows, we interpret the material of this section in terms of the natural transformation of diffunctors

$$C^<_{k,l}\colon \mathrm{Diff}^\bullet_k(\mathrm{Diff}^>_l) \longrightarrow \mathrm{Diff}^<_{k+l}, \quad C^>_{k,l}\colon \mathrm{Diff}^>_k(\mathrm{Diff}^>_l) \longrightarrow \mathrm{Diff}^>_{k+l}\,.$$

17.3. Homomorphisms representing gluing transformations. The isomorphism (15.10), together with the universality property of modules of jets, leads to the following chain of isomorphisms:

$$\operatorname{Diff}_k^> (\operatorname{Diff}_l^> P) \cong \operatorname{Hom}_A^> (\mathcal{J}_<^k(\mathfrak{M}), \operatorname{Diff}_l^> P)$$
$$\cong \operatorname{Hom}_A^> (\mathcal{J}_<^k(\mathfrak{M}), \operatorname{Hom}_A^> (\mathcal{J}_<^l(\mathfrak{M}), P)) \qquad (17.3)$$
$$\cong \operatorname{Hom}_A^> (_< \mathcal{J}_>^l(\mathfrak{M}) \otimes_A \mathcal{J}_<^k(\mathfrak{M}), P),$$

which implies that $_< \mathcal{J}_>^l(\mathfrak{M}) \otimes_A \mathcal{J}_<^k(\mathfrak{M}) = \, _< \mathcal{J}^l(\mathcal{J}_<^k(\mathfrak{M}); \mathfrak{M})$ is the representing A-module for the diffunctor $\operatorname{Diff}_k^> (\operatorname{Diff}_l^>)$.

If $P = \, _< \mathcal{J}_>^l(\mathfrak{M}) \otimes_A \mathcal{J}_<^k(\mathfrak{M})$, then to the identity homomorphism $I_{k,l}$ of the module $_< \mathcal{J}_>^l(\mathfrak{M}) \otimes_A \mathcal{J}_<^k(\mathfrak{M})$, according to (17.3), there corresponds the operator $\Delta \in \operatorname{Diff}_k^> (\operatorname{Diff}_l^> (_< \mathcal{J}_>^l(\mathfrak{M}) \otimes_A \mathcal{J}_<^k(\mathfrak{M})))$ such that

$$[\Delta(a)](b) = j_l^{\mathfrak{M}}(b) \otimes j_k^{\mathfrak{M}}(a) \in \, _< \mathcal{J}_>^l(\mathfrak{M}) \otimes_A \mathcal{J}_<^k(\mathfrak{M}).$$

Since, by definition, we have $C_{k,l}^> (\Delta) = \text{Д}_l^> \circ \Delta \in \operatorname{Diff}_{k+l}^> (_< \mathcal{J}_> \otimes_A \mathcal{J}_<)$, it follows that

$$\bigl(C_{k,l}^> (\Delta) \bigr)(a) = [\Delta(a)](1) = j_l^{\mathfrak{M}}(1) \otimes j_k^{\mathfrak{M}}(a).$$

Therefore, according to Proposition 15.2, the homomorphism representing the gluing transformation $C_{k,l}$ will be the A-homomorphism

$$\mathcal{J}_<^{k+l}(\mathfrak{M}) \to \, _< \mathcal{J}_>^l(\mathfrak{M}) \otimes_A \mathcal{J}_<^k(\mathfrak{M}) = \mathcal{J}_<^l(\mathcal{J}_<^k(\mathfrak{M}); \mathfrak{M}),$$
$$j_{k+l}^{\mathfrak{M}}(a) \mapsto j_l^{\mathfrak{M}}(1) \otimes j_k^{\mathfrak{M}}(a) = j_l^{\mathfrak{M}}(j_k^{\mathfrak{M}}(a)) \qquad (17.4)$$

(see 15.41).

17.4. The diffunctors $\operatorname{Df}_{(k,l)}$. Now we can describe the kernel of the gluing transformation $C_{k,l}^>$. Since $\operatorname{Ker} \text{Д}_l^> = D_{(l)}(Q)$, it follows that $\bigl(C_{k,l}^>(Q) \bigr)(\Delta) = 0$ if and only if $\operatorname{Im} \Delta \subset D_{(l)}(Q)$. The set of these operators will be denoted as follows:

$$\operatorname{Df}_{(l,k)}(Q) \overset{\text{def}}{=} \{ \Delta \in \operatorname{Diff}_k^> (\operatorname{Diff}_l^>(Q)) \mid \operatorname{Im} \Delta \subset D_{(l)}(Q) \}.$$

Thus,

$$\operatorname{Ker} C_{k,l}^>(Q) = \operatorname{Df}_{(l,k)}(Q). \qquad (17.5)$$

The space $\operatorname{Df}_{(l,k)}(Q)$ possesses a natural A-bimodule structure, in which the left and right multiplication are defined by

$$\text{(left)} \quad \Delta \mapsto a_{D_{(l)}(Q)} \circ \Delta, \qquad \text{(right)} \quad \Delta \mapsto \Delta \circ a_A.$$

The left and right A-modules of the A-bimodule $\operatorname{Df}_{(l,k)}(Q)$ are denoted by $\operatorname{Df}_{(l,k)}^<(Q)$ and $\operatorname{Df}_{(l,k)}^>(Q)$, respectively.

Exercise. Show that the identical transformation of change of module structure

$$\operatorname{Df}_{(l,k)}^<(Q) \to \operatorname{Df}_{(l,k)}^>(Q) \quad \text{and} \quad \operatorname{Df}_{(l,k)}^>(Q) \to \operatorname{Df}_{(l,k)}^<(Q)$$

is a differential operator of order $\leq k$ (see Exercise in 9.67).

Thus, we have constructed the new diffunctors

$$\mathrm{Df}^<_{(l,k)} \colon Q \mapsto \mathrm{Df}^<_{(l,k)}(Q) \quad \text{and} \quad \mathrm{Df}^>_{(l,k)} \colon Q \mapsto \mathrm{Df}^>_{(l,k)}(Q).$$

(Check this!)

In what follows, it will be convenient to describe these diffunctors by the following formulas:

$$
\begin{aligned}
\mathrm{Df}^<_{(l,k)} &= \mathrm{Diff}^<_k (D_{(l)} \subset \mathrm{Diff}^>_l) \\
\mathrm{Df}^>_{(l,k)} &= \mathrm{Diff}^>_k (D_{(l)} \subset \mathrm{Diff}^>_l).
\end{aligned}
\tag{17.6}
$$

Accordingly, their values on the module Q are expressed by the formulas

$$
\begin{aligned}
\mathrm{Df}^<_{(l,k)}(Q) &= \mathrm{Diff}^<_k (D_{(l)}(Q) \subset \mathrm{Diff}^>_l(Q)) \\
\mathrm{Df}^>_{(l,k)}(Q) &= \mathrm{Diff}^>_k (D_{(l)}(Q) \subset \mathrm{Diff}^>_l(Q)).
\end{aligned}
\tag{17.7}
$$

Such descriptions are useful in that they explicitly indicate various connections between different diffunctors. For example, we see directly that $\mathrm{Diff}^<_k (D_{(l)}(Q) \subset \mathrm{Diff}^>_l(Q)) \subset \mathrm{Diff}^<_k(D_{(l)}(Q))$, i.e., that $\mathrm{Df}^<_{(l,k)}(Q)$ is a submodule of the module $\mathrm{Diff}^<_k(D_{(l)}(Q))$ and we observe a natural embedding of diffunctors $\mathrm{Df}^<_{(l,k)} \to \mathrm{Diff}^<_k \circ D_{(l)}$. Similarly, the same description shows that $\mathrm{Df}^<_{(l,k)}(Q)$ is not a submodule of the A-module $\mathrm{Diff}^<_k (\mathrm{Diff}^>_l(Q))$, while the natural embedding

$$\mathrm{Df}^<_{(l,k)}(Q) \to \mathrm{Diff}^<_k (\mathrm{Diff}^>_l(Q))$$

is a differential operator. (Find the order of this operator!)

The above can be presented in the form of exact sequences of A-homomorphisms

$$
\begin{aligned}
0 \to \mathrm{Df}^<_{(l,k)}(Q) \to \mathrm{Diff}^\bullet_k(\mathrm{Diff}^>_l(Q)) \xrightarrow{C^<_{(k,l)}(Q)} \mathrm{Diff}^<_{k+l}(Q), \\
0 \to \mathrm{Df}^>_{(l,k)}(Q) \to \mathrm{Diff}^>_k(\mathrm{Diff}^>_l(Q)) \xrightarrow{C^>_{(k,l)}(Q)} \mathrm{Diff}^>_{k+l}(Q),
\end{aligned}
\tag{17.8}
$$

where $\mathrm{Df}^>_{(l,k)}(Q) = \mathrm{Ker}\, C^>_{(k,l)}(Q)$. Since the sequences (17.8) are exact for any A-module Q, we obtain exact sequences of diffunctors

$$
\begin{aligned}
0 \to \mathrm{Df}^<_{(l,k)} \to \mathrm{Diff}^\bullet_k(\mathrm{Diff}^>_l) \xrightarrow{C^<_{(k,l)}} \mathrm{Diff}^<_{k+l}, \\
0 \to \mathrm{Df}^>_{(l,k)} \to \mathrm{Diff}^>_k(\mathrm{Diff}^>_l) \xrightarrow{C^>_{(k,l)}} \mathrm{Diff}^>_{k+l} .
\end{aligned}
\tag{17.9}
$$

17.5. The diffunctors $D_{(l,k)}$ and related chain complexes. The diffunctor $D_{(l,k)}$ can be defined by the formula

$$Q \mapsto D_{(l,k)}(Q) \overset{\mathrm{def}}{=} D_{(k)}(D_{(l)}(Q) \subset \mathrm{Diff}^>_l(Q)), \tag{17.10}$$

which means that here we are considering differential operators $\Delta \colon A \to \mathrm{Diff}^>_l(Q)$ of order $\leq k$ such that $\Delta(1_A) = 0$ and $\mathrm{Im}\,\Delta \subset D_{(l)}(Q)$.

The module $D_{(l,k)}(Q)$, unlike $\mathrm{Df}_{(l,k)}(Q)$, is supplied with only one A-module structure; in it, the multiplication by the element $a \in A$ of the operator $\Delta \colon A \to \mathrm{Diff}_l^{>}(Q)$ is defined by the rule $(a, \Delta) \mapsto a_{D_{(l)}(Q)} \circ \Delta$. Here $a_{D_{(l)}(Q)}$ denotes the operator of multiplication by the element a in $D_{(l)}(Q)$. Since $D_{(l)}(Q)$ is a submodule of the module $\mathrm{Diff}_l^{<}(Q)$, it follows that $a_{D_{(l)}(Q)}$ is the restriction to $D_{(l)}(Q)$ of the operator $a_{\mathrm{Diff}_l^{<}(Q)}$ of multiplication by $a \in A$ of elements of the module $\mathrm{Diff}_l^{<}(Q)$. Therefore, $D_{(l,k)}(Q)$ is a submodule of the module $\mathrm{Df}_{(l,k)}^{<}(Q)$ and thereby we have defined the *embedding* diffunctor $D_{(l,k)} \to \mathrm{Df}_{(l,k)}^{<}$.

Further, let us denote by $D_{(k)}(\mathrm{Diff}_l^{>}(Q))$ the A-module that coincides with $D_{(k)}(\mathrm{Diff}_l^{>}(Q))$ as a vector space over $\Bbbk$ and in which multiplication by an element $a \in A$ is defined by the rule

$$(a, \Delta) \mapsto a_{\mathrm{Diff}_l^{<}(Q)} \circ \Delta.$$

Then, $D_{(k)}(\mathrm{Diff}_l^{>}(Q))$ can be regarded as a submodule of the module $\mathrm{Diff}_k^{\bullet}(\mathrm{Diff}_l^{>}(Q))$. Denote by $C_{(k,l)}^{\circ}(Q)$ the restriction of the gluing homomorphism $C_{(k,l)}^{<}(Q)$ to $D_{(k)}(\mathrm{Diff}_l^{>}(Q))$ and note that, using the same considerations as in the proof of (17.5), we can derive the equality

$$\mathrm{Ker}\, C_{(k,l)}^{\circ}(Q) = D_{(k,l)}(Q). \tag{17.11}$$

Note that $C_{(k,l)}^{\circ}(Q)$ maps $D_{(k)}(\mathrm{Diff}_l^{>}(Q))$ to $D_{(k+l)}^{<}(Q) \subset \mathrm{Diff}_{k+l}^{<}(Q)$. For this reason, the composition

$$D_{(k)}(\mathrm{Diff}_l^{>}(Q)) \xrightarrow{C_{(k,l)}^{\circ}(Q)} \mathrm{Diff}_{k+l}^{>}(Q) \xrightarrow{Д_{k+l}^{<}} Q$$

is trivial (is the map to zero). Now, taking all the above into account, we obtain a sequence of homomorphisms

$$0 \to D_{(l,k)}(Q) \to D_{(k)}(\mathrm{Diff}_l^{>}(Q)) \xrightarrow{C_{(k,l)}^{\circ}(Q)} \mathrm{Diff}_{k+l}^{<}(Q) \xrightarrow{Д_{k+l}^{<}} Q \to 0, \tag{17.12}$$

in which the composition of any two successive homomorphisms is trivial. In other words, the sequence (17.12) is a chain complex. Moreover, the nontrivial homology of this complex can occur only at the term $\mathrm{Diff}_{k+l}^{<}(Q)$. This is an obvious consequence of the fact that the homomorphism $Д_{k+l}^{<}$ is surjective, while $D_{(l,k)}(Q)$ is the kernel of the homomorphism $C_{(k,l)}^{\circ}(Q)$.

Note that the complex (17.12) is the result of applying the sequence

$$0 \to D_{(l,k)} \to D_{(k)} \circ \mathrm{Diff}_l^{>} \xrightarrow{C_{(k,l)}^{\circ}} \mathrm{Diff}_{k+l}^{<} \xrightarrow{Д_{k+l}^{<}} \mathrm{id}_{\mathrm{Mod}\,A} \to 0 \tag{17.13}$$

of natural transformations of diffunctors to the module Q, where $\mathrm{id}_{\mathrm{Mod}\,A}$ is the identity functor on the category $\mathrm{Mod}\,A$. Denote this sequence by $\mathrm{Sp}_{(k,l)}$. Then the complex (17.12) should be denoted by $\mathrm{Sp}_{(l,k)}(Q)$. This notation indicates that we are dealing with the functor $Q \mapsto \mathrm{Sp}_{(l,k)}(Q)$ from the category $\mathrm{Mod}\,A$ to category of chain complexes of A-homomorphisms.

This is an example of a type of diffunctor that differs from those studied previously. One other example of a diffunctor of this type is the sequence $\mathrm{Spen}_{(l,k)}$:

$$0 \to \mathrm{Df}^<_{(l,k)} \to \mathrm{Diff}^<_{(k)}(\mathrm{Diff}^>_l) \xrightarrow{C^<_{(k,l)}} \mathrm{Diff}^<_{k+l} \to 0. \qquad (17.14)$$

From this point of view, the chain complex (17.8) is the value of this diffunctor on the module Q. For this reason, it should denoted by $\mathrm{Spen}_{(l,k)}(Q)$.

Such "complex"-valued diffunctors, as well as other diffunctors, may be related by natural transformations. An example of such a transformation is the transformation $\mathrm{Sp}_{(l,k)} \to \mathrm{Spen}_{(l,k)}$ described by the following diagram:

$$
\begin{array}{ccccccc}
D_{(l,k)} & \longrightarrow & D_{(k)}(\mathrm{Diff}^>_l) & \xrightarrow{C^\circ_{k,l}} & \mathrm{Diff}^<_{k+l} & \xrightarrow{\text{Д}^<_2} & \mathrm{id}_{\mathrm{Mod}\,A} \\
\downarrow & & \downarrow & & \downarrow{\scriptstyle =} & & \downarrow \\
\mathrm{Df}^<_{(l,k)} & \longrightarrow & \mathrm{Diff}^\bullet_k(\mathrm{Diff}^>_l) & \xrightarrow{C^<_{k,l}} & \mathrm{Diff}^<_{k+l} & \longrightarrow & 0,
\end{array}
\qquad (17.15)
$$

in which the two vertical transformation on the left are the embeddings of diffunctors described above.

Exercise. 1. Show that the homology of the complex (17.12) is trivial if and only if the homomorphism $C^<_{(k,l)}(Q)$ is surjective.

2. Let $A = C^\infty(\mathbf{K})$ (see Sections 9.45 and 9.45), $Q = A$ and $k = l = 1$. Compute the homology of the complex (17.12) and ascertain that it is nontrivial.

17.6. Modules, representing the diffunctors $D_{(l,k)}$ and $\mathrm{Df}^<_{(l,k)}$. Now we can describe the modules representing the diffunctors $D_{(l,k)}$ and $\mathrm{Df}^<_{(l,k)}$ in a differentially closed category $\mathfrak{M}$. To do this, it is first necessary to describe the representing module of the diffunctor $D_{(k)}(\mathrm{Diff}^>_l)$. It can be calculated by using the isomorphism (15.10) and Proposition 15.20 in the following way:

$$
\begin{aligned}
D_{(k)}(\mathrm{Diff}^>_l(P)) &= \mathrm{Hom}^<_A(\Lambda^{(k)}_{\mathfrak{M}}, \mathrm{Diff}^>_l(P)) \\
&= \mathrm{Hom}^<_A(\Lambda^{(k)}_{\mathfrak{M}}, \mathrm{Hom}^>_A(\mathcal{J}^l_<(\mathfrak{M}), P)) \\
&= \mathrm{Hom}^<_A({}_<\mathcal{J}^l_>(\mathfrak{M}) \otimes_A \Lambda^{(k)}_{\mathfrak{M}}, P) = \mathrm{Hom}_A(\mathcal{J}^l_<(\Lambda^{(k)}_{\mathfrak{M}}; \mathfrak{M}), P),
\end{aligned}
$$

where the reduced jets $\Lambda^{(k)}_{\mathfrak{M}}$ are defined in Section 15.21. Thus, the module $\mathcal{J}^l_<(\Lambda^{(k)}_{\mathfrak{M}}; \mathfrak{M})$ is the representing object of the diffunctor $D_{(k)}(\mathrm{Diff}^>_l)$ in the category $\mathfrak{M}$.

Further, it follows directly from the proposition in Section 15.2 and formula (15.41), that the homomorphism

$$C^{(k,l)}_{\circ,\mathfrak{M}} \colon \mathcal{J}^{k+l}_<(\mathfrak{M}) \to \mathcal{J}^k_<(\Lambda^{(k)}_{\mathfrak{M}}; \mathfrak{M}), \quad j^{\mathfrak{M}}_{k+l}(a) \mapsto j^{\mathfrak{M}}_l(d_{(k)}a), \qquad (17.16)$$

represents the transformation of diffunctors $C^{\circ}_{(k,l)}$ in the category $\mathfrak{M}$. Besides, from the exact sequence of diffunctors

$$0 \to D_{(l,k)} \to D_{(k)} \circ \mathrm{Diff}_l^{>} \xrightarrow{C^{\circ}_{(k.l)}} \mathrm{Diff}_{k+l}^{<}$$

and from the corollary in Section 15.2, it follows that the representing module for the diffunctor $\mathrm{Df}_{(l,k)}^{<}$ exists and is isomorphic to $\mathrm{coker}\, C^{(k,l)}_{\circ,\mathfrak{M}}$. Denote it by $\Lambda^{(l,k)}_{\mathfrak{M}}$. Then

$$0 \leftarrow \Lambda^{(l,k)}_{\mathfrak{M}} \xleftarrow{\pi_{(l.k)}} \mathcal{J}^l_{<}(\Lambda^{(k)}_{\mathfrak{M}}; \mathfrak{M}) \xleftarrow{C^{(k.l)}_{\circ.\mathfrak{M}}} \mathcal{J}^{k+l}_{<}(A; \mathfrak{M}) \xleftarrow{i_{k+l}} A \leftarrow 0, \qquad (17.17)$$

where $i_{k+l}(a) = a j_{k+l}(1_A)$, is the cochain complex representing the complex of diffunctors (17.13).

Similarly, from the exact sequence of diffunctors

$$0 \to \mathrm{Df}_{(l,k)}^{<} \to \mathrm{Diff}_k^{\bullet}(\mathrm{Diff}_l^{>}) \xrightarrow{C_{(k.l)}} \mathrm{Diff}_{k+l}^{<}$$

and the corollary in Section 15.2, it follows that the module that represents the diffunctor $\mathrm{Df}_{(l,k)}^{<}$ exists and can be identified with $\mathrm{coker}\, C^{(k,l)}_{\mathfrak{M}}$, where

$$C^{(k,l)}_{\mathfrak{M}} \colon \mathcal{J}^{k+l}_{<}(\mathfrak{M}) \to \mathcal{J}^l_{<}(\mathcal{J}^k_{<}(\mathfrak{M}); \mathfrak{M}), \quad j^{\mathfrak{M}}_{k+l}(a) \mapsto j^{\mathfrak{M}}_l(j_k a),$$

is the homomorphism representing the transformation $C^{<}_{(k,l)}$ in the category $\mathfrak{M}$. This module will be denoted by $\mathfrak{L}^{(l,k)}_{\mathfrak{M}}$. Further, let us note that the natural embedding of diffunctors $D_{(l,k)} \to \mathrm{Df}_{(l,k)}^{<}$ induces a *natural* epimorphism $\mathfrak{L}^{(l,k)}_{\mathfrak{M}} \to \Lambda^{(l,k)}_{\mathfrak{M}}$.

Finally, comparing the diagrams given below, we can begin to understand why the diffunctors $D_{(l,k)}$ and $\mathrm{Df}_{(l,k)}$ exist "in nature".

$$\Lambda^{(l,k)}_{\mathfrak{M}} \xleftarrow{\pi_{(l,k)}} \mathcal{J}^l_{<}(\Lambda^{(k)}_{\mathfrak{M}}; \mathfrak{M}) \qquad \Lambda^2(M) \xleftarrow{\pi_{(1.1)}} \mathcal{J}^1_{<}(\Lambda^1(M)) \qquad (17.18)$$

The diagram on the right is a particular case of the one on the left when $k = l = 1$ and $\mathfrak{M} = \mathrm{GMod}\, C^{\infty}(M)$. On the other hand, it is one of the possible definitions of differential forms of second degree on a manifold M and of the standard exterior derivation d acting on differential forms of first degree. Why this is so will be explained below. Thus, elements of the module $\Lambda^{(l,k)}_{\mathfrak{M}}$ are the *higher analogs of differential forms of second degree*.

Exercise. The formula (17.10) defining the diffunctor $D_{(k,l)}$ may also be regarded as the definition of the embedding $i_{(l,k)} \colon D_{(l,k)} \to D_{(k)} \circ D_{(l)}$. Show that $\Lambda^{(l)}_{\mathfrak{M}} \otimes_A \Lambda^{(k)}_{\mathfrak{M}}$ is the representing module of the diffunctor

$D_{(k)} \circ D_{(l)}$ in the category $\mathfrak{M}$ and describe the homomorphism $\Lambda_{\mathfrak{M}}^{(l)} \otimes_A$ $\Lambda_{\mathfrak{M}}^{(k)} \to \Lambda_{\mathfrak{M}}^{(k,l)}$ representing the transformations $i_{(l,k)}$.

17.7. The diffunctor D_2 and the meaning of anti-commutativity.
In order to clarify the not too obvious geometrical meaning of the diffunctors $D_{(l,k)}$, let us consider the simplest case $k = l = 1$. In what follows, it will be convenient to simplify the notation somewhat by putting

$$D_{(1,1)} = D_2 \quad \Leftrightarrow \quad D_2 = D(D \subset \mathrm{Diff}_1^{<}).$$

If $\Delta \colon A \to \mathrm{Diff}_1^{<}(P)$ is a first-order differential operator, then

$$\Delta(a) \in \mathrm{Diff}_1^{<}(P) \quad \text{and} \quad (\Delta(a))(b) \in P, \quad a, b \in A.$$

Let us put $\langle a, b \rangle = (\Delta(a))(b)$. If $\Delta \in D(\mathrm{Diff}_1^{<}(P))$, then

$$\Delta(ab) = \Delta(a) \circ b + \Delta(b) \circ a.$$

If, moreover, $\Delta \in D_2(P)$, then $\Delta(a') \in D(P)$ for all $a' \in A$. This means that $(\Delta(ab))(1_A) = 0$ or, equivalently, that

$$(\Delta(a))(b) + (\Delta(b))(a) = \langle a, b \rangle + \langle b, a \rangle = 0.$$

Thus, the form $\langle a, b \rangle$ is skew-symmetric. Besides, it is obviously $\Bbbk$-bilinear and is a *biderivation*. This means that

$$\langle ab, c \rangle = a\langle b, c \rangle + b\langle a, c \rangle, \qquad \langle a, bc \rangle = b\langle a, c \rangle + c\langle a, b \rangle.$$

Indeed, the first of these equalities follows from the second one in view of the skew-symmetry of the form $\langle \cdot, \cdot \rangle$, while the second one, because of the inclusion $\Delta(a) \in D(P)$, can be obtained by the following calculation:

$$\langle a, bc \rangle = (\Delta(a))(bc) = b(\Delta(a))(c) + c(\Delta(a))(b) = b\langle a, c \rangle + c\langle a, b \rangle.$$

17.8 Lemma–Exercise. *Let $(a, b) \mapsto \langle a, b \rangle$, $a, b \in A$, be a $\Bbbk$-bilinear, skew-symmetric biderivation with values in P. Show that the operator*

$$\Delta \colon A \to D(D(P)), \qquad \Delta(a) \stackrel{\mathrm{def}}{=} \langle a, \cdot \rangle,$$

belongs to $D_2(P)$.

Note that the type of biderivation indicated in the lemma naturally constitutes an A-module, while the lemma itself, together with the previous constructions, establishes an isomorphism between this module and $D_2(P)$. This isomorphism clarifies the hidden meaning of the skew-symmetry of the biderivation $\langle \cdot, \cdot \rangle$ corresponding to the derivation

$$\Delta \in D(D \subset \mathrm{Diff}_1^{<}(P)) = D_2(P).$$

It consists in that Δ is a derivation with respect to the right as well as the left A-module structure of the A-bimodule $\mathrm{Diff}_1(P)$.

17.9. Bivectors. In the case when $A = P = C^\infty(M)$, we will simplify the notation, putting $D_2(M) \stackrel{\text{def}}{=} D_2(C^\infty(M))$. Elements of the $C^\infty(M)$-module $D_2(M)$ have a direct differential-geometrical interpretation. Namely, they can be naturally identified with bivector fields on the manifold M. Indeed, a bivector field on M, according to the accepted coordinate-free definition of tensors, is a $C^\infty(M)$-bilinear skew-symmetric form, defined on differential forms of first degree and taking values in the algebra $C^\infty(M)$. Let

$$(\omega_1, \omega_2) \mapsto \Theta(\omega_1, \omega_2), \quad \omega_1, \omega_2 \in \Lambda^1(M),$$

be such a form. Then the form

$$(f, g) \mapsto \langle f, g \rangle_\Theta \stackrel{\text{def}}{=} \Theta(df, dg), \quad f, g \in C^\infty(M),$$

is a skew-symmetric biderivation. Conversely, to any biderivation $\langle \cdot, \cdot \rangle$ of the type considered above there corresponds a bivector field, defined by the formula

$$\Theta\left(\sum_i f_i dg_i, \sum_j f_j' dg_j' \right) \stackrel{\text{def}}{=} \sum_{i,j} f_i f_j' \langle g_i, g_j' \rangle,$$

where $f_i, g_i, f_j', g_j' \in C^\infty(M)$. That this is well defined easily follows from the fact that the value of the function $\langle f', g' \rangle$ at some point $z \in M$ depends only on $d_z f'$ and $d_z g'$. In its turn, this follows from the fact that we have $\langle f', g' \rangle(z) = 0$ if $d_z g' = 0$. Since the operation $f \mapsto \langle f', f \rangle$ is a derivation in the algebra $C^\infty(M)$, it represents some vector field X on M. Therefore, $\langle f', g' \rangle(z) = \big(X(g') \big)(z) = X_z(g') = d_z g'(X_z) = 0$, as required.

Thus we see that elements of the module $D_2(P)$ may be represented as skew-symmetric biderivations with values in P and, in particular, as contravariant skew-symmetry tensors of second order, or bivectors, on the manifold M if $P = C^\infty(M)$.

In the case under consideration, the chain complex (17.12) acquires the form

$$0 \to D_2(P) \to D^<(\text{Diff}_1^> P) \xrightarrow{C_{1.1}^<} \text{Diff}_2^<(P) \xrightarrow{Д_2} P \to 0 \qquad (17.19)$$

and is called the *Spencer* Diff-*complex of second order of the module* P.

Exercise. Let π be a vector bundle over the manifold M and $P = \Gamma(\pi)$.
1. Express $\Delta \in D_2(P)$ in coordinates.
2. Show that the homology of the complex (17.19) is trivial.

17.10. Differential 2-forms. As we have already seen (Section 17.6), the representing object for the diffunctor $D_2 = D_{(1,1)}$ exists and is denoted by $\Lambda_{\mathfrak{M}}^{(1,1)}$. In what follows, it will be convenient to replace this notation by $\Lambda_{\mathfrak{M}}^2$ or by $\Lambda^2(M)$ when $A = C^\infty(M)$ and $\mathfrak{M} = \text{GMod}\, C^\infty(M)$. It is necessary to specify the complex (17.17) in order to prove the existence of the representing module $\Lambda_{\mathfrak{M}}^2$:

$$0 \leftarrow \Lambda_{\mathfrak{M}}^2 \xleftarrow{S_{1.1}} \mathcal{J}_<^1(\Lambda_{\mathfrak{M}}^{(1)}; \mathfrak{M}) \xleftarrow{S_{2.0}} \mathcal{J}_<^2(\mathfrak{M}) \xleftarrow{i_2} A \leftarrow 0, \qquad (17.20)$$

where we have put $S_{2,0} = C^{1,1}_{\circ,\mathfrak{M}}$ and $S_{1,1} = \pi_{(1,1)}$ (see (17.16)). Here we have $\Lambda^2_\mathfrak{M} \overset{\text{def}}{=} \operatorname{coker} S_{2,0}$ and $S_{2,0}(j^\mathfrak{M}_2 a) = j^\mathfrak{M}_1(da)$.

Below, we shall present a more convenient description of the module $\Lambda^2_\mathfrak{M}$ and, in particular, we will show that $\Lambda^2(M)$ consists of second-order differential forms on the manifold M in the sense as they are understood in classical differential geometry.

17.11. The exterior product. First of all, note that the map

$$\operatorname{Diff}^{>}_1 P \ni \Delta \mapsto \Delta - \Delta(1) \in D(P)$$

is an A-homomorphism. Thus we obtain a natural transformation of diffunctors $\operatorname{Diff}^{>}_1 \to D$. Taking its composition from the left with the diffunctor D, we arrive at the transformation

$$D^{<}(\operatorname{Diff}^{>}_1) \overset{\partial}{\longrightarrow} D(D) = D \circ D. \tag{17.21}$$

In more detail, its value on the module P can be described as follows. Let $\Delta \in D^{<}(\operatorname{Diff}^{>}_1(P))$. Then

$$\Delta(a) \in \operatorname{Diff}^{>}_1(P), \ a \in A, \quad \text{and} \quad \Delta(a) - \big(\Delta(a)\big)(1) \in D(P).$$

Therefore,

$$\partial(\Delta) \colon a \mapsto \Delta(a) - \Delta(a)(1). \tag{17.22}$$

Accordingly, the homomorphism

$$\partial^*_\mathfrak{M} \colon \Lambda^1_\mathfrak{M} \otimes_A \Lambda^1_\mathfrak{M} \to \mathcal{J}^1_{<}(\Lambda^1_\mathfrak{M}; \mathfrak{M}),$$

representing the transformation ∂ in the category $\mathfrak{M}$, can be described, according to Section 15.2 and (17.22), by the following formula:

$$\partial^*_\mathfrak{M}(d^\mathfrak{M}a' \otimes_A d^\mathfrak{M}a) = [j^\mathfrak{M}_1(a') - a'j^\mathfrak{M}_1(1_A)) \otimes^{>}_A d^\mathfrak{M}a . \tag{17.23}$$

Note that

$$\big(j^\mathfrak{M}_1(a') - a'j^\mathfrak{M}_1(1_A)\big) \otimes^{>}_A d^\mathfrak{M}a$$
$$= \big(j^\mathfrak{M}_1(1_A) \otimes^{>}_A a'd^\mathfrak{M}a\big) - a'j^\mathfrak{M}_1(1_A) \otimes^{>}_A d^\mathfrak{M}a$$
$$= j^\mathfrak{M}_1(a'd^\mathfrak{M}a) - a'j^\mathfrak{M}_1(d^\mathfrak{M}a). \tag{17.24}$$

Having in mind (17.23), as well as the fact that the module $\operatorname{Im} S_{2,0}$ is generated by elements of the form $j^\mathfrak{M}_1(d^\mathfrak{M}b)$, $b \in A$, we see that

$$\partial^*_\mathfrak{M}(d^\mathfrak{M}a' \otimes_A d^\mathfrak{M}a) + \partial^*_\mathfrak{M}(d^\mathfrak{M}a \otimes_A d^\mathfrak{M}a')$$
$$= j^\mathfrak{M}_1(d^\mathfrak{M}aa') - a'j^\mathfrak{M}_1(d^\mathfrak{M}a) - aj^\mathfrak{M}_1(d^\mathfrak{M}a') \in \operatorname{Im} S_{2,0}. \tag{17.25}$$

Therefore, $(S_{1,1} \circ \partial^*_\mathfrak{M})(d^\mathfrak{M}a' \otimes_A d^\mathfrak{M}a + d^\mathfrak{M}a \otimes_A d^\mathfrak{M}a') = 0$. On the other hand, the representing homomorphism of the composition of diffunctors

$$w_{1,1} \colon D_2 \to D^{<}(\operatorname{Diff}^{<}_1) \overset{\partial}{\longrightarrow} D(D)$$

is $(w_{1,1})^*_{\mathfrak{M}} = S_{1,1} \circ \mathfrak{d}^*_{\mathfrak{M}}$. Since $w_{1,1}$ is an embedding of diffunctors, the homomorphism

$$(w_{1,1})^*_{\mathfrak{M}} \colon \Lambda^1_{\mathfrak{M}} \otimes_A \Lambda^1_{\mathfrak{M}} \to \Lambda^2_{\mathfrak{M}}$$

is surjective and elements of the form $d^{\mathfrak{M}}a' \otimes_A d^{\mathfrak{M}}a + d^{\mathfrak{M}}a \otimes_A d^{\mathfrak{M}}a'$ generate its kernel. In other words, this kernel consists of the symmetric part of the tensor product $\Lambda^1_{\mathfrak{M}} \otimes_A \Lambda^1_{\mathfrak{M}}$, so that

$$\Lambda^2_{\mathfrak{M}} = \Lambda^1_{\mathfrak{M}} \otimes_A \Lambda^1_{\mathfrak{M}} / \Lambda^1_{\mathfrak{M}} \odot_A \Lambda^1_{\mathfrak{M}} = \Lambda^1_{\mathfrak{M}} \wedge_A \Lambda^1_{\mathfrak{M}}.$$

But the last equality is the standard definition of the *exterior square* of the module $\Lambda^1_{\mathfrak{M}}$. Now, since the exterior square coincides with

$$(w_{1,1})^*_{\mathfrak{M}} \colon \Lambda^1_{\mathfrak{M}} \otimes_A \Lambda^1_{\mathfrak{M}} \to \Lambda^2_{\mathfrak{M}} = \Lambda^1_{\mathfrak{M}} \wedge_A \Lambda^1_{\mathfrak{M}}, \qquad (17.26)$$

we come to the *conceptual definition* of the exterior product of 1-forms.

Definition. The *exterior product of first degree differential forms* is defined as the homomorphism $(w_{1,1})^*_{\mathfrak{M}}$, representing the transformation of diffunctors $w_{1,1} \colon D_2 \to D(D)$. In other words,

$$\omega_1 \wedge \omega_2 \overset{\text{def}}{=} (w_{1,1})^*_{\mathfrak{M}}(\omega_1 \otimes_A \omega_2), \quad \omega_1, \omega_2 \in \Lambda^1_{\mathfrak{M}}. \qquad (17.27)$$

Remark. At this point, the reader may want to ask what all these complications are needed when the standard descriptive definition is so simple? The answer is that the conceptual definition automatically carries over to various nonstandard situations, whereas it is difficult, if not impossible, to guess the correct generalization of the descriptive definition. Situations of this type arise, for example, in the differential calculus over graded commutative algebras, which will be studied in the concluding chapter of this book.

17.12. The exterior differential. We shall define the *exterior differential* acting on 1-forms by putting (see 17.18)

$$d^{\mathfrak{M}}_1 \overset{\text{def}}{=} d^{\mathfrak{M}}_{(1,1)} = S_{1,1} \circ j^{\mathfrak{M}}_1 \colon \Lambda^1_{\mathfrak{M}} \to \Lambda^2_{\mathfrak{M}}.$$

Having in mind (17.23), note that

$$
\begin{aligned}
j^{\mathfrak{M}}_1(a'd^{\mathfrak{M}}a) &= j^{\mathfrak{M}}_1(a') \otimes^>_A d^{\mathfrak{M}}a \\
&= \mathfrak{d}^*_{\mathfrak{M}}(d^{\mathfrak{M}}a' \otimes_A d^{\mathfrak{M}}a) + a'j^{\mathfrak{M}}_1(1_A) \otimes^>_A d^{\mathfrak{M}}a \\
&= \mathfrak{d}^*_{\mathfrak{M}}(d^{\mathfrak{M}}a' \otimes_A d^{\mathfrak{M}}a) + a'j^{\mathfrak{M}}_1(d^{\mathfrak{M}}a) \\
&= \mathfrak{d}^*_{\mathfrak{M}}(d^{\mathfrak{M}}a' \otimes_A d^{\mathfrak{M}}a) \mod (\operatorname{Im} S_{2,0}).
\end{aligned}
$$

Therefore, since $\operatorname{Im} S_{2,0} = \operatorname{Ker} S_{1,1}$, it follows that

$$
\begin{aligned}
d^{\mathfrak{M}}_1(a'd^{\mathfrak{M}}a) &= S_{2,0}(j^{\mathfrak{M}}_1(a'd^{\mathfrak{M}}a)) = S_{2,0}(\mathfrak{d}^*_{\mathfrak{M}}(d^{\mathfrak{M}}a' \otimes_A d^{\mathfrak{M}}a)) \\
&= (w_{1,1})^*_{\mathfrak{M}}(d^{\mathfrak{M}}a' \otimes_A d^{\mathfrak{M}}a) = d^{\mathfrak{M}}a' \wedge_A d^{\mathfrak{M}}a.
\end{aligned}
$$

Thus, we have established the analog of the well-known property of the exterior differential:

$$d_1^{\mathfrak{M}}(a'd^{\mathfrak{M}}a) = d^{\mathfrak{M}}a' \wedge_A d^{\mathfrak{M}}a. \qquad (17.28)$$

In particular, this implies that $d_1^{\mathfrak{M}} \circ d^{\mathfrak{M}} = 0$.

Exercise.   1. Show that the derivation

$$\eth_2 \colon a \mapsto d^{\mathfrak{M}}a \wedge d^{\mathfrak{M}}, \quad \text{where} \quad (d^{\mathfrak{M}}a \wedge d^{\mathfrak{M}})(b) \overset{\text{def}}{=} d^{\mathfrak{M}}a \wedge d^{\mathfrak{M}}b,$$

is the element of the module $D_2(\Lambda_{\mathfrak{M}}^2) = \mathrm{Hom}_A(\Lambda_{\mathfrak{M}}^2, \Lambda_{\mathfrak{M}}^2)$, which corresponds to the identity homomorphism $\mathrm{id}_{\Lambda_{\mathfrak{M}}^2}$. Accordingly, the biderivation

$$(a,b) \mapsto d^{\mathfrak{M}}a \wedge d^{\mathfrak{M}}b$$

is *universal* in the sense that for any $\Delta \in D_2(P)$, $P \in \mathrm{Ob}\,\mathfrak{M}$, there exists a unique homomorphism of A-modules $h \colon \Lambda_{\mathfrak{M}}^2 \to P$ such that

$$\big(\Delta(a)\big)(b) = h(d^{\mathfrak{M}}a \wedge d^{\mathfrak{M}}b).$$

  2. Describe the module of differential 2-form on the "cross", i.e., when $A = C^\infty(\mathbf{K})$ and $\mathfrak{M} = \mathrm{GMod}\,C^\infty(\mathbf{K})$.

17.13. The Spencer second-order jet complex. The results of the two previous sections can be reduced to the following commutative diagram.

$$0 \longleftarrow \Lambda_{\mathfrak{M}}^2 \overset{S_{1.1}}{\longleftarrow} \mathcal{J}_<^1(\Lambda_{\mathfrak{M}}^1; \mathfrak{M}) \overset{S_{2.0}}{\longleftarrow} \mathcal{J}_<^2(\mathfrak{M}) \overset{i_2}{\longleftarrow} A \longleftarrow 0 \qquad (17.29)$$

with vertical maps $= \! \uparrow j_0^{\mathfrak{M}}$, $\uparrow j_1^{\mathfrak{M}}$, $\uparrow j_2^{\mathfrak{M}}$, $\uparrow$

$$\Lambda_{\mathfrak{M}}^2 \overset{d_1^{\mathfrak{M}}}{\longleftarrow} \Lambda_{\mathfrak{M}}^1 \overset{d^{\mathfrak{M}}}{\longleftarrow} A \longleftarrow 0$$

By definition, its upper row is called the *Spencer jet complex of second order* of the category $\mathfrak{M}$. It can be nonexact only at the term $\mathcal{J}_<^2(\mathfrak{M})$. Its lower row is the initial fragment of the *de Rham complex* of the same category. This terminology is justified by the following statement.

Proposition. *If $A = C^\infty(M)$ and $\mathfrak{M} = \mathrm{GMod}\,C^\infty(M)$, then the diagram (17.29) becomes the same diagram, but consisting of jets and differential forms in the sense of standard differential geometry.*

◀ Since $\Lambda^2(M) = \Lambda^1(M) \wedge \Lambda^1(M)$ and a similar fact also holds for $\Lambda_{\mathfrak{M}}^2$, the identity of the modules $\Lambda^2(M)$ and $\Lambda_{\mathrm{GMod}\,C^\infty(M)}^2$ is a consequence of the previously established identity of the modules $\Lambda^1(M)$ and $\Lambda_{\mathrm{GMod}\,C^\infty(M)}^1$. Similarly, from the fact that the differentials d and $d^{\mathrm{GMod}\,C^\infty(M)}$ coincide, and from formulas (17.28), it follows that the standard exterior differentials d_1 on M and $d_1^{\mathrm{GMod}\,C^\infty(M)}$ coincide. The coincidence of the other elements of the diagram (17.29) with the standard ones was established earlier.   ▶

17.14. Substitution operations. All the known notions and operations of modern differential geometry that have to do with differential forms, as well as with all its other objects, are due in some way to various relations between diffunctors. For example, we have seen above that this is how the multiplicative structure in jets, the exterior differential on 1-forms, and other such objects arise. Moreover, in this way, we have discovered some new (not known previously) objects $\Lambda_{\mathfrak{M}}^{(k,l)}$ and the *natural* differential operators $d_{(k,l)}$. We shall illustrate this fact once again by demonstrating the substitution operation of vector fields into differential forms.

To any derivation $X \in D(A)$, we can associate the transformation of diffunctors $i^X \colon D \to D_2$. Its value $i_P^X \colon D(P) \to D_2 P$ on the module P can be conveniently defined by means of the biderivation $\langle \cdot, \cdot \rangle_Z$, $Z \in D(P)$, corresponding to $i_P^X(Z)$ (Lemma 17.8):

$$\left(i_P^X(Z) \right)(a,b) = \langle a,b \rangle_Z \overset{\text{def}}{=} X(a)Z(b) - X(b)Z(a), \quad a,b \in A, \quad Z \in D(P).$$

Now let $i_X \colon \Lambda_{\mathfrak{M}}^2 \to \Lambda_{\mathfrak{M}}^1$ be the homomorphism representing the transformation of diffunctors i^X and let $\omega \in \Lambda_{\mathfrak{M}}^2$. Then the 1-form $i_X(\omega) \in \Lambda_{\mathfrak{M}}^1$ will be called the *substitution of the derivation X into the 2-form ω*. In explicit notation, the substitution operation is expressed by the formula

$$i_X(a' d^{\mathfrak{M}} a \wedge d^{\mathfrak{M}} b) = a'(X(a) d^{\mathfrak{M}} b - X(b) d^{\mathfrak{M}} a) \qquad (17.30)$$

which is a consequence of Exercise 1 from Section 17.12.

17.15. The diffunctors D_σ and Df_σ. The pairs of diffunctors

$$(D_{(k)}, \mathrm{Df}_{(k)} \overset{\text{def}}{=} \mathrm{Diff}_k) \text{ and } (D_{(k,l)}, \mathrm{Df}_{(k,l)})$$

constructed earlier are the first terms of an infinite sequence of similar diffunctors. These new pairs of diffunctors correspond to sequences $\sigma = (\sigma_1, \ldots, \sigma_m)$ of positive integers and are denoted by $(D_\sigma, \mathrm{Df}_\sigma)$. Here the values of the diffunctor Df_σ are A-bimodules, while the symbols $\mathrm{Df}_\sigma^<(P)$ and $\mathrm{Df}_\sigma^>(P)$ denote the "left" and "right" A-modules of the bimodule $\mathrm{Df}_\sigma(P)$. Thus the diffunctors $\mathrm{Df}_\sigma^<$ and $\mathrm{Df}_\sigma^>$ are taken into consideration. Here we have the embedding $D_\sigma \subset \mathrm{Df}_\sigma^<$.

The pair $(D_\sigma, \mathrm{Df}_\sigma)$ is constructed inductively according to the scheme used above to define the functors $D_{(k,l)}$ and $\mathrm{Df}_{(k,l)}$. Namely, let

$$(\sigma, k) = (\sigma_1, \ldots, \sigma_m, k).$$

Now we can define

$$\mathrm{Df}_{(\sigma,k)}^< \overset{\text{def}}{=} \mathrm{Diff}_k^< (D_\sigma \subset \mathrm{Df}_\sigma^>), \quad \mathrm{Df}_{(\sigma,k)}^> \overset{\text{def}}{=} \mathrm{Diff}_k^> (D_\sigma \subset \mathrm{Df}_\sigma^>)$$

$$D_{(\sigma,k)} \overset{\text{def}}{=} D_{(k)} (D_\sigma \subset \mathrm{Df}_\sigma^>). \qquad (17.31)$$

The meaning of these formulas is exactly the same as that of (17.10), (17.6). It is easy to verify that the pair of diffunctors defined in this way satisfies the conditions formulated above, which were needed to continue the inductive procedure. (Check this!)

The construction (17.31) may be generalized by defining the diffunctors denoted by the symbol $\Phi_\tau(D_\sigma \subset \mathrm{Df}_\sigma^>)$, where Φ_τ is one of the diffunctors $D_\tau, \mathrm{Df}_\tau^\leq$ or $\mathrm{Df}_\tau^>$, while $\tau = (\tau_1, \ldots, \tau_l)$. The meaning of this symbol becomes clear by induction on the length l of the sequence τ, assuming that σ is fixed. For $l = 1$, this symbol coincides with $\Phi_{(\sigma,k)}$ provided $\tau = (k)$. The inductive step is described by following formula.

$$\Phi_{(k,\tau)}(D_\sigma \subset \mathrm{Df}_\sigma^>) \overset{\text{def}}{=} \Phi_{(k)}(D_\tau(D_\sigma \subset \mathrm{Df}_\sigma^>) \subset \mathrm{Df}_\tau^>(D_\sigma \subset \mathrm{Df}_\sigma^>)).$$

The existence of the diffunctors $\Phi_\tau(D_\sigma \subset \mathrm{Df}_\sigma^>)$, in particular, allows to define the embedding of diffunctors

$$D_{(\sigma,k)} = D_\sigma(D_{(k)} \subset \mathrm{Diff}_k^>) \to D_\sigma^<(\mathrm{Diff}_k^>), \qquad (17.32)$$

which, according to the scheme 15.28, leads to the important natural differential operator that we shall discuss in the next section.

17.16. The modules representing D_σ and Df_σ. The representing objects for the diffunctors D_σ and Df_σ in any differentially closed category $\mathfrak{M}$ exist and are denoted by $\Lambda_\mathfrak{M}^\sigma$ and $\mathfrak{L}_\mathfrak{M}^\sigma$, respectively. This is proved in essentially the same way (see above) as for the diffunctors $D_{(l,k)}$ and $\mathrm{Df}_{(l,k)}$. However, the proof requires some new technical details that lie outside the scope of this book. Therefore, the existence of these objects will be proved later in a simpler (in this respect) particular case, when the sequence σ consists only of ones. In the framework of this approach, we shall obtain classical differential forms on the manifold M together with the exterior differential and all the rest, choosing, for the main category $\mathfrak{M}$, the category $\mathrm{GMod}\, C^\infty(M)$.

On the other hand, the definition of exterior differential itself and its higher analogs is not complicated at all, provided we assume that the existence of the representing modules $\Lambda_\mathfrak{M}^\sigma$ has been established. Indeed, the embedding of diffunctors (17.32) yields the following generalization of the diagram (17.18):

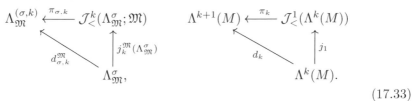

$$(17.33)$$

The homomorphism $\pi_{\sigma,k}$ in the left diagram is representing for the embedding of diffunctors (17.32). In its turn, it allows to define the differential

$$d_{\sigma,k}^\mathfrak{M} \overset{\text{def}}{=} \pi_{\sigma,k} \circ j_k^\mathfrak{M}(\Lambda_\mathfrak{M}^\sigma),$$

which is a differential operator of order k that can be naturally regarded as the *higher analog of the exterior differential*. This is motivated by the fact that the right diagram, by definition, is a particular case of the left one, namely when both sequences σ and (σ, k) consist only of ones and $\mathfrak{M} = \mathrm{GMod}\, C^\infty(M)$. In the diagram, we used one of the standard notations for differential forms and exterior differentials on the manifold M in order to stress that the objects obtained are actually well-known ones. This will be verified later.

Concluding this section, it remains to note that the appropriate composition of the differentials $d_\sigma^{\mathfrak{M}}$ yield the de Rham cochain complex and its higher analogs. Namely, let $\sigma = (\sigma_1, \sigma_2, \dots, \sigma_k, \dots)$ be an infinite sequence of positive integers and $\sigma(l) = (\sigma_1, \sigma_2, \dots, \sigma_l)$. Let us put

$$d_{\sigma,k}^{\mathfrak{M}} \overset{\text{def}}{=} d_{\sigma(k+1),\sigma(k)}^{\mathfrak{M}} \colon \Lambda_{\mathfrak{M}}^{\sigma(k)} \to \Lambda_{\mathfrak{M}}^{\sigma(k+1)}. \tag{17.34}$$

The sequence corresponding to the pair $(\sigma, \mathfrak{M})$

$$0 \to A \xrightarrow{d_{\sigma,0}^{\mathfrak{M}}} \Lambda_{\mathfrak{M}}^{\sigma(1)} \xrightarrow{d_{\sigma,1}^{\mathfrak{M}}} \Lambda_{\mathfrak{M}}^{\sigma(2)} \xrightarrow{d_{\sigma,2}^{\mathfrak{M}}} \cdots \xrightarrow{d_{\sigma,k-1}^{\mathfrak{M}}} \Lambda_{\mathfrak{M}}^{\sigma(k)} \xrightarrow{d_{\sigma,k}^{\mathfrak{M}}} \cdots, \tag{17.35}$$

denoted by $\mathrm{dF}_\sigma^{\mathfrak{M}}$, is a cochain complex. Below this will be proved for sequences $\sigma = (1, 1, \dots, 1, \dots)$. Moreover, we shall prove that this complex is identical to the classical de Rham cochain complex on the manifold M in the case $\mathfrak{M} = \mathrm{GMod}\, C^\infty(M)$.

Note that if a (finite) sequence τ majorizes a finite sequence σ, then we have an embedding of diffunctors $D_\sigma \to D_\tau$ and $\mathrm{Df}_\sigma \to \mathrm{Df}_\tau$. This is easily verified by using the inductive definition (17.31) and the obvious embeddings $D_k \to D_l$ and $\mathrm{Diff}_k \to \mathrm{Diff}_l$, which exist if $k \leq l$. These embeddings are represented by certain surjective homomorphism $\Lambda_\tau^{\mathfrak{M}} \to \Lambda_\sigma^{\mathfrak{M}}$, $\mathfrak{L}_\tau^{\mathfrak{M}} \to \mathfrak{L}_\sigma^{\mathfrak{M}}$. If an infinite sequence τ majorizes a similar sequence σ, then, obviously, $\tau(m)$ majorizes $\sigma(m)$ for any m. In this case, the family of representing homomorphisms $\Lambda_{\tau(m)}^{\mathfrak{M}} \to \Lambda_{\sigma(m)}^{\mathfrak{M}}$ defines the surjective homomorphism of complexes $\mathrm{dF}_\tau^{\mathfrak{M}} \to \mathrm{dF}_\sigma^{\mathfrak{M}}$. In its turn, the inverse limit $\mathrm{dF}_\infty^{\mathfrak{M}}$ of this system of homomorphisms is the *infinite analog of the de Rham complex*. All of this shows, once again, how little we know of the "universe" of natural structures of the differential calculus, in particular, of differential geometry.

17.17. The diffunctors D_k. Now let us consider in more detail the diffunctors D_σ and Df_σ for the sequences $\sigma^k \overset{\text{def}}{=} (1, 1, \dots, 1)$ (k times) and let us simplify the notation by putting $\mathrm{Df}_k = \mathrm{Df}_{\sigma^k}$, $D_k = D_{\sigma^k}$ (do not confuse D_k with $D_{(k)}$!). The inductive definition (17.31) in this notation has the following form:

$$D_{k+1} = D(D_k \subset \mathrm{Df}_k^{>}), \quad \mathrm{Df}_{k+1} = \mathrm{Diff}_1(D_k \subset \mathrm{Df}_k^{>}). \tag{17.36}$$

Recall that $\mathrm{Df}_{k+1}(P)$ is an A-bimodule and its left and right A-module structures are denoted by $\mathrm{Df}_{k+1}^{<}(P)$ and $\mathrm{Df}_{k+1}^{>}(P)$, respectively. More precisely, if the operator $\Delta \colon A \to \mathrm{Df}_k^{>}(P)$ belongs to $\mathrm{Df}_{k+1}(P)$, then the result of its multiplication from the *left* by $a \in A$ is the operator $a_{\mathrm{Df}_k^{<}(P)} \circ \Delta$ or, which is the same, the operator $a_{D_k(P)} \circ \Delta$. The result of its multiplication from the *right* by $a \in A$ is the operator $\Delta \circ a_A$.

Let Q be an A-bimodule, while $Q_<$ and $Q_>$ are the corresponding left and right A-modules. The operators $I^< \colon Q_< \to Q_>$ and $I^> \colon Q_> \to Q_<$, which are, as maps of sets, identity maps, are called *change of structure* operators. Recall also that if $p \in P$, then the symbol p_A denotes the homomorphism $A \to P$, defined by the rule $p_A(a) = ap$.

Lemma. *For any $k \geq 0$:*

(1) *The embedding $D_{k+1}(P) \to \mathrm{Df}^<_{k+1}(P)$ is a homomorphism of A-modules.*

(2) *The change of structure operators in the A-bimodule $\mathrm{Df}_{k+1}(P)$ are differential operators of first order.*

(3) *The projection*

$$\mathrm{Df}^>_{k+1}(P) \ni \Delta \xrightarrow{\Pi^>} \Delta - \Delta(1) \in D_{k+1}(P)$$

is a homomorphism of A-modules.

(4) *If $\Delta \in \mathrm{Df}_{k+1}(P)$, then the decomposition*

$$\Delta = \Delta(1) + (\Delta - \Delta(1))$$

establishes an isomorphism of A-modules

$$D_k(P) \oplus D_{k+1}(P) \cong \mathrm{Df}^<_{k+1}(P).$$

◄ These statements are easily proved by induction. For $k = 0$, they are obvious, since $D_1 = D$ and $\mathrm{Df}_1 = \mathrm{Diff}_1$. Further, assuming that they hold for $k \geq 0$, let us show that they are valid for $k + 1$ as well.

The statement (1) obviously follows from the definition of left multiplication.

Let $I^< \colon \mathrm{Df}^<_{k+1}(P) \to \mathrm{Df}^>_{k+1}(P)$ be the operator changing the left structure in the A-module to its right structure. Now if we assume that $\Delta \colon A \to \mathrm{Df}^>_k(P)$ is a $\Bbbk$-linear map, then a straightforward calculation shows that

$$\left(\delta_{a_1,\dots,a_l}(I^<)\right)(\Delta) = (-1^l)\delta_{a_1,\dots,a_l}(\Delta).$$

In particular, if $\Delta \in \mathrm{Df}_{k+1}(P)$, then Δ is a differential operator of first order or less and, therefore,

$$\left(\delta_{a_1,a_2}(I^<)\right)(\Delta) = \delta_{a_1,a_2}(\Delta) = 0.$$

Since Δ is arbitrary, we have $\delta_{a_1,a_2}(I^<) = 0$, i.e., $I^<$ is a first order differential operator. This same argument can also be carried over, word for word, to case of the operator $I^>$.

Finally, since $a^>(\Delta) = \Delta \circ a$ (recall that the operator of right multiplication by a in $\mathrm{Df}^>_{k+1}(P)$ is denoted by $a^>$), it follows that

$$\Pi^>(a^>\Delta) = \Delta \circ a - \Delta(a).$$

Now if $\Delta \in \mathrm{Df}_{k+1}(P)$, then we also have $\Delta \in D(D_k(P))$ and, therefore,

$$\Delta \circ a - \Delta(a) = a\Delta.$$

This proves statement (3), while statement (4) is obvious. ▶

Exercise. Let $(\Delta_0, \Delta) \in D_k(P) \oplus D_{k+1}(P)$. Let us put

$$a * (\Delta_0, \Delta) \overset{\text{def}}{=} \big(a\Delta_0 + \Delta(a), a\Delta\big), \quad a \in A.$$

Show that the operation of multiplication (defined above) by elements of the algebra A supplies $D_k(P) \oplus D_{k+1}(P)$ with a second A-module structure and the module obtained in this way is isomorphic to the module $\mathrm{Df}_{k+1}^{>}(P)$.

17.18. Polyderivations. As we have already seen in Section 17.7, elements of the module $D_2(P)$ can be naturally interpreted as P-valued skew-symmetric biderivations, i.e., as bivectors in differential-geometrical terminology. Below, we will show that the elements of the module $D_k(P)$, $k > 2$, may also be interpreted similarly.

Arguing by induction, which we can begin from the previously considered case $k = 2$, we assume that to any element $\nabla \in D_l(P)$, $l \leq k$ we can assign the skew-symmetric l-form

$$(a_1, \ldots, a_l) \mapsto \langle a_1, \ldots, a_l \rangle_\nabla \in P, \; a_1, \ldots, a_l \in A,$$

which is $\Bbbk$-linear derivation in each argument. To do this, let $\Delta \in D_{k+1}(P)$. Since $\Delta \colon A \to \mathrm{Df}_k^{>} P$ is a derivation, we have

$$\Delta(ab) = a_{\mathrm{Df}_k^{>}(P)} \cdot \Delta(b) + b_{\mathrm{Df}_k^{>}(P)} \cdot \Delta(a) = \Delta(b) \circ a + \Delta(a) \circ b. \quad (17.37)$$

On the other hand, $\Delta(ab) \in D_k(P)$, since $\operatorname{Im} \Delta \in D_k(P)$. Therefore, $\Delta(ab) \colon A \to \mathrm{Df}_{k-1}^{>}(P)$ is a derivation and so $\big(\Delta(ab)\big)(1) = 0$. In view of (17.37), this is equivalent to the equality $\big(\Delta(b)\big)(a) + \big(\Delta(a)\big)(b) = 0$. This condition can be conveniently rewritten in the form

$$[a, b]_\Delta + [b, a]_\Delta = 0, \quad (17.38)$$

where $[a, b]_\Delta \overset{\text{def}}{=} \big(\Delta(a)\big)(b)$. Note that

$$(a, b) \mapsto [a, b]_\Delta, \quad a, b \in A,$$

is a $\Bbbk$-bilinear skew-symmetric form with values in $D_{k-1}(P)$ and, moreover, it is a biderivation.

Finally, let us put

$$\langle a_1, a_2, a_3, \ldots, a_{k+1} \rangle_\Delta \overset{\text{def}}{=} [\ldots [[\Delta(a_1)](a_2)] \ldots](a_{k+1}). \quad (17.39)$$

The P-valued form (17.39) is obviously k-linear and is a $\Bbbk$-linear derivation in each argument. On the other hand, we have

$$\begin{aligned} \langle a_1, a_2, a_3, \ldots, a_{k+1} \rangle_\Delta &= \langle a_2, \ldots, a_{k+1} \rangle_{\Delta(a_1)} \\ \langle a_3, \ldots, a_{k+1} \rangle_{(\Delta(a_1))(a_2)} &= \langle a_3, \ldots, a_{k+1} \rangle_{[a_1, a_2]_\Delta}. \end{aligned} \quad (17.40)$$

By the induction hypothesis, the second expression in this chain of equalities is skew-symmetric in the whole family of arguments, except the first. Besides, in view of (17.37), the last expression is skew-symmetric in the first two arguments. From these two facts, it obviously follows that the forms $\langle \cdot, \ldots, \cdot \rangle_\Delta$ are skew-symmetric.

The skew-symmetric $\Bbbk$-linear k-forms that are at the same time derivations in each argument, are usually called P-valued k-*derivations of the algebra* A or *polyderivations*, if it is not necessary to specify the *multiplicity* k. All such k-derivations obviously constitute an A-module that will be denoted by $\mathfrak{D}_k(P)$.

Thus, the formula (17.39) may be regarded as the definition of the homomorphism

$$i_{D,\mathfrak{D}} \colon D_k(P) \to \mathfrak{D}_k(P), \quad \Delta \mapsto \langle \cdot, \ldots, \cdot \rangle_\Delta. \tag{17.41}$$

Moreover, to each P-valued k-derivation we can assign an element of the module $D_k(P)$ simply by reversing the previous arguments (see the Lemma-Exercise from Section 17.7). This shows that $i_{D,\mathfrak{D}}$ is an isomorphism.

17.19. A bit of philosophy. Possibly, at this point the reader may want to ask why do we need these complicated manipulations with the diffunctors D_k, while the equivalent notion of k-derivation is quite elementary. The general answer to this question is that this notion, as well as any other *descriptive* notion, does not explicitly contain any information on connections with other natural constructions of the differential calculus. In other words, working only with descriptions, it is difficult, if not impossible, to discover the structure of relationships of the corresponding notions, i.e., to understand what may be called the *algebra of their calculus*. For example, it is hardly possible to guess the existence of higher analogs of differential forms and of the de Rham complexes starting from the standard definitions. What is more, the descriptive definitions often turn out to be "incorrect" in the sense that they cannot be generalized to various singular, infinite-dimensional, and other nonstandard situations. Such is, for example, is the well-known definition of tensors. Concerning the concrete diffunctors D_k, it will be shown below that they lead directly to the construction of Spencer complexes and, as a consequence, to differential forms and de Rham complexes over arbitrary commutative algebras.

17.20. Polyvectors. A polyderivation over the algebra $C^\infty(M)$ with values in geometrical modules has a geometrical interpretation as a *polyvector*, or *polyvector field* on a manifold M. Recall that according to the standard definition, by a k-vector on M we mean a $C^\infty(M)$-linear skew-symmetric k-form whose arguments are differential forms of first order on M. Here we assume that the values of such a form are sections of a vector bundle π or, as a rule, simply functions.

This interpretation is merely the expression of the fact that to each k-derivation of the type mentioned above we can assign a k-vector and vice versa. The corresponding construction for $k = 2$ was already carried out in Section 17.9. Now we shall study this situation in a wider context.

Let

$$\alpha(\omega_1, \ldots, \omega_k) \in P, \quad \omega_1, \ldots, \omega_k \in \Lambda^1_{\mathfrak{M}}(A), \qquad (17.42)$$

be an A-linear skew-symmetric polylinear form with values in $P \in \mathrm{Ob}\,\mathfrak{M}$. Then

$$\chi_\alpha(a_1, \ldots, a_k) \overset{\text{def}}{=} \alpha(d^{\mathfrak{M}} a_1, \ldots, d^{\mathfrak{M}} a_k), \quad a_1, \ldots, a_k \in A,$$

is a P-valued k-derivation of the algebra A. The family of all k-forms (17.42) obviously constitutes an A-module that will be denoted as $\mathfrak{P}_{k,\mathfrak{M}}(P)$. Then the map

$$\chi \colon \mathfrak{P}_{k,\mathfrak{M}}(P) \to \mathfrak{D}_k(P), \quad \alpha \mapsto \chi_\alpha,$$

is a homomorphism of A-modules. Since the module $\Lambda^1_{\mathfrak{M}}(A)$ is generated by the differentials $d^{\mathfrak{M}} a$, $a \in A$, the form χ_α is trivial if and only if the form α is trivial. Therefore, the homomorphism χ is injective, but can it be inverted?

To answer this question, consider a k-derivation $\beta \in \mathfrak{D}_k(P)$. If the form $\alpha = \chi^{-1}(\beta)$ exists, then we necessarily have

$$\alpha(d^{\mathfrak{M}} a_1, \ldots, d^{\mathfrak{M}} a_k) = \beta(a_1, \ldots, a_k).$$

Here, in order to define the value of $\alpha(\omega_1, \ldots, \omega_k)$, we must represent each 1-form ω_i as

$$\omega_i = \sum_{j=1}^{m_i} b^i_j d^{\mathfrak{M}} a^i_j, \quad a^i_j, b^i_j \in A, \qquad (17.43)$$

and put

$$\alpha(\omega_1, \ldots, \omega_k) \overset{\text{def}}{=} \sum_{j_1, \ldots, j_k} b^1_{j_1} \ldots b^k_{j_k} \beta(a^1_{j_1}, \ldots, a^k_{j_k}). \qquad (17.44)$$

The fact that this is well defined must be verified, i.e., we must show that the right-hand side of the definition (17.44) does not depend on how the form ω_i is expressed in (17.43). This is equivalent to saying that, for each $i \in \{1, \ldots, k\}$, this right-hand side does not depend on the representation of the form ω_i under the condition that for all other forms ω_j such representations are fixed. In view of the skew-symmetry of summands of (17.44), we can assume that $i = 1$. Thus, regarding that the representations (17.43) for the forms ω_j, $j > 1$, as fixed, let us consider the derivation

$$a \overset{X}{\mapsto} \sum_{j_2, \ldots, j_k} b^2_{j_2} \ldots b^k_{j_k} \beta(a, a^2_{j_2}, \ldots, a^k_{j_k}).$$

In view of universality of the module $\Lambda^1_{\mathfrak{M}}(A)$, there exists a unique homomorphism $h_X \colon \Lambda^1_{\mathfrak{M}}(A) \to P$ such that $X = h_X \circ d^{\mathfrak{M}}$. Here

$$h_X(\omega) = \sum_j b_j X(a_j),$$

where $\omega = \sum_j b_j d^{\mathfrak{M}} a_j$. This shows that $h_X(\omega)$ does not depend on the representation of the form ω as linear combination of differentials. To conclude the proof, it remains to note that $h_X(\omega_1)$ exactly coincides with the right-hand side of the equality (17.44).

Thus, we have established natural isomorphisms

$$D_k(P) \cong \mathfrak{D}_k(P) \cong \mathfrak{P}_k(P). \qquad (17.45)$$

This gives us three different views of the same notion. Which of them is more convenient depends on the context. For example, the natural symplectic structure on the cotangent space T^*M of a manifold M may also be understood as the Poisson bracket on it, as well as a Poisson bivector. As we showed in Chapter 10, its analog for an arbitrary commutative algebra can be naturally obtained in the form of a biderivation, i.e., of the bracket $\{\cdot, \cdot\}$ on the algebra of symbols $S_*(A)$, defined in Section 10.2. On the other hand, it is more convenient to work with the corresponding Poisson bivector by using of Schouten brackets [13].

17.21. The algebra of polyderivations. Similarly to its symmetric analogs (see Section 16.7), A-valued polyderivations, as well as polyvectors, can be multiplied. The product of a k-vector α by an l-vector β is the $(k+l)$-vector, denoted by $\alpha \wedge \beta$, and defined by the following formula:

$$[\alpha \wedge \beta](\omega_1, \ldots, \omega_{k+l}) \stackrel{\text{def}}{=} \sum_{|I|=k} \alpha(\omega_{i_1}, \ldots, \omega_{i_k}) \cdot \beta(\omega_{j_1}, \ldots, \omega_{j_l}), \qquad (17.46)$$

where the sequences $I = \{i_1, \ldots, i_k\}$ and $J = \{j_1, \ldots, j_l\}$ are given in increasing order and $I \cup J = \{1, 2, \ldots, k+l\}$. Thus the direct sum

$$\mathfrak{P}_*(A) = \sum_{k \geq 0} \mathfrak{P}_k(A)$$

acquires an algebra structure. Moreover, formula (17.46) is also meaningful when $\beta \in \mathfrak{D}_l(P)$. In this case, $\alpha \wedge \beta \in \mathfrak{P}_{k+l}(P)$.

Exercise. Show that

1. $\beta \wedge \alpha = (-1)^{kl} \alpha \wedge \beta$, if $\alpha \in \mathfrak{P}_k(A), \beta \in \mathfrak{P}_l(A)$;

2. $\mathfrak{P}_*(A)$ is an associative algebra, while $\mathfrak{P}_*(P)$ is a $\mathfrak{P}_*(A)$-module with respect to the multiplication (17.46).

The natural isomorphisms (17.45) allow us to carry over the (graded) algebra structure of $\mathfrak{P}_*(A)$ to

$$D_*(A) = \sum_{k \geq 0} D_k(A)$$

(respectively, to $\mathfrak{D}_*(A) = \sum_{k \geq 0} \mathfrak{D}_k(A)$), as well as to regard

$$D_*(P) = \sum_{k \geq 0} D_k(P)$$

(respectively, $\mathfrak{D}_*(P) = \sum_{k \geq 0} \mathfrak{D}_k(P)$) as a $D_*(A)$-module (respectively, a $\mathfrak{D}_*(A)$-module).

17.22. The diffunctors $D_{k,l}$. The interpretation of elements of the module $D_k(P)$ as k-derivations allows us to define the diffunctors

$$D_{k,l} \overset{\text{def}}{=} D_k(D_l \subset \mathrm{Df}_l^{>}).$$

By definition, elements of the module $D_{k,l}(P)$ are k-derivations β of the algebra A with values in the module $\mathrm{Df}_l^{>}(P)$ such that we have $\beta(a_1, \ldots, a_k) \in D_l(P)$.

Proposition. *The diffunctors $D_{k,l}$ and D_{k+l} are equivalent.*

◀ We must indicate a natural isomorphism between the modules $D_{k,l}(P)$ and $D_{k+l}(P)$. To do this, we will show that any element $\beta \in D_{k,l}$ is uniquely defined by a $(k+l)$-derivation and vice versa.

By definition, $\beta(a_1, \ldots, a_k) \in D_l(P)$. Consider the form

$$\nabla_\beta(a_1, \ldots a_{k+l}) \overset{\text{def}}{=} \langle a_{k+1}, \ldots, a_{k+l} \rangle_{\beta(a_1, \ldots, a_k)} \qquad (17.47)$$

(see definition (17.39) for $\Delta = \beta(a_1, \ldots, a_k)$). This form is skew-symmetric in the first k and last l arguments and is a derivation in each of them. Besides,

$$\beta(a_1, \ldots, a_{k-1}, a_k a_{k+1})$$
$$= \beta(a_1, \ldots, a_{k-1}, a_k) \circ a_{k+1} + \beta(a_1, \ldots, a_{k-1}, a_{k+1}) \circ a_k, \quad (17.48)$$

since, in particular, β is a derivation in the last argument. On the other hand,

$$\beta(a_1, \ldots, a_{k-1}, a_k a_{k+1}) \in D_l(P) = D^{<}(D_{l-1}(P) \subset \mathrm{Df}_{l-1}^{>}(P))$$

Therefore, we have $\beta(a_1, \ldots, a_{k-1}, a_k a_{k+1})(1_A) = 0$ and, applying the operators from (17.48) to 1_A, we obtain

$$\beta(a_1, \ldots, a_{k-1}, a_k)(a_{k+1}) + \beta(a_1, \ldots, a_{k-1}, a_{k+1})(a_k) = 0. \qquad (17.49)$$

Finally, from (17.40) and (17.49), it follows that

$$\langle a_{k+1}, a_{k+2}, \ldots, a_{k+l} \rangle_{\beta(a_1,\ldots,a_{k-1},a_k)}$$
$$+ \langle a_k, a_{k+2}, \ldots, a_{k+l} \rangle_{\beta(a_1,\ldots,a_{k-1},a_{k+1})}$$
$$= \langle a_{k+2}, \ldots, a_{k+l} \rangle_{\beta(a_1,\ldots,a_{k-1},a_k)(a_{k+1})}$$
$$+ \langle a_{k+2}, \ldots, a_{k+l} \rangle_{\beta(a_1,\ldots,a_{k-1},a_{k+1})(a_k)} = 0$$

or, equivalently,

$$\nabla_\beta(a_1, \ldots, a_{k-1}, a_k, a_{k+1}, a_{k+2}, \ldots, a_{k+l})$$
$$+ \nabla_\beta(a_1, \ldots, a_{k-1}, a_{k+1}, a_k, a_{k+2}, \ldots, a_{k+l}) = 0. \quad (17.50)$$

Thus, the form ∇_β is skew-symmetric in the k-th and $(k+1)$-st arguments. This, together with the properties of the skew-symmetric form ∇_β, proves its skew-symmetry in all the arguments.

Thus, we have constructed a map

$$D_{k,l}(P) \to D_{k+l}(P), \quad \beta \mapsto \nabla_\beta,$$

which is obviously a homomorphism of A-modules. Conversely, the equality (17.47) may be regarded as the definition of the form β, assuming that the form ∇_β is given. Then, inverting the order of the steps of the proof of the formula, starting from (17.48) and ending at (17.50), it is easy to see that the form β defined in this way is indeed an element of the module $D_{k,l}(P)$. ▶

17.23 Exercise. In Section 17.15, we inductively defined the diffunctors $\Phi_\tau(D_\sigma \subset \mathrm{Df}_\sigma^>)$. Show that the diffunctor $D_\tau(D_\sigma \subset \mathrm{Df}_\sigma^>)$ defined in this way is equivalent to $D_{k,l}$ if the sequences τ and σ consist of k and l ones, respectively.

17.24. Existence of differential forms. Now everything is ready to prove the existence of representing objects $\Lambda_{\mathfrak{M}}^k$ for the diffunctors D_k in a differentially closed category of A-modules $\mathfrak{M}$, i.e., to prove the existence of modules of differential forms. To do this, it is natural to use induction on k, having in mind that for $k = 1, 2$ these modules are already constructed (see Sections 17.10 and 17.6). Therefore, it remains to show how to construct the module $\Lambda_{\mathfrak{M}}^{k+1}$ under the assumption that the modules $\Lambda_{\mathfrak{M}}^m$ are already constructed for $m \leq k$, $k \geq 2$. Do do this, by the corollary from Section 15.3, it suffices to represent the diffunctor D_{k+1} as the kernel of some transformation of diffunctors for which the existence of representing modules has been established.

The first thing that we shall need is the following transformation of diffunctors, which we will denote by $\mathsf{t}_{k,l}$:

$$D_k(\mathrm{Diff}_l^>) = D_{k-1}(D(\mathrm{Diff}_l^>) \subset \mathrm{Diff}_1^>(\mathrm{Diff}_l^>))$$
$$\xrightarrow{\mathsf{t}_{k,l}} D_{k-1}(\mathrm{Diff}_1^>(\mathrm{Diff}_l^>)). \quad (17.51)$$

Here the expression $D_m(\Phi(\mathrm{Diff}_n^>(P)))$ stands for the family of m-derivation with values in $\Phi(\mathrm{Diff}_n^>(P))$ supplied with the A-module structure induced by left multiplication in the A-bimodule $\mathrm{Diff}_n(P)$.

Further, let us consider the *Spencer transformation* S_{k+l}^k defined by the following commutative diagram:

$$D_k(\mathrm{Diff}_l^>) \xrightarrow{\quad S_{k+l}^k \quad} D_{k-1}(\mathrm{Diff}_{l+1}^>) \qquad (17.52)$$

with $t_{k,l}$ down-left to $D_{k-1}(\mathrm{Diff}_1^>(\mathrm{Diff}_l^>))$ and $D_{k-1}^<(C_{1,l}^>)$ up-right.

We stress that the A-module structure in the lower term of the diagram is induced by the left multiplication in Diff_l.

Proposition. *The following equality holds*:

$$\mathrm{Ker}\, S_{k+l}^k = D_{k-1}(D_{(l,1)} \subset \mathrm{Df}_{(l,1)}^>),$$

in particular, $\mathrm{Ker}\, S_{k+1}^k = D_{k+1}$.

◄ First note that $\mathrm{Ker}\, C_{1,l}^> = \mathrm{Diff}_1^>(D_{(l)} \subset \mathrm{Diff}_l^>)$. Therefore,

$$\mathrm{Ker}\, D_{k-1}(C_{1,l}^>) = D_{k-1}(\mathrm{Diff}_1^>(D_{(l)} \subset \mathrm{Diff}_l^>)) = D_{k-1}(\mathrm{Df}_{(l,1)}^>)$$

(see (17.10)). But since $t_{k,l}$ is an embedding, while $\mathrm{Df}_{(l,1)}^>$ is a submodule of the module $\mathrm{Diff}_1^>(\mathrm{Diff}_l^>)$, it follows from (17.51) that

$$\mathrm{Ker}\, S_{k+l}^k = (\mathrm{Im}\, t_{k,l}) \cap (\mathrm{Ker}\, D_{k-1}(C_{1,l}^>))$$
$$= D_{k-1}((D(\mathrm{Diff}_l^>) \cap \mathrm{Df}_{(l,1)}^>) \subset \mathrm{Df}_{(l,1)}^>) = D_{k-1}(D_{(l,1)} \subset \mathrm{Df}_{(l,1)}^>).$$

This proves the first assertion, while the second one follows from the relation

$$D_{(1,1)} = D_2,\ \mathrm{Df}_{(1,1)}^> = \mathrm{Df}_2^> \quad \text{and} \quad D_{k-1}(D_2 \subset \mathrm{Df}_2^>(P)) = D_{k+1}$$

(see the Proposition from Section 17.22). ►

The second assertion of the proposition just proved above may be described by the exact sequence of diffunctors

$$0 \to D_{k+1} \to D_k(\mathrm{Diff}_1^>) \xrightarrow{S_{k+1}^k} D_{k-1}(\mathrm{Diff}_2^>). \qquad (17.53)$$

The homomorphism

$$S_{k,\mathfrak{M}}^{k+l} \colon \mathcal{J}^{l+1}(\Lambda_{\mathfrak{M}}^{k-1}; \mathfrak{M}) \to \mathcal{J}^l(\Lambda_{\mathfrak{M}}^k; \mathfrak{M}), \qquad (17.54)$$

which represents the Spencer transformation (17.52) is called the *Spencer operator* (in the category $\mathfrak{M}$). In these terms, the corollary from Section 15.3 applied to the exact sequence (17.53), proves the required fact.

Corollary. *The A-module $\Lambda_{\mathfrak{M}}^{k+1} = \mathrm{coker}\, S_{k,\mathfrak{M}}^{k+1}$ is the representing object of the diffunctor D_{k+1} in the category $\mathfrak{M}$.*

Accordingly, the exact sequence of homomorphisms

$$0 \leftarrow \Lambda_{\mathfrak{M}}^{k+1} \leftarrow \mathcal{J}^1(\Lambda_{\mathfrak{M}}^k; \mathfrak{M}) \xleftarrow{S_{k \cdot \mathfrak{M}}^{k+1}} \mathcal{J}^2(\Lambda_{\mathfrak{M}}^{k-1}; \mathfrak{M}) \qquad (17.55)$$

is the representation of the exact sequence of diffunctors (17.53).

17.25. The representing module for the diffunctor $\mathfrak{D}_k$. A technically simpler, but conceptually less satisfactory way of defining differential forms is to use the equivalence of the diffunctors D_k and $\mathfrak{D}_k$. Indeed, as will be shown below, the representing module for $\mathfrak{D}_k$ may be constructed as the range of the universal k-derivation, simply by generalizing the construction of the differential 1-form from Section 14.4 to the case $k > 1$. In view of the equivalence $i_{D,\mathfrak{D}}$ of these diffunctors (see (17.41)), the representing module for $\mathfrak{D}_k$ is isomorphic to the representing module for D_k and so can serve as the definition of the module of differential k-forms.

We begin by recalling that, according to its definition, the tensor product $A^{\otimes_k m} \stackrel{\text{def}}{=} A \otimes_k \dots \otimes_k A$ (m times) is the universal object for k-linear m-forms on the algebra A understood as a k-vector space. This means that any such form φ with values in a k-vector space V can be represented in a unique way as follows

$$\varphi(a_1, \dots, a_m) = h_\varphi(a_1 \otimes_k \dots \otimes_k a_m),$$

where $h_\varphi \colon A^{\otimes_k m} \to V$ is a k-linear map. In what follows, we regard $A^{\otimes_k m}$ as a k-vector space and assign to it the A-module $A \otimes_k A^{\otimes_k m}$, in which multiplication by an element $a \in A$ is defined by the rule

$$(a, a' \otimes_k a_1 \otimes_k \dots \otimes_k a_m) \mapsto aa' \otimes_k a_1 \otimes_k \dots \otimes_k a_m.$$

Let $P \in \mathrm{Ob}\,\mathfrak{M}$ and $\mathfrak{F}_m(P)$ be the vector space of k-linear m-forms on the module P. If $\varphi \in \mathfrak{F}_m(P)$, then $a\varphi \in \mathfrak{F}_m(P)$, where

$$(a\varphi)(a_1, \dots, a_m) \stackrel{\text{def}}{=} a\varphi(a_1, \dots, a_m).$$

This operation of multiplication by elements of the algebra A transforms $\mathfrak{F}_m(P)$ into an A-module. Here, as can be easily seen, the m-form

$$\Phi_m \colon (a_1, \dots, a_m) \mapsto 1_A \otimes_k a_1 \otimes_k \dots \otimes_k a_m \in A \otimes_k A^{\otimes_k m}$$

is universal in the sense that for any form $\varphi \in \mathfrak{F}_m(P)$ there exists a unique A-homomorphism $\widetilde{h}_\varphi \colon A \otimes_k A^{\otimes_k m} \to P$ such that

$$\varphi(a_1, \dots, a_m) = \widetilde{h}_\varphi(1_A \otimes_k a_1 \otimes_k \dots \otimes_k a_m). \qquad (17.56)$$

Further, in the A-module $A \otimes_k A^{\otimes_k m}$, we consider the submodule R generated by all possible elements of the two following types:

$1_\mathbf{R}$. $1_A \otimes_k a_1 \otimes_k \dots \otimes_k a_m + (-1)^\sigma 1_A \otimes_k a_{\sigma(1)} \otimes_k \dots \otimes_k a_{\sigma(m)}$, where σ is a permutation, while $(-1)^\sigma$ is its sign.

$2_\mathbf{R}$. $1_A \otimes_k a_1 \otimes_k \dots \otimes_k a_i a_i' \otimes_k \dots \otimes_k a_m$
$\qquad -a_i \otimes_k a_1 \otimes_k \dots \otimes_k a_i' \otimes_k \dots \otimes_k a_m$
$\qquad\qquad -a_i' \otimes_k a_1 \otimes_k \dots \otimes_k a_i \otimes_k \dots \otimes_k a_m, \; 1 \le i \le m.$

Proposition. *The quotient module*[1] $Л^m(A) \overset{\text{def}}{=} A \otimes_{\Bbbk} A^{\otimes_{\Bbbk} m} / R$ *is the representing module for the diffunctor* $\mathfrak{D}_k$ *in the category* Mod$A$.

◄ First, for convenience, let us introduce the following notation:

$$Л^m(A) \ni a\partial a_1|\partial a_2|\dots|\partial a_m \overset{\text{def}}{=} a \otimes_{\Bbbk} a_1 \otimes_{\Bbbk} a_2 \otimes_{\Bbbk} \dots \otimes_{\Bbbk} a_m \mod \mathcal{R},$$

assuming that $\partial a_1|\dots|\partial a_m = 1_A \cdot \partial a_1|\dots|\partial a_m$. From the fact that elements of the form $1_\mathbf{R}$ and $2_\mathbf{R}$ belong to the submodule R, it follows that

$$(a_1,\dots,a_m) \mapsto \partial a_1|\dots|\partial a_m \qquad (17.57)$$

is an m-derivation of the algebra A with values in the module $Л_m(A)$. Moreover, any m-derivation

$$(a_1,\dots,a_m) \mapsto \varphi(a_1,\dots,a_m) \in P$$

with values in an A-module P is, in particular, a $\Bbbk$-linear m-form on the algebra A. Further, since the form φ is skew-symmetric and is a derivation in each argument, it follows that the homomorphism $\tilde{h}_\varphi$ (see (17.56)) annihilates elements of the form $1_\mathbf{R}$ and $2_\mathbf{R}$ and, therefore, annihilates the whole submodule R. Hence, $\tilde{h}_\varphi$ can be represented as the composition

$$A \otimes_{\Bbbk} A^{\otimes_{\Bbbk} m} \overset{\text{pr}}{\longrightarrow} Л^m(A) \overset{h_\varphi}{\longrightarrow} P\,,$$

where pr is the quotient map, while h_φ is an A-homomorphism such that

$$h_\varphi(a \cdot \partial a_1|\dots|\partial a_m) = a\varphi(a_1,\dots,a_m). \qquad ► \qquad (17.58)$$

As to the case of an arbitrary differentially closed category of A-modules $\mathfrak{M}$, the existence of an A-module representing in $\mathfrak{M}$ the diffunctor $\mathfrak{D}_k$ is obtained by an appropriate factorization of the module $Л^m(A)$ (see the Proposition above). This module will be denoted by $Л^m_{\mathfrak{M}}(A)$. In this situation, the homomorphism

$$i_\mathfrak{D}^\mathfrak{M} \colon Л^m_{\mathfrak{M}}(A) \to \Lambda^m_{\mathfrak{M}}(A)\,,$$

which represents the equivalence of diffunctors $i_{D,\mathfrak{D}} \colon D_k \to \mathfrak{D}_k$ in the category $\mathfrak{M}$, is called the *canonical* isomorphism of modules $Л^m_{\mathfrak{M}}(A)$ and $\Lambda^m_{\mathfrak{M}}(A)$.

17.26. Universal polyderivation. The formula (17.58) shows that the polyderivation (17.57) is *universal*. Let us consider the polyderivation

$$(a_1,\dots,a_m) \mapsto \partial^\mathfrak{M} a_1|\dots|\partial^\mathfrak{M} a_m, \qquad (17.59)$$

where

$$\partial^\mathfrak{M} a_1|\dots|\partial^\mathfrak{M} a_m \overset{\text{def}}{=} \text{pr}_\mathfrak{M}(\partial a_1|\dots|\partial a_m),$$

---

[1] The Cyrillic letter Л, which resembles the Greek Λ, reads as the Latin L.

while $\mathrm{pr}_{\mathfrak{M}} \colon \Pi^m(A) \to \Pi^m_{\mathfrak{M}}(A)$ is the natural projection expressing the universality of the polyderivation (17.57); then this polyderivation will be universal in the category $\mathfrak{M}$. The range of the universal polyderivation (17.59) may be carried over to $\Lambda^m_{\mathfrak{M}}(A)$ by means of the isomorphism $i^*_{D,\mathfrak{D}}$. This can be done directly as follows.

Having in mind the isomorphism

$$D_m(\Lambda^m_{\mathfrak{M}}(A)) = \mathrm{Hom}_A(\Lambda^m_{\mathfrak{M}}(A), \Lambda^m_{\mathfrak{M}}(A)),$$

let us consider the element $\Delta_{m,\mathfrak{M}} \in D_m(\Lambda^m_{\mathfrak{M}}(A))$, which under this isomorphism corresponds to the identity map of the module $\Lambda^m_{\mathfrak{M}}(A)$. The polyderivation corresponding to it in the notation of Section 17.18 is given by

$$(a_1, \dots, a_m) \mapsto \langle a_1, \dots, a_m \rangle_{\Delta_{m,\mathfrak{M}}}.$$

Its universality follows from the fact that, under the isomorphism $i^*_{D,\mathfrak{D}}$, it corresponds to the identity homomorphism of the representing module $\Pi^k_{\mathfrak{M}}(A)$, which in its turn corresponds to the universal polyderivation (17.59) (see (17.58)). In other words, we have the equality

$$\langle a_1, \dots, a_m \rangle_{\Delta_{m,\mathfrak{M}}} = i^{\mathfrak{M}}_{\mathfrak{D}}(\partial^{\mathfrak{M}} a_1 | \dots | \partial^{\mathfrak{M}} a_m). \qquad (17.60)$$

17.27. The exterior product. Note that by (15.10), the A-module $\Pi^k_{\mathfrak{M}}(A) \otimes_A \Pi^l_{\mathfrak{M}}(A)$ represents the diffunctor $\mathfrak{D}_k \circ \mathfrak{D}_l$ in any differentially closed category of A-modules $\mathfrak{M}$. The equivalence $i_{D,\mathfrak{D}}$ assigns to the natural embedding of diffunctors

$$D_{k+l} = D_{k,l} = D_k(D_l \subset \mathrm{Df}^>_l) \longrightarrow D_k \circ D_l$$

the embedding

$$\nu_{k,l} \colon \mathfrak{D}_{k+l} \to \mathfrak{D}_k \circ \mathfrak{D}_l.$$

Namely, if $\varphi \in \mathfrak{D}_{k+l}$, then $\psi = \nu_{k,l}(\varphi)$ is defined by the rule

$$\psi(a_1, \dots, a_k) \colon (a_{k+1}, \dots a_{k+l}) \mapsto \varphi(a_1, \dots, a_k, a_{k+1}, \dots a_{k+l}). \qquad (17.61)$$

According to Section 15.2, the transformation $\nu_{k,l}$ is represented in the category $\mathfrak{M}$ by the corresponding homomorphism of representing modules

$$\lambda_{k,l} \colon \Pi^k_{\mathfrak{M}}(A) \otimes_A \Pi^l_{\mathfrak{M}}(A) \to \Pi^{k+l}_{\mathfrak{M}}(A).$$

More precisely,

$$\lambda_{k,l} \colon (\partial a_1 | \dots | \partial a_k) \otimes_A (\partial a_{k+1} | \dots | \partial a_{k+l})$$
$$\mapsto \partial a_1 | \dots | \partial a_k | \partial a_{k+1} | \dots | \partial a_{k+l}. \qquad (17.62)$$

Indeed, in view of the the isomorphism

$$\mathfrak{D}_k(\mathfrak{D}_l(\Pi^{k+l}_{\mathfrak{M}}(A)) = \mathrm{Hom}_A(\Pi^k_{\mathfrak{M}}(A), \mathfrak{D}_l(\Pi^{k+l}_{\mathfrak{M}}(A)))$$

the transformation $\lambda_{k,l}$ assigns to the element $\partial a_1 | \dots | \partial a_k \in \Pi^k_{\mathfrak{M}}(A)$ the l-derivation

$$\Pi^l_{\mathfrak{M}}(A) \ni \partial a_{k+1} | \dots | \partial a_{k+l} \mapsto \partial a_1 | \dots | \partial a_k | \partial a_{k+1} | \dots | \partial a_{k+l} \in \Pi^{k+l}_{\mathfrak{M}}(A).$$

Then, if $\varphi \in \mathfrak{D}_{k+l}(P)$ and

$$\varphi(a_1, \ldots, a_k, a_{k+1}, \ldots a_{k+l}) = F(\partial a_1 | \ldots | \partial a_k | \partial a_{k+1} | \ldots | \partial a_{k+l}),$$

where $F \in \mathrm{Hom}_A(\Pi_{\mathfrak{M}}^{k+l}(A), P) = \mathfrak{D}_{k+l}(P)$, then the homomorphism $F \circ \lambda_{k,l}$, due to the existence of the isomorphism

$$\mathfrak{D}_k(\mathfrak{D}_l(P)) = \mathrm{Hom}_A(\Pi_{\mathfrak{M}}^k(A), \mathfrak{D}_l(P))$$

understood as an element of the module $\mathrm{Hom}_A(\Pi_{\mathfrak{M}}^k(A), \mathfrak{D}_l(P))$, obviously coincides with $\nu_{k,l}(\varphi)$ (see 17.61 above).

Finally, using the canonical isomorphism

$$i_{\mathfrak{D}}^{\mathfrak{M}} \colon \Pi_{\mathfrak{M}}^n(A) \to \Lambda_{\mathfrak{M}}^n(A), \quad n = k, l,$$

let us carry over the homomorphism $\lambda_{k,l}$ to differential forms:

$$\lambda_{k,l} \colon \Lambda_{\mathfrak{M}}^k(A) \otimes_A \Lambda_{\mathfrak{M}}^l(A) \to \Lambda_{\mathfrak{M}}^{k+l}(A).$$

17.28. Definition. The *exterior product* of differential forms $\omega \in \Lambda_{\mathfrak{M}}^k(A)$ and $\rho \in \Lambda_{\mathfrak{M}}^l(A)$ is defined as the element

$$\omega \wedge \rho \stackrel{\mathrm{def}}{=} \lambda_{k,l}(\omega \otimes_A \rho) \in \Lambda_{\mathfrak{M}}^{k+l}(A).$$

Let us introduce the useful *intermediate* notation

$$da_1^{\mathfrak{M}} | \ldots | da_m^{\mathfrak{M}} = i_{\mathfrak{D}}^{\mathfrak{M}}(\partial a_1 | \ldots | \partial a_m) \in \Lambda_{\mathfrak{M}}^k(A).$$

Then, from the definition of the representing homomorphism from Section (15.2), it follows that

$$(da_1^{\mathfrak{M}} | \ldots | da_k^{\mathfrak{M}}) \wedge (da_{k+1}^{\mathfrak{M}} | \ldots | da_{k+l}^{\mathfrak{M}}) =$$
$$da_1^{\mathfrak{M}} | \ldots | da_k^{\mathfrak{M}} | da_{k+1}^{\mathfrak{M}} | \ldots | da_{k+l}^{\mathfrak{M}}. \quad (17.63)$$

The right-hand side of this formula does not depend on the choice of the position of the symbol $\wedge$ between the symbols da_i in its left-hand side. This obviously implies the *associativity of the exterior product*, since the module $\Lambda_{\mathfrak{M}}^m(A)$ is generated by elements of the form $d^{\mathfrak{M}} a_1 | \ldots | d^{\mathfrak{M}} a_m$.

Note that the symbol $d^{\mathfrak{M}} a_1 | \ldots | d^{\mathfrak{M}} a_m$ inherits the properties of the symbol $\partial a_1 | \ldots | \partial a_m$. In particular, it is skew-symmetric w.r.t. the arguments of $d^{\mathfrak{M}} a_i$. Therefore, (17.63) implies that

$$(d^{\mathfrak{M}} a_1 | \ldots | d^{\mathfrak{M}} a_k) \wedge (d^{\mathfrak{M}} a_{k+1} | \ldots | d^{\mathfrak{M}} a_{k+l})$$
$$= (-1)^{kl} (d^{\mathfrak{M}} a_{k+1} | \ldots | d^{\mathfrak{M}} a_{k+l}) \wedge (d^{\mathfrak{M}} a_1 | \ldots | d^{\mathfrak{M}} a_k). \quad (17.64)$$

For the same reason as above, this implies the skew-symmetry of the exterior product: $\lambda_{l,k} = (-1)^{kl} \lambda_{k,l}$.

17.29. The algebra of differential forms. The exterior product operation transforms the direct sum

$$\Lambda_{\mathfrak{M}}^* = \Lambda_{\mathfrak{M}}^*(A) \stackrel{\mathrm{def}}{=} \bigoplus_{k \geq 0} \Lambda_{\mathfrak{M}}^k(A)$$

into a *graded* associative algebra. From the usual point of view, multiplication in this algebra is skew-commutative. However, in the "graded sense" this algebra is commutative. The exact meaning of this notion will be given in Chapter 21. As to now, we shall only indicate the remarkable fact that the differential calculus may also be developed over graded commutative algebras, for example, over the algebra of differential forms itself. In particular, we can construct the algebra of differential forms over the algebra of differential forms $\Lambda_{\mathfrak{M}}^*(A)$ and, continuing this process, arrive at a new mathematical object, namely the algebra of iterated differential forms.

Notation. Further, for the two most important categories, we shall use the following simplified notation:

$$\Lambda^i(A) \stackrel{\text{def}}{=} \Lambda_{\text{Mod}A}^i, \quad \Lambda_\Gamma^i(A) \stackrel{\text{def}}{=} \Lambda_{\text{GMod}A}^i$$

and similarly for $i = *$.

17.30. The isomorphism $\lambda_m^{\mathfrak{M}}$. Recall that the exterior degree of an A-module P, denoted by $P^{\wedge m}$, is the quotient module of its tensor degree $P^{\otimes m}$ by the submodule R_m generated by all possible elements of the form

$$p_{\sigma(1)} \otimes_A \cdots \otimes_A p_{\sigma(m)} - (-1)^\sigma p_1 \otimes_A \cdots \otimes_A p_m, \quad p_i \in P,$$

where σ is a permutation of the indices $1, \dots, m$. Obviously, $P^{\wedge m}$ is the universal object for skew-symmetric A-polylinear forms on P. The image of the element $p_1 \otimes_A \cdots \otimes_A p_m$ under the factorization homomorphism $P^{\otimes m} \to P^{\wedge m}$ is denoted by $p_1 \wedge \dots \wedge p_m$ and is called the *exterior product* of the elements $p_1, \dots, p_m$.

Proposition. *The map*

$$\lambda_m^{\mathfrak{M}} \colon \left(\Lambda_{\mathfrak{M}}^1(A)\right)^{\wedge m} \to \Lambda_{\mathfrak{M}}^m(A),$$

given by $\lambda_m^{\mathfrak{M}}(d^{\mathfrak{M}}a_1 \wedge \dots \wedge d^{\mathfrak{M}}a_m) = d^{\mathfrak{M}}a_1 | \dots | d^{\mathfrak{M}}a_m$, *is an isomorphism.*

◄ Let us prove this statement by induction, first noting that for $m = 1$ it is tautological. Now let us assume that it is valid for some $m > 1$ and let us consider the $(m + 1)$-form on $\Lambda_{\mathfrak{M}}^1(A)$

$$(\omega_0, \omega_1, \dots, \omega_m) \mapsto \wedge_{1,m}^{\mathfrak{M}}\left(\omega_0 \otimes_A \left(\lambda_m^{\mathfrak{M}}(\omega_1 \wedge \dots \wedge \omega_m)\right)\right), \quad \omega_i \in \Lambda_{\mathfrak{M}}^1(A),$$

with values $\Lambda_{\mathfrak{M}}^{m+1}(A)$. Using the inductive hypothesis and the definition of exterior product $\wedge_{1,m}^{\mathfrak{M}}$, we see that

$$(d^{\mathfrak{M}}a_0, d^{\mathfrak{M}}a_1, \dots, d^{\mathfrak{M}}a_m)$$
$$3 \mapsto \lambda_{1,m}^{\mathfrak{M}}\left(d^{\mathfrak{M}}a_0 \otimes_A (d^{\mathfrak{M}}a_1 | \dots | d^{\mathfrak{M}}a_m)\right) = d^{\mathfrak{M}}a_0 | d^{\mathfrak{M}}a_1 | \dots | d^{\mathfrak{M}}a_m.$$

Moreover, this form is skew-symmetric and A-linear in each argument. Therefore, in view of the previously mentioned universality of the exterior degree $\left(\Lambda_{\mathfrak{M}}^1(A)\right)^{\wedge(m+1)}$, there exists a homomorphism

$$\lambda_{m+1}^{\mathfrak{M}} \colon \left(\Lambda_{\mathfrak{M}}^1(A)\right)^{\wedge(m+1)} \to \Lambda_{\mathfrak{M}}^{m+1}(A),$$

taking $d^{\mathfrak{M}} a_0 \wedge d^{\mathfrak{M}} a_1 \wedge \ldots \wedge d^{\mathfrak{M}} a_m$ to $d^{\mathfrak{M}} a_0 | d^{\mathfrak{M}} a_1 | \ldots | d^{\mathfrak{M}} a_m$.

On the other hand,

$$(a_0, a_1, \ldots, a_m) \mapsto d^{\mathfrak{M}} a_0 \wedge d^{\mathfrak{M}} a_1 \wedge \ldots \wedge d^{\mathfrak{M}} a_m$$

is a $(m+1)$-derivation of the algebra A with values in the A-module $\left(\Lambda^1_{\mathfrak{M}}(A)\right)^{\wedge(m+1)}$. Therefore, having in mind the universality of the $(m+1)$-derivation

$$(a_0, a_1, \ldots, a_m) \mapsto d^{\mathfrak{M}} a_0 | d^{\mathfrak{M}} a_1 | \ldots | d^{\mathfrak{M}} a_m,$$

we see that there is a homomorphism

$$\nu^{m+1}_{\mathfrak{M}} \colon \Lambda^{m+1}_{\mathfrak{M}}(A) \to \left(\Lambda^1_{\mathfrak{M}}(A)\right)^{\wedge(m+1)},$$

taking $d^{\mathfrak{M}} a_0 | d^{\mathfrak{M}} a_1 | \ldots | d^{\mathfrak{M}} a_m$ to $d^{\mathfrak{M}} a_0 \wedge d^{\mathfrak{M}} a_1 \wedge \ldots \wedge d^{\mathfrak{M}} a_m$. This shows that $\lambda^{\mathfrak{M}}_{m+1}$ and $\nu^{m+1}_{\mathfrak{M}}$ are mutually inverse isomorphisms. ▶

It will be useful for the reader to compare the above proof with the one given in Section 17.10 for $m = 2$ and based on a direct calculation.

Since the isomorphism $\lambda^{\mathfrak{M}}_m$ identifies the modules $\Lambda^m_{\mathfrak{M}}(A)$, $\left(\Lambda^1_{\mathfrak{M}}(A)\right)^{\wedge m}$, and $\Lambda^m_{\mathfrak{M}}(A)$, it follows that the temporary symbol $d^{\mathfrak{M}} a_0 | d^{\mathfrak{M}} a_1 | \ldots | d^{\mathfrak{M}} a_m$ can be replaced by the standard one, i.e., $d^{\mathfrak{M}} a_0 \wedge d^{\mathfrak{M}} a_1 \wedge \ldots \wedge d^{\mathfrak{M}} a_m$.

17.31. The Kahler differentials. It should be especially noted that in algebraic geometry, the module of differential k-forms over a commutative algebra, usually denoted by Ω^k, is defined as $\left(\Omega^1\right)^{\wedge k}$, while Ω^1, as the module of *Kahler differentials*. From our point of view, $\Omega^1 = \Lambda^1_{\mathrm{Mod}A}$. This approach looks quite natural from the historical perspective, however, it has the drawback that it misses certain important natural connections of the calculus of differential forms with other structures of the differential calculus and, above all, with the Spencer complexes, which will be discussed below. And this approach cannot be carried over directly to tensors of other types and leads to definite difficulties in passing to graded algebras. Under it, there remain "shadows" of the dependence of differential forms on the choice of the category of modules.

18

The (Co)Chain Complexes Coming from the Spencer Sequence

In this chapter, we continue our study of the Spencer sequence and the related diffunctors. We discover that the Spencer Diff-sequence is a chain complex and that the Spencer jet complex is a cochain complex. Further, along the same lines, we obtain the de Rham cochain complex and continue our study of differential forms, moving in the direction of their conceptual definition.

18.1. The diffunctors of Spencer complexes. The exact sequence of diffunctors (17.53) is the initial segment of the *Spencer* Diff-*sequence* of order $k + 1$, which is formed by the Spencer transformations

$$S_{k+1}^r, \ 0 \leq r \leq k + 1,$$

(see (17.52)). This whole sequence is of order m, it is denoted by Spen_m, and has the following form:

$$0 \to D_m \to D_{m-1}^<(\mathrm{Diff}_1^>) \xrightarrow{S_m^{m-1}} D_{m-2}^<(\mathrm{Diff}_2^>) \xrightarrow{S_m^{m-2}} \ldots$$

$$\ldots \xrightarrow{S_m^{l+1}} D_l(\mathrm{Diff}_{m-l}^>) \xrightarrow{S_m^l} D_{l-1}(\mathrm{Diff}_{m-l+1}^>) \xrightarrow{S_m^{l-1}} \ldots \qquad (18.1)$$

$$\ldots \xrightarrow{S_m^2} D_1^<(\mathrm{Diff}_{m-1}^>) \xrightarrow{S_m^1} \mathrm{Diff}_m^< \to 0.$$

Theorem. *The Spencer* Diff-*sequence is a chain complex.*

◀ For $m = 0$, we assume that $D_0 = \mathrm{id}$ (the identity functor), so that the sequences Spen_0 and Spen_1 are of the form

$$0 \to \mathrm{id} \to 0 \quad \text{and} \quad 0 \to D = D_1 \to \mathrm{Diff}_1 \to 0,$$

© Springer Nature Switzerland AG 2020
J. Nestruev, *Smooth Manifolds and Observables*, Graduate Texts
in Mathematics 220, https://doi.org/10.1007/978-3-030-45650-4_18

respectively. Since there is nothing to prove in these two cases, we assume that $m > 1$.

We must show that $\operatorname{Im} S_m^{l+1} \subset \operatorname{Ker} S_m^l$. To this end, it suffices to show that $\operatorname{Im} D_l^<(C_{1,m-l-1}^\circ) \subset \operatorname{Ker} S_m^l$, since $\operatorname{Im} S_m^{l-1} \subset \operatorname{Im} D_l^<(C_{1,m-l-1}^\circ)$ (see (17.52)). To do this, we shall use the fact that

$$D_{l+1} = D_{l-1}(D_2 \subset \mathrm{Df}_2^>)\,,$$

so that

$$D_{l+1}(\operatorname{Diff}_{m-l-1}^>) = D_{l-1}(D_2(\operatorname{Diff}_{m-l-1}^>) \subset \mathrm{Df}_2^>(\operatorname{Diff}_{m-l-1}^>))\,. \qquad (18.2)$$

Here

$$\begin{aligned} D_2(\operatorname{Diff}_{m-l-1}^>)) &= D(D(\operatorname{Diff}_{m-l-1}^>)) \subset \operatorname{Diff}_1^>(\operatorname{Diff}_{m-l-1}^>))),\\ \mathrm{Df}_2^>(\operatorname{Diff}_{m-l-1}^>)) &= \operatorname{Diff}_1^>(D(\operatorname{Diff}_{m-l-1}^>)) \subset \operatorname{Diff}_1^>(\operatorname{Diff}_{m-l-1}^>))) \end{aligned} \qquad (18.3)$$

On the other hand, by definition (see (17.52))

$$S_m^{l+1} = D_l(C_{1,m-l-1}^>) \circ \mathsf{t}_{l+1,m-l-1}\,.$$

Since $\mathsf{t}_{l+1,m-l-1}$ is an embedding, we can assume that the transformation S_m^{l+1} is induced by the action of the gluing transformation $C_{1,m-l-1}^>$ on the right-hand side of the equality (18.2). Taking into account the fact that $C_{1,m-l-1}^>$ transforms the diffunctor $D(\operatorname{Diff}_{m-l-1}^>)$ into $D_{(m-l)}$, and $\operatorname{Diff}_1^>(\operatorname{Diff}_{m-l-1}^>)$ into $\operatorname{Diff}_{(m-l)}^>$, we see that, by (18.3), it induces the transformations

$$\begin{aligned} D_2(\operatorname{Diff}_{m-l-1}^>)) &\to D(D_{(m-l)} \subset \operatorname{Diff}_{m-l}^>) = D_{(m-1,1)},\\ \mathrm{Df}_2^>(\operatorname{Diff}_{m-l-1}^>)) &\to \operatorname{Diff}_1^>(D_{(m-l)} \subset \operatorname{Diff}_{m-l}^>) = \mathrm{Df}_{(m-l,1)}^>\,. \end{aligned}$$

Finally, in view of (18.2), these transformations induce the transformation of functors

$$D_{l+1}(\operatorname{Diff}_{m-l-1}^>) \to D_{l-1}(D_{(m-1,1)} \subset \mathrm{Df}_{(m-l,1)}^>) = \operatorname{Ker} S_m^l$$

(see the Proposition from Section 17.24).   ▶

The projection $Д_{m,P} \colon \operatorname{Diff}_m^< P \to P$ can be regarded as the value on the module P of the natural transformation of diffunctors $Д_m \colon \operatorname{Diff}_m^< \to \operatorname{Diff}_0^<$ (see (15.13).

Exercise. Show that the composition of the transformations of diffunctors $Д_m \circ S_m^1$ vanishes.

This fact allows us to obtain the *extended Spencer Diff-sequence* by replacing the fragment $\operatorname{Diff}_m^< \to 0$ in (18.1) by $\operatorname{Diff}_m^< \xrightarrow{Д_m^<} \operatorname{Diff}_0^< \to 0$.

Thus, the Spencer Diff-sequence Spen_m determines a chain-complex-valued diffunctor (see Section 17.5). Its value on the A-module Q is denoted by $\operatorname{Spen}_m(Q)$ and is called the *Spencer Diff-complex of the module Q*.

18.2 Lemma–Exercise. *Show that for $r \geq 0$ the following diagram commutes:*

$$
\begin{array}{ccc}
D_k(\text{Diff}_{l+r}^{>}) & \xrightarrow{S_{k+l+r}^{l}} & D_{k-1}(\text{Diff}_{l+r+1}^{>}) \\
\uparrow & & \uparrow \\
D_k(\text{Diff}_{l}^{>}) & \xrightarrow{S_{k+l}^{l}} & D_{k-1}(\text{Diff}_{l+1}^{>})
\end{array}
\qquad (18.4)
$$

In view of (18.4), the natural embeddings of diffunctors

$$
D_l(\text{Diff}_{m-l}) \to D_l(\text{Diff}_{m-l+r}), \quad 0 \leq l \leq m,
$$

define an embedding of the Spencer sequence Spen_m into Spen_{m+r}, denoted by $\text{sp}_{m,m+r}$, which is a homomorphism of chain complexes. The direct limit of the chain of embeddings

$$
\text{Spen}_0 \xrightarrow{\text{sp}_{0,1}} \cdots \xrightarrow{\text{sp}_{m-1,m}} \text{Spen}_m \xrightarrow{\text{sp}_{m,m+1}} \text{Spen}_{m+1} \xrightarrow{\text{sp}_{m+1,m+2}} \cdots, \qquad (18.5)
$$

denoted by Spen_*, is called the *infinite Spencer* Diff-*sequence*. It is a chain-complex-valued diffunctor, *filtered* by the chain-complex-valued diffunctors Spen_m.

18.3. Spencer Diff-complexes. Applying the sequence of diffunctors (18.1) to an A-module P, we obtain a finite or infinite *Spencer* Diff-*complex*, denoted by $\text{Spen}_m(P)$ or $\text{Spen}_*(P)$, respectively. The complex $\text{Spen}_*(P)$ is filtered by the subcomplexes $\text{Spen}_m(P)$. The *extended* version of the complex $\text{Spen}_m(P)$ ends with the fragment

$$
\text{Diff}_m^{<} P \xrightarrow{\text{Л}_{m,P}^{<}} P \to 0, \quad 0 \leq m \leq \infty.
$$

It should be noted that the differentials $S_m^l = S_m^l(P)$ of the chain complexes $\text{Spen}_m(P)$, $0 \leq m \leq \infty$, are A-homomorphisms. Therefore, the homology groups of these chain complexes are A-modules, and so one can speak of their supports in $|A|$. If P is a projective module over the algebra $C^\infty(M)$, then the zero-dimensional homology group of the complex $\text{Spen}_m(P)$ is isomorphic to P, while the other groups are trivial. If P is not projective, then the homology groups of nonzero dimension, generally speaking, are nontrivial and their support consists of those points of the space $|A|$ where projectivity fails. In particular, for manifolds with singularities, this support coincides with the set of singular points. Moreover, by means of Spencer Diff-complexes, we can construct important homology invariants of differential operators.

Namely, let $\Delta \in \text{Diff}_k(P,Q)$. The homomorphism

$$
h_*^\Delta \colon \text{Diff}^{>} P \to \text{Diff}^{>} Q
$$

(see Section 15.10) obviously induces a homomorphism of sequences

$$
\text{sp}_\Delta \colon \text{Spen}_* P \to \text{Spen}_* Q,
$$

which is a homomorphism of A-modules that shifts the filtration by k, i.e.,

$$\mathrm{sp}_\Delta(\mathrm{Spen}_m(P)) \subset \mathrm{Spen}_{m+k}(Q), \ m \geq 0. \tag{18.6}$$

Exercise. Show that sp_Δ is a morphism (chain map) of complexes.

Further, denote the homology group of the complex $\mathrm{Spen}_*(R)$ at the term $D_i(\mathrm{Diff}^> R)$ by $H_i(\mathrm{Spen}_*(R))$. Being a chain map, sp_Δ induces a homomorphism of the homology group

$$H_i(\mathrm{sp}_\Delta)\colon H_i(\mathrm{Spen}_*(P)) \to H_i(\mathrm{Spen}_*(Q)). \tag{18.7}$$

Since sp_Δ is a homomorphism of A-modules, then so are the maps $H_i(\mathrm{sp}_\Delta)$. This is a homology invariant of the operator Δ. Moreover, one may consider the complexes $\mathrm{Ker}(\mathrm{sp}_\Delta)$, $\mathrm{Im}(\mathrm{sp}_\Delta)$, and $\mathrm{coker}(\mathrm{sp}_\Delta)$. Their homology is also an invariant of the operator Δ. Homology groups of this kind play an important role in the contemporary geometrical theory of differential equations (see [3]).

18.4. Spencer δ-complexes. A standard construction of the homological algebra relates a certain spectral sequence to the filtration of the complex $\mathrm{Spen}_*(P)$ by the subcomplexes $\mathrm{Spen}_m(P)$. Its detailed study lies outside of the scope of this book, and here we will only describe its initial term. By definition, it consists of the quotient complexes

$$\mathrm{Sml}_m(Q) \overset{\mathrm{def}}{=} \mathrm{Spen}_m(Q)/\mathrm{Spen}_{m-1}(Q),$$

which are called *Spencer $d\delta$-complexes* or *δ-Diff-complexes*. The homology groups of these complexes are called the *δ-homology groups of the module Q of order m*. They are A-modules and constitute the terms of the above mentioned spectral sequence that follow the initial term.

In view of (18.6), the homomorphism sp_Δ induces a homomorphism of quotient complexes

$$\mathrm{sd}_\Delta^m\colon \mathrm{Sml}_m(P) \to \mathrm{Sml}_{m+k}(Q)$$

and therefore, a homomorphism of δ-homology

$$H_i(\mathrm{sd}_\Delta^m)\colon H_i(\mathrm{Sml}_m(P)) \to H_i(\mathrm{Sml}_{m+k}(Q)).$$

Obviously, the maps $H_i(\mathrm{sd}_\Delta^m)$ are homomorphisms of A-modules. As in Section 18.3, the groups $H_i(\mathrm{Sml}_m(Q))$ are invariants of the module Q, while the A-homomorphisms $H_i(\mathrm{sd}_\Delta^m)$ and the homology of the chain complexes $\mathrm{Ker}(\mathrm{sd}_\Delta^m)$, $\mathrm{Im}(\mathrm{sd}_\Delta^m)$, and $\mathrm{coker}(\mathrm{sd}_\Delta^m)$ are invariants of the operator Δ.

Exercise. Let $A = C^\infty(M)$, $Q = \Gamma(\pi)$, and $m > 0$. Show that the homology groups $H_i(\mathrm{Sml}_m(Q))$, $i > 0$, are trivial and deduce from this that $H_0(\mathrm{Spen}_m(Q)) = Q$ and $H_i(\mathrm{Spen}_m(Q)) = 0$, $i > 0$.

18.5. Spencer jet cochain complexes. Since the representing objects for the diffunctors $\mathrm{Diff}_k^<$ and D_l in a differentially closed category exist, it follows that, according to Section 17.16, the representing objects for $D_l(\mathrm{Diff}_k^>)$

also exist. Therefore, the complex representing the extended diffunctor Spen_m has the following form:

$$0 \to A \xrightarrow{i_m^{\mathfrak{M}}} \mathcal{J}_<^m(A; \mathfrak{M}) \xrightarrow{S_{1,\mathfrak{M}}^m} \mathcal{J}_<^{m-1}(\Lambda_{\mathfrak{M}}^1; \mathfrak{M}) \xrightarrow{S_{2,\mathfrak{M}}^m} \dots$$

$$\dots \xrightarrow{S_{l,\mathfrak{M}}^m} \mathcal{J}_<^{m-l}(\Lambda_{\mathfrak{M}}^l; \mathfrak{M}) \xrightarrow{S_{l+1,\mathfrak{M}}^m} \mathcal{J}_<^{m-l-1}(\Lambda_{\mathfrak{M}}^{l+1}; \mathfrak{M}) \xrightarrow{S_{l+2,\mathfrak{M}}^m} \dots \qquad (18.8)$$

$$m \dots \xrightarrow{S_{m-1,\mathfrak{M}}^m} \mathcal{J}_<^1(\Lambda_{\mathfrak{M}}^{m-1}; \mathfrak{M}) \xrightarrow{S_{m,\mathfrak{M}}^m} \Lambda_{\mathfrak{M}}^m \to 0,$$

where $i_m^{\mathfrak{M}}(a) = a j_m^{\mathfrak{M}}(1_A)$.

Theorem. *The sequence (18.8) is a cochain complex known as the* Spencer jet cochain complex of order m in the category $\mathfrak{M}$. *It is denoted by* $\mathrm{Spen}_{\mathfrak{M}}^m(A)$. ▶

Being representative objects of diffunctors for Spencer Diff-complexes, the Spencer jet complexes possess the naturality property. The cohomology groups of these complexes constitute *Spencer jet cohomology*. The cohomology group of the extended Spencer jet complex $\mathrm{Spen}_{\mathfrak{M}}^m$ at the term $\mathcal{J}_<^l(\Lambda_{\mathfrak{M}}^k; \mathfrak{M})$ will be denoted by $H^{m-i}(\mathrm{Spen}_{\mathfrak{M}}^m)$ (see Section 18.5). Since the initial fragment of the cochain complex $\mathrm{SP}_{\mathfrak{M}}^m$ has the form

$$0 \to A \xrightarrow{i_m^{\mathfrak{M}}} \mathcal{J}_<^m(A; \mathfrak{M}) \xrightarrow{S_{1,\mathfrak{M}}^m} \mathcal{J}_<^{m-1}(\Lambda_{\mathfrak{M}}^1; \mathfrak{M}),$$

it follows that $H^0(\mathrm{Spen}_{\mathfrak{M}}^m) \supset \mathrm{Im}\, i_m^{\mathfrak{M}} \cong A$, and the cohomology class corresponding to the element $a \in A$ is represented by the cocycle $a j_m^{\mathfrak{M}}(1_A)$ (see (18.8)).

18.6. Spencer differentials. The above conceptual definition of the A-homomorphisms $S_{l,\mathfrak{M}}^m$, also called *Spencer differentials*, is not too convenient from the practical point of view, and how this shortcoming can be overcome will be explained below.

Proposition. *Let $a \in A$, $\omega \in \Lambda_{\mathfrak{M}}^{k-1}$ and $l = m - k$. Then*

$$S_{k,\mathfrak{M}}^{k+l}(a j_{l+1}^{\mathfrak{M}}(\omega)) = a j_l^{\mathfrak{M}}(d^{\mathfrak{M}}\omega). \qquad (18.9)$$

◀ Recall that the homomorphism $S_{l+1,\mathfrak{M}}^{k+l}$ represents the transformation of diffunctors S_{k+l}^{l+1} defined as the composition of transformations

$$D_k(\mathrm{Diff}_l^>) \xrightarrow{\mathsf{t}_{k,l}} D_{k-1}(\mathrm{Diff}_1^>(\mathrm{Diff}_l^>)) \xrightarrow{D_{k-1}(C_{1,l}^>)} D_{k-1}(\mathrm{Diff}_{l+1}^>)$$

(see (17.52)). Therefore, $S_{l+1,\mathfrak{M}}^{k+l}$ is the composition of the homomorphisms representing the transformations $D_{k-1}(C_{1,l}^>)$ and $\mathsf{t}_{k,l}$, respectively. Let us describe them separately.

First of all, note that the diffunctor $D_{k-1}(\mathrm{Diff}_1^>(\mathrm{Diff}_l^>))$ in the category $\mathfrak{M}$ is represented by the A-module $\mathcal{J}_<^l(\mathcal{J}_<^1(\Lambda_{\mathfrak{M}}^{k-1}; \mathfrak{M}); \mathfrak{M})$. Indeed, if $P \in \mathrm{Ob}\,\mathfrak{M}$, then, in view of the isomorphism (15.10), and the results of

Sections 15.20 and 17.3, we see that

$$
\begin{aligned}
D_{k-1}(\mathrm{Diff}_1^>(\mathrm{Diff}_l^>(P))) &= \mathrm{Hom}_A(\Lambda_{\mathfrak{M}}^{k-1}, \mathrm{Diff}_1^>(\mathrm{Diff}_l^>(P))) \\
&= \mathrm{Hom}_A(\Lambda_{\mathfrak{M}}^{k-1}, \mathrm{Hom}_A(_<\mathcal{J}_>^l(\mathfrak{M}) \otimes_A \mathcal{J}_<^1(\mathfrak{M}), P)) \\
&= \mathrm{Hom}_A(_<\mathcal{J}_>^l(\mathfrak{M}) \otimes_A {}_<\mathcal{J}_>^1(\mathfrak{M}) \otimes_A \Lambda_{\mathfrak{M}}^{k-1}, P)
\end{aligned}
$$

It remains to notice that

$$
<\mathcal{J}>^l(\mathfrak{M}) \otimes_A {}_<\mathcal{J}_>^1(\mathfrak{M}) \otimes_A \Lambda_{\mathfrak{M}}^{k-1} = \mathcal{J}_<^l(\mathcal{J}_<^1(\Lambda_{\mathfrak{M}}^{k-1}; \mathfrak{M}); \mathfrak{M}).
$$

Further, in view of the fact that the transformation $D_{k-1}(C_{1,l}^>)$ is induced by the transformation $C_{1,l}^>$, the homomorphism

$$
\mathcal{J}_<^{l+1}(\Lambda_{\mathfrak{M}}^{k-1}; \mathfrak{M}) \longrightarrow \mathcal{J}_<^l(\mathcal{J}_<^1(\Lambda_{\mathfrak{M}}^{k-1}; \mathfrak{M}); \mathfrak{M}),
$$

representing this transformation is induced by the homomorphism representing $C_{1,l}^>$. Therefore, from the results of Section 17.3, we see that

$$
\begin{aligned}
j_{l+1}^{\mathfrak{M}}(\omega) = j_{l+1}^{\mathfrak{M}}(1) \otimes_A^> \omega \mapsto j_l^{\mathfrak{M}}(1) \otimes_A^> j_1^{\mathfrak{M}}(1) \otimes_A^> \omega \\
= j_l^{\mathfrak{M}}(1) \otimes_A^> j_1^{\mathfrak{M}}(\omega) = j_l^{\mathfrak{M}}(j_1^{\mathfrak{M}}(\omega)),
\end{aligned} \tag{18.10}
$$

where $\omega \in \Lambda_{\mathfrak{M}}^{k-1}$ (see 15.41).

In its turn, the transformation $\mathsf{t}_{k,0}$ is induced by the embedding $D_k \to D_{k-1}(\mathrm{Diff}_1^>)$, represented by the homomorphism

$$
S_{k,\mathfrak{M}}^k \colon \mathcal{J}_<^1(\Lambda_{\mathfrak{M}}^{k-1}; \mathfrak{M}) \to \Lambda_{\mathfrak{M}}^k, \quad j_1^{\mathfrak{M}}(\omega) \mapsto d^{\mathfrak{M}}(\omega).
$$

Recall (see (17.33) and Section 17.16) that by definition

$$
d^{\mathfrak{M}} = d_{k-1}^{\mathfrak{M}} \overset{\mathrm{def}}{=} S_{k,\mathfrak{M}}^k \circ j_1^{\mathfrak{M}}.
$$

Thus, the homomorphism representing $\mathsf{t}_{k,l}$

$$
\mathcal{J}_<^l(\mathcal{J}_<^1(\Lambda_{\mathfrak{M}}^{k-1}; \mathfrak{M}); \mathfrak{M}) \to \mathcal{J}_<^l(\Lambda_{\mathfrak{M}}^k; \mathfrak{M})
$$

takes $j_l^{\mathfrak{M}}(j_1^{\mathfrak{M}}(\omega))$ to $j_l^{\mathfrak{M}}(d^{\mathfrak{M}}(\omega))$. Together with (18.10), this shows that

$$
S_{k,\mathfrak{M}}^{k+l}(j_{l+1}^{\mathfrak{M}}(\omega)) = j_l^{\mathfrak{M}}(d^{\mathfrak{M}}\omega).
$$

But since $S_{k,\mathfrak{M}}^{k+l}$ is an A-homomorphism, this implies (18.9).   ▶

18.7. Spencer δ-complexes. The *Spencer $j\delta$-complex*, or *δ-jet complex* in a differentially closed category $\mathfrak{M}$ of A-modules is the complex representing the complex-valued diffunctor $\mathrm{Sml}_{\mathfrak{m}}^{\mathfrak{M}}$ (see Section 18.4). The number m is called its *order*. According to our general principles, this cochain complex, denoted by $\mathrm{Csm}_{\mathfrak{M}}^m$, is isomorphic to the complex which is the kernel of the projection $\mathrm{Spen}_{\mathfrak{M}}^m \to \mathrm{Spen}_{\mathfrak{M}}^{m-1}$. The cohomology groups of this cochain complex constitute the *Spencer δ-cohomology of order m in the category $\mathfrak{M}$* or simply the Spencer δ-cohomology if the rest is clear from the context. Since the Spencer coboundary operators are A-module homomorphisms, Spencer δ-cohomology groups, just as Spencer cohomology groups, possess an A-module structure.

Spencer δ-cohomology is very useful for the computation of Spencer cohomology. Indeed, the Spencer δ-cohomology is the first term of the spectral sequence that *represents* the spectral sequence diffunctor mentioned in Section 18.4. According to the definition, there is a short exact sequence of cochain complexes

$$0 \to \mathrm{Csm}_{\mathfrak{M}}^m \to \mathrm{Spen}_{\mathfrak{M}}^m \to \mathrm{Spen}_{\mathfrak{M}}^{m-1} \to 0$$

and the corresponding long exact cohomology sequence will be

$$\cdots \to H^i(\mathrm{Csm}_{\mathfrak{M}}^m) \to H^i(\mathrm{Spen}_{\mathfrak{M}}^m) \to H^i(\mathrm{Spen}_{\mathfrak{M}}^{m-1}) \to \ldots \qquad (18.11)$$

However, they are also interesting in themselves. For instance, if some δ-cohomology group in a geometrical category is not a projective A-module, then the algebra A itself is not smooth (see the proposition 20.31). Moreover, the points at which this module is not projective are singular points of the spectrum of the algebra A itself. For example, we will see that such is the "crossing point" of the cross $\mathbf{K}$.

Remark. It is clear that the Spencer (δ-)homology and cohomology groups are very delicate invariants of singular points of smooth manifolds and algebraic varieties. Nevertheless, this very natural approach to the description of such singular points remains practically unstudied.

As is easy to see, the δ-complex $\mathrm{Csm}_{\mathfrak{M}}^m$ is a subcomplex of the Spencer complex $\mathrm{Spen}_{\mathfrak{M}}^m$, whose terms generated by elements of the form

$$\delta_{m,\mathfrak{M}}^{a_1,\ldots,a_m} \cdot j_m(\omega), \ a_1,\ldots,a_m \in A, \ \omega \in \Lambda_{\mathfrak{M}}^s(A), \ s \in \mathbb{N}.$$

In particular, for a smooth category $\mathfrak{M}$ the complex $\mathrm{Csm}_{\mathfrak{M}}^m$ has the form

$$0 \to \mathrm{Csml}_{\mathfrak{M}}^m(A) \xrightarrow{\delta_{1,\mathfrak{M}}^m} \mathrm{Csml}_{\mathfrak{M}}^{m-1}(A) \otimes_A \Lambda_{\mathfrak{M}}^1(A) \xrightarrow{\delta_{2,\mathfrak{M}}^m} \ldots$$

$$\xrightarrow{\delta_{m-k,\mathfrak{M}}^m} \mathrm{Csml}_{\mathfrak{M}}^k(A) \otimes_A \Lambda_{\mathfrak{M}}^{m-k}(A) \xrightarrow{\delta_{m-k+1,\mathfrak{M}}^m} \mathrm{Csml}_{\mathfrak{M}}^{k-1}(A) \otimes_A \Lambda_{\mathfrak{M}}^{m-k+1}(A)$$

$$\ldots \xrightarrow{\delta_{m-1,\mathfrak{M}}^m} \mathrm{Csml}_{\mathfrak{M}}^1(A) \otimes_A \Lambda_{\mathfrak{M}}^{m-1}(A) \xrightarrow{\delta_{m,\mathfrak{M}}^m} \Lambda_{\mathfrak{M}}^m(A) \to 0. \qquad (18.12)$$

Here the action of the differential $\delta_{m-k+1,\mathfrak{M}}^m$ is described by the following formula

$$\delta_{m-k+1,\mathfrak{M}}^m(\delta_{k,\mathfrak{M}}^{a_1,\ldots,a_k} \cdot j_k^{\mathfrak{M}}(\omega)) = \sum_{l=1}^{k} \delta_{k-1,\mathfrak{M}}^{a_1,\ldots,\widehat{a_l},\ldots,a_k} \cdot j_{k-1}^{\mathfrak{M}}(d^{\mathfrak{M}} a_l \wedge \omega) \quad (18.13)$$

Indeed, $\delta_{m-k+1,\mathfrak{M}}^m(\delta_{k,\mathfrak{M}}^{a_1,\ldots,a_k}) = \left(S_{m-k+1,\mathfrak{M}}^m\right)\big|_{\mathrm{Csml}_{\mathfrak{M}}^k(A)}$, while $S_{m-k+1,\mathfrak{M}}^m$ is described by the formula (18.9). Besides,

$$\delta_{k,\mathfrak{M}}^{a_1,\ldots,a_k} = \sum_{|I| \leq k} (-1)^I a_I j_k^{\mathfrak{M}}(a_J),$$

where $I = \{i_1, \ldots, i_k\} \subset \{1, 2, \ldots, k\}$, $J = \{1, 2, \ldots, k\} \setminus I$. Therefore,

$$\delta^m_{m-k+1,\mathfrak{M}}(\delta^{a_1,\ldots,a_k}_{k,\mathfrak{M}} \cdot j^{\mathfrak{M}}_k(\omega)) = \sum_{|I| \leq k} (-1)^I a_I j^{\mathfrak{M}}_{k-1}(d(a_J\omega)) =$$

$$\sum_{|I| \leq k} (-1)^I a_I j^{\mathfrak{M}}_{k-1}(a_J d\omega) + \sum_{|I| \leq k} (-1)^I a_I \sum_{j_s \in J} j^{\mathfrak{M}}_{k-1}(a_{J_s} da_{j_s} \wedge \omega), \quad (18.14)$$

where $J_s = J \setminus \{j_s\}$. Note that $\delta^{a_1,\ldots,a_k}_{k-1,\mathfrak{M}} = 0$ and so we see that

$$\sum_{|I| \leq k} (-1)^I a_I j^{\mathfrak{M}}_{k-1}(a_J d\omega) = \delta^{a_1,\ldots,a_k}_{k-1,\mathfrak{M}} \cdot j^{\mathfrak{M}}_k(d\omega)) = 0$$

as well as

$$\sum_{|I| \leq k} (-1)^I a_I \sum_{j_s \in J} j^{\mathfrak{M}}_{k-1}(a_{J_s} da_{j_s} \wedge \omega) = \sum_{I \subset I_l} (-1)^I a_I \sum_{l \in J_l} j^{\mathfrak{M}}_{k-1}(a_{J_l} da_l \wedge \omega),$$

where $I_l = \{1, \ldots, \hat{l}, \ldots, k\}$, $J_l = I_l \setminus I$. It remains to note that

$$\sum_{I \subset I_l} (-1)^I a_I \sum_{l \in J_l} j^{\mathfrak{M}}_{k-1}(a_{J_l} da_l \wedge \omega) = \sum_{l=1}^{k} \delta^{a_1,\ldots,\widehat{a_l},\ldots,a_k}_{k-1,\mathfrak{M}} \cdot j^{\mathfrak{M}}_{k-1}(d^{\mathfrak{M}}a_l \wedge \omega).$$

Using the fact that the homomorphism $sc^k_{\mathfrak{M}}$ (formula (16.8)) is an isomorphism in the smooth case (see Section 16.11), it is useful to rewrite formula (18.13) in more expressive form

$$\delta^m_{m-k+1,\mathfrak{M}}((d^{\mathfrak{M}}a_1 \odot \cdots \odot d^{\mathfrak{M}}a_k) \otimes_A \omega) =$$

$$\sum_{i=1}^{k} (d^{\mathfrak{M}}a_1 \odot \cdots \odot \widehat{d^{\mathfrak{M}}a_i} \odot \cdots \odot d^{\mathfrak{M}}a_k) \otimes_A (d^{\mathfrak{M}}a_i \wedge \omega). \quad (18.15)$$

Note that the complexes $\mathrm{Csm}^0_{\mathfrak{M}}$ and $\mathrm{Csm}^1_{\mathfrak{M}}$ have the form $0 \to A \to 0$ and $0 \to \Lambda^1_{\mathfrak{M}} \xrightarrow{=} \Lambda^1_{\mathfrak{M}} \to 0$ and so their cohomology groups do not carry any useful information about the algebra A and the category $\mathfrak{M}$, unlike the complexes $\mathrm{Csm}^m_{\mathfrak{M}}$, $m \geq 2$.

18.8. The de Rham complex. The commutative diagram

$$\mathcal{J}^l_<(\Lambda^k_{\mathfrak{M}}; \mathfrak{M}) \xrightarrow{S^{k+l}_{k+1,\mathfrak{M}}} \mathcal{J}^{l-1}_<(\Lambda^{k+1}_{\mathfrak{M}}; \mathfrak{M}) \xrightarrow{S^{k+l}_{k+2,\mathfrak{M}}} \mathcal{J}^{l-2}_<(\Lambda^{k+2}_{\mathfrak{M}}; \mathfrak{M}) \quad (18.16)$$

$$\uparrow j^{\mathfrak{M}}_l \qquad\qquad \uparrow j^{\mathfrak{M}}_{l-1} \qquad\qquad \uparrow j^{\mathfrak{M}}_{l-2}$$

$$\Lambda^k_{\mathfrak{M}} \xrightarrow{\ d^{\mathfrak{M}}_k\ } \Lambda^{k+1}_{\mathfrak{M}} \xrightarrow{\ d^{\mathfrak{M}}_{k+1}\ } \Lambda^{k+2}_{\mathfrak{M}}$$

is a direct consequence of Proposition 18.9. It shows that

$$j^{\mathfrak{M}}_{l-2} \circ d^{\mathfrak{M}}_{k+1} \circ d^{\mathfrak{M}}_k = 0, \text{ since } S^{k+l}_{k+2,\mathfrak{M}} \circ S^{k+l}_{k+1,\mathfrak{M}} = 0.$$

But the map $j^{\mathfrak{M}}_r$ is injective, hence $d^{\mathfrak{M}}_k \circ d^{\mathfrak{M}}_{k+1} = 0$.

Definition. The sequence

$$0 \to A = \Lambda^0_{\mathfrak{M}} \xrightarrow{d^{\mathfrak{M}}_0} \Lambda^1_{\mathfrak{M}} \xrightarrow{d^{\mathfrak{M}}_1} \cdots \xrightarrow{d^{\mathfrak{M}}_{k-1}} \Lambda^k_{\mathfrak{M}} \xrightarrow{d^{\mathfrak{M}}_k} \Lambda^k_{\mathfrak{M}} \xrightarrow{d^{\mathfrak{M}}_{k+1}} \cdots \qquad (18.17)$$

is a cochain complex, called the *de Rham complex of the algebra A in the category $\mathfrak{M}$* and denoted by $\mathrm{dR}_{\mathfrak{M}}(A)$. The cohomology of this complex is called the *de Rham cohomology of the algebra A in the category $\mathfrak{M}$*.

Note also that the family of maps

$$j^{\mathfrak{M}}_{m-k} \colon \Lambda^k_{\mathfrak{M}} \to \mathcal{J}^{m-k}_<(\Lambda^k_{\mathfrak{M}}; \mathfrak{M}), \quad 0 \le k \le m,$$

defines a chain morphism of the complexes $\mathrm{dR}_{\mathfrak{M}}(A) \to \mathrm{Spen}^m_{\mathfrak{M}}$. This follows from the commutativity of the diagram (18.16).

18.9. The exterior differential and the exterior product. A remarkable fact is that the exterior differential is a *graded* derivation of the exterior product in a sense that will be explained below. Moreover, using exterior multiplication, we can give a simpler and practically more convenient description of this differential. To do this, let us use the interpretation of the diffunctor D_{k+1} as $D_{k,1}$ (see Section 17.22) and recall that $d^{\mathfrak{M}}_k$ is defined as the composition

$$\Lambda^k_{\mathfrak{M}} \xrightarrow{j^{\mathfrak{M}}_1} \mathcal{J}^1_<(\Lambda^k_{\mathfrak{M}}; \mathfrak{M}) \xrightarrow{S^k_{k,\mathfrak{M}}} \Lambda^{k+1}_{\mathfrak{M}},$$

in which the homomorphism $S^k_{k,\mathfrak{M}}$ represents the embedding of diffunctors

$$D_{k+1} = D_{k,1} = D_k(D \subset \mathrm{Diff}^>_1) \xrightarrow{t_{k+1,0}} D_k(\mathrm{Diff}^>_1),$$

i.e., $S^k_{k,\mathfrak{M}} = h^{\mathfrak{M}}_{d^{\mathfrak{M}}_k}$ (see Section 15.9). This embedding assigns to each P-valued $(k+1)$-derivation $\varphi(a_0, a_1, \ldots, a_k)$ a $\mathrm{Diff}^>_1$ P-valued k-derivation

$$(a_1, \ldots, a_k) \mapsto \psi(a_1, \ldots, a_k) \in \mathrm{Diff}^>_1 P,$$

such that

$$\psi(a_1, \ldots, a_k)(a) \overset{\mathrm{def}}{=} \varphi(a, a_1, \ldots, a_k). \qquad (18.18)$$

This is equivalent to the fact that under the canonical isomorphism $D_k(\mathrm{Diff}^>_1(P)) = \mathrm{Hom}_A(\Lambda^k_{\mathfrak{M}}, \mathrm{Diff}^>_1 P)$, the k-derivation ψ corresponds to the homomorphism

$$d^{\mathfrak{M}} a_1 \wedge \ldots \wedge d^{\mathfrak{M}} a_k \mapsto \{a \mapsto \varphi(a, a_1, \ldots, a_k)\} \in \mathrm{Diff}^>_1 P.$$

In its turn, to the operator $\{a \mapsto \varphi(a, a_1, \ldots, a_k)\}$, the canonical isomorphism

$$\mathrm{Hom}_A(\Lambda^k_{\mathfrak{M}}, \mathrm{Diff}^>_1 P) = \mathrm{Hom}_A(\mathcal{J}^1_<(\Lambda^k_{\mathfrak{M}}; \mathfrak{M}), P)$$

assigns the homomorphism

$$j^{\mathfrak{M}}_1(a) \otimes^>_A (d^{\mathfrak{M}} a_1 \wedge \ldots \wedge d^{\mathfrak{M}} a_k) \mapsto \varphi(a, a_1, \ldots, a_k)$$

or, which is the same,

$$j_1^{\mathfrak{M}}(ad^{\mathfrak{M}}a_1 \wedge \ldots \wedge d^{\mathfrak{M}}a_k) \mapsto \varphi(a, a_1, \ldots, a_k). \qquad (18.19)$$

In particular, let us now suppose that $P = \Lambda_{\mathfrak{M}}^{k+1}$ while $\varphi = \varphi_{\mathrm{id}}$ is the $(k+1)$-derivation that corresponds to $\mathrm{id}_{\Lambda_{\mathfrak{M}}^{k+1}}$ under the canonical isomorphism $D_{k+1}(\Lambda_{\mathfrak{M}}^{k+1}) = \mathrm{Hom}_A(\Lambda_{\mathfrak{M}}^{k+1}, \Lambda_{\mathfrak{M}}^{k+1})$. Obviously,

$$\varphi_{\mathrm{id}}(a, a_1, \ldots, a_k) = d^{\mathfrak{M}}a \wedge d^{\mathfrak{M}}a_1 \wedge \ldots \wedge d^{\mathfrak{M}}a_k$$

(see Section 17.30). It follows from the proposition in Section 15.2 that the homomorphism $S_{k,\mathfrak{M}}^k \in \mathrm{Hom}_A(\mathcal{J}^1(\Lambda_{\mathfrak{M}}^k; \mathfrak{M}), \Lambda_{\mathfrak{M}}^{k+1}) = D_k(\mathrm{Diff}_1^>(\Lambda_{\mathfrak{M}}^{k+1}))$ is the image of the $(k+1)$-derivation φ_{id} under the composition of homomorphisms

$$\mathrm{Hom}_A(\Lambda_{\mathfrak{M}}^{k+1}, \Lambda_{\mathfrak{M}}^{k+1}) = D_{k+1}(\Lambda_{\mathfrak{M}}^{k+1}) \overset{t_{k+1,0}}{\longrightarrow}$$

$$\overset{t_{k+1,0}}{\longrightarrow} D_k(\mathrm{Diff}_1^>(\Lambda_{\mathfrak{M}}^{k+1})) = \mathrm{Hom}_A(\mathcal{J}^1(\Lambda_{\mathfrak{M}}^k; \mathfrak{M}), \Lambda_{\mathfrak{M}}^{k+1}), \qquad (18.20)$$

where the equality sign stands for the canonical isomorphisms. Therefore, from (18.19), we see that

$$S_{k,\mathfrak{M}}^k \colon j_1^{\mathfrak{M}}(ad^{\mathfrak{M}}a_1 \wedge \ldots \wedge d^{\mathfrak{M}}a_k) \mapsto d^{\mathfrak{M}}a \wedge d^{\mathfrak{M}}a_1 \wedge \ldots \wedge d^{\mathfrak{M}}a_k.$$

But since $d_k^{\mathfrak{M}} = S_{k,\mathfrak{M}}^k \circ j_1^{\mathfrak{M}}$, it follows that

$$d_k^{\mathfrak{M}}(ad^{\mathfrak{M}}a_1 \wedge \ldots \wedge d^{\mathfrak{M}}a_k) = d^{\mathfrak{M}}a \wedge d^{\mathfrak{M}}a_1 \wedge \ldots \wedge d^{\mathfrak{M}}a_k. \qquad (18.21)$$

This is the required description of the differential $d_k^{\mathfrak{M}}$, which is quite sufficient, since the differential forms $ad^{\mathfrak{M}}a_1 \wedge \ldots \wedge d^{\mathfrak{M}}a_k$ are additive generators of the module $\Lambda_{\mathfrak{M}}^k$.

Further, as a rule, we omit the index k and write $d^{\mathfrak{M}}$ instead of $d_k^{\mathfrak{M}}$. Similarly, just as in the case of smooth manifolds, we call the form $\omega \in \Lambda_{\mathfrak{M}}^k$ *closed* if $d^{\mathfrak{M}}\omega = 0$, and *exact* if $\omega = d^{\mathfrak{M}}\rho$.

Corollary. (a) *Any exact form* $\omega \in \Lambda_{\mathfrak{M}}^k$ *is the sum of forms*

$$d^{\mathfrak{M}}a_1 \wedge \ldots \wedge d^{\mathfrak{M}}a_k.$$

(b) $d^{\mathfrak{M}}(ad^{\mathfrak{M}}\omega) = d^{\mathfrak{M}}a \wedge d^{\mathfrak{M}}\omega.$

◄ Assertion (a) is a direct consequence of formulas (18.21) and the fact that the forms $a_1 \wedge d^{\mathfrak{M}}a_2 \wedge \ldots \wedge d^{\mathfrak{M}}a_k$ are additive generators of $\Lambda_{\mathfrak{M}}^{k-1}$. According to this same assertion, the form $d^{\mathfrak{M}}\omega$ is the sum of the forms $d^{\mathfrak{M}}a_0 \wedge d^{\mathfrak{M}}a_1 \wedge \ldots \wedge d^{\mathfrak{M}}a_k$, and assertion (b) again directly follows from formulas (18.21). ►

Exercise. Is the form $d_{\mathrm{alg}}e^x - e^x d_{\mathrm{alg}} \in \Lambda_{\mathrm{alg}}^1$ considered in Section 14.10 closed?

18.10 Proposition. *Let* $\omega \in \Lambda_{\mathfrak{M}}^k$ *and* $\rho \in \Lambda_{\mathfrak{M}}^l$. *Then*

$$d^{\mathfrak{M}}(\omega \wedge \rho) = d^{\mathfrak{M}}\omega \wedge \rho + (-1)^k \omega \wedge d^{\mathfrak{M}}\rho. \qquad (18.22)$$

◄ Since the forms

$$ad^{\mathfrak{M}}a_1 \wedge \ldots \wedge d^{\mathfrak{M}}a_m = ad^{\mathfrak{M}}(a_1 d^{\mathfrak{M}}a_2 \wedge \ldots \wedge d^{\mathfrak{M}}a_m)$$

are additive generators of $\Lambda_{\mathfrak{M}}^m$, we can assume that $\omega = ad^{\mathfrak{M}}\omega_1$ and $\rho = a'd^{\mathfrak{M}}\rho_1$. In this case, we have $\omega \wedge \rho = aa'd^{\mathfrak{M}}\omega_1 \wedge d^{\mathfrak{M}}\rho_1$. Here the form $d^{\mathfrak{M}}\omega_1 \wedge d^{\mathfrak{M}}\rho_1$ is exact (Corollary (a) from the previous section). Having in mind this corollary and formula (18.21), we see that

$$\begin{aligned}
d^{\mathfrak{M}}(\omega \wedge \rho) &= d^{\mathfrak{M}}(aa') \wedge d^{\mathfrak{M}}\omega_1 \wedge d^{\mathfrak{M}}\rho_1 \\
&= a'd^{\mathfrak{M}}a \wedge d^{\mathfrak{M}}\omega_1 \wedge d^{\mathfrak{M}}\rho_1 + ad^{\mathfrak{M}}a' \wedge d^{\mathfrak{M}}\omega_1 \wedge d^{\mathfrak{M}}\rho_1 \\
&= (d^{\mathfrak{M}}a \wedge d^{\mathfrak{M}}\omega_1) \wedge (a'd^{\mathfrak{M}}\rho_1) + (-1)^k (ad^{\mathfrak{M}}\omega_1) \wedge (d^{\mathfrak{M}}a' \wedge d^{\mathfrak{M}}\rho_1) \\
&= d^{\mathfrak{M}}\omega \wedge \rho + (-1)^k \omega \wedge d^{\mathfrak{M}}\rho. \quad \blacktriangleright
\end{aligned}$$

Formula (18.22) is an example of the *graded Leibnitz rule*, while the exterior differential $d^{\mathfrak{M}}$ is a *graded derivation of degree 1* of the algebra of differential forms.

18.11. Substituting vector fields into differential forms. As we mentioned before, in standard differential geometry, a differential form of degree k is understood as a skew-symmetric $C^\infty(M)$-polylinear function $\omega(X_1, \ldots, X_k)$ of vector arguments $X_i \in D(M)$. In these terms, the important operation $i_X \colon \Lambda^k(M) \to \Lambda^{k-1}(M)$ of substituting vector fields into differential forms has a simple description:

$$i_X(\omega)(X_1, \ldots, X_{k-1}) \overset{\text{def}}{=} \omega(X, X_1, \ldots, X_{k-1}).$$

Here we assume that $i_X|_{\Lambda^0} = 0$. (Recall that $\Lambda^0 = C^\infty(M)$.) Stated in that way, however, this operation is not convenient for generalizations to polyvectors and commutative algebras of general form. As we saw in Section 17.14, the conceptually correct approach consists of regarding i_X as a homomorphism that represents a certain transformation of diffunctors $i^X \colon D_{k-1} \to D_k$. We will describe this approach now, in the general situation of polyderivations.

Let $\nabla \in \mathfrak{D}_k(A)$ (see Section 17.18) and $\Delta \in \mathfrak{D}_l(P)$. Using the notation of Section 9.58, and having in mind that $n = k + l$, let us put

$$i_l^\nabla(\Delta)(a_1, \ldots, a_{k+l}) \overset{\text{def}}{=} \sum_{I+J=I^n, |I|=k} (-1)^{I,J} \nabla(a_I) \Delta(a_J), \qquad (18.23)$$

where $(-1)^{I,J}$ stands for the sign of the permutation of indices

$$\{1, \ldots, k+l\} \mapsto \{i_1, \ldots, i_k, j_1, \ldots, j_l\};$$

here $I = \{i_1, \ldots, i_k\}$, $J = \{j_1, \ldots, j_l\}$. It is obvious that $i_l^\nabla(\Delta) \in \mathfrak{D}_{k+l}(P)$, while $i_l^\nabla \colon \mathfrak{D}_l \to \mathfrak{D}_{k+l}$ is a transformation of diffunctors. It is also easy to see that the homomorphism $i_\nabla^{\mathfrak{M},l} \colon \Lambda_{\mathfrak{M}}^{k+l} \to \Lambda_{\mathfrak{M}}^k$, which represents this transformation in some differentially closed category $\mathfrak{M} \subset \mathrm{Mod}\, A$, is given by the formula

$$i_\nabla^{\mathfrak{M},l}(d^{\mathfrak{M}}a_1 \wedge \ldots \wedge d^{\mathfrak{M}}a_{k+l}) = \sum_{I+J=I^{k+l},\,|J|=l} (-1)^{I,J}\nabla(a_I)d^{\mathfrak{M}}(a_J), \qquad (18.24)$$

where $d^{\mathfrak{M}}(a_J) \stackrel{\text{def}}{=} d^{\mathfrak{M}}a_{j_1} \wedge \ldots \wedge d^{\mathfrak{M}}a_{j_l}$ if $J = (j_1, \ldots, j_l)$. By definition, we can write $i_\nabla^{\mathfrak{M},l} = 0$ if $l < 0$. Finally, bringing together the homomorphisms $\{i_\nabla^{\mathfrak{M},l}\}$, we obtain a "global" *substitution operator* or *inner product*

$$i_\nabla^{\mathfrak{M}} \stackrel{\text{def}}{=} \bigoplus_{l \geq 0} i_\nabla^{\mathfrak{M},l} : \Lambda_{\mathfrak{M}}^* \to \Lambda_{\mathfrak{M}}^*. \qquad (18.25)$$

Below, we will use the simplified notation $i_\nabla = i_\nabla^{\text{Mod}\,A}$ and $i_\nabla^\Gamma = i_\nabla^{\text{GMod}\,A}$. In particular, if $\nabla = X \in D(A)$, we use the standard notation i_X.

18.12. Exercise.

1. Show that

$$i_X^{\mathfrak{M}}(\omega \wedge \rho) = i_X^{\mathfrak{M}}(\omega) \wedge \rho + (-1)^m \omega \wedge i_X^{\mathfrak{M}}(\rho), \ \omega \in \Lambda_{\mathfrak{M}}^k. \qquad (18.26)$$

2. Let $X_1, \ldots, X_k \in D(A)$. Show that

$$i_{X_k}^{\mathfrak{M}} \circ i_{X_{k-1}}^{\mathfrak{M}} \circ \cdots \circ i_{X_1}^{\mathfrak{M}} = i_{X_1 \wedge \ldots \wedge X_k}^{\mathfrak{M}}. \qquad (18.27)$$

In particular, $i_X \circ i_Y = -i_Y \circ i_X$.

18.13. Naturality of differential forms. Let $\varphi \colon A \to B$ be a homomorphism of commutative algebras, while $\mathfrak{M}$ and $\mathcal{K}$ are φ-connected categories (see Section 15.24) of A- and B-modules, respectively. Then according to the approach of Section 15.25, there must exist a homomorphism of A-modules $\varphi^* = \varphi^*_{\mathfrak{M},\mathcal{K}} \colon \Lambda_{\mathfrak{M}}^k \to \Lambda_{\mathcal{K}}^k$, representing the diffunctor D_k in the categories $\mathfrak{M}$ and $\mathcal{K}$, respectively. As pointed out in that section, to this end it is necessary to describe the transformation $T_{D_k}^\varphi$. This is easy to do if we interpret D_k as the polyderivation diffunctor, i.e., as $\mathfrak{D}_k$. Indeed, from this point of view, $T_{D_k}^\varphi$ simply assigns to any k-derivation of the algebra B

$$(b_1, \ldots, b_k) \mapsto \psi(b_1, \ldots, b_k) \in Q, \quad \text{where} \quad Q \in \text{Ob}\,\mathcal{K},$$

the k-derivation of the algebra A

$$(a_1, \ldots, a_k) \mapsto \psi(\varphi(a_1), \ldots, \varphi(a_k)) \in Q_\varphi.$$

Using the isomorphisms $\lambda_{\mathfrak{M}}^k$ and $\lambda_{\mathcal{K}}^k$ (see Section 17.30), the required map may be alternatively defined as the map $\left(\Lambda_{\mathfrak{M}}^1\right)^{\wedge k} \to \left(\Lambda_{\mathcal{K}}^1\right)^{\wedge k}$ which is the k-th exterior product of the previously defined (see the diagram (15.50)) map

$$\varphi^*_{\mathfrak{M},\mathcal{K}} \colon \Lambda_{\mathfrak{M}}^1 \to \Lambda_{\mathcal{K}}^1.$$

Under this approach, it becomes obvious that $\varphi^* = \varphi^*_{\mathfrak{M},\mathcal{K}} \colon \Lambda_{\mathfrak{M}}^* \to \Lambda_{\mathcal{K}}^*$ is a homomorphism of exterior algebras, i.e.,

$$\varphi^*(\omega \wedge \rho) = \varphi^*(\omega) \wedge \varphi^*(\rho), \ \omega, \rho \in \Lambda_{\mathfrak{M}}^*. \qquad (18.28)$$

18.14 Exercise. Show that the two above definitions of the map φ^* are equivalent.

According to either one of these definitions, we have

$$\varphi^*(d^{\mathfrak{M}}a_1 \wedge \ldots \wedge d^{\mathfrak{M}}a_k) = d^{\mathcal{K}}(\varphi(a_1)) \wedge \ldots \wedge d^{\mathcal{K}}(\varphi(a_k)).$$

Therefore, in view of (18.21), we can write:

$$
\begin{aligned}
\varphi^*(d^{\mathfrak{M}}_k(ad^{\mathfrak{M}}a_1 \wedge \ldots \wedge d^{\mathfrak{M}}a_k)) &= \varphi^*(d^{\mathfrak{M}}a \wedge d^{\mathfrak{M}}a_1 \wedge \ldots \wedge d^{\mathfrak{M}}a_k) \\
&= d^{\mathcal{K}}(\varphi(a)) \wedge d^{\mathcal{K}}(\varphi(a_1)) \wedge \ldots \wedge d^{\mathcal{K}}(\varphi(a_k)) \\
&= d^{\mathcal{K}}_k(\varphi(a)d^{\mathcal{K}}(\varphi(a_1)) \wedge \ldots \wedge d^{\mathcal{K}}(\varphi(a_k))) \\
&= d^{\mathcal{K}}_k(\varphi^*(ad^{\mathfrak{M}}a_1 \wedge \ldots \wedge d^{\mathfrak{M}}a_k)),
\end{aligned}
$$

i.e., $\varphi^*(d^{\mathfrak{M}}_k(\omega)) = d^{\mathcal{K}}_k(\varphi^*(\omega))$ where ω is a differential form given by

$$\omega = ad^{\mathfrak{M}}a_1 \wedge \ldots \wedge d^{\mathfrak{M}}a_k.$$

But such forms additively generate $\Lambda^k_{\mathfrak{M}}$, and so it follows that

$$\varphi^* \circ d^{\mathfrak{M}}_k = d^{\mathcal{K}}_k \circ \varphi^*, \tag{18.29}$$

i.e., we have proved the *naturality of the exterior differential.*

18.15. Example. It is remarkable that the operation of substituting vector fields into differential forms can be naturally carried over to relative vector fields. Namely, let $X \colon A \to B$ be such a field with respect to the algebra homomorphism $\varphi \colon A \to B$, i.e., a $\Bbbk$–homomorphism satisfying the relative Leibnitz rule

$$X(ab) = \varphi(a)X(b) + \varphi(b)X(a),$$

and let $\mathfrak{M}$ and $\mathcal{K}$ be φ-connected categories of A- and B-modules, respectively, (see Section 15.24). In a non-conceptual way, this can be done simply by putting

$$
\begin{aligned}
i_X&(ad^{\mathcal{K}}a_1 \wedge \ldots \wedge d^{\mathcal{K}}a_k) \\
&= \varphi^*(a) \sum_{i=1}^k (-1)^{i-1} X(a_i)d^{\mathcal{K}}\varphi^*(a_1) \wedge \ldots \\
&\quad \ldots \wedge d^{\mathcal{K}}\varphi^*(a_{i-1}) \wedge d^{\mathcal{K}}\varphi^*(a_{i+1}) \wedge \ldots \wedge d^{\mathcal{K}}\varphi^*(a_k).
\end{aligned}
\tag{18.30}
$$

(see (18.25)). However, this definition requires checking that i_X is well defined. To avoid this, let us modify the approach of Section 18.11 so that it will include the relative case. We can show that

$$i_X(\omega \wedge \rho) = i_X(\omega) \wedge \varphi^*(\rho) + (-1)^i \varphi^*(\omega) \wedge i_X(\rho). \tag{18.31}$$

If M, N are smooth manifolds, $A = C^\infty(N)$, $B = C^\infty(M)$, and $F \colon M \to N$ is a smooth map, then in the standard notation formula (18.31) acquires the form:

$$i_X(\omega \wedge \rho) = i_X(\omega) \wedge F^*(\rho) + (-1)^i F^*(\omega) \wedge i_X(\rho), \tag{18.32}$$

18.16. Description of the algebra of differential forms. The exact description of the algebra $\Lambda^*_{\mathfrak{M}}(A)$ depends on how the algebra A itself is presented. Below we indicate the main way of doing this in terms of the generators of this algebra.

Proposition. *Let the algebra A be generated by the elements $\{a_\alpha\}_{\alpha \in \aleph}$. Then*

(a) *the algebra $\Lambda^*_{\mathfrak{M}}(A)$, regarded as an A-module, is generated by elements of the form $d^{\mathfrak{M}} a_{\alpha_1} \wedge \ldots \wedge d^{\mathfrak{M}} a_{\alpha_k}$, $k \in \mathbb{N}$, $\alpha_i \in \aleph$;*

(b) *if the algebra homomorphism $\varphi \colon A \to B$ is surjective and the categories $\mathfrak{M}$ and $\mathcal{K}$ are φ-connected, the algebra homomorphism $\varphi^*_{\mathfrak{M},\mathcal{K}} \colon \Lambda^*_{\mathfrak{M}}(A) \to \Lambda^*_{\mathcal{K}}(B)$ is also surjective;*

(c) *the kernel of the homomorphism $\varphi^*_{\mathfrak{M},\mathcal{K}}$ contains the ideal $\mathcal{I}^\varphi_{\mathfrak{M},\mathcal{K}}$ of the algebra $\Lambda^*_{\mathfrak{M}}(A)$ generated by $\mathrm{Ker}\,\varphi \cup \mathrm{Ker}(\varphi^*_{\mathfrak{M},\mathcal{K}}|_{\Lambda^1_{\mathfrak{M}}})$. Here the homogeneous components Λ^k_φ of the graded quotient algebra*

$$\Lambda^*_\varphi \stackrel{\mathrm{def}}{=} \Lambda^*_{\mathfrak{M}}(A)/\mathcal{I}^\varphi_{\mathfrak{M},\mathcal{K}}$$

acquire the structure of a B-module, while the B-module Λ^k_φ is isomorphic to $\Lambda^k_{\mathcal{K}}(B)$ if it belongs to the category $\mathcal{K}$. The latter always holds if $\mathfrak{M} = \mathrm{Mod}\,A$, $\mathcal{K} = \mathrm{Mod}\,B$.

◄ (a) Since $\Lambda^k_{\mathfrak{M}}(A) = (\Lambda^1_{\mathfrak{M}})^{\wedge k}(A)$, it suffices to show that the A-module $\Lambda^1_{\mathfrak{M}}(A)$ is generated by the differentials $d^{\mathfrak{M}} \alpha_\alpha$, $\alpha \in \aleph$. But this obviously follows from the fact that the algebra A, regarded as a $\Bbbk$-module (i.e. $\Bbbk$-vector space), is generated by elements of the form $a_{\alpha_1} \ldots a_{\alpha_m}$, $\alpha_i \in \aleph$, whose differentials are linear combinations of the differentials $d^{\mathfrak{M}} \alpha_\alpha$ with coefficients from A.

(b) Note that the algebra B is generated by the elements of $\varphi(a_\alpha)$ since φ is surjective. Therefore, in view of item (a), (18.29), and (18.28), the algebra $\Lambda_{\mathcal{K}}$ is generated (as a B-module) by elements of the form

$$d^{\mathcal{K}}(\varphi(a_{\alpha_1})) \wedge \ldots \wedge d^{\mathcal{K}}(\varphi(a_{\alpha_k})) = \varphi^*(d^{\mathfrak{M}} a_{\alpha_1}) \wedge \ldots \wedge \varphi^*(d^{\mathfrak{M}} a_{\alpha_k})$$
$$= \varphi^*(d^{\mathfrak{M}} a_{\alpha_1} \wedge \ldots \wedge d^{\mathfrak{M}} a_{\alpha_k}). \tag{18.33}$$

(c) The first part of this statement is obvious. Note further that the ideal $\mathcal{I}^\varphi_{\mathfrak{M},\mathcal{K}}$ is homogeneous, i.e.,

$$\mathcal{I}^\varphi_{\mathfrak{M},\mathcal{K}} = \sum_{k \in \mathbb{N}} \mathcal{I}^{\varphi,k}_{\mathfrak{M},\mathcal{K}}, \quad \mathcal{I}^{\varphi,k}_{\mathfrak{M},\mathcal{K}} \stackrel{\mathrm{def}}{=} \mathcal{I}^\varphi_{\mathfrak{M},\mathcal{K}} \cap \Lambda^k_{\mathfrak{M}}(A),$$

and let us consider the quotient algebra $\Lambda^*_\varphi = \Lambda^*_{\mathfrak{M}}(A)/\mathcal{I}^\varphi_{\mathfrak{M},\mathcal{K}}$. Therefore,

$$\Lambda^*_\varphi = \sum_{k \in \mathbb{N}} \Lambda^k_\varphi, \quad \text{where} \quad \Lambda^k_\varphi \stackrel{\mathrm{def}}{=} \Lambda^k_{\mathfrak{M}}(A)/\mathcal{I}^{\varphi,k}_{\mathfrak{M},\mathcal{K}}.$$

Since $\operatorname{Ker}\varphi \cdot \Lambda^k_{\mathfrak{M}}(A) \subset \mathcal{I}^{\varphi,k}_{\mathfrak{M},\mathcal{K}}$, it follows that $\operatorname{Ker}\varphi \cdot \Lambda^k_\varphi = 0$, and so Λ^k_φ may be regarded as an $(A/\operatorname{Ker}\varphi)$-module, or, equivalently, as a B-module. But, by definition, $\mathcal{I}^{\varphi,1}_{\mathfrak{M},\mathcal{K}} = \operatorname{Ker}(\varphi^*_{\mathfrak{M},\mathcal{K}}|_{\Lambda^1_{\mathfrak{M}}})$, and we obtain an isomorphism $\Lambda^1_\varphi = \Lambda^1_\mathcal{K}(B)$ of $\Bbbk$-vector spaces which, for the reasons indicated, may be regarded as an isomorphism of B-modules. Thus, we can identify the image $[d^{\mathfrak{M}}a_\alpha]$ of the element $d^{\mathfrak{M}}a_\alpha$ under the natural projection $\Lambda^1_{\mathfrak{M}} \to \Lambda^1_\varphi$ with $d^\mathcal{K}b_\alpha$, where $b_\alpha = \varphi(a_\alpha)$. Hence,

$$(b_{\alpha_1}, \ldots, b_{\alpha_k}) \mapsto [d^{\mathfrak{M}}a_{\alpha_1}] \wedge \ldots \wedge [d^{\mathfrak{M}}a_{\alpha_k}] \in \Lambda^k_\varphi, \quad b_{\alpha_i} = \varphi(a_{\alpha_i}),$$

is a k-derivation of the algebra B that we will denote by $\psi(b_1, \ldots, b_k)$. Besides, the homomorphism $\varphi^*_{\mathfrak{M},\mathcal{K}}$ induces the projection

$$[\varphi]\colon \Lambda^k_\varphi \to \Lambda^k_\mathcal{K}, \quad [d^{\mathfrak{M}}a_1] \wedge \ldots \wedge [d^{\mathfrak{M}}a_k] \mapsto d^\mathcal{K}(\varphi(a_1)) \wedge \ldots \wedge d^\mathcal{K}(\varphi(a_k)).$$

Now if $\Lambda^k_\varphi \in \operatorname{Ob}\mathcal{K}$, then the $\Bbbk$-derivations ψ can be obtained from the universal $\Bbbk$-derivation in the category $\mathcal{K}$ by means of a B-homomorphism $[\psi]\colon \Lambda^k_\mathcal{K} \to \Lambda^k_\varphi$ such that

$$d^\mathcal{K}b_1 \wedge \ldots \wedge d^\mathcal{K}b_k \mapsto [d^{\mathfrak{M}}a_1] \wedge \ldots \wedge [d^{\mathfrak{M}}a_k], \quad \text{if} \quad b_i = \varphi(a_i).$$

This immediately implies $[\varphi]$ and $[\psi]$ are homomorphisms inverse to each other. ▶

18.17. Examples. In the examples presented below, we show how the proposition from the previous section allows to concretely describe the differential forms of various algebras. Here we use the simplified notation $\Lambda^i(A)$ for $\Lambda^i_{\operatorname{Mod}A}$ and $\Lambda^i_\Gamma(A)$ for $\Lambda^i_{\operatorname{GMod}A}$ and omit the indices $\mathfrak{M}$ and $\mathcal{K}$ in $\mathcal{I}^\varphi_{\mathfrak{M},\mathcal{K}}$ and in $\mathcal{I}^{\varphi,k}_{\mathfrak{M},\mathcal{K}}$.

I. Let $A = \Bbbk[x_1, \ldots, x_n]$ and $\mathfrak{M} = \operatorname{Mod}A$. Then the A-module $\Lambda^*_{\mathfrak{M}}(A)$ is free, and the forms that read

$$d^{\mathfrak{M}}x_{i_1} \wedge \ldots \wedge d^{\mathfrak{M}}x_{i_k}, \ 1 \leq i_1 < \cdots < i_k \leq n,$$

constitute its free basis. The A-modules $\Lambda^*_{\mathfrak{M}}(A)$ are geometrical if $\Bbbk$ is a field of zero characteristic, and nongeometrical otherwise.

II. Let $A = \mathbb{F}_2[x_1, \ldots, x_n]$. Recall that the geometrization $\Gamma(A)$ of the algebra A is the Boolean algebra B_n of subsets of a set of n elements. It is freely generated by the elements $\bar{x}_i = x_i \mod \operatorname{Inv}(A)$, $i = 1, \ldots, n$, satisfying the condition $\bar{x}_i^2 = \bar{x}_i$. Therefore, the algebra of geometrical differential forms $\Lambda^*_\Gamma(A)$, which, as an A-module, is the geometrization of the algebra $\Lambda^*(A)$ described above turns out to be none other than the algebra of skew-symmetric polynomials in the variables $d\bar{x}_i$, $i = 1, \ldots, n$, with coefficients in B_n.

III. Let $A = \Bbbk[x]$ and $B = \Bbbk[x]/(x^m)$. The algebra B is generated by elements of the form

$$\sum_{i=0}^{m-1} \lambda_i \theta^i, \quad \text{where } \theta = x \mod x^m, \ \theta^m = 0.$$

Then the proposition from the previous section shows that $\Lambda^i(B) = 0$ if $i > 1$, while the A-module $\mathcal{I}^{\varphi,1}$ is generated by the elements $mx^{m-1}dx$ and $x^m dx$. If we assume, for the sake of simplicity, that $m1_{\Bbbk} \neq 0$, it will follow that the B-module $\Lambda^1(B)$ is generated by the differential $d\theta$ and obeys the relation $\theta^{m-1}d\theta = 0$. In other words, the forms $\theta^i d\theta$, $0 \leq i \leq m-2$, constitute a basis of $\Lambda^1(B)$ regarded as a vector space over $\Bbbk$. Since the ideal $\mathrm{Inv}(B)$ is generated by the element θ, the geometrization of the B-module $\Lambda^1(B)$, i.e, the B-module $\Lambda^1_{\Gamma}(B)$ is a one-dimensional space (over $\Bbbk$) $\Bbbk \cdot d^{\Gamma}\theta$. This example is interesting in that it shows that differential forms of higher degrees may be nontrivial while the spectrum of the algebra itself is "zero-dimensional".

IV. $A = \Bbbk[x_1, x_2]$ and $B = \Bbbk[x_1, x_2]/(x_1^2 + x_2^2 - 1)$ (the algebra of functions on the "$\Bbbk$-circle"). If the characteristic of the field $\Bbbk$ is not 2, then the A-module $\mathcal{I}^{\varphi,1}$ is generated by the 1-forms

$$\omega_i = f(x_1, x_2)dx_i, \ i = 1, 2, \ \omega = x_1 dx_1 + x_2 dx_2 = 1/2 \cdot df(x_1, x_2),$$

where $f = f(x_1, x_2) = x_1^2 + x_2^2 - 1$. These forms together with the form $\rho = x_1 dx_2 - x_2 dx_1$ generate the A-module $\Lambda^1(A)$, which can be seen from the fact that $dx_1 = x_1 \omega - x_2 \rho - f dx_1$ and similarly for dx_2. This implies that $\Lambda^1(B) = B \cdot \varphi^*(\rho)$ is a free one-dimensional module.

In its turn, the A-module $\mathcal{I}^{\varphi,2}$ is generated by the forms

$$f dx_1 \wedge dx_2 = dx_1 \wedge \omega_2, \quad x_1 dx_i \wedge dx_2 = -dx_2 \wedge \omega, \quad x_2 dx_1 \wedge dx_2 = dx_1 \wedge \omega.$$

But since

$$dx_1 \wedge dx_2 = x_1(x_1 dx_1 \wedge dx_2) + x_2(x_2 dx_1 \wedge dx_2) - f dx_1 \wedge dx_2 \in \mathcal{I}^{\varphi,2},$$

we have $\Lambda^2(B) = 0$.

If the characteristic of the field $\Bbbk$ is 2, then $df = 0$, and therefore, the A-module $\mathcal{I}^{\varphi,1}$ is generated by the 1-forms $f dx_i$, $i = 1, 2,$, while $\mathcal{I}^{\varphi,2}$ is generated by the 2-form $f dx_1 \wedge dx_2$. Thus, in this case, the B-modules $\Lambda^1(B)$ and $\Lambda^2(B)$ are free, while $\{d\varphi(x_1), d\varphi(x_2)\}$ and $\{d\varphi(x_1) \wedge d\varphi(x_2)\}$ constitute their free bases, respectively.

Finally, the modules $\Lambda^i(B)$ are trivial for obvious reasons if $i > 2$, independently of the characteristic of the field $\Bbbk$, and since all the other modules $\Lambda^i(B)$, $0 \leq i \leq 2$, are free, they are geometrical if and only if the algebra B is geometrical.

V. $A = \Bbbk[x, y]$ and $B = \Bbbk[x, y]/(xy)$ ("$\Bbbk$-cross"). The A-module $\mathcal{I}^{\varphi,1}$ is generated by the 1-forms

$$xy dx, \quad xy dy, \quad x dy + y dx = d(xy).$$

From this it is easy to derive that $\Lambda^1(B)$, regarded as a $\Bbbk$-vector space, is the direct sum

$$\Lambda^1(B) = (\Bbbk[\bar{x}] \cdot d\bar{x}) \oplus (\Bbbk[\bar{y}] \cdot d\bar{y}) \oplus (\Bbbk \cdot \rho),$$
$$\text{where} \quad \bar{x} = \varphi(x), \bar{y} = \varphi(y), \rho = \bar{x} d\bar{y} = -\bar{y} d\bar{x},$$

while the B-module structure is ruled by the relations $\bar{x}\bar{y} = 0$, $\bar{x}\rho = \bar{y}\rho = 0$. This implies that

$$\Lambda_x \stackrel{\text{def}}{=} (\Bbbk[\bar{x}] \cdot d\bar{x}) \oplus (\Bbbk \cdot \rho) \text{ and } \Lambda_y \stackrel{\text{def}}{=} (\Bbbk[\bar{y}] \cdot d\bar{y}) \oplus (\Bbbk \cdot \rho)$$

are submodules of $\Lambda^1(B)$ so that $\Lambda^1(B) = \Lambda_x + \Lambda_y$ and $\Lambda_x \cap \Lambda_y = (\Bbbk \cdot \rho)$.

It is obvious that $|B| = \{(\lambda, 0)\}_{\lambda \in \Bbbk} \cup \{(0, \mu)\}_{\mu \in \Bbbk} \subset \Bbbk^2$ (the union of the "x-axis" and the "y-axis" $\Bbbk^2$). A direct calculation shows that the $\Bbbk$-vector space $\Lambda^1(B)_h$, $h \in |B|$, is one-dimensional if $h \neq (0, 0)$ and that the B-module $\Lambda^1(B)$ is geometrical if $\Bbbk$ is a field of zero characteristic. Here the $\Bbbk$-vector space $\Lambda^1(B)_{(0,0)}$ is two-dimensional, while the images of the forms $d\bar{x}$ and $d\bar{y}$ constitute its basis. This implies, in particular, that the B-module $\Lambda^1(B)$ is not projective and so the algebra B is not smooth.

On the other hand,

$$\rho \in \bar{x}\Lambda^1(B) \cap \bar{y}\Lambda^1(B) \in \text{Inv} \Lambda^1_\Gamma(B),$$

so that the form ρ is not geometrical. In its turn, this implies that the form $d\rho = d\bar{x} \wedge d\bar{y}$ is also nongeometrical, and therefore $\Lambda^i_\Gamma(B) = 0$, $i > 1$.

VI. The considerations developed above can be applied word for word to describe the geometrical differential forms on the cross $\mathbf{K}$ (see Section 9.45). In particular, $\Lambda^i_\Gamma(\mathbf{K}) = 0, i > 1$, while $\Lambda^1_\Gamma(\mathbf{K})$ consists of forms expressed as $f(\bar{x})d\bar{x}$ and $g((y))d\bar{y}$. Here

$$\bar{y} \cdot f(\bar{x})d\bar{x} = 0 = \bar{x} \cdot g((y))d\bar{y}.$$

Exercise. Let $P = C^\infty(\mathbf{K})_h$. Show that the A-module $\text{Diff}^>_1 P$ is not geometrical and deduce from this fact that $\Lambda^2_\Gamma(\mathbf{K}) = 0$.

18.18. Description of cosymbols. Recall that the group of cosymbols $\text{Csml}^{\mathfrak{M}}_k(P) \subset \mathcal{J}^k(P; \mathfrak{M})$ of an A-module P has the additive generators

$$\delta_{a_1,\ldots,a_k}(j^{\mathfrak{M}}_k)(p), \quad a_1, \ldots, a_k \in A, \quad p \in P.$$

Here it is important to note that the operator

$$\delta_{a_1,\ldots,a_k}(j^{\mathfrak{M}}_k) \colon P \to \mathcal{J}^k(P; \mathfrak{M})$$

is of order zero and so is an A-module homomorphism, both with respect to the right and the left A-module structure of the bimodule $\mathcal{J}^k(P; \mathfrak{M})$. In other words, we have the relation

$$\delta_{a_1,\ldots,a_k}(j^{\mathfrak{M}}_k)(ap) = a\delta_{a_1,\ldots,a_k}(j^{\mathfrak{M}}_k)(p), \quad a \in A, \quad p \in P. \tag{18.34}$$

Despite its simplicity, it is precisely this relation which allows one to give an exact description of cosymbols for a wide class of A-modules (see the example in the next section). Furthermore, the following fact simplifies this description.

Exercise. Let the algebra A be generated by the elements a^0_i, while the A-module P is generated by the elements p^0_j. Show that the elements of the form $\delta_{a^0_{i_1},\dots,a^0_{i_k}}(j^{\mathfrak{M}}_k)(p^0_j)$ generate the submodule $\mathrm{Csml}^{\mathfrak{M}}_m(P)$ of the A-module $\mathcal{J}^k(P;\mathfrak{M})$. (Compare with item (a) of the proposition in Section 18.16.)

18.19. Exercise-Example: cosymbols for the cross. Describe the module $\mathrm{Csml}^{\Gamma}_m(\mathbf{K})$ of geometrical symbols of the algebra $C^{\infty}(\mathbf{K})$.

19

Differential Forms: Classical and Algebraic Approach

The last three chapters of this book differ, both in the style of exposition and the character of their contents, from the previous ones. The topics considered in each of them could be developed into separate books, perhaps into several volumes, or form the basis of further research.

The present chapter deals with differential forms and their applications. We see how differential forms can be conceptually redefined and then naturally generalized to become an inherent and very important part of the differential calculus over commutative algebras. The applications include the definition and study of symplectic, Poisson, and contact manifolds, Monge–Ampère equations, the Lie derivative, infinitesimal symmetries, derivations of vector bundles, Killing fields, Jacobi brackets, etc.

19.1. Relationship of classical differential forms with $\mathrm{dR}_{\mathfrak{M}}(A)$. First of all, let us state and prove the following

Proposition. *If* $A = C^\infty(M)$ *and* $\mathfrak{M} = \mathrm{GMod}\,A$, *then the cochain complex* $\mathrm{dR}_{\mathfrak{M}}(A)$ *is identical to the standard de Rham cochain complex (of the given manifold M) defined in Section 18.8.*

◄ Recall that the $C^\infty(M)$-module of differential forms of degree k on the manifold M is denoted by $\Lambda^k(M)$. As we noted previously (see Section 14.11), we have $\Lambda^1(M) = \Lambda^1_{\mathfrak{M}}$ (the natural identification). On the other hand, we know that $\Lambda^k(M) = \left(\Lambda^1(M)\right)^{\wedge k}$. Besides, we have $\Lambda^k_{\mathfrak{M}} = \left(\Lambda^1_{\mathfrak{M}}\right)^{\wedge k}$ (by the proposition in Section 17.30). This allows us to identify $\Lambda^k_{\mathfrak{M}}$ with $\Lambda^k(M)$.

© Springer Nature Switzerland AG 2020
J. Nestruev, *Smooth Manifolds and Observables*, Graduate Texts
in Mathematics 220, https://doi.org/10.1007/978-3-030-45650-4_19

Further, we see that in the notation of Section 14.11, $d_{\text{geom}} = d_0^{\mathfrak{M}}$, so that the theorem from the same section allows us to identify $d_0^{\mathfrak{M}} : C^\infty(M) \to \Lambda_{\mathfrak{M}}^1$ with the standard differential $d = d_{\text{geom}} : C^\infty(M) \to \Lambda^1(M)$. Moreover, the formula (18.21) shows that the action of the differential $d_k^{\mathfrak{M}}$ can be expressed in terms of the differential $d^{\mathfrak{M}} = d_0^{\mathfrak{M}}$. Since a similar formula is valid for the standard exterior differential $d_k : \Lambda^k(M) \to \Lambda^{k+1}(M)$ on M, this allows us to identify $d_k^{\mathfrak{M}}$ with d_k.   ▶

Let us stress that the rules of behavior of differential forms in differential geometry are simply the manifestation of the general naturality conditions that we studied in Section 18.13. In that context, we considered homomorphisms $\varphi = F^*$, where $F : M \to N$ is a smooth map and the categories of geometrical modules played the role of the categories $\mathfrak{M}$ and $\mathcal{K}$. Here the image of the differential form $\omega \in \Lambda^k(N)$ under the homomorphism F^* is denoted by $F^*(\omega)$ and called its *inverse image under the map F*. Besides, by the *restriction of the form $\rho \in \Lambda^k(M)$ to a submanifold $L \subset M$* we mean the form $i^*(\rho) \in \Lambda^k(L)$, where $i : L \to M$ is the inclusion map. Traditionally, it is denoted by $\rho|_L$.

Exercise. Describe the algebra of differential forms on the cross $\mathbf{K}$ as this was done in example V from Section 18.17.

19.2. Standard description of differential forms on manifolds. Differential forms on a manifold M are standardly defined in differential geometry as skew-symmetric $C^\infty(M)$-polylinear functions
$$\omega(X_1, \ldots, X_k) \in C^\infty(M)$$
of vector fields $X_i \in D(M)$ (see Section 16.13). As we have already noted,
$$\omega(X_1, \ldots, X_k) \stackrel{\text{def}}{=} (i_{X_k} \circ \cdots \circ i_{X_1})(\omega),.$$
In particular, in view of (18.24), this implies that
$$(df_1 \wedge \ldots \wedge df_k)(X_1, \ldots, X_k) = \det \|X_i(f_j)\|. \tag{19.1}$$
And, in general, all the operations with differential forms are defined and described in a similar way. For instance, the definition of the exterior product looks as follows:
$$(\omega \wedge \rho)(X_1, \ldots, X_{k+l})$$
$$\stackrel{\text{def}}{=} \sum_I (-1)^{(I,\bar{I})} \omega(X_{i_1}, \ldots, X_{i_k}) \rho(X_{j_1}, \ldots, X_{j_l}), \tag{19.2}$$
where $I = \{i_1, \ldots, i_k\} \subset \{1, \ldots, k+l\}$, $\bar{I} = \{1, \ldots, k+l\} \setminus I$ and $(-1)^{(I,\bar{I})}$ is the sign of the corresponding permutation. Similarly,
$$d\omega(X_1, \ldots, X_{k+1}) \stackrel{\text{def}}{=} \sum_i (-1)^{i-1} X_i \omega(X_1, \ldots, \hat{X}_i, \ldots, X_{k+1})$$
$$+ \sum_{i<j} (-1)^{i+j} \omega([X_i, X_j], X_1, \ldots, \hat{X}_i, \ldots, \hat{X}_j, \ldots, X_{k+1}). \tag{19.3}$$

Exercise. Prove formulas (19.2) and (19.3).

19.3. Examples. I. If $\dim M = n$, then the vector bundle associated with the $C^\infty(M)$-module $\Lambda^n(M)$ is one-dimensional. We can see this, for example, from the fact that locally any n-form on M can be written as $f(x_1, \ldots, x_n) \, dx_1 \wedge \ldots \wedge dx_n$. If this module is free, then its basis consists of one element, which is called the *volume form* of the manifold M. In that case, the manifold is called *orientable* and *nonorientable* otherwise. If ω is a volume form and $\rho \in \Lambda^n(M)$, then $\rho = f\omega$, $f \in C^\infty(M)$. If ρ is another volume form, then the function f is nonzero everywhere. Now if, moreover, f is positive, then the forms ω and ρ are called equivalent. The *orientation of the manifold* M is an equivalence class of its volume forms. Obviously, there exists exactly two orientations. They are opposite to each other in the sense that if one of them is given by the volume form ω, then the other one is given by $-\omega$.

II. Let us take the n-sphere S^n to be hypersurface $\{x_1^2 + \ldots + x_{n+1}^2 = 1\}$ in Euclidean space $\mathbb{R}^{n+1}$. The form $\omega_{n+1} = dx_1 \wedge \ldots \wedge dx_{n+1}$ is obviously a volume form on the manifold $\mathbb{R}^{n+1}$. It is also easy to see that the restriction of the form

$$\omega = i_V(\omega_{n+1}), \quad \text{where } V = \sum_{i=1}^{n+1} x_i \frac{\partial}{\partial x_i}$$

to the sphere S^n is the sphere's volume form. Denote it by ρ.

The central symmetry

$$T \colon (x_1, \ldots, x_{n+1}) \mapsto (-x_1, \ldots, -x_{n+1})$$

leaves the sphere S^n invariant, as well as the vector field V. But since $T^*(x_i) = -x_i$, it follows that $T^*(\omega_{n+1}) = (-1)^{n+1}\omega_{n+1}$. Therefore, $T^*(\omega) = (-1)^{n+1}\omega$ and so $\hat{T}^*(\rho) = (-1)^{n+1}\rho$, where $\hat{T} = T|_{S^n}$. Thus $\hat{T}$ preserves the orientation of the sphere if n is odd and reverses it if n is even. Further, gluing together antipodal points of the sphere, we obtain the n-dimensional projective space $\mathbb{R}P^n$. Now if $F \colon S^n \to \mathbb{R}P^n$ is the gluing map, then $F \circ \hat{T} = F$. If $\nu \in \Lambda^n(\mathbb{R}P^n)$ is the volume form, then $F^*(\nu)$ is the volume form on S^n, since F is a local diffeomorphism. But $\hat{T}^*(F^*(\nu)) = (F \circ \hat{T})^*(\nu) = F^*(\nu)$ and, therefore, in that case, $\hat{T}$ preserves the orientation of the sphere. But we have seen that this is possible only if n is odd, and so it follows that projective spaces of even dimension have no volume form. Thus the manifolds $\mathbb{R}P^{2n}$ are nonorientable, while $\mathbb{R}P^{2n+1}$ are orientable.

Below we present examples of two very important geometrical structures defined by means of differential forms.

19.4. Symplectic manifolds. A pair (M^{2n}, Ω), $\Omega \in \Lambda^2(M)$, is said to be a *symplectic manifold* if $d\Omega = 0$, while the form Ω itself is nondegenerate. This means that the map

$$\gamma_\Omega \colon D(M) \to \Lambda^1(M), \quad X \mapsto -i_X(\Omega),$$

is an isomorphism of $C^\infty(M)$-modules. The classical example of a symplectic manifold is the total space of the cotangent bundle T^*M. In that case, $\Omega = d\rho$, where $\rho \in \Lambda^1(T^*N)$ is the universal 1-form (see Section 14.5). The form Ω that appears in the definition of symplectic manifold is customarily called *symplectic*.

The above definition of symplectic manifold can be carried over, word for word, to the case of an arbitrary commutative algebra. Here, however, we must take care in choosing the category $\mathfrak{M}$. It must satisfy the condition that the module $\Lambda^1_{\mathfrak{M}}$ is isomorphic to the module $D(M)$. For instance, for the algebra $A = C^\infty(M)$, the category GMod$A$ works, but the category Mod$A$ (see Section 14.11) does not. Thus, having accepted this definition, we can speak of *symplectic algebras* or, more precisely, of *symplectic triples* $(A, \Omega, \mathfrak{M})$.

19.5 Proposition. *The formula*

$$\{f, g\} \overset{\text{def}}{=} \Omega(\gamma_\Omega^{-1}(df), \gamma_\Omega^{-1}(dg)), \quad f, g \in C^\infty(M), \tag{19.4}$$

supplies each symplectic manifold with a Poisson bracket.

◄ The fact that this bracket is skew-symmetric is obvious, while the Jacobi identity, as we will show, follows from the fact that the form Ω is closed. By the notation $X_f \overset{\text{def}}{=} \gamma_\Omega^{-1}(df)$, we stress that the vector field $\gamma_\Omega^{-1}(df)$ is Hamiltonian with respect to the bracket (19.4). In this notation, we can write

$$\{f, g\} = \Omega(X_f, X_g) = i_{X_f}(dg) = X_f(g)$$

Since the form Ω is closed, we have

$$0 = d\Omega(X_f, X_g, X_h) = X_f(\Omega(X_g, X_h)) - \Omega([X_f, X_g], X_h) + \text{cycle}$$
$$= X_f(\{g, h\}) - [X_f, X_g](h) + \text{cycle} = \{f, \{g, h\}\} - [X_f, X_g](h) + \text{cycle}$$

Further, since

$$[X_f, X_g](h) = X_f(X_g(h)) - X_g(X_f(h)) = \{f, \{g, h\}\} - \{g, \{f, h\}\},$$

we see that

$$0 = d\Omega(X_f, X_g, X_h) = 3(\{f, \{g, h\}\} + \text{cycle}). \quad ►$$

The Poisson bracket related to symplectic manifolds as indicated above will be called *symplectic*. Such a bracket is uniquely defined by its nondegeneracy property. This property can be stated in different ways, among which the most convenient for our aims is the following.

To any Poisson bracket (respectively, any *Poisson manifold*, i.e., a smooth manifold supplied with a Poisson bracket) corresponds the derivation

$$\chi \colon C^\infty(M) \to D(M), \quad f \mapsto X_f.$$

If the corresponding homomorphism $h_\chi \colon \Lambda^1(M) \to D(M)$ is an isomorphism, then the given bracket (respectively, the Poisson manifold) is called *nondegenerate*. In the case of a symplectic Poisson bracket, the definition immediately implies that the composition

$$D(M) \xrightarrow{\gamma_\Omega} \Lambda^1(M) \xrightarrow{\chi} D(M)$$

is the identity map, so that $\chi = \gamma_\Omega^{-1}$ is an isomorphism.

Conversely, let $\{\cdot,\cdot\}$ be a nondegenerate Poisson bracket on M, while X_f is the Hamiltonian vector field determined by that bracket. Since h_χ is an isomorphism, it follows that the $C^\infty(M)$-module $D(M)$ is generated by the fields X_f and we can define the 2-form $\Omega_\chi \in \Lambda^2(M)$ by putting

$$\Omega_\chi(f'X_f, g'dg) = f'g'\{X_f, X_g\}.$$

After that is done, by reversing the proof of Proposition 19.4, we see that $d\Omega(X_f, X_g, X_h) = 0$, which implies that $d\Omega = 0$.

In conclusion, let us note that the definition of nondegenerate Poisson brackets given above can be carried over directly to arbitrary commutative algebras. One only has to correctly choose the appropriate category $\mathfrak{M}$ of modules over the main algebra.

19.6 Exercises.   1. Let $X, Y \in D(M)$ and $[X, Y] = 0$. Show that

$$\{f, g\} \stackrel{\text{def}}{=} X(f)Y(g) - Y(f)X(g)$$

is a Poisson bracket which is not symplectic if $\dim M > 2$.

2. Show that $\Omega^{\wedge n}$ is a volume form if Ω is a symplectic form on a $2n$-dimensional manifold.

3. Show that Cartesian product of symplectic manifolds is also a symplectic manifold.

4. Show that the Poisson bracket introduced in Section 10.13 is nondegenerate.

5. Let $A = \Bbbk[x_1, \ldots, x_{2n}]$ and $\mathfrak{M} = \mathrm{Mod}\,A$. Describe all the symplectic triples of the form $(A, \Omega, \mathrm{Mod}\,A)$. One obvious solution is provided by the form $\Omega_n = dx_1 \wedge dx_2 + \cdots + dx_{2n-1} \wedge dx_{2n}$.

6. Under the conditions of the previous item, let $\Bbbk = \mathbb{F}_p$. Describe all the symplectic triples of the form $(A, \Omega, \mathrm{GMod}\,A)$.

19.7. Contact manifolds: origins. Contact geometry, or the geometry of contact manifolds, became a complete theory dealing with (nonlinear) first-order differential equations for functions in one variable in the work of Sophus Lie. From the contemporary point of view, such an equation $\mathcal{E}$ has the form $F = 0$, where $F \in C^\infty(J^1 P)$, while P is a one-dimensional projective $C^\infty(M)$-module. The solution of the equation $\mathcal{E}$ in the usual understanding is an element $p \in P$ such that $j_1(p)^*(F) = 0$ (see Section

14.20). From the geometrical point of view, $\mathcal{E}$ is a hypersurface in J^1P whose solutions are the elements $p \in P$ such that $\text{Im } j_1(p) \subset \mathcal{E}$.

19.8 Exercise. In the coordinates on J^1P specified in Section 14.20, describe the condition $j_1(p)^*(F) = 0$ and verify that it is actually a first-order differential equation in the classical sense.

With the manifold J^1P, one can naturally associate the *contact module*

$$\mathcal{C}\Lambda \overset{\text{def}}{=} \{\rho \in \Lambda^1(J^1P) \mid j_1(p)^*(\rho) = 0, \forall p \in P\}$$

and the *contact distribution*

$$\mathcal{C} \overset{\text{def}}{=} \{X \in D(J^1P) \mid i_X(\rho) = 0, \forall \rho \in \mathcal{C}\Lambda\}.$$

Since $\mathcal{C}$ is a submodule of the module $D(J^1P)$, the vector bundle associated with $\mathcal{C}$ is a subbundle of the tangent bundle $T(J^1P)$. In particular, its fiber $\mathcal{C}_z$ at the point $z \in J^1P$ is a subspace of the tangent space $T_z J^1P$, while the family $z \mapsto \mathcal{C}_z$ is a *distribution* on the manifold J^1P in the sense in which that term is used in differential geometry.

19.9 Exercises.   1. Show that in the coordinates on J^1P specified in Section 14.20, the contact distribution $\mathcal{C}$ is locally generated by the vector fields

$$\frac{\partial}{\partial x_i} + p_i \frac{\partial}{\partial u}, \quad \frac{\partial}{\partial p_i}, \quad i = 1, \ldots, n.$$

2. In the same coordinates, let

$$\omega = du - \sum_{i=1}^{n} p_i dx_i.$$

Show that the (local) section s of the bundle J^1P will have the form $s = j_1(p)$ if and only if $s^*(\omega) = 0$.

3. Show that $\omega \wedge (d\omega)^{\wedge n}$ is a volume form on J^1P, i.e.,

$$\left. \left(\omega \wedge (d\omega)^{\wedge n} \right) \right|_z \neq 0, \forall z \in J^1P.$$

A consequence of these exercises is the fact that the contact module $\mathcal{C}\Lambda$ is projective and one-dimensional. In its turn, this implies that the module $\mathcal{C}$ is also projective and its dimension is $2n$. In other words, the tangent subspace $\mathcal{C}_z$ of its associated geometrical distribution is of dimension $2n$. The characteristic property of the contact distribution $\mathcal{C}$ will be stated following a useful digression.

Example. Note that $J^1P = J^1M$ if $P = C^\infty(M)$. On J^1M, there exists a global analog of the differential form ω from the previous exercises, because $J^1M = T^*M \oplus \mathbb{R}$; namely, it is the differential form

$$\omega_M = \pi_{\mathbb{R}}^*(du) - \pi_{T^*M}(\rho),$$

where ρ is the universal 1-form on T^*M, u is the coordinate on $\mathbb{R}$, while π_{T^*M} and $\pi_{\mathbb{R}}$ are the projections of J^1M on T^*M and $\mathbb{R}$, respectively.

19.10. Digression: curvature of distributions. A *distribution* on the manifold M is a submodule $\mathcal{D}$ of the module $D(M)$. The quotient module $\varkappa = \varkappa_D \overset{\text{def}}{=} D(M)/\mathcal{D}$ is said to be *normal* for the distribution $\mathcal{D}$, while the $\varkappa$-valued form

$$R_{\mathcal{D}}(X,Y) \overset{\text{def}}{=} [X,Y] \mod \mathcal{D}, \quad X, Y \in \mathcal{D},$$

is called its *curvature*. This form is obviously skew-symmetric and, as is easily seen, is $C^{\infty}(M)$-bilinear. With the curvature form, we naturally associate the *curvature map*

$$\hat{R}_{\mathcal{D}} \colon \mathcal{D} \to \operatorname{Hom}_{C^{\infty}(M)}(\mathcal{D}, \varkappa_{\mathcal{D}}),$$
$$\hat{R}_{\mathcal{D}}(X) \colon Y \mapsto R_{\mathcal{D}}(X,Y), \quad X, Y \in \mathcal{D}. \tag{19.5}$$

The distribution $\mathcal{D}$ is called *regular* if the modules $\mathcal{D}$ and $\varkappa_D$ are projective. As a rule, in differential geometry, only regular distributions are considered.

Geometrically, the distribution $\mathcal{D}$ is characterized by the family of vector spaces $\mathcal{D}_z$, $z \in M$, constituting $|\mathcal{D}|$ (see Section 12.11). Since $\mathcal{D} \subset D(M)$, we can define the natural map $\gamma_z = \gamma_{z,\mathcal{D}} \colon \mathcal{D}_z \to D(M)_z = T_z M$. The geometrical portrait of the distribution $\mathcal{D}$ is obtained from the family of subspaces $\operatorname{Im} \gamma_{z,\mathcal{D}} \subset T_z M$. It is precisely in this way that it is convenient to visualize any distribution (see Figures 19.1 and 19.2). If γ_z is an embedding, then z is a *regular point* of the distribution $\mathcal{D}$. If not, it is said to be a *singular point*. If $\mathcal{D}$ is regular, then all the points of the manifold M are regular and the geometrical image of the distribution $\mathcal{D}$ is a subbundle of the tangent bundle. The *dimension* (or *rank*) *of a regular distribution* is defined as the dimension of that subbundle. In the general case, the manifold M is divided into domains of regular points complemented by the nowhere dense set of singular points.

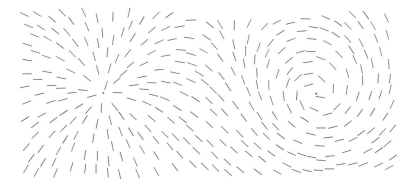

Figure 19.1. One-dimensional distribution with two singular points.

The approach to the theory of distributions briefly sketched in this section is presented here in a form that can be automatically carried over to

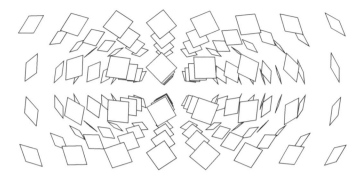

Figure 19.2. Two-dimensional distribution.

arbitrary commutative algebras. The definitions and constructions of this section were precisely motivated by this consideration. Besides, this allows to develop the theory of distributions with singularities even in the classical context of smooth manifolds, which would be quite arduous within the framework of the old purely geometrical approach. For example, we obtain an interesting class of distributions with singularities by stipulating that the module $\mathcal{D}$ be projective, at the same time omitting the requirement that the normal module $\gamma_{\mathcal{D}}$ be projective.

Remark. Just as in 19.7, we can construct the distribution $\mathcal{C}_k$ on the manifold $J^k P$; this distribution is the higher analog of a contact structure (see the next section) and plays a fundamental role in the geometrical theory of differential equations.

Exercises.   1. Verify that the definitions and constructions of this section can be generalized to arbitrary commutative algebras simply by replacing $D(M)$ by $D(A)$.

   2. Give examples of distributions with singularities for which $\mathcal{D}$ is projective.

   3. Show that the distribution $\mathcal{D}$ is regular if and only if all the maps $\gamma_{z,\mathcal{D}}$ are injective.

19.11. More exercises. Let $\omega \in \Lambda^1(M)$ and
$$\alpha_\omega \colon D(M) \to C^\infty(M),\ X \mapsto i_X \omega\,.$$

With the form ω, let us associate the distribution $\mathcal{D}_\omega \stackrel{\text{def}}{=} \operatorname{Ker}\alpha_\omega$. Then, obviously, $\varkappa_\omega \stackrel{\text{def}}{=} \varkappa_{\mathcal{D}_\omega} \cong \operatorname{Im}\alpha_\omega \subset C^\infty(M)$. Therefore, we can assume that the curvature form $R_\omega \stackrel{\text{def}}{=} R_{\mathcal{D}_\omega}$ is $C^\infty(M)$-valued. Show that

   1. the distribution $\mathcal{D}_\omega$ is regular if and only if the form ω is *nonsingular*, i.e., $\omega_z \neq 0$, $\forall z \in M$;

2. $\alpha_m(R_\omega(X,Y)) = d\omega(X,Y)$, $X,Y \in \mathcal{D}_\omega$; in other words, the curvature form R_ω is the restriction of the form $d\omega$ to the distribution $\mathcal{D}_\omega$;

3. the construction of the distribution $\mathcal{D}_\omega$ can be carried over automatically to arbitrary commutative algebras.

19.12. Contact manifolds: abstract algebraic approach. The form $\omega \in \Lambda^1(N)$ on a manifold N of dimension $2n+1$ is said to be *contact* if $\omega \wedge (d\omega)^{\wedge n}$ is a volume form on N. It is easy to see that all the covectors ω_z, $z \in M$, are nonzero. If $\omega' = f\omega$, then

$$\omega' \wedge (d\omega')^{\wedge n} = f\omega \wedge (df \wedge \omega + f d\omega)^{\wedge n} = f^{n+1}\omega \wedge (d\omega)^{\wedge n}.$$

Therefore, $\omega' \wedge (d\omega')^{\wedge n}$ is also a volume form on N provided that the function f does not vanish anywhere. In that case, the contact forms are considered to be equivalent. A one-dimensional projective submodule $\mathcal{C}\Lambda \subset \Lambda^1(N)$ is referred to as a *contact structure* (or as a so-called *Cartan structure*) if it is locally generated by contact forms. In these terms, a *contact manifold* is a manifold with a contact structure on it, i.e., the pair $(N, \mathcal{C}\Lambda)$. If N possesses a global contact form ω, then we can write $\mathcal{C}\Lambda = C^\infty(N) \cdot \omega$.

Example. The well-known Darboux lemma asserts that for any contact form there exist local *canonical coordinates*, in which the form can be expressed as

$$dx_{2n+1} - \sum_{i=1}^{n} x_{n+i} dx_i.$$

In other words, all contact forms of the same dimension are locally equivalent.

Exercise. Indicate some contact form on the 3-sphere.

The definition of contact manifold may be convenient from the computational point of view, however, we will use the conceptual definition, which is as follows.

Definition. A regular distribution $\mathcal{D}$ is said to be *contact* if the curvature map $\hat{R}_\mathcal{D}$ is locally an isomorphism. In this case $(M, \mathcal{D})$ is said to be a *contact manifold*.

In order to verify that this definition is equivalent to the generally accepted one, note that if $\hat{R}_\mathcal{D}$ is an isomorphism, then the dimension of the projective module $\text{Hom}_{C^\infty(M)}(\mathcal{D}, \varkappa_\mathcal{D})$ (see (19.5)) equals $\dim \mathcal{D}$ and, therefore, $\varkappa_\mathcal{D}$ is a one-dimensional projective module. Now since the modules $\mathcal{D}$ and $\text{Hom}_{C^\infty(M)}(\mathcal{D}, C^\infty(M))$ are isomorphic, the module $\varkappa_\mathcal{D}$ is isomorphic to $C^\infty(M))$. Therefore, locally there exists a 1-form $\omega \in \Lambda^1(M)$ such that $\mathcal{D} = \mathcal{D}_\omega$ (see Section 19.11) and so $d\omega(X,Y) = -\omega([X,Y])$ for $X,Y \in \mathcal{D}$. But since the conditions $\omega(Z) = 0$ and $Z \in \mathcal{D}$ are equivalent, this implies that $d\omega|_{\mathcal{D}} = f\varkappa_{\mathcal{D}}$, where $f \in C^{\infty}(M)$ is nonzero everywhere. This shows that ω is a contact form. But $\hat{R}_{\mathcal{D}}$ is a local isomorphism, and by arguing as above, we conclude that $\mathcal{D}$ is locally a contact form.

19.13. Example: the secret of the Monge–Ampère equation. Recall that the general two-dimensional scalar *Monge–Ampère equation* reads

$$N(u_{xx}u_{yy} - u_{xy}^2) + Au_{xx} + Bu_{xy} + Cu_{yy} + D = 0, \qquad (19.6)$$

where N, A, B, C, D are certain functions of the variables x, y, u, u_x, u_y. Concrete instances of this class of equations were studied by many authors, beginning with Monge's 1784 paper. The intimate connection of this equation with contact geometry was discovered by Sophus Lie, who, in particular, established that this class is invariant under contact transformations, i.e., under diffeomorphisms of contact manifolds preserving the contact structure. Below, we indicate the previously unnoticed main cause of this, thereby once again demonstrating the importance of "conceptually correct" definitions.

The curvature form of the distribution $\mathcal{D}$ allows us to distinguish the class of *self-adjoint endomorphisms* of the module $\mathcal{D}$. Namely, an endomorphism $A \colon \mathcal{D} \to \mathcal{D}$ is called *self-adjoint* if

$$R_{\mathcal{D}}(AX, Y) = R_{\mathcal{D}}(X, AY), \ \forall X, Y \in \mathcal{D}.$$

Self-adjoint endomorphisms constitute a *Jordan algebra*[1], denoted $\mathrm{Snd}(\mathcal{D})$, in which multiplication is defined by the formula

$$S * T \overset{\text{def}}{=} \frac{S \circ T + T \circ S}{2},$$

and $\mathrm{id}_{\mathcal{D}}$ is the unit. It is also obvious that $\mathrm{Snd}(\mathcal{D})$ is a submodule of the $C^{\infty}(J^1(P))$-module $\mathrm{End}(\mathcal{D})$.

Now let us consider the contact distribution $\mathcal{C}$ on $J^1 P$ from Section 19.7. We shall say that an element $p \in P$ is *invariant with respect to the endomorphism* $T \in \mathrm{Snd}(\mathcal{C})$ if the vector field $T(X)$, $X \in \mathcal{C}$, is tangent to the image of the section $j_1(p)$ whenever the vector field X is tangent to it. The search for elements of the module P invariant w.r.t. T yields a system of first-order differential equations, because T is a homomorphism, while j_1 is an operator of first order. In particular, let us put $\dim M = 2$ and pass to the classical notation, taking $(x = x_1, y = x_2, u, p = p_1, q = p_2)$ to be the standard local chart on $J^1 P$ (see Section 14.20). In these coordinates,

---

[1] A Jordan algebra is a nonassociative algebra over a field whose multiplication satisfies the following axioms:
  1. $xy = yx$ (commutative law),
  2. $(xy)(xx) = x(y(xx))$ (Jordan identity).

the basis of the locally free module $\mathcal{C}$ is given by the vector fields

$$\frac{\partial}{\partial_x} + p\frac{\partial}{\partial_u}, \quad \frac{\partial}{\partial_y} + q\frac{\partial}{\partial_u}, \quad \frac{\partial}{\partial_p}, \quad \frac{\partial}{\partial_q},$$

and the endomorphism T with the matrix

$$\begin{pmatrix} B & -2A & 0 & -2N \\ 2C & -B & 2N & 0 \\ 0 & 2D & B & 2C \\ -2D & 0 & -2A & -B \end{pmatrix}$$

in this basis is self-adjoint. After that, the reader will easily verify that equation (19.6) exactly expresses the invariance condition of the function $u = u(x, y)$ w.r.t. the operator T as defined above. It is useful to note that $T^2 = \Delta \cdot I$, where I is the unit matrix and $\Delta = B^2 - 4AC + 4ND$. Note that equation (19.6) is elliptic (respectively, parabolic or hyperbolic) if $\Delta < 0$ (respectively, $\Delta = 0$ or $\Delta > 0$).

The submodule $\mathcal{A}$ of the algebra $\mathrm{Snd}(\mathcal{C})$ generated by T and $\mathrm{id}_{\mathcal{C}}$ is, moreover, a Jordan subalgebra of this algebra. Obviously, the invariance condition for the element p w.r.t. the endomorphisms $T' \in \mathcal{A}$ is equivalent to its invariance w.r.t. T. The condition of $\mathcal{A}$-invariance is conceptually preferable to the condition of T-invariance, since it does not depend on the choice of local coordinates on $J^1 P$. The initial endomorphism T can be locally normalized so that the endomorphism $T_0 = fT$ will satisfy the condition $T_0^2 = \epsilon \, \mathrm{id}_{\mathcal{C}}$, where $\epsilon = \pm 1, 0$. This, of course, is possible in those domains of the manifold $J^1 P$ where Δ does not change sign. Finally, all the above can be carried over to any five-dimensional contact manifold $(N, \mathcal{D})$ by setting the problem of finding so-called Legendre manifolds invariant w.r.t. the given two-dimensional (as a projective module) $C^\infty(N)$-module of the Jordan subalgebra $\mathcal{A} \subset \mathrm{Snd}(\mathcal{D})$. (In this situation, Legendre manifolds are two-dimensional submanifolds $L \subset N$ such that $R_{\mathcal{D}}|_L = 0$.)

19.14. Remarks. 1. The invariance condition described above implies that the solutions of the Monge–Ampère equation are so-called ζ-complex curves. By definition, these are curves over the algebras of ζ-complex "numbers" $x + y\zeta, x, y \in \mathbb{R}$, for which $\zeta^2 = \pm 1$ or 0 depending on the type of equation under consideration; thus, for an elliptic Monge–Ampère equation, $\zeta^2 = -1$, and so its solutions are complex curves in the ordinary sense.

2. The interpretation of the Monge–Ampère equations given above indicates new approaches to their solution, their classification, and so on.

3. The fact that the Monge–Ampère equation gives a description in coordinates of the problem of finding invariant elements $p \in P$ in the sense indicated above has a deep meaning in the theory of geometrical singularities of multivalued solutions of nonlinear differential equations. Here we can only mention that these equations coincide with equations describing singularities of their solutions of certain types, and the singularities themselves are classified by the algebras of ζ-complex numbers.

19.15. The Lie derivative: general idea. Any vector field X on a manifold can be interpreted as the velocity of a certain *flow*. One can also imagine that the flow captures and carries a certain geometrical quantity $\mathfrak{G}$, deforming it along the way. On the intuitive level, the *Lie derivative* of this quantity along the vector field X, denoted by $L_X(\mathfrak{G})$, is the velocity of its deformation under the action of the flow. We can intuitively think of the flow as a family of diffeomorphisms $F_t \colon M \to M$, $F_0 = \mathrm{id}_M$ depending on "time." However, in order to "observe" it, it is necessary to pass to the family of transformations of the observables $F_t^* \colon C^\infty(M) \to C^\infty(M)$. Once this is done, it is natural to define the velocity of the field $\{F_t\}$ as the derivative

$$X = \frac{dF_t^*}{dt}\Big|_{t=0}, \quad X(f) \overset{\text{def}}{=} \frac{dF_t^*(f)}{dt}\Big|_{t=0}, \quad f \in C^\infty(M).$$

The next manipulation shows that X is indeed a vector field on M:

$$\begin{aligned} X(fg) &= \frac{dF_t^*(fg)}{dt}\Big|_{t=0} = \frac{[dF_t^*(f)F_t^*(g)]}{dt}\Big|_{t=0} \\ &= \frac{dF_t^*(f)}{dt}\Big|_{t=0} \cdot g + f \cdot \frac{dF_t^*(g)}{dt}\Big|_{t=0} = X(f)g + fX(g). \end{aligned}$$

As an example, let us compute the Lie derivative of $L_X(Y)$ of a field $Y \in D(M)$. The image $Y_t = F_t(Y)$ of the field Y under the diffeomorphism F_t is defined by the relation $F_t^* \circ Y_t = Y \circ F_t^*$. Let us take the derivatives of both sides of this equation, having in mind that, by definition, we have $L_X(Y) = dY_t/dt|_{t=0}$. Since

$$d\big(F_t^* \circ Y_t\big)\big|_{t=0} = d(F_t^*)/dt\big|_{t=0} \circ Y + \mathrm{id}_{C^\infty(M)} \circ dY_t/dt\big|_{t=0} = X \circ Y + L_X(Y),$$

it follows that

$$X \circ Y + L_X(Y) = \frac{Y \circ F_t^*}{dt}\Big|_{t=0} = Y \circ X.$$

Thus, we obtain

$$L_X(Y) = [Y, X]. \tag{19.7}$$

This formula clarifies the conceptual meaning of the commutator of vector fields. In particular, it implies the remarkable relation $L_Y(X) = -L_X(Y)$, which is not at all obvious a priori from the definition of the Lie derivative. Besides, it allows to define the Lie derivative of vector fields for arbitrary commutative algebras, when the heuristic considerations described above lose their meaning.

Sometimes, in order to avoid misunderstandings, we use the notation L_X^Ψ, where the index Ψ indicates the nature of the quantity to which the Lie derivative is applied. For example, L_X^D denotes the Lie derivative of a vector field, while L_X^Λ denotes the Lie derivative of differential forms.

Exercises.   1. Using 19.7, show that $[L_X^D, L_Y^D] = L_{[X,Y]}^D$.

2. From the general idea underlying the notion of Lie derivative, carry out a heuristic argument showing that $[L_X^\Psi, L_Y^\Psi] = L_{[X,Y]}^\Psi$.

19.16. The Lie derivative of differential forms. If $\omega \in \Lambda^k(M)$, then the family $F_t^*(\omega)$ is a deformation of the differential form ω under the action of the flow F_t. Therefore, according to the informal definition,

$$L_X(\omega) = \left.\frac{dF_t^*(\omega)}{dt}\right|_{t=0}, \quad \omega \in \Lambda^k(M). \tag{19.8}$$

However, this definition, although it clarifies the meaning of the Lie derivative, does not indicate any method of its practical computation and further manipulations. Hence, it is necessary to find the analog of formula (19.7) for differential forms. To do this, we will use the result of the following simple exercise.

Exercise. From the definition 19.8, show that

1. $L_X(f) = X(f)$, $f \in C^\infty(M)$;

2. $L_X(\omega \wedge \rho) = L_X(\omega) \wedge \rho + \omega \wedge L_X(\rho)$, $\omega, \rho \in \Lambda^*(M)$;

3. $d \circ L_X = L_X \circ d$.

These formulas imply that

$$L_X(f dg_1 \wedge \ldots \wedge dg_k) = L_X(f) dg_1 \wedge \ldots \wedge dg_k$$
$$+ \sum_i dg_1 \wedge \ldots \wedge dX(g_i) \wedge \ldots \wedge dg_k.$$

Therefore, the Lie derivative L_X is entirely determined by the three properties listed in the above exercise.

Further, let us note that

$$L_X(f) = X(f) = (i_X \circ d)(f),$$
$$L_X(dg) = dL_X(g) = dX(g) = (d \circ i_X)(dg).$$

From this, one might suppose that $L_X = d \circ i_X + i_X \circ d$. This is indeed the case; to prove that formula, it suffices to verify that the operator defined in that way satisfies relations 1–3 in the above exercise. The first of them is obvious, the third immediately follows from $d^2 = 0$. The second one, as the calculation below shows, can be established by successive applications of formulas (18.22) and (18.26) under the assumption that ω belongs to

$\Lambda^k(M)$:

$$(d \circ i_X + i_X \circ d)(\omega \wedge \rho)$$
$$= d(i_X(\omega) \wedge \rho + (-1)^k \omega \wedge i_X(\rho)) + i_X(d\omega \wedge \rho + (-1)^k \omega \wedge d\rho))$$
$$= d(i_X(\omega)) \wedge \rho + (-1)^{k-1} i_X(\omega) \wedge d\rho + (-1)^k d\omega \wedge i_X(\rho) + \omega \wedge d(i_X(\rho))$$
$$\quad + i_X(d\omega) \wedge \rho + (-1)^{k+1} d\omega \wedge i_X(\rho) + (-1)^k i_X(\omega) \wedge d\rho + \omega \wedge i_X(d\rho)$$
$$= (d \circ i_X + i_X \circ d)(\omega) \wedge \rho + \omega \wedge (d \circ i_X + i_X \circ d)(\rho). \tag{19.9}$$

Thus the previous heuristic considerations lead us to the following definition.

Definition. Let $X \in D(A)$. The operator

$$L_X = d^{\mathfrak{M}} \circ i_X + i_X \circ d^{\mathfrak{M}}$$

is called the *Lie derivative of differential forms in the category* $\mathfrak{M}$.

As before, a direct consequence of this definition is the formula

$$L_X(a \, da_1 \wedge \ldots \wedge da_k) = X(a) da_1 \wedge \ldots \wedge da_k$$
$$+ \sum_i a \, da_1 \wedge \ldots \wedge dX(a_i) \wedge \ldots \wedge da_k. \tag{19.10}$$

19.17. Relative Lie derivative. The previous definition can be immediately generalized to relative vector fields. Namely, if $X \colon A \to B$ is such a field with respect to the algebra homomorphism $\varphi \colon A \to B$, while $\mathfrak{M}$ and $\mathcal{K}$ are φ-connected categories, then we can put

$$L_X \overset{\text{def}}{=} d^{\mathfrak{M}} \circ i_X + i_X \circ d^{\mathcal{K}} \colon \Lambda_{\mathcal{K}}^*(A) \to \Lambda_{\mathfrak{M}}^*(B). \tag{19.11}$$

Exercise. Verify the validity of the following formulas

1. $d^{\mathfrak{M}} \circ L_X = L_X \circ d^{\mathcal{K}}$;

2. $L_X(a) = X(a)$, $a \in A$;

3. $L_X(\omega \wedge \rho) = L_X(\omega) \wedge \varphi^*(\rho) + \varphi^*(\omega) \wedge L_X(\rho)$, $\omega, \rho \in \Lambda^*(A)$;

4. $L_{aX} = a L_X + d^{\mathfrak{M}} a \wedge i_X$, $(d^{\mathfrak{M}} a \wedge i_X)(\omega) \overset{\text{def}}{=} d^{\mathfrak{M}} a \wedge i_X(\omega)$.

Recall that a map of smooth sets (see Section 7.13). $F \colon M \to N$ is called smooth if $F^*(C^\infty(N)) \subset C^\infty(M)$.

Proposition. *Let $F_t \colon M \to N$, $-\epsilon < t < \epsilon$, be a family of smooth maps of smooth sets and let $X = \frac{dF_t^*}{dt}|_{t=0}$ be the corresponding relative vector field. Then*

$$L_X(\omega) = \frac{dF_t^*(\omega)}{dt}\Big|_{t=0}, \quad \omega \in \Lambda_\Gamma^*(C^\infty(N)).$$

◀ The proof directly follows from the fact that the operators L_X and $\frac{dF_t^*}{dt}|_{t=0}$ acting on $\Lambda_\Gamma^*(C^\infty(N))$ coincide on forms of degree ≤ 1 and the

first three formulas of the previous exercise are satisfied for both of them.
▶

19.18 Proposition. *Let* $X, Y \in D(A)$, $a \in A$ *and* $\omega \in \Lambda_{\mathfrak{M}}^{k}$. *Then the following relations hold:*

(a) $[L_X, L_Y] = L_{[X,Y]}$,

(b) $[L_X, i_Y] = i_{[X,Y]}$.

◀ Using formula (19.10) these two relations can be proved by a direct calculation. A more economical proof consists in the obvious verification of these relations for forms of degree ≤ 1. Then, in order to generalize this to higher degree forms, it suffices to check that the operators $[L_X, L_Y]$ and $[L_X, i_Y]$ act on the exterior product of differential from according to the same rules as the operators $L_{[X,Y]}$ and $i_{[X,Y]}$, respectively. But this last fact follows from the result of the exercise in Section 19.17.   ▶

19.19. Let us define the *Lie derivative of a differential form* by the following formula:

$$L_X(\omega)(X_1, \ldots, X_k) =$$
$$X(\omega(X_1, \ldots, X_k)) + \sum_i \omega(X_1, \ldots, [X_i, X], \ldots, X_k). \quad (19.12)$$

It is easily checked by induction (using the relation (see (19.18))) that

$$i_{X_1} \circ L_X = L_X \circ i_{X_1} - [L_X, i_{X_1}] = L_X \circ i_{X_1} - i_{[X, X_1]}.$$

Namely, if $\omega \in \Lambda_{\mathfrak{M}}^{k}$, then this relation implies that

$$L_X(\omega)(X_1, \ldots, X_k) = ((i_{X_1} \circ L_X)(\omega))(X_2, \ldots, X_k)$$
$$= L_X(i_{X_1}(\omega))(X_2, \ldots, X_k) + \omega([X_1, X], X_2, \ldots, X_k). \quad (19.13)$$

For $k = 1$, the last relation immediately yields the required result. Assuming further that the formula that we are proving holds for forms of degree $k \geq 1$, we obtain

$$L_X(i_{X_1}(\omega))(X_2, \ldots, X_k) = X(i_{X_1}(\omega))(X_2, \ldots, X_k)$$
$$+ \sum_{2 \leq i \leq k} i_{X_1}(\omega))(X_2, \ldots, [X_i, X], \ldots, X_k) = X(\omega(X_1, X_2, \ldots, X_k))$$

$$+ \sum_{2 \leq i \leq k} (\omega(X_1, X_2, \ldots, [X_i, X], \ldots, X_k).$$

Comparing this last relation to formula (19.13), we conclude the proof.

19.20. Infinitesimal symmetries: general idea. The Lie derivative is part of the natural language of infinitesimal symmetries of mathematical structures and objects of various nature. On the purely intuitive level, a symmetry of some object $\mathcal{O}$ is simply a transformation (of a specific type) of the object into itself, usually having a specific term associated to it. For example, a symmetry of some algebraic object is usually called an automorphism, while the symmetry of manifold is its diffeomorphism. The symmetries of a given object $\mathcal{O}$ obviously constitute a group, which it is natural to denote by $\operatorname{Sym} \mathcal{O}$.

Now let $F_t \colon \mathcal{O} \to \mathcal{O}$, $F_0 = \operatorname{id}_{\mathcal{O}}$, be a family of symmetries depending on "time." Then it would be appropriate to call $dF_t/dt|_{t=0}$ an *infinitesimal symmetry* of the object $\mathcal{O}$. Of course, this very rough idea must be clarified. For example, the notion of derivative for a family of diffeomorphisms $F_t \colon M \to M$ makes no sense because the difference $F_{t+\Delta t} - F_t$ is meaningless. But it will acquire a meaning if we recall that $M = |C^\infty(M)|$ and $F_t = |H_t|$, where H_t is an automorphism of the algebra $C^\infty(M)$. Indeed, in this case the classical formula

$$\frac{dH_t}{dt} \stackrel{\text{def}}{=} \lim_{\Delta t \to 0} \frac{1}{\Delta t}(H_{t+\Delta t} - H_t)$$

makes sense, first of all because the set of automorphisms of the algebra $C^\infty(M)$ constitute a vector space over $\mathbb{R}$. Assuming that the following limit

$$\Theta \stackrel{\text{def}}{=} \frac{dH_t}{dt}\bigg|_{t=0} = \lim_{\Delta t \to 0} \frac{1}{\Delta t}(H_{\Delta t} - \operatorname{id}_{C^\infty(M)})$$

exists, it is not hard to see that $\Theta \colon C^\infty(M) \to C^\infty(M)$ is an $\mathbb{R}$-linear map and $\Theta(fg) = \Theta(f)g + f\Theta(g)$. Thus, the heuristic considerations developed above lead us to the idea of defining the *infinitesimal symmetry of a manifold* M or the *algebra* $C^\infty(M)$ as the vector field $\Theta \in D(M)$. In this form, the idea can be carried over immediately to the general case of commutative algebras, whereas this cannot be done from the heuristic considerations that led to it. Thus it is natural to call any "vector field" $X \in D(A)$ an *infinitesimal symmetry* of the algebra A. Comparing the ideas of infinitesimal symmetry (IS) with the idea of Lie derivative (see Section 19.15), we come to the conclusion that a field $X \in D(M)$ should be referred to as an IS of the quantity $\mathfrak{G}$ provided that $L_X(\mathfrak{G}) = 0$. For example, the field X is the IS of the field $Y \in D(M)$ if $[X, Y] = 0$, or, respectively, the IS of the form $\omega \in \Lambda^k(M)$ if $L_X(\omega) = 0$. In particular, if Ω is a symplectic form, then it is an IS of the symplectic manifold (M, Ω). Or if X is the velocity field of a moving liquid, while $\omega = dx \wedge dy \wedge dz$ is the Euclidean volume form, then X is an IS of ω if and only if the liquid is incompressible.

Exercises. Show that

(1) a Hamiltonian field X_f is the IS of a symplectic manifold (M, Ω);

(2) $\operatorname{div} X = 0$ if and only if X is the IS of the Euclidean volume form.

These examples and similar ones show that infinitesimal symmetries appear in different guises in various circumstances in differential geometry, mechanics, and physics. For example, a Killing field in Riemannian geometry (see Example V in Section 19.21) is an IS of Riemann manifolds.

Besides, such examples show that the notion of infinitesimal symmetry, as well as the notion of Lie derivative, automatically carry over to objects of the differential calculus in commutative algebras.

19.21. Specifying the definitions of infinitesimal symmetry. The exact definition of the notion of infinitesimal symmetry in various concrete contexts is not always obvious and may require additional computations. From this point of view, we must first determine what the ordinary notion of infinitesimal symmetry is in the given context, which, as a rule, is not difficult. Then we must differentiate the family of such symmetries depending on "time" and declare that the obtained result is an infinitesimal symmetry. The following examples illustrate this procedure.

Examples. I. *Poisson manifolds.* From the intuitive geometrical point of view, a symmetry of a Poisson manifold $(M, \{\cdot, \cdot\})$ is a diffeomeorphism F preserving the bracket $\{\cdot, \cdot\}$. But, as we have seen, it is more appropriate to speak of the symmetry of the Poisson algebra $(C^\infty(M), \{\cdot, \cdot\})$, i.e., of isomorphisms $H \colon C^\infty(M) \to C^\infty(M)$ that preserve this bracket. This clearly means that $H(\{f, g\}) = \{H(f), H(g)\}$. Further, following our procedure, let us consider the family of symmetries H_t, $H_0 = \mathrm{id}\,|_{C^\infty(M)}$, and formally differentiate this definition of symmetry w.r.t. t:

$$\frac{dH_t(\{f,g\})}{dt} = \left\{ \frac{dH_t(f)}{dt}, H_t(g) \right\} + \left\{ H(f), \frac{H_t(g)}{dt} \right\}.$$

The last equation for $X = d/dt|_{t=o}$ acquires the form

$$X(\{f,g\}) = \{X(f), g\} + \{f, X(g)\}$$

and should be regarded as a symmetry of the given Poisson manifold. Moreover, this "computational definition" can be immediately carried over to arbitrary Poisson algebras. Namely, *a derivation $X \in D(A)$ is said to be an infinitesimal symmetry of the Poisson algebra $(A, \{\cdot, \cdot\})$, if*

$$X(\{a_1, a_2\}) = \{X(a_1), a_2\} + \{a_1, X(a_2)\}, \ \forall a_1, a_2 \in A. \tag{19.14}$$

Note that Hamiltonian fields X_a, $a \in A$, are infinitesimal symmetries of the given Poisson algebra, since in the present case the condition (19.14) is identical to the Jacobi identity for the elements a, a_1, a_2.

II. *Distributions.* From the geometrical point of view, a symmetry of a distribution $(M, \mathcal{D})$ is a diffeomorphism $F \colon M \to M$ which preserves the module $\mathcal{D}$. This means that the image $F(Y)$ of any field $Y \in \mathcal{D}$ also belongs to $\mathcal{D}$. Further, differentiating, the inclusion $F_t(Y) \in \mathcal{D}$, where F_t, $F_0 = \mathrm{id}\,|_M$, is the family of symmetries of the distribution $\mathcal{D}$, we obtain the inclusion $dF_t(Y)/dt|_{t=0} \in \mathcal{D}$. But $dF_t(Y)/dt = L_X(Y) = [Y, X]$, where $X = dF_t^*/dt|_{t=0}$ (see Section 19.15). Thus, a vector field $X \in D(M)$ such that $[Y, X] \in \mathcal{D}$ should be called an *infinitesimal symmetry of the distribution $\mathcal{D}$* provided $Y \in \mathcal{D}$ or, symbolically, if $[X, \mathcal{D}] \subset \mathcal{D}$. An infinitesimal symmetry of a contact distribution is called a *contact field*.

As before, these definitions can be automatically carried over to arbitrary commutative algebras. (Do that!)

III. *Tensors.* A symmetry of a covariant tensor T of rank k on a manifold M, understood as a $C^\infty(M)$-polylinear function $T(X_1, \ldots, X_m)$ on $D(M)$ (see Section 16.13) can be naturally considered as being a diffeomorphism $F \colon M \to M$ such that

$$F^*(T(X_1, \ldots, X_m)) = T(F(X_1), \ldots, F(X_m)).$$

Differentiating this equality w.r.t. t for the family of symmetries F_t and using, as in the previous example, the fact that

$$dF_t(Y)/dt = L_X(Y) = [Y, X], \text{ where } X = \left.\frac{dF_t^*}{dt}\right|_{t=0},$$

we obtain the following definition of an infinitesimal symmetry X for the tensor T:

$$X(T(X_1, \ldots, X_m) = \sum_{i=1}^{m} T(X_1, \ldots, [X_i, X], \ldots, X_m). \qquad (19.15)$$

Moreover, as is easily seen, similar considerations lead to the formula

$$L_X(T)(X_1, \ldots, X_m) = X(T(X_1, \ldots, X_m)) -$$
$$- \sum_{i=1}^{m} T(X_1, \ldots, [X_i, X], \ldots, X_m). \qquad (19.16)$$

The definition (19.15) automatically carries over to arbitrary smooth algebras, while the formula (19.15) in that case may be taken for the definition of the Lie derivative of the tensor T.

Recall that a nondegenerate symmetric covariant tensor $g(X_1, X_2)$ of rank 2 is said to be a *pseudo-Riemannian metric* on the manifold M. In the literature, the infinitesimal symmetries of this tensor are called *Killing fields*.

VI. *Modules.* It is natural to understand a symmetry of a vector bundle π over a manifold M as a diffeomorphism F of the manifold which is covered by a morphism of the bundle π to itself, the latter being an isomorphism on each fiber. The algebraic paraphrase of this is a pair $(\bar{H}, H)$ consisting of an automorphism H of the algebra $C^\infty(M)$ and an automorphism $\bar{H}$ of an $\mathbb{R}$-vector space P, where P is a $C^\infty(M)$-module, the two automorphisms being related by the formula $\bar{H}(fp) = H(f)\bar{H}(p)$, $f \in C^\infty(M)$, $p \in P$. We will obtain an infinitesimal version of this by differentiating at $t = 0$ the family of symmetries $(\bar{H}_t, H_t)$ under the assumption $\bar{H}_0 = \mathrm{id}_P$, $H_0 = \mathrm{id}_{C^\infty(M)}$ together with the connecting relation $\bar{H}_t(fp) = H_t(f)\bar{H}_t(p)$. As a result, we obtain the formula

$$\bar{X}(fp) = X(f)p + f\bar{X}(p), \qquad (19.17)$$

where $\bar{X} \stackrel{\text{def}}{=} d\bar{H}_t/dt|_{t=0}$, $X \stackrel{\text{def}}{=} dH_t/dt|_{t=0}$. Here X, as we have already seen, is the vector field $\bar{X} \in \text{Diff}_1(P, P)$, which follows directly from the relation (19.17).

The previous argument shows that in the general algebraic context, a pair

$$(\bar{X}, X), \ \bar{X} \in \text{Diff}_1(P, P), \ X \in D(A)$$

obeying the relation

$$\bar{X}(ap) = X(a)p + a\bar{X}(p), \quad a \in A, \ p \in P. \tag{19.18}$$

should be called an infinitesimal symmetry of the A-module P.

In the geometrical context, such a pair is known as a Der-*operator* or a *derivation of the vector bundle* π. Such operators arise in different situations of differential geometry. For example, such are *covariant derivatives* in the theory of connections.

19.22 Exercises. 1. Give an example of a Poisson manifold/algebra whose infinitesimal symmetry is not exhausted by Hamiltonian fields.

2. Let (N, ω) be a contact manifold (see Section 19.12). Show that a field $X \in D(N)$ is contact if and only if $L_X(\omega) = f\omega$ for some function $f \in C^\infty(N)$.

3. Describe Killing fields in the Euclidean metric

$$g = \sum_{i=1}^{n} dx_i \odot dx_i \ \left(= \sum_{i=1}^{n} dx_i^2\right).$$

19.23. The Lie algebra of infinitesimal symmetries. The set $\text{sym}\,\mathcal{O}$ of infinitesimal symmetries of some object $\mathcal{O}$ has a natural Lie algebra structure. This can be seen by applying the general procedure for finding infinitesimal symmetries from Section 19.21. Finite symmetries of some object $\mathcal{O}$ obviously form a group under composition. If H_t and G_t are families of symmetries of an object $\mathcal{O}$ depending on "time" and defined over some algebra $C^\infty(M)$, then $H_t \circ G_t$ is also such a family. Therefore, $d(H_t \circ G_t)/dt|_{t=0}$ is also an IS of such an object if $H_0 = G_0 = \text{id}_{C^\infty(M)}$. Now if $X = dH_t/dt|_{t=0}$, $Y = dG_t/dt|_{t=0}$, then

$$d(H_t \circ G_t)/dt|_{t=0} = (dH_t/dt \circ G_t + H_t \circ dG_t/dt)|_{t=0} = X + Y,$$

i.e., $X + Y \in \text{sym}\,\mathcal{O}$. This shows how an additive structure arises in $\text{sym}\,\mathcal{O}$, thereby transforming $\text{sym}\,\mathcal{O}$ into an Abelian group.

Similarly, the multiplication of infinitesimal symmetries by numbers is introduced by the relation $dH_{\lambda t}/dt|_{t=0} = \lambda X$, $\lambda \in \mathbb{R}$. Further, the relation

$$d(G_s \circ H_t \circ G_s^{-1})/dt|_{t=0} = G_s \circ X \circ G_s^{-1}$$

shows that $G_s \circ X \circ G_s^{-1}$ is a family of infinitesimal symmetries depending on s. Finally, the relation

$$d(G_s \circ X \circ G_s^{-1})/ds|_{s=0} = [Y, X]$$

implies that $[Y, X]$ is also an IS. This is how the Lie bracket is introduced in sym $\mathcal{O}$.

The fact that the operations defined above satisfy the axioms defining the notion of Lie algebra is also proved by differentiating the appropriate expressions. We leave this verification to the reader. In cases where the IS of the quantity $\mathfrak{G}$ is expressed by the formula $L_X(\mathfrak{G}) = 0$, the fact that the infinitesimal symmetries constitute a Lie algebra is a consequence of the formula $[L_X, L_Y] = L_{[X,Y]}$ that we established above for vector fields and differential forms (see Sections 19.15 and 19.18).

19.24. Example: description of contact fields. The definition of contact manifold that we have been using has the advantage, among other things, of yielding a simple description of its infinitesimal symmetries, i.e., contact vector fields. The Lie algebra of these fields for a contact manifolds $(M, \mathcal{C})$ will be denoted by Cont $\mathcal{C}$. Obviously, Cont $\mathcal{C}$ is an $\mathbb{R}$-vector space. Here it should be noted that the field fZ, $f \in C^\infty(M)$, is not a contact field in general even if the field Z is contact.

The $\mathbb{R}$-linear map $\mathfrak{c}$: Cont $\mathcal{C} \to \varkappa = $ Cont $/\mathcal{C}$

$$\text{Cont } \mathcal{C} \ni Z \quad \mapsto \quad (Z \bmod \mathcal{C}) \in \varkappa$$

plays the key role in the description of contact fields. First of all, it is injective. Indeed, Ker $\mathfrak{c} = ($Cont $\mathcal{C}) \cap \mathcal{C}$. Hence, if $Z \in $ Ker $\mathfrak{c}$, then $[Z, \mathcal{C}] \subset \mathcal{C}$, because the field Z is contact. Now since Z belongs to $\mathcal{C}$, the inclusion $[Z, \mathcal{C}] \subset \mathcal{C}$ is equivalent to the equality $\hat{R}_{\mathcal{C}}(Z) = 0$ (см. (19.5)). But since $\hat{R}_{\mathcal{C}}$ is a local isomorphism, it follows that $Z = 0$.

If the $C^\infty(M)$-module $\varkappa$ is free, then the map $\mathfrak{c}$ will also be surjective. To verify this, let us assign, to each field $Z \in D(M)$, the homomorphism of $C^\infty(M)$-modules

$$\phi_Z: \mathcal{C} \to \varkappa, \quad \phi_Z(X) = [X, Z](\bmod \mathcal{C}).$$

Recall also that the curvature map

$$\hat{R}_{\mathcal{C}}: \mathcal{C} \to \text{Hom}_{C^\infty(M)}(\mathcal{C}, \varkappa)$$

is an isomorphism if the module $\varkappa$ is free. Therefore, we can find a field $Y \in \mathcal{C}$ such that $\hat{R}_{\mathcal{C}}(Y) = \phi_Z$. But $\hat{R}_{\mathcal{C}}(Y) = -\phi_Y$, so that we will have $\phi_Z - \phi_Y = \phi_{Z-Y} = 0$. Now the last equality is equivalent to

$$[X, Z - Y] \in \mathcal{C}, \ \forall X \in \mathcal{C},$$

and so we can conclude that $Z - Y \in $ Cont $\mathcal{C}$. Thus, the map $\mathfrak{c}$ is an isomorphism of vector spaces if the module $\varkappa = \varkappa_{\mathcal{C}}$ is free. But, since in the general case the module $\varkappa$ is projective, it follows that $\mathfrak{c}$ is a local isomorphism.

It is also important to note that $\mathfrak{c}$ allows to carry over the $C^\infty(M)$-module structure from $\varkappa$ to Cont $\mathcal{C}$. Hence Cont $\mathcal{C}$ is a submodule of the module $\varkappa$. Finding conditions for the coincidence of Cont $\mathcal{C}$ and $\varkappa$ is a topological problem, but this is hardly a topic to be discussed here.

If $Z \in$ Cont $\mathcal{C}$ and $\mathfrak{c}(Z) = \nu$, then ν is said to be the *generating section of the contact field Z*, and is denoted by X_ν. Since the module $\varkappa$ is one-dimensional, the local coordinates of its elements can be identified with functions; we then use the notation X_f, $f \in C^\infty(M)$. In its turn, the algebra Lie structure in Cont $\mathcal{C}$ is transferred by $\mathfrak{c}$ to $\varkappa$, because $\mathfrak{c}$ is a local isomorphism, while the map $\varkappa \ni \nu \mapsto X_\nu \in D(M)$ is a first-order differential operator (see the exercise below). The bracket on $\varkappa$ defined in this way is denoted by $\{\cdot, \cdot\}_\mathcal{C}$, so that by definition, we have

$$[X_\mu, X_\nu] = X_{\{\mu,\nu\}_\mathcal{C}}, \quad \mu, \nu \in \varkappa.$$

This bracket is a first-order differential operator in each argument. In the literature such a bracket is often called the *Jacobi bracket*.

19.25 Exercises.   1. Show that the field

$$Z = [f, [g, \chi]](\nu) = X_{fg\nu} - f X_{g\nu} - g X_{f\nu} + fg X_\nu,$$

where $f, g \in C^\infty(M)$, $\mu, \nu \in \varkappa$ is a contact field that belongs to $\mathcal{C}$ and deduce from this that the map $\varkappa \ni \nu \mapsto X_\nu \in D(M)$ is a first-order differential operator.

2. Express the field X_ν in the canonical coordinates of Section 14.20, assuming that $\partial/\partial u$ is a basis vector field of $\varkappa$.

3. Express the bracket $\{\cdot, \cdot\}_\mathcal{C}$ in the same coordinates.

4. Let $\nu_0 \in \varkappa$ be the (local) basis of $\varkappa$ and define (locally) $X_f \overset{\text{def}}{=} X_{f\nu_0}$. Consider the subalgebra

$$C^\infty(M)_0 \overset{\text{def}}{=} \{f \in C^\infty(M) \mid X_{\nu_0}(f) = 0\} \subset C^\infty(M).$$

Show that $\{C^\infty(M)_0, C^\infty(M)_0\}_\mathcal{C} \subset C^\infty(M)_0$ and prove that the restriction of the bracket $\{\cdot, \cdot\}_\mathcal{C}$ to the subalgebra $C^\infty(M)_0$ transforms it into a Poisson algebra.

19.26. Conclusion. At first glance, this chapter might seem to be merely a technical compilation of algebraic generalizations of many classical notions from differential geometry, mechanics, physics, and differential equations. But it is actually much more—it opens the way to many concrete applications of the theory developed in this book to these and other branches of the physico-mathematical sciences, in particular in practically important situations when the classical theory of smooth manifolds is not applicable because smoothness requirements no longer hold and singu-larities appear. It turns out that the differential calculus on commutative algebra, as developed here, is the tool that works in these situations just as well as the classical theory works in the smooth case.

20

Cohomology

20.1. In the previous chapters, starting from conceptually defined differential forms, we were able to define the de Rham and Spencer complexes (Chapter 18) and construct the main functors of the differential calculus (Chapters 17 and 19). The next natural step in this direction is to study the cohomology of these complexes, which carries a wealth of useful information.

In this chapter, we present a series of examples and constructions that show how these cohomology theories may be effectively applied, and we demonstrate the techniques used in working with them.

20.2. De Rham cohomology of commutative algebras. Let $\mathfrak{M}$ be a differentially closed category of A-modules. Denote by

$$H^i_{\mathrm{dR}}(\mathfrak{M}) \stackrel{\mathrm{def}}{=} \frac{\operatorname{Ker} d^{\mathfrak{M}}_i}{\operatorname{Im} d^{\mathfrak{M}}_{i-1}}$$

the i-th de Rham cohomology group of the algebra A in the category $\mathfrak{M}$ (see Section 18.8). In the most important cases, we simplify this notation to

$$H^i_{\mathrm{dR}}(A) \stackrel{\mathrm{def}}{=} H^i_{\mathrm{dR}}(\operatorname{Mod} A), \quad H^i_{\Gamma\,\mathrm{dR}}(A) \stackrel{\mathrm{def}}{=} H^i_{\mathrm{dR}}(\operatorname{GMod} A)$$

Proposition 18.10 allows to carry over, in the standard way, the exterior multiplication operation of differential forms (see Section 17.18) to the cohomology groups.

$$[\omega] \wedge [\rho] \stackrel{\mathrm{def}}{=} [\omega \wedge \rho], \ \omega \in \Lambda^k_{\mathfrak{M}}, \ \rho \in \Lambda^l_{\mathfrak{M}},$$

© Springer Nature Switzerland AG 2020
J. Nestruev, *Smooth Manifolds and Observables*, Graduate Texts
in Mathematics 220, https://doi.org/10.1007/978-3-030-45650-4_20

where $[\nu] \in H^m(\mathfrak{M})$ denotes the cohomology class of the closed form ν. This multiplication is obviously skew-commutative, i.e., $[\rho] \wedge [\omega] = (-1)^{kl} [\omega] \wedge [\rho]$. Therefore, it transforms the direct sum

$$H^*_{\mathrm{dR}}(\mathfrak{M}) = \bigoplus_{i \geq 0} H^i_{\mathrm{dR}}(\mathfrak{M})$$

into a skew-commutative algebra, or, equivalently, into a $\mathbb{Z}$-graded commutative algebra ($\mathbb{Z}$-graded commutative algebras will be discussed in the next chapter, see 21.6). Since $d_0^{\mathfrak{M}}(1_A) = 0$, the cohomology class $[1_A]$ is well defined and is the unit of this algebra.

Let $\varphi \colon A \to B$ be a homomorphism of commutative algebras, let $\mathfrak{M}$ and $\mathcal{K}$ be φ-connected differentially closed categories of A- and B-modules, respectively. The naturality condition of the exterior differential (see Section 18.13) is tantamount to the fact that the family of maps $\varphi_* \colon \Lambda^i_{\mathfrak{M}} \to \Lambda^i_{\mathcal{K}}$, $i \geq 0$, determines a homomorphism of the corresponding de Rham complexes that we denote by

$$\varphi_{\mathrm{dR}} = \varphi_{\mathrm{dR}}^{\mathfrak{M}, \mathcal{K}} \colon \mathrm{dR}_{\mathfrak{M}}(A) \to \mathrm{dR}_{\mathcal{K}}(B).$$

Accordingly, the cohomology group homomorphism induced by φ_{dR} is denoted by

$$\varphi^i_{\mathrm{dR}} \colon H^i_{\mathrm{dR}}(\mathfrak{M}) \to H^i_{\mathrm{dR}}(\mathcal{K}), \quad \varphi^*_{\mathrm{dR}} \colon H^*_{\mathrm{dR}}(\mathfrak{M}) \to H^*_{\mathrm{dR}}(\mathcal{K}).$$

Since $\varphi_* \colon \Lambda^*_{\mathfrak{M}} \to \Lambda^*_{\mathcal{K}}$ is a homomorphism of algebras of differential forms, it follows that

$$\varphi^*_{\mathrm{dR}} \colon H^*_{\mathrm{dR}}(\mathfrak{M}) \to H^*_{\mathrm{dR}}(\mathcal{K}) \tag{20.1}$$

is also a homomorphism of algebras with unit. In particular, if $A = B$ and $\varphi = \mathrm{id}_A$, then any subcategory $\mathcal{K}$ of the category $\mathfrak{M}$ is tautologically φ-connected with it and conversely. Therefore, in this case, it is natural to refer to the homomorphism (20.1) as the *comparison homomorphism* of the de Rham cohomology of the algebra A in the categories $\mathfrak{M}$ and $\mathcal{K}$. For example, we have the most obvious comparison homomorphism $H^*_{\mathrm{dR}}(A) \to H^*_{\Gamma \, \mathrm{dR}}(A)$.

We conclude this section with the following statement, which has now become obvious:

Proposition. *The de Rham cohomology groups of the algebra $C^\infty(M)$ in the category of geometrical modules coincide with the ordinary de Rham cohomology groups of the manifold M, i.e.,,*

$$H^*_{\Gamma \, \mathrm{dR}}(C^\infty(M)) = H^*(M).$$

◄ This automatically follows from the isomorphism (constructed earlier) between the algebras $\Lambda^*_{\mathrm{GMod}\, C^\infty(M)}$ and $\Lambda^*(M)$ that identifies $d_{\mathrm{GMod}\, C^\infty(M)}$ and $d = d_M$. ►

20.3. Exercises. Compute the de Rham cohomology of the algebra A in the categories $\mathrm{Mod}A$ and $\mathrm{GMod}A$ if

1. $A = \mathbb{C}(x)$ and $A = \mathbb{R}(x)$ (the field of rational functions on $\mathbb{C}$ and $\mathbb{R}$, respectively);

2. $A = \mathbb{F}_2(x)$ (the field of rational functions on the field $\mathbb{F}_2$);

3. $A = S^{-1}(\mathbb{C}^\infty(\mathbb{R}^2)$, where S is the multiplicative set generated by the coordinate functions x and y;

4. $A = \mathbb{C}^\infty(S^1)$ (use the fact that A is isomorphic to the algebra of periodic functions on $\mathbb{R}^1$);

5. $A = \mathbb{F}_2[x,y]$ (use the fact that A is a module over the algebra $\mathbb{F}_2[x^2, y^2]$);

6. $A = \mathbb{F}_2[x,y]/(x^2 + y^2 - 1)$;

7. $A = C^\infty(\mathbf{K})$ (only in the category $\mathrm{GMod}A$), where $\mathbf{K}$ is coordinate cross in $\mathbb{R}^2$.

20.4. Residue theory. It is useful to have a procedure that would allow to determine whether or not a closed form determines a nonzero de Rham cohomology class. Below, we explain how it is possible, for certain algebras, to construct such a procedure on the basis of an algebraic analog of Leray's residue theory.

Let A be an algebra without zero divisors, let $\mathfrak{M}$ be a differentially closed category over A, and let $\Lambda_\mathfrak{M}^*$ be the algebra of differential forms in that category.

Leray condition. Suppose $s \in A$ is an irreversible nonnilpotent element and $S = \{s^n\}_{n \geqslant 0}$. Recall (see 15.30) that $\Lambda_{S^{-1}\mathfrak{M}}^* = S^{-1}\Lambda_\mathfrak{M}^*$. To simplify the notation, the image of $\Lambda_\mathfrak{M}^*$ under the localization map $S^{-1} \colon \Lambda_\mathfrak{M}^* \to \Lambda_{S^{-1}\mathfrak{M}}^*$ will be denoted by Λ^*. Set

$$\Lambda_s^* \overset{\text{def}}{=} \Lambda^*/(s\Lambda^* + ds \wedge \Lambda^*).$$

Strictly speaking, the right-hand side of the previous equality should have been written as $\Lambda^*/(S^{-1}(s)\Lambda^* + d_{S^{-1}\mathfrak{M}}S^{-1}(s) \wedge \Lambda^*)$, but, in order to simplify the notation and make the formulas more readable, we will write simply s instead of $S^{-1}(s)$, dropping the subscript of the exterior derivation operator. The image of a differential form $\omega \in \Lambda_s^*$ under the factorization map $\Lambda^* \to \Lambda_s^*$ is called the *restriction of ω to the ideal* $S^{-1}(sA) \subset S^{-1}A$ and is denoted by $\omega|_s$. The choice of the term "restriction" in this situation will become clear in the following example.

Example. Suppose that $s \in \Bbbk[x_1, \dots, x_n]$ is an irreducible polynomial, $A \subset \bar{A} \overset{\text{def}}{=} \Bbbk(x_1, \dots, x_n)$ is the algebra of all fractional-rational functions not containing s in the denominator, and $\mathfrak{M} = \mathrm{Mod}\,A$. In this case, we have $S^{-1}A = \bar{A}$, $\Lambda^* = \Lambda_{\mathrm{Mod}\,A}^*$, $\Lambda_s^* = \Lambda_{\mathrm{Mod}\,A/\nu_s}^*$, where ν_s is the principal ideal generated by s. (Check this.) From the geometrical point of view, the elements of Λ_s^* may be regarded as the restrictions of the forms from Λ^* to the submanifold in $\Bbbk^n$ determined by the equation $s = 0$.

Below we assume that for the form $ds \in \Lambda^*$, the *Leray condition* holds, i.e., the following complex

$$0 \to A \xrightarrow{\wedge ds} \Lambda^1 \xrightarrow{\wedge ds} \ldots \xrightarrow{\wedge ds} \Lambda^{i-1} \xrightarrow{\wedge ds} \Lambda^i \xrightarrow{\wedge ds} \ldots$$

is acyclic.

For simplicity, we set $p = \infty$ if $\operatorname{char} \Bbbk = 0$.

Exercise. Under the conditions of the previous example, verify whether the Leray condition holds for the polynomial $s = x^2 + y^2 - 1 \in \Bbbk[x, y]$.

20.5 Proposition. *Let $\omega \in \Lambda^*$ and $\varphi = \omega/s^n$, $n < p$. Suppose that there exists a form $\alpha \in \Lambda^*$ such that $d\varphi = \alpha/s^n$; then there exist forms ψ, $\theta \in \Lambda^*$ such that $\varphi = ds/s^n \wedge \psi + \theta/s^{n-1}$.*

◀ Since $d\varphi = \alpha/s^n$, it follows that

$$nds \wedge \omega = sd\omega - s\alpha \tag{20.2}$$

Therefore, we have $ds \wedge d\omega = ds \wedge \alpha$. By the Leray condition, there exists a form $\theta \in \Lambda^*$ such that $d\omega = ds \wedge n\theta + \alpha$. Substituting the last relation in (20.2), we obtain $nds \wedge \omega = nds \wedge s\omega$. Again using the Leray condition, we find a form $\psi \in \Lambda^*$ such that $\omega = s\theta + ds\psi$. ▶

Corollary. *If the assumptions of Proposition 20.5 hold and $n > 1$, then there exist forms β, $\gamma \in \Lambda^*$ such that $\varphi = d(\beta/s^{n-1}) + \gamma/s^{n-1}$.* ▶

20.6 Proposition. *Let $\omega \in \Lambda^*$ and let the form $\varphi = \omega/s$ be closed, then there exist forms ψ, $\theta \in \Lambda^*$ such that $\varphi = ds/s \wedge \psi + \theta$. The restrictions $\psi|_s$, $\theta|_s$ of the forms ψ, θ on S are uniquely determined, and $\psi|_s$ is closed. The forms $\psi|_s$ are called the* residue forms *of φ on S and are denoted by $\operatorname{res}[\varphi]_s$.*

◀ 1. The the existence of the forms φ, θ follows from Proposition 20.5.

2. Uniqueness of $\psi|_s$. It suffices to prove that $\psi|_s = 0$ whenever $\varphi = 0$. If $ds/s \wedge \psi + \theta = 0$, then $ds \wedge \theta = 0$ and by the Leray condition there exists a form $\theta' \in \Lambda_p^*$ such that $\theta = ds \wedge \theta'$. Hence,

$$ds \wedge \psi + s\theta = ds \wedge (\psi + s\theta').$$

Using the lemma once more, we obtain $\psi + s\theta' = ds \wedge \theta''$, where $\theta'' \in \Lambda_p^{l-2}$ so that $\psi|_s = 0$.

The form $ds/s \wedge \theta = ds/s \wedge \varphi$ is closed, therefore by the previous argument $\theta|_s = \operatorname{res}[ds/s \wedge \theta]_s$ is uniquely defined.

3. The form $\psi|_s$ is closed. Since $d\varphi = ds/s \wedge d\psi + d\theta = 0$, it follows that $d\psi|_s$ is the zero residue form on S and, by item 2 of the proof, is equal to zero. ▶

For the statement and proof of this proposition, we followed F. Pham. If in the conditions of the previous proposition we have $\varphi \in \Lambda_s^l$, then, as is easily seen, $\mathrm{res}[\varphi]_s = 0$. Indeed, in that case, we may put $\theta = \varphi$ and $\psi = 0$. Then $\mathrm{res}[\varphi]_s = \psi|_s = 0$.

Obviously, if $\varphi \in \Lambda^*$, then $\mathrm{res}[\varphi]_s = 0$.

20.7 Proposition. *Let* $\varphi = \omega/s$, $\omega \in \Lambda^*$, ω *be an exact form, and* char $\Bbbk = 0$, *then* $\mathrm{res}[\varphi]_s$ *is also exact.*

◄ By definition, $\omega/s = d(\alpha'/s^n)$, $\alpha' \in \Lambda^*$. If $n > 1$, then, by the corollary to Proposition 20.5, there exist forms β, $\gamma \in \Lambda^*$ such that

$$\frac{\alpha'}{s^n} = d\left(\frac{\beta}{s^{n-1}}\right) + \frac{\gamma}{s^{n-1}}.$$

Therefore, $\omega/s = d(\gamma/s^{n-1})$. Hence, by induction on n, we can find a form $\alpha \in \Lambda^*$ such that $\omega/s = d(\alpha/s)$. By Proposition 20.5, there exist forms ψ, $\theta \in \Lambda^*$ such that $\alpha/s = ds/s \wedge \psi + \theta$. Thus

$$\frac{\omega}{s} = d\left(\frac{\alpha}{s}\right) = -\frac{ds}{s} \wedge d\psi + d\theta,$$

and $\mathrm{res}[\omega/s]_s = -d\psi|_s$. ►

20.8 Proposition. *Let* $\varphi = \omega/s^n$, $\omega \in \Lambda^*$, $0 \leqslant n < p$, *be a closed form. Then there exists a form* $\alpha \in \Lambda^*$ *such that* φ *is cohomologous to* α/s.

◄ If $n = 0$, then $\alpha = s\omega$. The proof of this proposition for the case $n > 1$ is similar to the proof of Proposition 20.7. ►

20.9. Residues in the case of characteristic zero. Let char $\Bbbk = 0$. If $\varphi \in \Lambda_{S^{-1}\mathfrak{M}}^*$ is a closed form, then, by Proposition 20.8, there exists a form $\alpha \in \Lambda^*$ such that φ is cohomologous to α/s. The cohomology class of the form $\mathrm{res}[\alpha/s]_s$ is uniquely determined by Proposition 20.6. It is called the *residue class* of φ, and is denoted by $\mathrm{Res}[\varphi]_s$.

Consider a cohomology class $\Theta \in H^l(S^{-1}\mathfrak{M})$. Let φ, ψ be arbitrary elements of Θ. Then, by Proposition 20.7, we have $\mathrm{Res}[\varphi]_s = \mathrm{Res}[\psi]_s$. Thus the $\Bbbk$-homomorphism

$$\mathrm{Res}_s \colon H^l(S^{-1}\mathfrak{M}) \to H^{l-1}(\Lambda_s^*), \quad \mathrm{Res}_s(\Theta) = \mathrm{Res}[\varphi]_s, \ \varphi \in \Theta,$$

is well defined.

20.10 Theorem. *The sequence*

$$0 \longrightarrow H^l(\Lambda^*) \xrightarrow{\ i\ } H^l(S^{-1}\mathfrak{M}) \xrightarrow{\ \mathrm{Res}_s\ } H^{l-1}(\Lambda_s^*) \tag{20.3}$$

is exact.

◄ Obviously, $\mathrm{Im}\, i \subset \mathrm{Ker}\, \mathrm{Res}_s$. Let φ be an arbitrary representative of a cohomology class $\Theta \in H^l(S^{-1}\mathfrak{M})$. By definition, the kernel of the factorization homomorphism $\Lambda^* \to \Lambda_s^*$ is generated by s and ds. This implies that there exist forms $\alpha,\ \beta,\ \gamma,\ \delta \in \Lambda^*$ such that

$$\varphi = \frac{ds}{s} \wedge (d\alpha + s\beta + ds \wedge \gamma) + \delta.$$

This means that the form φ is cohomologous to $ds \wedge \beta + \delta$. Therefore $\Theta \in \operatorname{Im} i.$ ▶

Remark. Let $R \subset A$ be a multiplicative subset, and $s \notin R$. Suppose $R^{-1}\colon \Lambda^*_{S^{-1}\mathfrak{M}} \to R^{-1}\Lambda^*_{S^{-1}\mathfrak{M}} = \Lambda^*_{R^{-1}S^{-1}\mathfrak{M}}$ is a monomorphism. Then the Leray condition holds for the element $d_{R^{-1}S^{-1}\mathfrak{M}} s \in R^{-1}\Lambda^*$, and the residue homomorphism

$$\operatorname{Res}_{R^{-1}s}\colon H^l(R^{-1}S^{-1}A; R^{-1}S^{-1}\mathfrak{M}) \to H^{l-1}(\Lambda^*_{R^{-1}s})$$

can be constructed. The diagram

$$
\begin{array}{ccc}
H^l(S^{-1}A; S^{-1}\mathfrak{M}) & \xrightarrow{\ \operatorname{Res}_s\ } & H^{l-1}(\Lambda^*_s) \\
\downarrow & & \downarrow \\
H^l(R^{-1}S^{-1}A; R^{-1}S^{-1}\mathfrak{M}) & \xrightarrow{\ \operatorname{Res}_{R^{-1}s}\ } & H^{l-1}(\Lambda^*_{R^{-1}s})
\end{array}
$$

is commutative. This means that it is not too difficult to generalize most of the constructions described above to algebraic geometry, specifically to schemes and sheaves.

Example: discrete valuation rings. A ring A with unit and without zero divisors is said to be a *discrete valuation ring* if there exists an element $s \in A$ such that any ideal of the ring A is generated by some degree of the element s. In other words, all the ideals of the ring A are principal ideals generated by powers of one and the same element s, which is defined up to multiplication by an invertible element. By definition, in A there exists a unique maximal ideal $\nu = \{sa,\ a \in A\}$.

An example of a discrete valuation ring is the ring of formal power series $\Bbbk[[x]]$ in one variable.

Each element $a \in A$ may be written in the form $\alpha s^n,\ n \geqslant 0$, where α is an invertible element. As above, let us consider the multiplicative set $S = \{s^n\}_{n \geqslant 0}$ and note that the fraction field $\bar{A}$ of the ring A is nothing but $S^{-1}A$. For the category $\mathfrak{M}$, let us take the category $\operatorname{Mod} A$ of all A-modules. In that case, we have $\Lambda^* = \Lambda^*(A)$ and $S^{-1}\operatorname{Mod} A = \operatorname{Mod} \bar{A}$. To simplify the notation, we put $\Lambda^*_{S^{-1}\operatorname{Mod} A} = \Lambda^*(\bar{A})$.

20.11 Exercises. 1. Prove that Λ^*_s does not depend on the choice of the generating element s and so we have the right to use the notation Λ^*_ν.

2. Prove that Λ^*_ν is isomorphic to $\Lambda^*(A/\nu)$.

20.12. Definition. Let $\Omega = \{\omega_j\}$ be a finite subset of Λ^*. A finite subset $G = \{g_i\}$ is called a *ν-system of* Ω if the following conditions hold:
   (1) the elements dg_i are linearly independent;
   (2) if dG is a subalgebra of L generated by dg_i, then $\Omega \subset dG$.

It is not difficult to prove the following two statements.

20.13 Proposition. *The forms dg_i are linearly independent over $\bar{A}$ iff the elements g_i are algebraically independent over the field of constants $\Bbbk$.*

◀ The simple proof of this proposition is left to the reader.   ▶

20.14 Proposition. *For any finite subset of $\Omega = \{\omega_j\} \subset \Lambda^*$, there exists a ν-system.*   ▶

20.15 Proposition. $ds/s \notin \Lambda^*$.

◀ Assume the converse. Then there exist elements $f_i, g_i \in A$ such that

$$\frac{ds}{s} = \sum_{i=1}^{k} f_i dg_i \qquad (20.4)$$

By Propositions 20.13, 20.14, the elements g_i are algebraically independent over $\Bbbk$. On the other hand, it follows from the last condition that the elements $\{s, g_1, \ldots, g_k\}$ are algebraically dependent over $\Bbbk$.

Let $F(x, y_1, \ldots, y_k) \in \Bbbk[x, y_1, \ldots, y_k]$ be an irreducible polynomial over $\Bbbk$ such that $F(s, g_1, \ldots, g_k) = 0$. It is easily shown that (20.4) is true iff the polynomial $F(x, y_1, \ldots, y_k)$ can be presented in the form

$$F(x, y_1, \ldots, y_k) = xF'(x, y_1, \ldots, y_k) + C,$$

where $F'(x, y_1, \ldots, y_k) \in \Bbbk[x, y_1, \ldots, y_k]$, $0 \neq C \in \Bbbk$. Then

$$sF'(s, g_1, \ldots, d_k) + C = 0,$$

and the element s is invertible in the algebra A:

$$\frac{1}{s} = -\frac{F'(s, g_1, \ldots, d_k)}{C}.$$

This contradiction proves the proposition.   ▶

20.16 Proposition. *For any finite subset of $\Omega = \{\omega_j\} \subset \Lambda^*$, there exists a ν-system containing s.*

◀ Let $\{g_i\}$ be any ν-system for the set Ω. If the set $\{s, g_i\}$ is algebraically independent over $\Bbbk$, then it is just the required ν-system. Otherwise, there exists a nontrivial constraint

$$\alpha_0 ds + \sum_{i=1}^{k} \alpha_i dg_i = 0, \ \alpha_0 \neq 0, \ \alpha_i \in A.$$

Each of the nonzero coefficients α_i may be presented in the form $\alpha_i = \beta_i s^{r_i}$, $\beta_i \in A$. Let $r = \min_{i>0}\{r_i\}$ and l be an integer such that $r_l = r$. The assumption $r_0 < r_l$ contradicts Proposition 20.15, and thus the set $G \backslash \{g_l\}$ is the required ν-system.   ▶

20.17 Proposition. *The Leray condition holds for the form ds.*

◄ Let $\omega \in \Lambda^*$. By the previous proposition, for any ω there exists a ν-system $\{s, g_i\}$. Obviously, if $ds \wedge \omega = 0$, then

$$\omega = \sum_{(i_1,\ldots,i_k)} h_{(i_1,\ldots,i_k)} ds \wedge dg^{i_1} \wedge \cdots \wedge dg^{i_k}. \quad ▶$$

Now to any closed form $\varphi \in \Lambda^*(\bar{A})$ we can assign its residue $\mathrm{res}[\varphi]_s$. Let s' be another generating element of the ideal ν. For ds', the Leray condition also obviously holds.

20.18 Proposition. $\mathrm{res}[\varphi]_s = \mathrm{res}[\varphi]_{s'}$.

◄ Indeed, $s = \alpha s'$, where α is an invertible element in A. Hence,

$$\frac{ds}{s} \wedge \psi + \theta = \frac{d(\alpha s')}{\alpha s'} \wedge \psi + \theta = \frac{ds'}{s'} \wedge \psi + \frac{d\alpha}{\alpha} \wedge \psi + \theta. \quad ▶$$

Thus, $\mathrm{res}[\varphi]_s$ does not depend on the choice of the element s generating the ideal ν and we can speak of the residue form $\mathrm{res}[\varphi]_\nu$ of the closed form φ on the ideal ν and of the residue class $\mathrm{Res}[\theta]_\nu$ of the cohomology class $\theta \in H^*(\bar{A})$.

Bringing together all the above concerning discrete valuation rings, we come to the conclusion that for such rings we have Theorem 20.10, and the exact sequence (20.3) takes the form

$$0 \longrightarrow H^l(A) \overset{i}{\longrightarrow} H^l(\bar{A}) \overset{\mathrm{Res}_\nu}{\longrightarrow} H^{l-1}(A/\nu). \quad (20.5)$$

The example that follows is interesting from two points of view. First, in it we consider a non-Noetherian algebra. Second, this example shows that the validity of the Leray condition depends on the choice of the "coefficient" category $\mathfrak{M}$.

Let A be a non-Noetherian local algebra, i.e., an algebra with unique maximal ideal ν. Assume that ν is principal, let s be a generator of ν, let $A_s = S^{-1}A$, and let $\bar{A}$ be the field of fractions of the algebra A. Denote by $\mathfrak{M}$, $\mathfrak{M}_s$ the complete subcategories of $\mathrm{Mod}\,A$, $\mathrm{Mod}\,A_s$, respectively, such that objects of $\mathfrak{M}$ and $\mathfrak{M}_s$ have no torsion. It is easy to prove that the categories $\mathfrak{M}$ and $\mathfrak{M}_s$ are differentially closed. The natural embeddings $A \subset A_s \subset \bar{A}$ induce the homomorphisms

$$\Lambda^*_{\mathfrak{M}} \overset{i}{\to} \Lambda^*_{\mathfrak{M}_s} \overset{j}{\to} \Lambda^*_{\mathcal{M}(\bar{A})}.$$

Since $\Lambda^*_{\mathfrak{M}}$ and $\Lambda^*_{\mathfrak{M}_s}$ are torsion-free, the homomorphisms i and j have no kernels.

20.19 Proposition. *Under the above described conditions, the Leray condition holds.*

◄ Consider any differential form

$$\omega = \sum_{(i_1,\ldots,i_k)} h_{(i_1,\ldots,i_k)} dg^{i_1} \wedge \cdots \wedge dg^{i_k}.$$

Put

$$\bar{R} \stackrel{\text{def}}{=} \Bbbk(s, h_{(i_1,\ldots,i_k)}, \ldots, g^1, \ldots) \text{ and } R \stackrel{\text{def}}{=} \bar{R} \cap A.$$

Therefore, R is a discrete valuation ring, whence the Leray condition holds for $\Lambda^*_{\text{Mod } R}$. As we know, the homomorphism $\alpha\colon \Lambda^*_{\text{Mod } R} \to \Lambda^*_{\text{Mod } \bar{R}}$ has no kernel. The extension $\bar{R} \subset \bar{A}$ is separable, therefore the homomorphism $\alpha\colon \Lambda^*_{\text{Mod } \bar{R}} \to \Lambda^*_{\text{Mod } \bar{A}}$ also has no kernel. Let

$$\omega' = \sum_{(i_1,\ldots,i_k)} h_{(i_1,\ldots,i_k)} dg^{i_1}_{\text{Mod } R} \wedge \cdots \wedge dg^{i_k}_{\text{Mod } R}.$$

If $ds \wedge \omega = 0$, then $d_{\text{Mod } R} s \wedge \omega' = 0$. By the Leray condition for the complex $\Lambda^*_{\text{Mod } R}$, there exists a form $\lambda' \in \Lambda^*_{\text{Mod } R}$ such that $\omega' = d_{\text{Mod } R} s \wedge \lambda'$. Set $\lambda = \beta \circ \alpha(\lambda')$. Then $\omega = \beta \circ \alpha(\omega') = ds \wedge \lambda$. ▶

20.20. Residues in the case of positive characteristic. Suppose that $\operatorname{char} \Bbbk = p > 0$. For $n \equiv 0 \bmod p$, Propositions 20.5, 20.8 do not hold. For this reason, the construction of residue forms and residue classes described above does not work. Instead of it, it is possible to construct a spectral sequence whose terms play a similar role. The interested reader can find the details of this construction in [26].

Concluding this excursion in residue theory, note that multiple (iterated) residues may be constructed in a similar way.

20.21. Relative cohomology. The naturally defined restriction map $\Lambda^*(M) \to \Lambda^*(N)$ is surjective if the submanifold $N \subset M$ is closed (see Sections 4.11, 4.12). Its kernel is obviously a subcomplex of the de Rham complex of manifold M and is denoted by

$$\cdots \to \Lambda^{(i-1)}(M, N) \to \Lambda^i(M, N) \to \Lambda^{i+1}(M, N) \to \cdots$$

It is the complex of *relative differential forms of the manifold M modulo N*. Its cohomology groups are denoted by

$$H^*(M, N) = \sum_i H^i(M, N).$$

Elements of $\Lambda^*(M, N)$ are called *relative differential forms*.

In the general algebraic context, such a situation arises when there is a surjective homomorphism of commutative algebras $\varphi\colon A \to B$, and the categories $\mathfrak{M}$ and $\mathcal{K}$ are φ-connected. The kernel of the homomorphism φ_{dR} (see Section (20.1)) will be denoted by

$$\text{dR}(\mathfrak{M}, \mathcal{K}) = \sum_i \text{dR}^i(\mathfrak{M}, \mathcal{K}), \tag{20.6}$$

and is called the *complex of φ-relative differential forms (in the category $\mathfrak{M}$)*. Sometimes, we shall also use the notation $\text{dR}(\varphi)$, which is more convenient in certain contexts. In particular, we should note the case of the comparison of categories, especially when $\mathcal{K}$ is a subcategory of $\mathfrak{M}$. A typical example is $\mathfrak{M} = \mathrm{Mod}\, A$, $\mathcal{K} = \mathrm{GMod}\, A$. Accordingly, $(\varphi, \mathfrak{M}, \mathcal{K})$-*relative cohomology groups* of the cochain complex 20.6 are denoted by

$$H_{\mathrm{dR}}^{*}(\mathfrak{M}, \mathcal{K}) = \sum_{i} H_{\mathrm{dR}}^{i}(\mathfrak{M}, \mathcal{K}). \tag{20.7}$$

Thus we have obtained a *short exact sequence of complexes*

$$0 \to \mathrm{dR}(\mathfrak{M}, \mathcal{K}) \xrightarrow{\varphi_{\mathrm{dR}}^{\mathfrak{M}.\mathcal{K}}} \mathrm{dR}(\mathfrak{M}) \xrightarrow{\varphi_{\mathrm{dR}}} \mathrm{dR}(\mathcal{K}) \to 0 \tag{20.8}$$

It is known from basic cohomology theory that to any short exact sequence of complexes one can canonically associate a *long exact sequence of cohomology groups* which, in the case of the sequence 20.8, will be of the form

$$\cdots \to H_{\mathrm{dR}}^{i-1}(\mathcal{K}) \xrightarrow{\partial^{i-1}} H_{\mathrm{dR}}^{i}(\mathfrak{M}, \mathcal{K}) \xrightarrow{H(\varphi_{\mathrm{dR}}^{\mathfrak{M}.\mathcal{K}})} H_{\mathrm{dR}}^{i}(\mathfrak{M}) \xrightarrow{H(\varphi_{\mathrm{dR}})}$$

$$\xrightarrow{H(\varphi_{\mathrm{dR}})} H_{\mathrm{dR}}^{i}(\mathcal{K}) \xrightarrow{\partial^{i}} H_{\mathrm{dR}}^{i+1}(\mathcal{K}) \to \cdots \tag{20.9}$$

Here $H(\varphi_{\mathrm{dR}}^{\mathfrak{M},\mathcal{K}})$ and $H(\varphi_{\mathrm{dR}})$ are homomorphisms of cohomology groups induced by the homomorphisms of the complexes $\varphi_{\mathrm{dR}}^{\mathfrak{M},\mathcal{K}}$ and φ_{dR}, respectively, while ∂^{i} is the *connecting homomorphism*. Recall that the connecting homomorphism is defined as follows. Let $[\theta] \in H_{\mathrm{dR}}^{i}(\mathcal{K})$, $\theta \in dR^{i}(\mathcal{K})$. Then, since 20.8 is exact, we have $\theta = \varphi_{\mathrm{dR}}^{\mathfrak{M},\mathcal{K}}(\theta')$, $\theta' \in dR^{i}(\mathfrak{M})$. But $d^{\mathcal{K}}(\theta) = 0$, and so $\varphi_{\mathrm{dR}}(d^{\mathfrak{M}}(\theta')) = (d^{\mathfrak{M}}(\varphi_{\mathrm{dR}}(\theta')) = 0$, in view of the exactness of (20.8), we obtain $d^{\mathfrak{M}}(\theta') \in \mathrm{Im}(\varphi_{\mathrm{dR}}^{\mathfrak{M},\mathcal{K}})$. Finally, we put

$$\partial^{I}([\theta]) \stackrel{\mathrm{def}}{=} [F] \in H_{\mathrm{dR}}^{i+1}(\mathfrak{M}, \mathcal{K}), \quad \text{where} \quad d^{\mathfrak{M}}(\theta') = \varphi_{\mathrm{dR}}^{\mathfrak{M},\mathcal{K}}(F).$$

The reader will easily check that the cohomology class $[F]$ does not depend on the choice of the differential forms used in the above construction.

The long exact sequence (20.9) describes the relationship between the cohomology of a complex and the cohomology of its kernel and image under certain maps. For this reason, sequences of this type are one of the main technical instruments used for the practical computation of cohomology, and not only cohomology.

20.22. Example: the algebraic Newton–Leibnitz formula. Here we present a simply example of the application of the sequence (20.9) to cohomological integration theory. The necessity of such a theory becomes apparent in the framework of standard differential geometry, where integration in the sense of measure theory is not possible for a number of reasons. From this point of view, in this section we shall discuss the simplest particular case of the general Stokes formula, namely the Newton–Leibnitz formula.

Let $\Bbbk$ be a field of zero characteristic. Then $|\Bbbk[x]| = \Bbbk$ (as sets). Also let $\lambda, \mu \in \Bbbk, \lambda \neq \mu$, and let $\mathfrak{I} = \{\lambda\} \cup \{\mu\}$ be a two-point set in $|\Bbbk[x]|$. The algebra of $\Bbbk$-valued functions on $\mathfrak{I}$ can be naturally identified with the algebra of ordered pairs $(\alpha, \beta) \in \Bbbk^2, \lambda \mapsto \alpha, \mu \mapsto \beta$ with coordinate-wise addition and multiplication. Denote this algebra by $\Bbbk_{[\lambda,\mu]}$, thereby stressing that we regard $\mathfrak{I}$ as an *oriented "interval"* $[\lambda, \mu]$ in $|\Bbbk[x]|$.

Now let us describe the exact sequence 20.9 in the category of geometrical modules for the surjection $\varphi \colon \Bbbk[x] \to \Bbbk_{[\lambda,\mu]}$, using the simplified notation $H^i(\Bbbk[x])$, $H^i(\Bbbk_{[\lambda,\mu]})$ and $H^i(\Bbbk[x]), \Bbbk_{[\lambda,\mu]})$ for its terms. It is obvious that all the nontrivial terms are grouped together in the fragment

$$0 \to H^0(\Bbbk[x]) \xrightarrow{H_{\mathrm{dR}}(\varphi)} H^0(\Bbbk_{[\lambda,\mu]}) \xrightarrow{\partial} H^1(\Bbbk[x], \Bbbk_{[\lambda,\mu]}) \to 0.$$

Here $H^0(\Bbbk[x])$ consists of constants and is therefore canonically isomorphic to $\Bbbk$, while $H^0(\Bbbk_{[\lambda,\mu]})$ is *canonically* isomorphic to $\Bbbk^2$. Besides, $[H_{\mathrm{dR}}(\varphi)](\nu) = (\nu, \nu)$. Therefore, $H^1(\Bbbk[x], \Bbbk_{[\lambda,\mu]}) \cong \operatorname{coker} H_{\mathrm{dR}}(\varphi)$. It is remarkable that among these isomorphisms there exist exactly two canonical ones. Namely, $\partial(\alpha, \beta) = \beta - \alpha$ and $\partial(\alpha, \beta) = \alpha - \beta$. On the other hand, according to the definition, we have $\partial(\alpha, \beta) = [df(x)]$, where the polynomial $f(x) \in \Bbbk[x]$ satisfies $f(\lambda) = \alpha$ and $f(\mu) = \beta$. Therefore, it is natural to regard the cohomology class of the 1-form $df(x)$ as a *relative form*, either as the number $f(\mu) - f(\lambda)$, or as $f(\lambda) - f(\mu)$, depending on the orientation of the interval $\mathfrak{I}$. For this reason, it is appropriate to denote the relative cohomology class $[df(x)]$ by the symbol $\int_\lambda^\mu df(x)$. This can be written in the form of the Newton–Leibnitz formula (over the field $\Bbbk$)

$$\int_\lambda^\mu f'(x)\, dx = f(\mu) - f(\lambda)$$

which under the cohomological approach is transformed from a theorem to the definition of the integral.

Exercise. Investigate the sequence 20.9 for the algebra $\Bbbk[x]$ in the case when in $|\Bbbk[x]|$, instead of two, there is an arbitrary finite number of points.

20.23. Homotopy formula. One of the most effective methods for calculating cohomology is based on the use homotopy operators. Historically, this method appeared in the old algebraic topology quite naturally. In terms of de Rham cohomology on manifolds or sets, it looks as follows.

Smooth maps of smooth sets $F, G \colon M \to N$ are homotopic if there exists a smooth family $F_t \colon M \to N, 0 \leq t \leq 1$ such that $F = F_0$ and $G = F_1$.

Proposition. (Homotopy theorem.) *If two maps F and G are homotopic, then the corresponding cohomology group homomorphisms*

$$H^i(F^*), H^i(G^*) \colon H^i(N) \to H^i(M)$$

coincide.

◄ Below the symbol F_t^* is understood as the homomorphism of algebras (corresponding to the map F_t) of the geometrical differential forms

$\Lambda^*(N) \to \Lambda^*(M)$. Recall that

$$X_{F_t} = \left. \frac{dF_t^*}{dt} \right|_{t=0}$$

is a vector field w.r.t. the map $F = F_0$. Similarly,

$$X_{F_s} = \left. \frac{dF_t^*}{dt} \right|_{t=s}$$

is an F_s-relative vector field. The desired homotopy operators are direct consequences of the generalized Newton–Leibnitz formula

$$F^* - G^* = F_1^* - F_0^* = \int_0^1 \frac{dF_t^*}{dt} dt, \tag{20.10}$$

which in its turn follows from the usual formula for functions, because $(F_t^*)_z \colon \Lambda(N)_z^* \to \Lambda^*(M)$, $z \in N$, is a map of finite-dimensional vector spaces, while F_t may be regarded as a family of such maps continuously depending on the parameter z. Now, using the proposition from Section 19.17 and the fact that the differentials d_M and d_N do not depend on t, we see that

$$\int_0^1 \frac{dF_t^*}{dt} dt = \int_0^1 (i_{X_t} \circ d_N + d_M \circ i_{X_t}) dt$$

$$= \left(\int_0^1 i_{X_t} dt \right) \circ d_N + d_M \circ \left(\int_0^1 i_{X_t} dt \right). \tag{20.11}$$

As the result, we come to the *homotopy formula*

$$F^* - G^* = \mathfrak{H} \circ d_N + d_M \circ \mathfrak{H}, \quad \text{where} \quad \mathfrak{H} \overset{\text{def}}{=} \int_0^1 i_{X_t} dt; \tag{20.12}$$

applying it to the closed form $\omega \in \Lambda^i(N)$ we see that

$$F^*(\omega) - G^*(\omega) = d_M(\mathfrak{H}(\omega)).$$

This implies that the image of the cohomology class of a closed form does not change when we pass to a homotopic map.   ▶

20.24. A few words about algebraic topology. Contemporary algebraic topology arose from concrete problems in the differential calculus. Thus, Poincaré, developing the theory of integration on manifolds, proposed to use their triangulation, thereby reducing the problem to a known Euclidean case. This in its turn led to the use of simplicial complexes, simplicial (co)homology, and so on. Poincaré was the first to use modern terminology and apply homotopy considerations to compute what was basically de Rham cohomology of Euclidean spaces. Below, we present a few examples of computations of this kind. Also see Section 20.30, where it is indicated how one can define cohomology with arbitrary coefficients on the basis of de Rham cohomology. More generally, the subject matter of this chapter may be regarded as a first step in the return of algebraic topology to its forgotten sources in the more general context of the differential calculus over commutative algebras.

20.25. Homotopy operators. The importance of homotopy techniques for the computation of (co)homology deserves to be distinguished by a special definition.

Definition. Let $K_\alpha = \{K_\alpha^i, d_\alpha\}$, $\alpha = 1, 2$, be cochain complexes. The cochain maps $\psi_\alpha \colon K_1 \to K_2$, $\alpha = 1, 2$, are said to be *homotopic* if there exists a map of cochain groups $\mathfrak{H} \colon K_1 \to K_2$, $\mathfrak{H}(K_1^i) \subset K_2^{i-1}$ such that

$$\psi_1 - \psi_2 = d_1 \circ \mathfrak{H} + \mathfrak{H} \circ d_2.$$

Here the map $\mathfrak{H}$ is called a *homotopy operator* or simply a *homotopy*.

From this algebraic point of view, formula (20.12) relates geometric homotopies (which are easy to see only for $\mathbb{R}$-algebras) to algebraic homotopies.

Exercise.   1. Generalize the notion of geometrical homotopy to maps of pairs (manifold, submanifold) and prove the analog of the Homotopy Theorem (see 20.23) for relative de Rham cohomology.

2. Let us call a commutative diagram of homomorphisms

$$
\begin{array}{ccc}
A & \xrightarrow{\varphi} & A' \\
\alpha \downarrow & & \downarrow \alpha' \\
B & \xrightarrow{\psi} & B'
\end{array}
\tag{20.13}
$$

a *morphism of* α to α'. Generalize the notion of algebraic homotopy to morphisms of surjections of commutative algebras over the same field and prove the Homotopy Theorem for relative de Rham cohomology.

20.26. Examples. I. *Poincaré Lemma.* Poincaré himself stated this lemma as asserting that any closed differential form of positive dimension in n-dimensional Euclidean space $\mathbb{R}^n$ is exact. To prove this, he considered the family of homotheties $F_t = \{x \mapsto tx, x = (x_1, \ldots, x_n), 0 \leq t \leq 1\}$. Further the argument goes as follows. First notice that $F_0 = \zeta \circ \psi$, where $\psi \colon R^n \to \{0\}$ is the map to the singleton $\{0\}$, while $\zeta \colon \{0\} \to \mathbb{R}^n$ is obviously an embedding. But since $F_1 = \mathrm{id}_{\mathbb{R}^n}$, the left-hand side of the homotopy formula reads $\mathrm{id}_{\Lambda^*(\mathbb{R}^n)} - \psi^* \circ \zeta^*$. Now the homomorphism $\psi^* \colon \Lambda^i(\mathbb{R}^n) \to \Lambda^i(\{0\})$, $i > 0$, is obviously trivial, and so the homotopy formula tells us that $\omega = d\mathfrak{H}(\omega)$, $\omega \in \Lambda^i(\mathbb{R}^i)$, if $d\omega = 0$.

II. By introducing the notion of *deformation retract*, the above proof of the Poincaré lemma carries over almost word for word to a much more general situation. In topology, a subspace W is said to be a *deformation retract* of $U \supset W$ if there exists a family of continuous maps

$$F_t, : U \to W, 0 \le t \le 1,$$

such that $F_0|_W = \mathrm{id}_W$, $\mathrm{Im}(F_0) = W$, $F_1 = \mathrm{id}_U$. For example, $\{0\}$ is a deformation retract of $\mathbb{R}^n$.

The left-hand side of the homotopy formula for cochains of the complex K of "ordinary" cohomology of U has the form $\mathrm{id}_K - F_0^*$ since $F_1 = \mathrm{id}_U$. This means that F_0^* induces the identity map of the cohomology $H^*(K)$ of the complex K. Further, let $\bar{F}_0 : U \to W$ be the restriction of the map F_0 to U. Then $F_0 = \iota \circ \bar{F}_0$, where $\iota : W \to U$ is the inclusion map and so $F_0^* = \bar{F}_0^* \circ \iota^*$. This implies that the cochain map ι^* induces the following injective map in cohomology: $H^*(U) \to H^*(W)$, while $\bar{F}_0^*$ (in the opposite direction) is surjective. On the other hand, $\iota^* \circ \bar{F}_0^* = \mathrm{id}_W^*$, since $\bar{F}_0 \circ \iota = \mathrm{id}_W$, which clearly implies the surjectivity of the cohomology homomorphism $H^*(W) \to H^*(U)$ induced by the cochain map ι^* and the injectivity of the cohomology map $H^*(U) \to H^*(W)$ induced by $\bar{F}_0^*$. In this way, we obtain an isomorphism between the cohomology groups of the given space U and those of its deformation retract W.

The above argument carries over easily to the general algebraic context. For example, for de Rham cohomology, we can consider a surjective homomorphism $\varphi : A \to B$, $\varphi|_B = \mathrm{id}\, B$, of the algebra A on its subalgebra B, together with φ-connected categories $\mathfrak{M}$ and $\mathcal{K}$ of A- and B-modules, respectively, and a given (algebraic) homotopy $\mathfrak{H} : \Lambda_{\mathfrak{M}}^i(A) \to \Lambda_{\mathcal{K}}^{i-1}(B)$ connecting id_A and φ. It is useful to note that the requirements imposed on f and B may be replaced by a single one only, namely $\varphi^2 = \varphi$ if we put $B = \mathrm{Im}\,\varphi$. From the same considerations as above, the homotopy formula corresponding to these data establishes an isomorphism between the cohomology algebras $H_{\mathfrak{M}}^*(A)$ and $H_{\mathcal{K}}^*(B)$. This motivates us to refer to the projection φ as a deformation retract in the general algebraic situation. Here, however, we lose the easily understandable geometrical mechanism for constructing the homotopy $\mathfrak{H}$, which must be searched for, depending on the nature of the algebra A. Finally, let us note that the above is applicable to any natural complex of the differential calculus, including the Spencer complex.

III. Let us present several typical examples of deformation retracts. First of all, they are given by fiber bundles whose fibers are *contractible*. The latter means that such a space (fiber) has the singleton as its retract. The image of any section of such a fiber bundle is a deformation retract of its total space. For instance, it is the zero section of any vector bundle, the central circle of the Möbius strip (see Figure 6.2 on page 67) or of the solid torus (Figure 20.1). The axis of an infinite solid cone (Figure 20.1) is its deformation retract, and the crossing point is that of the cross. Note also that the perpendicular projection of the infinite solid cone onto its axis gives a simple example of a Serre fibration (i.e., a fibration satisfying the

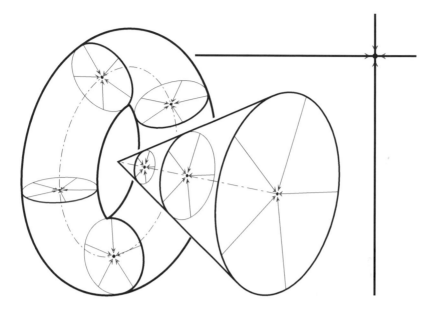

Figure 20.1. Three deformation retracts.

so-called homotopy lifting property) whose fibers are homotopy equivalent, but not homeomorphic, so that this fibration is not a fiber bundle.

IV. *The classical Newton–Leibnitz formula.* Each of the endpoints of the interval $I = [a, b]$ is its deformation retract, hence we have $H^0(I) = \mathbb{R}$ and $H^i(I) = 0$, $i > 0$, so that the argumentation in 20.22 carries over, words for word, to the algebra $C^\infty(I)$ and to the relative cohomology of the pair $(I, \partial I)$. Thus we have shown that the classical Newton–Leibnitz formula may be regarded as the definition of the integral

$$\int_a^b df(x), \ \ f(x) \in C^\infty(I).$$

20.27. The suspension isomorphism. It is well known that, in algebraic topology, there are some very useful constructions carried out on any topological space, such as the *suspension* and the *loop space*. The first of these is interesting in that it allows to compare the cohomology groups of spaces of different dimensions (we already saw this for retractions). Recall that the suspension ΣX over the space X is obtained from the product $X \times I$, $I = [0, 1]$, when its top ($X \times \{1\}$) and bottom ($X \times \{0\}$) are contracted to singletons u_1 and u_0, so that $BX = (X \times \{1\}) \bigcap (X \times \{0\})$ is contracted to the set $\{u_1, u_0\}$ (see Figure 20.2). In that case, we have the following natural isomorphism

$$H^i(X \times I, BX; R) \cong H^i(\Sigma X, \{u_1, u_0\}; R)$$

between the "ordinary" relative cohomology groups with coefficients in any Abelian group R.

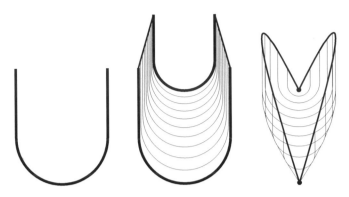

Figure 20.2. Suspension.

Below, for a smooth set X, we compute the relative de Rham cohomology groups $H^i(X \times I, BX)$. To this end, let us consider the following fragment of the exact sequence for the pair $(X \times I, BX)$ (see 20.9):

$$H^i(X \times I) \xrightarrow{\imath} H^i(BX) \xrightarrow{\partial} H^{i+1}(X \times I, BX) \xrightarrow{\jmath} H^{i+1}(X \times I). \quad (20.14)$$

But $X \equiv X \times \{\epsilon\}$, $\epsilon \in I$, is obviously a deformation retract of $X \times I$, so that we have the canonical isomorphism $H^j(X \times I) \cong H^j(X)$. Therefore, having in mind this isomorphism, we have $H^I(BX) \cong H^j(X) \times H^j(X)$ and $\imath(\vartheta) = (\vartheta, \vartheta)$. Moreover, the homomorphism $\jmath$ is trivial, since the restriction of a cocycle representing $\vartheta \in H^j(X \times I, BX)$ on $X \times \{\epsilon\}$, $\epsilon = 0, 1$, is trivial by definition, and, on the other hand, $H^j(X \times I) \cong X \times \{\epsilon\}$, $\epsilon = 0, 1$. Now

$$H^{i+1}(X \times I, BX) \cong \operatorname{coker} \partial,$$

while

$$\operatorname{coker} \partial \cong H^i(BX)/\operatorname{Im} \imath \cong [H^i(X) \times H^i(X)]/\Delta \cong H^i(X),$$

where $\Delta \cong \operatorname{Im} \imath \cong H^i(X)$ is the diagonal in $H^i(X) \times H^i(X)$. In other words, $H^{i+1}(X \times I, BX)$ can be naturally identified with the antidiagonal in $H^i(X) \times H^i(X)$, which by definition consists of pairs $(\vartheta, -\vartheta)$, $\vartheta \in H^i(X)$. Thus, the *suspension isomorphism*

$$\Uparrow_X : H^i(X) \to H^{i+1}(X \times I, BX),$$

in view of the isomorphisms specified above, has the form $\vartheta \mapsto (\vartheta, -\vartheta)$, which is not very informative. The reader can remedy this situation by doing the following exercise.

20.28 Exercises.   1. Noting that the exterior product of a relative cohomology class by an absolute one is a relative class of the same pair or a surjection, prove that $\Uparrow_X (\vartheta) = \operatorname{pr}_1^*(dx) \wedge \operatorname{pr}_2^*(\vartheta)$, pr_1 and pr_2

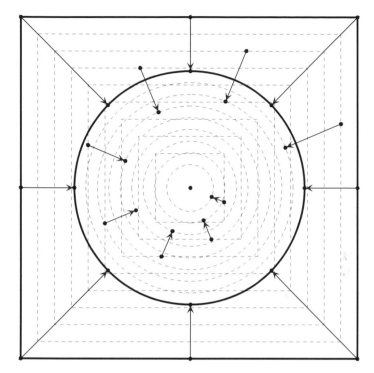

Figure 20.3. Homotopy from the square to the disk.

are the projections of $X \times I$ onto I and X, respectively, while dx is understood as the relative cohomology class of the pair $(I, \partial I)$.

2. Consider the cube $\nabla^k = I_1 \times, \ldots, \times I_k$, where I_i is a copy of the closed interval I with the standard coordinate x_i. Verify that $H^i(\nabla^k, \partial \nabla^k) = 0$ if $i \neq k$, and

$$H^k(\nabla^k, \partial \nabla^k) = \mathbb{R} \cdot d\bar{x}_1 \wedge \cdots \wedge d\bar{x}_k, \quad \bar{x}_i = \mathrm{pr}_i^*(x_i),$$

where $\mathrm{pr}_i \colon \nabla^k \to I_i$ is the natural projection.

3. Let $D^k = \{\sum_{i=1}^k x_i^2 \leq 1\} \subset \mathbb{R}^k$ be the k-dimensional disk. Show that $H^i(D^k, \partial D^k) \cong H^i(\nabla^k, \partial \nabla^k)$, using the representation of the pairs shown in Figure 20.3 and the visually obvious geometrical homotopies that the picture suggests.

4. Compute the cohomology groups of the k-dimensional sphere S^k, regarding it as the boundary of the disk D^{k+1}.

5. Iterating the suspension construction k times, one obtains the isomorphism

$$\Uparrow_U^k : H^i(U) \to H^{i+k}(U \times \nabla^k), U \times \partial \nabla^k),$$
$$\vartheta \mapsto \operatorname{pr}_\nabla^*(\bar{x}_1 \wedge \cdots \wedge d\bar{x}_k) \wedge \operatorname{pr}_U^* \vartheta, \qquad (20.15)$$

where $\operatorname{pr}_\nabla^*$ and pr_U^* are the natural projections of $U \times \nabla^k)$ on ∇^k and U, respectively. Verify this.

Now if we replace, using item 3 of the Exercise, ∇^k by D^k in (20.15), then we obtain the so-called Thom isomorphism for the Cartesian product by the disk. In the general case, this isomorphism takes place for fibrations with fiber the disk, and plays an important role in the theory of characteristic classes.

Exercise. In the example from Section 20.22, if we replace the algebra $\Bbbk[x]$ by $\Bbbk[x_1, \ldots, x_n]$, then the result does not change. Verify this directly, or by constructing the homotopy needed to prove this fact.

20.29. Homotopy complex: general idea. There is a remarkable analogy between homotopy operators and cochain maps. Namely, in the notation of Section 20.23, the relation $d_1 \circ \psi - \psi \circ d_2 = 0$ defines a cochain map. Similarly, it is natural to interpret the relation $d_1 \circ \mathfrak{H} + \mathfrak{H} \circ d_2 = 0$, where $\mathfrak{H} : K_1 \to K_2$, $\mathfrak{H}(K_1^i) \subset K_2^{i-1}$, as a cochain map of degree -1. This map obviously induces the following map on the cohomology level: $H^i(\mathfrak{H}^i) : H^i(K_1) \to H^{i-1}(K_2)$. At first glance it would seem that there is no sense in comparing cohomology groups of different dimensions. However, in connection with the suspension isomorphism, when the cohomology groups of the suspension over a space are mapped to another space, such a situation arises. Similarly, the map of a space to the suspensions over another space induces a homomorphism of the cohomology groups of the second space to those of the first one, raising the dimension by 1. Moreover, having in mind the isomorphism of iterated suspension or the Thom isomorphism, we see that the comparison of cohomology groups of different dimensions make sense, and so it is natural to include in our considerations the "higher" analogs of homotopy operators, which change the degree of cochains by a fixed integer m.

If we assume that $\mathfrak{H} : K_1 \to K_2$, $\mathfrak{H}(K_1^i) \subset K_2^{i+m}$, and put

$$\eth \overset{\text{def}}{=} d_1 \circ \mathfrak{H} - (-1)^m \mathfrak{H} \circ d_2,$$

then, obviously, $\eth^2 = 0$ and so we have obtained a *homotopical complex*. Here the fact that ψ is a cochain means that $\eth\psi = 0$, i.e., that ψ is a cocycle of degree zero of that complex. Similarly, the fact that cochain maps ψ_1 are ψ_2 homotopic means that $\psi_1 - \psi_2 = \eth\mathfrak{H}$, i.e., ψ_1 and ψ_2 lie in the same cohomology class of the homotopic complex.

The general scheme described above, however, needs to be developed, because it leaves unanswered the question of the nature of the maps

$\mathfrak{H}\colon K_1 \to K_2$ under consideration. For instance, they can simply be linear maps of vector spaces or differential operators of some kind. In the next section, in order to illustrate this aspect, we shall consider the de Rham complex of smooth manifolds or smooth sets.

20.30. De Rham cohomology with arbitrary coefficients. A sufficiently complete understanding of the meaning and the role of de Rham cohomology is a thing of the future, when there will be enough "experimental material" obtained from computations and applications. But at this point, we already have and can suspect a lot of unexpected and interesting developments. For example, de Rham cohomology can be nontrivial for algebras A with a finite spectrum $|A|$. Or it seems quite natural to study the structure of those algebras related by homomorphisms that induce isomorphisms in de Rham cohomology or, possibly, in *differential (co)homology* of other types, say, Spencer cohomology.

In this connection, we must stress that the language of the differential calculus in commutative algebras allows to carry over to this new context the entire standard apparatus of algebraic topology. For example, using it, one can show that the cohomology of a simplicial complex S with real coefficients is isomorphic to the de Rham cohomology of the algebra of *smooth functions on* S in the category of geometrical modules over that algebra. By definition, that algebra consists of functions on S whose restriction to any closed simplex is smooth in the ordinary sense. Moreover, in the corresponding de Rham complex, we will be able to distinguish the *integer subcomplex*, whose cohomology will be identical to the ordinary cohomology of S with integer coefficients. In its turn, this will allow us to define the cohomology of S with arbitrary coefficients. In other words, the historical process that began with the combinatorial topology of simplicial complexes and much later led to the well-known de Rham theorem can be transformed into the cornerstone of the de Rham cohomology of commutative algebras even on the basis of this familiar material.

20.31 Proposition. *In the geometrical category* $\mathfrak{M} = \mathrm{GMod}\ C^\infty(M)$, *the Spencer δ-cohomology of a smooth manifold M, i.e., the cohomology groups of the cochain complexes* $\mathrm{Csm}_{\mathfrak{M}}^m$ *(see Section 18.7) and the cohomology groups of the Spencer jet complexes* $\mathrm{Spen}_{\mathfrak{M}}^m$ *are trivial if $m \geq 1$. Besides,*

$$H^0(\mathrm{Csm}_{\mathfrak{M}}^0) = H^0(\mathrm{Spen}_{\mathfrak{M}}^m) = C^\infty(M),\ m \geq 0.$$

◄ Since the differentials of the Spencer jet complexes are homomorphisms of $C^\infty(M)$-modules, it suffices to prove the required statement locally. If $x = \{(x_1, \ldots, x_n)\}$ are local coordinates in some domain $U \subset M$, then the $C^\infty(U)$-module

$$\mathrm{Csml}_\Gamma^k(C^\infty(U)) \bigotimes_{C^\infty(U)} \Lambda_\Gamma^{m-k}(C^\infty(U))$$

is generated by elements of the form

$$f(x)(dx_{i_1} \odot \cdots \odot dx_{i_k}) \otimes_{C^\infty(U)} (dx_{j_1} \wedge \ldots \wedge dx_{j_{m-k}}), \; f(x) \in C^\infty(U). \quad (20.16)$$

This expression is rather ambiguous, since the symbols dx_i in the product $dx_{i_1} \odot \cdots \odot dx_{i_k}$ denote 1-cosymbols, whereas they appear in the product $dx_{j_1} \wedge \ldots \wedge dx_{j_{m-k}}$ as 1-forms. To stress this difference, we denote these dx_i by different symbols, e.g., $\bar{x}_i = dx_i$ for 1-cosymbols and $d\bar{x}_j = dx_j$ for 1-forms. In this notation, elements of (20.16) at some point $z \in U$ can be written in the form

$$c\bar{x}_{i_1} \ldots \bar{x}_{i_k} \cdot d\bar{x}_{j_1} \wedge \ldots \wedge d\bar{x}_{j_{m-k}} \; c \in \mathbb{R},$$

i.e., as differential forms over the polynomial algebra $\mathbb{R}[\bar{x}_1, \ldots, \bar{x}_n]$. Here the forms corresponding to elements of the complex $\mathrm{Csm}_\Gamma^m(C^\infty(M))$ at the point z constitute a subcomplex, which is invariant w.r.t. the "Poincaré homotopy" (see the first example in Section 20.26). Thus the required assertion concerning the cohomology groups of the complexes $\mathrm{Csm}_\Gamma^m(C^\infty(M))$ is a consequence of the Poincaré lemma.

As to the cohomology groups of the complexes $\mathrm{Spen}_\Gamma^m(C^\infty(M))$, they can be easily computed from the long exact cohomology sequence corresponding to the short exact sequence of complexes

$$0 \to \mathrm{Csm}_\Gamma^m(C^\infty(M)) \to \mathrm{Spen}_\Gamma^m(C^\infty(M)) \to \mathrm{Spen}_\Gamma^{m-1}(C^\infty(M)) \to 0.$$

Indeed, this long exact sequence consists of fragments of the form

$$0 \to H^i(\mathrm{Csm}_\Gamma^m(C^\infty(M))) \to H^i(\mathrm{Spen}_\Gamma^m(C^\infty(M))) \to$$
$$\to H^i(\mathrm{Spen}_\Gamma^{m-1}(C^\infty(M))) \to H^{i+1}(\mathrm{Csm}_\Gamma^m(C^\infty(M))) \to 0,$$

in which the groups $H^i(\mathrm{Csm}_\Gamma^m(C^\infty(M)))$ and $H^{i+1}(\mathrm{Csm}_\Gamma^m(C^\infty(M)))$ are trivial if, as we noted before, $m \geq 1$. This obviously implies that the groups $H^i(\mathrm{Spen}_\Gamma^m(C^\infty(M)))$ and $H^i(\mathrm{Spen}_\Gamma^{m-1}(C^\infty(M)))$ are isomorphic. Therefore,

$$H^i(\mathrm{Spen}_\Gamma^m(C^\infty(M))) = H^i(\mathrm{Spen}_\Gamma^0(C^\infty(M))),$$

while the complex $\mathrm{Spen}_\Gamma^0(C^\infty(M))$ has the form $0 \to C^\infty(M) \to 0$ and hence has the cohomology groups indicated above. ▶

20.32 Exercises.   1. Let P be a geometrical projective module over the geometrical algebra A. Let also $P^{\odot l}$ (resp., $P^{\wedge k}$) be the l-th symmetric (resp., the k-th exterior) degree of P. Then the elements

$$(p_{i_1} \odot \cdots \odot p_{i_l}) \otimes_A (p_{j_1} \wedge \ldots \wedge p_{j_k}), \quad p_{i_s}, p_{j_r} \in P.$$

generate the A-module $P^{\odot l} \otimes_A P^{\wedge k}$. Show that

$$\mathrm{SL}^m(P) \overset{\mathrm{def}}{=} \sum_{l+k=m} P^{\odot l} \bigotimes_A P^{\wedge k}$$

is a cochain complex whose differentials are the following A-homo-morphisms

$$\delta^m_{m-k,P}\colon P^{\odot(m-k+1)} \bigotimes_A P^{\wedge k-1} \to P^{\odot(m-k)} \bigotimes_A P^{\wedge k},$$

$$(p_{i_1} \odot \cdots \odot p_{i_{m-k+1}}) \bigotimes_A (p_{j_1} \wedge \ldots \wedge p_{j_{k-1}})$$

$$\mapsto \sum_{s=1}^{m-k+1} (p_{i_1} \odot \cdots \widehat{p_{i_s}} \odot \cdots \odot p_{i_{m-k+1}}) \bigotimes_A (p_{i_s} \wedge p_{j_1} \wedge \ldots \wedge p_{j_k}).$$

(20.17)

Here we can assume, by definition, that $P^{\odot 0} = P^{\wedge 0} = A$.

2. Prove that $H^i(\mathrm{SL}^m(P)) = 0$ if $i \geq 1$, and that $H^0(\mathrm{SL}^m(P)) = A$ by indicating the analog of the "Poincaré homotopy" in this case (see the proof of Proposition 20.31).

3. Verify that for the smooth category $\mathfrak{M}$ over the algebra A, we have $H^i(\mathrm{Spen}_{\mathfrak{M}}^m(A)) = 0$ if $m \geq 1$, and $H^0(\mathrm{Spen}_{\mathfrak{M}}^m(A)) = A$.

20.33. Problem: Spencer cohomology of the cross K. In this book, we have repeatedly studied the properties of **K**. For the cross, the computation of Spencer cohomology reduces to that of the corresponding Spencer δ-cohomology via the short exact sequences (18.11) or the appropriate spectral sequence. We propose the following problem: calculate the geometrical Spencer δ-cohomology of the cross directly, and show that it is nontrivial because **K** has a singularity, namely its singular point. The solution of this problem, in our opinion, is worthy of publication in a serious journal.

20.34. Adjoint operators. In this section and the next ones, we assume that we are dealing with a differentially closed category $\mathfrak{M}$. (In our exposition, we follow [15].) To simplify the notation, the symbol $\mathfrak{M}$ in the notation for diffunctors and their representing objects will be omitted. Let P be an A-module, and let us consider the sequence of maps

$$0 \to \mathrm{Diff}^>(P, A) \xrightarrow{d^\circ} \mathrm{Diff}^>(P, \Lambda^1) \xrightarrow{d^\circ} \mathrm{Diff}^>(P, \Lambda^2) \xrightarrow{d^\circ} \ldots, \qquad (20.18)$$

where, as in Section 9.68, we have $\mathrm{Diff}^>(P, \Lambda^i) = \bigcup_{n \geq 0} \mathrm{Diff}_n^>(P, \Lambda^i)$ and $d^\circ(\Delta) \stackrel{\text{def}}{=} d \circ \Delta$, $\Delta \in \mathrm{Diff}^>(P, \Lambda^i)$. According to the definition of the right A-module structure in $\mathrm{Diff}^>(P, \Lambda^i)$, the map d° is an A-homomorphism and, in view of the equality $d \circ d = 0$, the sequence 20.18 is a cochain complex. Denote the cohomology module of this complex at the term $\mathrm{Diff}^>(P, \Lambda^n)$ by $\widehat{P}_n$, $n \geq 0$.

With each differential operator $\Delta \in \mathrm{Diff}^>(P, Q)$, we can associate the following morphism of complexes:

$$\ldots \xrightarrow{d^\circ} \mathrm{Diff}^>(Q, \Lambda^{k-1}) \xrightarrow{d^\circ} \mathrm{Diff}^>(Q, \Lambda^k) \xrightarrow{d^\circ} \ldots \qquad (20.19)$$

$$\downarrow \tilde{\Delta} \qquad\qquad\qquad \downarrow \tilde{\Delta}$$

$$\ldots \xrightarrow{d^\circ} \mathrm{Diff}^>(P, \Lambda^{k-1}) \xrightarrow{d^\circ} \mathrm{Diff}^>(P, \Lambda^k) \xrightarrow{d^\circ} \ldots,$$

where

$$\tilde{\Delta}(\nabla) \overset{\mathrm{def}}{=} \nabla \circ \Delta \in \mathrm{Diff}^>(P, \Lambda^k), \ \ \nabla \in \mathrm{Diff}^>(Q, \Lambda^k). \qquad (20.20)$$

From the commutativity of the diagram (20.19) it follows that we have a map of the cohomology of the complex $\mathrm{Diff}^>(Q, \Lambda^*)$ to the cohomology of the complex $\mathrm{Diff}^>(P, \Lambda^*)$.

Definition. The operator $\Delta_n^* \colon \widehat{Q}_n \to \widehat{P}_n$ generated by the operator $\Delta \in \mathrm{Diff}^>(P, Q)$ is called the $(n\text{-th})$ *adjoint operator* of the operator Δ.

Denote by $[X]$ the cohomology class of the operator $X \in \mathrm{Diff}^>(P, \Lambda^n)$, $d^\circ(X) = 0$. Then, by definition,

$$\Delta_n^*([X]) = [X \circ \Delta].$$

20.35 Proposition. (1) *If* $\Delta \in \mathrm{Diff}_k(P, Q)$, *then* $\Delta_n^* \in \mathrm{Diff}_k(\widehat{Q}_n, \widehat{P}_n)$.
(2) *For any operators* $\Delta \in \mathrm{Diff}^>(P, Q)$ *and* $\nabla \in \mathrm{Diff}^>(Q, R)$, *the following equality holds*

$$(\nabla \circ \Delta)_n^* = \Delta_n^* \circ \nabla_n^*. \qquad (20.21)$$

◄ (1) Denoting, as above, the operator of right multiplication by $a \in A$ by the symbol $a^>$, we obtain

$$\delta_a(\Delta_n^*)([X]) = \Delta_n^*(a^>[X]) - a^>\Delta_n^*([X]) = [X \circ a \circ \Delta] - [X \circ \Delta \circ a]$$
$$= (a \circ \Delta)_n^*([X]) - (\Delta \circ a)_n^*([X]) = -\delta_a(\Delta)_n^*([X]).$$

In other words, $\delta_a(\Delta_n^*) = -\delta_a(\Delta)_n^*$. Thus,

$$\delta_{a_0, \ldots, a_k}(\Delta_n^*) = \delta_{a_0, \ldots, a_k}(\Delta)_n^* = 0.$$

(2)

$$(\nabla \circ \Delta)_n^*([X]) = [X \circ \nabla \circ \Delta] = \Delta_n^*([X \circ \nabla]) = \Delta_n^*(\nabla_n^*([X])). \quad ►$$

Examples. 1. Let a denote an element of the algebra A, and at the same time, the operator of left multiplication by this element in the A-module P. Then $a_n^*([\nabla]) = [\nabla \circ a] = a^>[\nabla]$, i.e., $a_n^* = a$.

2. To any element $p \in P$, we can assign the operator denoted by the same letter and given by $p \colon A \to P$, $a \mapsto pa$, $a \in A$. Then $p_n^*([\nabla]) = [\nabla \circ p]$, $p_n^* \in \mathrm{Hom}_A(\widehat{P}_n, \widehat{A}_n)$. Thus there exists a natural pairing

$$\langle \cdot, \cdot \rangle \colon P \otimes \widehat{P}_n \to \widehat{A}_n, \ \langle p, \widehat{p} \rangle = p_n^*(\widehat{p}), \ \widehat{p} \in \widehat{P}_n.$$

20.36. Integral forms. Let

$$\ldots \longrightarrow P^{k-1} \xrightarrow{\Delta_k} P^k \xrightarrow{\Delta_{k+1}} P^{k+1} \longrightarrow \ldots$$

be a complex of differential operators. Replacing the modules P^k by $\widehat{P_n^k}$ and taking into account the equalities $\Delta_k^* \circ \Delta_{k+1}^* = (\Delta_{k+1} \circ \Delta_k)^* = 0$, we obtain the complex

$$\ldots \longleftarrow \widehat{P_n^{k-1}} \xleftarrow{\Delta_k^*} \widehat{P_n^k} \xleftarrow{\Delta_{k+1}^*} \widehat{P_n^{k+1}} \longleftarrow \ldots,$$

which is called $(n$-$th)$ *dual* to the given one. The complex dual to the de Rham complex is called the *complex of integral forms* and is denoted as follows:

$$0 \longleftarrow \Sigma_0^n \xleftarrow{\delta} \Sigma_1^n \xleftarrow{\delta} \ldots \xleftarrow{\delta} \Sigma_k^n \xleftarrow{\delta} \Sigma_{k+1}^n \xleftarrow{\delta} \ldots,$$

where $\Sigma_i^n = \widehat{\Lambda_n^i}$ and $\delta = d^*$. The module $\Sigma_0^n = \widehat{A_n}$ is called the $(n$-$th)$ *module of volume forms* or the $(n$-$th)$ *Berezinian* and is denoted by $\mathcal{B}_n$ or simply by $\mathcal{B}$ if the value of n is fixed.

The differential operator $Д\colon \mathrm{Diff}^>(\Lambda^k) \to \Lambda^k$, $Д(\nabla) = \nabla(1)$ (see Section 15.8) induces a cohomology homomorphism

$$\int \colon \mathcal{B}_n \to H^n(\Lambda^*),$$

since to any volume form $\omega \in \mathcal{B}_n$ there corresponds a de Rham cohomology class, namely $\int \omega$. This is yet another version of the operation of integration.

Obviously, $\int[\nabla] = \nabla(1) \mod d\Lambda^{n-1}$, $\nabla \in \mathrm{Diff}(\Lambda^n)$.

20.37 Proposition. $\int \delta\omega = 0$, *where* $\omega \in \Sigma_1$.

◄ Putting $\omega = [\nabla]$, we obtain $\delta\omega = [\nabla \circ d]$. Therefore,

$$\int \delta\omega = [(\nabla \circ d)(1)] = 0. \blacktriangleright$$

20.38 Proposition. *For any* $\Delta \in \mathrm{Diff}(P, Q)$, $p \in P$ *and* $\widehat{q} \in \widehat{Q}_n$, *the following equality* ("integration by parts formula") *holds*

$$\int \langle \Delta(p), \widehat{q} \rangle = \int \langle p, \Delta^*(\widehat{q}) \rangle. \qquad (20.22)$$

◄ Let $\widehat{q} = [\nabla]$, where $\nabla \colon Q \to \Lambda^n$. Then,

$$\int \langle \Delta(p), \widehat{q} \rangle = \int [\nabla \circ \Delta(p)] = \int [\nabla \circ \Delta \circ p]$$

$$= \int \langle p, [\nabla \circ \Delta] \rangle = \int \langle p, \Delta^*(\widehat{q}) \rangle. \qquad \blacktriangleright$$

Note that in order to find the volume form modules $\mathcal{B}_n$, it is necessary to compute the cohomology of the complex (20.18) for $P = A$, i.e., of the following complex:

$$0 \to \mathrm{Diff}^> A \xrightarrow{d^\circ} \mathrm{Diff}^> \Lambda^1 \xrightarrow{d^\circ} \dots \xrightarrow{d^\circ} \mathrm{Diff}^> \Lambda^n \xrightarrow{d^\circ} \dots , \qquad (20.23)$$

Exercise. Let $A = C^\infty(\mathbb{R}^n)$. Prove that in the category of geometrical A-modules, we have

$$\mathcal{B}_i = \begin{cases} 0, & \text{if } i \neq n; \\ \Lambda^n, & \text{if } i = n. \end{cases}$$

The result of this exercise justifies the term "volume form module" for $\mathcal{B}_n = \Sigma_0^n = \widehat{A}_n$ and explains why we have called the map $\mathcal{B}_n \to H^n(\Lambda^*)$ "integration." The origins of the term "Berezinian," will be explained in the next chapter. The free module $\mathcal{B}_n = \Lambda^n$ is one-dimensional and its generator has the form $dx_1 \wedge dx_2 \wedge \dots \wedge dx_n$, where $x_1, \dots, x_n$ are the coordinates in $\mathbb{R}^n$. When we pass from the coordinates x_i to the coordinates v_j, the new generator is obtained from the old one via multiplication by the determinant $\det J(\frac{x}{v})$ of the Jacobi matrix $J(\frac{x}{v})$.

Terra incognita. As far as we know, no one has tried to apply the constructions of the algebraic theory of linear differential operators to p-adic functions (and to p-adic analysis). Apparently, in this direction, one can count on interesting results, which should also have applications to "p-adic mathematical physics." Besides a simple description of p-adic differential operators, there is a particular interest in the study of differential and integral forms, of the abovementioned Spencer complexes, and of the cohomological approach to integration. In particular, this might lead to understand the conceptual nature of the Feynman integral.

We hope that our book will attract the attention of researchers to these questions.

21

Differential Operators over Graded Algebras

Throughout the whole book, commutative algebras and modules over them were in the center of our considerations. In the seventies of the previous century, thanks to the pioneering work of Felix Berezin, it became clear that algebras with "anticommuting variables" are not only interesting *per se*, but also have important applications to physics and can readily be studied. Such algebras were called *graded commutative* (or *supercommutative*).

We have already met with some examples of these algebras—first of all, the algebras of differential forms and the de Rham cohomology of commutative algebras. Here we shall expand the list of examples and explain what the theory of differential operators for such algebras is like. The proof of many statements in this case will be a simple paraphrase of similar statements for commutative algebras; we shall omit these proofs, leaving them to the reader as exercises. The aim of this chapter is not only to give a systematic description of the differential calculus in the case of commutative graded algebras, but to show its usefulness.

21.1. Definition. Let G be a commutative (Abelian) semigroup with unit (which may sometimes be a group). An associative algebra with unit $\mathcal{A}$ over the field $\Bbbk$ is called *G-graded*, if it can be represented in the form

$$\mathcal{A} = \oplus_{g \in G} A^g,$$

where the A^g satisfy $A^g \cdot A^q \subset A^{g+q}$ for all $g, q \in G$. If $a \in A^g$, then g is called the *grading* of a and is denoted by $\omega(a)$, while a is called a *homogeneous element of grading g*.

© Springer Nature Switzerland AG 2020
J. Nestruev, *Smooth Manifolds and Observables*, Graduate Texts
in Mathematics 220, https://doi.org/10.1007/978-3-030-45650-4_21

In what follows, when dealing with elements of graded algebras or with the other graded objects defined below, we shall assume that they are homogeneous unless the converse is explicitly indicated.

21.2. Example. Second-order matrices. Consider the simplest non-commutative algebra—the algebra $\mathrm{Mat}(A, 2)$ of second-order matrices over a noncommutative algebra A of zero characteristic. Any such matrix can be uniquely presented as the sum of two matrices of certain special forms:

$$\begin{pmatrix} x & y \\ z & w \end{pmatrix} = \begin{pmatrix} a & b \\ -b & a \end{pmatrix} + \begin{pmatrix} c & d \\ d & -c \end{pmatrix}.$$

Denote the set of matrices of the first type by M^0, and of the second type, by M^1. Let $X_1, X_2 \in M^0$, $Y_1, Y_2 \in M^1$. It is easy to check that

$$X_1 + X_2 \in M^0, \ Y_1 + Y_2 \in M^1,$$
$$X_1 \cdot X_2 \in M^0, \ Y_1 \cdot Y_2 \in M^0, \ X_1 \cdot Y_1 \in M^1.$$

This means that

$$\mathrm{Mat}(A, 2) = M^0 \oplus M^1$$

is a $\mathbb{Z}_2$-graded algebra. Note that if A is the field of real numbers, then M^0 is isomorphic to the field of complex numbers.

We can go further: divide each of the components M^0 and M^1 into two pieces and take $\mathbb{Z}_2 \oplus \mathbb{Z}_2$ for the grading semigroup. More precisely, let us put

$$M^{0,0} = \left\{ \begin{pmatrix} a & 0 \\ 0 & a \end{pmatrix} \right\}, \ M^{1,1} = \left\{ \begin{pmatrix} 0 & a \\ -a & 0 \end{pmatrix} \right\},$$
$$M^{0,1} = \left\{ \begin{pmatrix} 0 & a \\ a & 0 \end{pmatrix} \right\}, \ M^{1,0} = \left\{ \begin{pmatrix} a & 0 \\ 0 & -a \end{pmatrix} \right\}.$$

Exercise. Check that

$$\mathrm{Mat}(A, 2) = \oplus_{g \in \mathbb{Z}_2 \oplus \mathbb{Z}_2} M^g$$

is a $(\mathbb{Z}_2 \oplus \mathbb{Z}_2)$-graded algebra .

21.3. Example. Quaternions. The algebra of quaternions $\mathbb{H}$ may be defined as the $\mathbb{R}$-algebra generated by the four elements $1, i, j, k$, whose products are given by the following table:

$\cdot$	1	i	j	k
1	1	i	j	k
i	i	-1	k	$-j$
j	j	$-k$	-1	i
k	k	j	$-i$	-1

(21.1)

As in the case of the algebra $\mathrm{Mat}(A, 2)$, there are two ways of representing of the set of quaternions as a graded algebra.

In the first one, we take $\mathbb{Z}_2$ for the grading group G and set

$$\mathbb{H}^0 = \mathbb{R} \oplus \mathbb{R} \cdot i, \quad \mathbb{H}^1 = \mathbb{R} \cdot j \oplus \mathbb{R} \cdot k.$$

Then $\mathbb{H} = \mathbb{H}^0 \oplus \mathbb{H}^1$. This decomposition of the quaternion algebra in a direct sum is compatible with the grading and allows to regard it as a $\mathbb{Z}_2$-graded algebra. (Check this!)

In the second version, for the grading semigroup G, we take the direct sum $\mathbb{Z}_2 \oplus \mathbb{Z}_2$ and define the direct summands as follows:

$$\mathbb{H}^{(0,0)} = \mathbb{R} \cdot 1, \quad \mathbb{H}^{(1,1)} = \mathbb{R} \cdot i,$$
$$\mathbb{H}^{(1,0)} = \mathbb{R} \cdot j, \quad \mathbb{H}^{(0,1)} = \mathbb{R} \cdot k.$$

Then $\mathbb{H} = \bigoplus_{g \in \mathbb{Z}_2 \oplus \mathbb{Z}_2} \mathbb{H}^g$ and the multiplication of homogeneous quaternions is compatible with the grading, so that we may regard $\mathbb{H}$ as $\mathbb{Z}_2 \oplus \mathbb{Z}_2$-graded. (Check this also!)

21.4. Definitions. 1. Let $\mathcal{A} = \bigoplus_{g \in G} A^g$ be a G-graded $\Bbbk$-algebra. The left $\mathcal{A}$-module $\mathcal{P} = \bigoplus_{g \in G} P^g$ is called G-*graded*, if $A^g \cdot P^q \subset P^{g+q}$ for all $g, q \in G$. Similarly, the right $\mathcal{A}$-module $\mathcal{P} = \bigoplus_{g \in G} P^g$ is called G-*graded* if $P^q \cdot A^g \subset P^{g+q}$ for all $g, q \in G$. If $p \in P^g$, then g is called the *grading* of p and is denoted by $\omega(p)$, while p is said to be a *homogeneous element of grading* g.

In what follows, unless indicated otherwise, elements of the graded algebras under consideration will be assumed homogeneous.

2. Let $\mathcal{P} = \bigoplus_{g \in G} P^g$ be a G-graded left (right) $\mathcal{A}$-module and let the family $Q^g \subset P^g$, $g \in G$ be such that $\mathcal{Q} = \bigoplus_{g \in G} Q^g$ is again a left (right) G-graded $\mathcal{A}$-module. Modules of this form will be called *left* (*right*) *graded submodules* of the G-graded $\mathcal{A}$-module $\mathcal{P}$.

Exercise. Verify that the *quotient module* $\mathcal{P}/\mathcal{Q} \overset{\text{def}}{=} \bigoplus_{g \in G} P^g/Q^g$ is indeed a left (right) G-graded $\mathcal{A}$-module.

3. If $\mathcal{P}$ and $\mathcal{Q}$ are G-graded $\mathcal{A}$-modules, then the $\Bbbk$-homomorphism $\phi \colon \mathcal{P} \to \mathcal{Q}$ is said to be G-*graded of degree* g if $\phi(P^{g_1}) \subset Q^{g_1+g}$ for any $g_1 \in G$. In what follows, speaking of morphisms of graded modules, we shall consider only graded morphisms. Denote the $\Bbbk$-module of all $\Bbbk$-homomorphisms from $\mathcal{P}$ to $\mathcal{Q}$ by $\operatorname{Hom}_{\Bbbk}^g(\mathcal{P}, \mathcal{Q})$. We shall denote the operator of multiplication by the element $a \in \mathcal{A}$ by the same letter a. If $\phi \in \operatorname{Hom}_{\Bbbk}^g(\mathcal{P}, \mathcal{Q})$, then, obviously, $\phi \circ a$ and $a \circ \phi$ belong to $\operatorname{Hom}_{\Bbbk}^{g+\omega(a)}(\mathcal{P}, \mathcal{Q})$. This means that the module

$$\operatorname{Hom}_{\Bbbk}(\mathcal{P}, \mathcal{Q}) = \bigoplus_{g \in G} \operatorname{Hom}_{\Bbbk}^g(\mathcal{P}, \mathcal{Q}) \tag{21.2}$$

possesses natural left as well as right $\mathcal{A}$-module structures. These structures are compatible: $a \circ (\phi \circ b) = (a \circ \phi) \circ b$, so that $\operatorname{Hom}_{\Bbbk}(\mathcal{P}, \mathcal{Q})$ is supplied with a bimodule structure over $\mathcal{A}$.

4. The tensor product over $\Bbbk$ of a left G-graded $\mathcal{A}$-module $\mathcal{P}$ and right G-graded $\mathcal{A}$-module $\mathcal{Q}$ is a $G \oplus G$-graded $\Bbbk$-module:

$$\mathcal{P} \otimes_\Bbbk \mathcal{Q} = \bigoplus_{g,r\in G} P^g \otimes_\Bbbk Q^r.$$

The multiplication operation of elements of the tensor product $\mathcal{P} \otimes_\Bbbk \mathcal{Q}$ by elements of the algebra $\mathcal{A}$ from the left $a(p \otimes_\Bbbk q) \overset{\text{def}}{=} ap \otimes_\Bbbk q$ and from the right $(p \otimes_\Bbbk q)a \overset{\text{def}}{=} p \otimes_\Bbbk qa$ are compatible, so that $\mathcal{P} \otimes_\Bbbk \mathcal{Q}$ is a bimodule over $\mathcal{A}$. It is convenient to think of it as being G-graded:

$$\mathcal{P} \otimes_\Bbbk \mathcal{Q} = \bigoplus_{g\in G}(\mathcal{P} \otimes_\Bbbk \mathcal{Q})^g, \ (\mathcal{P} \otimes_\Bbbk \mathcal{Q})^g \overset{\text{def}}{=} \bigoplus_{r+t=g} P^r \otimes_\Bbbk Q^t.$$

21.5. The parity form. Now suppose that the Abelian grading semigroup G is supplied with a *parity form*, i.e., a $\mathbb{Z}$-bilinear symmetric map $\langle \cdot, \cdot \rangle$: $G \times G \to \mathbb{Z}_2$. Here we assume that char $\Bbbk \neq 2$. An important and often arising particular case of a parity form is the form that can be constructed from the *parity homomorphism* $\rho\colon G \to \mathbb{Z}_2$, $\langle g_1, g_2 \rangle_\rho = \rho(g_1) \cdot \rho(g_2)$. Parity forms constructed via the parity homomorphism will be called *decomposable*.

If the grading semigroup G possesses a parity form $\langle \cdot, \cdot \rangle$ and $\mathcal{U}, \mathcal{V}$ are arbitrary G-graded objects, then to any homogeneous elements $u \in \mathcal{U}$, $v \in \mathcal{V}$, we can assign the element $\langle u, v \rangle \in \mathbb{Z}_2$ by setting

$$\langle u, v \rangle \overset{\text{def}}{=} \langle \omega(u), \omega(v) \rangle.$$

If the parity form was constructed from the corresponding homomorphism, then it is sometimes convenient to use the notation $|u|$ instead of $\rho(\omega(u))$. In that case $\langle u, v \rangle = |u| \cdot |v|$.

21.6. Definition. Let a parity form $\langle \cdot, \cdot \rangle$ be given. A G-graded algebra $\mathcal{A}$ is said to be *graded commutative* (we stress—not just commutative, but graded commutative), if $ab = (-1)^{\langle a,b \rangle} ba$ for any homogeneous elements $a, b \in \mathcal{A}$.

A homogeneous element $a \in \mathcal{A}$ is called *even*, if $\langle a, b \rangle = 0$ for all homogeneous $b \in \mathcal{A}$, and *odd*, if $\langle a, a \rangle = 1$. The fact that an element a is odd means that $a^2 = 0$. If the parity form $\langle \cdot, \cdot \rangle$ is decomposable, then the evenness of an element $a \in \mathcal{A}$ is equivalent to the condition $|a| = 0$, and its oddness, to $|a| = 1$, while the equality $ab = -ba$ holds iff both a and b are odd. In the general case, there may exist elements that are neither odd nor even (examples will appear below).

If $\mathcal{P}$ is supplied with both a left and a right module structure over a graded commutative algebra $\mathcal{A}$, then these structures are called *compatible* if $ap = (-1)^{\langle a,p \rangle} pa$ for any homogeneous elements $a \in A$, $p \in P$. In that case $\mathcal{P}$ is said to be a *two-sided* module over $\mathcal{A}$. Of course, to any right $\mathcal{A}$-module, we can assign a left one by setting $ap \overset{\text{def}}{=} (-1)^{\langle a,p \rangle} pa$ for any

$a \in \mathcal{A}$, $p \in \mathcal{P}$, and conversely. In what follows, unless the contrary is explicitly stated, all the modules are assumed two-sided.

21.7. Definition. Let a G-graded algebra $\mathcal{A}$ be commutative, and $\mathcal{P}$, $\mathcal{Q}$ be G-graded modules over $\mathcal{A}$. A $\Bbbk$-homomorphism $\phi \colon \mathcal{P} \to \mathcal{Q}$ with grading $g = \omega(\phi)$ is said to be an $\mathcal{A}$-*homomorphism* (or a *homomorphism over* $\mathcal{A}$), if for any $a \in \mathcal{A}$, $p \in \mathcal{P}$, we have

$$\phi(ap) = (-1)^{\langle a, \phi \rangle} a\phi(p). \tag{21.3}$$

The set of all $\mathcal{A}$-homomorphisms with fixed grading $g \in G$ from $\mathcal{P}$ to $\mathcal{Q}$ possesses a natural A^0-module structure (A^0 being the 0-component of the graded algebra $\mathcal{A}$) that we denote by $\mathrm{Hom}_{\mathcal{A}}^g(\mathcal{P}, \mathcal{Q})$. Recall that the operator of multiplication by an element $a \in \mathcal{A}$ is denoted by the same letter a. If $\phi \in \mathrm{Hom}_{\mathcal{A}}^g(\mathcal{P}, \mathcal{Q})$, then, obviously, $\phi \circ a$ and $a \circ \phi$ will belong to $\mathrm{Hom}_{\mathcal{A}}^{g+\omega(a)}(\mathcal{P}, \mathcal{Q})$. This means that the $\Bbbk$-module

$$\mathrm{Hom}_{\mathcal{A}}(\mathcal{P}, \mathcal{Q}) = \bigoplus_{g \in G} \mathrm{Hom}_{\mathcal{A}}^g(\mathcal{P}, \mathcal{Q}) \tag{21.4}$$

possesses natural left and right $\mathcal{A}$-module structures. From condition (21.3), it follows that these two structures coincide and the *module* $\mathrm{Hom}_{\mathcal{A}}(\mathcal{P}, \mathcal{Q})$ *of homomorphisms from* $\mathcal{P}$ *to* $\mathcal{Q}$ is a two-sided $\mathcal{A}$-module. For any G-graded $\mathcal{A}$-module $\mathcal{P}$, we obviously have

$$\mathrm{Hom}_{\mathcal{A}}(\mathcal{A}, \mathcal{P}) = \mathcal{P}. \tag{21.5}$$

21.8. Tensor product of graded modules. Let $\mathcal{P}, \mathcal{Q}$ be G-graded $\mathcal{A}$-modules. In the direct product $\mathcal{P} \times \mathcal{Q}$, let us consider the submodule $\mathcal{U}$ generated by elements of the form

$$(ap_1 + bp_2, q_1) - (ap_1, q_1) - (bp_2, q_1),$$
$$(p_1, aq_1 + bq_2) - (-1)^{\langle a, p_1 \rangle}(ap_1, q_1) - (-1)^{\langle b, p_1 \rangle}(bp_1, q_2),$$
$$a, b \in \mathcal{A}, \; p_1, p_2 \in \mathcal{P}, \; q_1, q_2 \in \mathcal{Q}.$$

The quotient module $(\mathcal{P} \times \mathcal{Q})/\mathcal{U}$ is called the *tensor product* of $\mathcal{P}$ and $\mathcal{Q}$; it is denoted by $\mathcal{P} \otimes_{\mathcal{A}} \mathcal{Q}$, while the image of (p, q) under the corresponding factorization homomorphism is denoted by $p \otimes q$.

Define the grading of an element $p \otimes q$ by putting $\omega(p \otimes q) \overset{\mathrm{def}}{=} \omega(p) + \omega(q)$. Then

$$\mathcal{P} \otimes_{\mathcal{A}} \mathcal{Q} = \bigoplus_{g \in G} (\mathcal{P} \otimes_{\mathcal{A}} \mathcal{Q})^g,$$

where $(\mathcal{P} \otimes_{\mathcal{A}} \mathcal{Q})^g = \{p \otimes q \mid \omega(p) + \omega(q) = g\}$. The formula $a(p \otimes q) = ap \otimes q$ supplies $\mathcal{P} \otimes_{\mathcal{A}} \mathcal{Q}$ with the structure of a left $\mathcal{A}$-module. Since, obviously,

$$A^r \cdot (\mathcal{P} \otimes_{\mathcal{A}} \mathcal{Q})^g \subset (\mathcal{P} \otimes_{\mathcal{A}} \mathcal{Q})^{g+r},$$

it follows that $\mathcal{P} \otimes_{\mathcal{A}} \mathcal{Q}$ is a G-graded $\mathcal{A}$-module. For any G-graded $\mathcal{A}$-module $\mathcal{P}$, we have the following obvious isomorphism:

$$\mathcal{A} \otimes_{\mathcal{A}} \mathcal{P} = \mathcal{P}. \tag{21.6}$$

Exercises.   1. Modify the above construction of tensor product in the case when $\mathcal{P}$ is a left, while $\mathcal{Q}$ is a right G-graded $\mathcal{A}$-module.

2. A map G of graded $\mathcal{A}$-modules $\alpha \colon \mathcal{P} \times \mathcal{Q} \to \mathcal{T}$ is called *bilinear in the grading g* if for any $a_1, a_2, b_1, b_2 \in \mathcal{A}$, $p_1, p_2 \in \mathcal{P}$, $q_1, q_2 \in \mathcal{Q}$, we have

$$\begin{aligned}
\alpha(a_1 p_1 + a_2 p_2, b_1 q_1 + b_2 q_2) = {}&(-1)^{\langle b_1, p_1 \rangle + \langle a_1, \alpha \rangle + \langle b_1, \alpha \rangle} a_1 b_1 \alpha(p_1, q_1) \\
&+ (-1)^{\langle b_2, p_1 \rangle + \langle a_1, \alpha \rangle + \langle b_2, \alpha \rangle} a_1 b_2 \alpha(p_1, q_2) \\
&+ (-1)^{\langle b_1, p_2 \rangle + \langle a_2, \alpha \rangle + \langle b_1, \alpha \rangle} a_2 b_1 \alpha(p_2, q_1) \\
&+ (-1)^{\langle b_2, p_2 \rangle + \langle a_2, \alpha \rangle + \langle b_2, \alpha \rangle} a_2 b_1 \alpha(p_2, q_2),
\end{aligned}$$

where, as before, in the parity form, we write α instead of g. Prove that for bilinear maps $\mathcal{P} \times \mathcal{Q} \to \mathcal{T}$, where $\mathcal{T}$ is an arbitrary G-graded $\mathcal{A}$-module, the tensor product $\mathcal{P} \otimes_{\mathcal{A}} \mathcal{Q}$ is a universal object, i.e., the map α may be presented as the composition of the factorization map

$$\mathcal{P} \times \mathcal{Q} \to (\mathcal{P} \times \mathcal{Q})/\mathcal{U} = \mathcal{P} \otimes_{\mathcal{A}} \mathcal{Q}$$

and the homomorphism $h_\alpha \colon \mathcal{P} \otimes_{\mathcal{A}} \mathcal{Q} \to \mathcal{T}$.

3. Consider the map

$$\eta \colon \operatorname{Hom}_{\mathcal{A}}(\mathcal{P} \otimes_{\Bbbk} \mathcal{Q}, \mathcal{U}) \to \operatorname{Hom}_{\Bbbk}(\mathcal{P}, \operatorname{Hom}_{\mathcal{A}}(\mathcal{Q}, \mathcal{U})) \tag{21.7}$$

defined by the following formula:

$$\eta(h)(p)(q) = h(p \otimes_{\Bbbk} q), \quad h \in \operatorname{Hom}_{\mathcal{A}}(\mathcal{P} \otimes_{\Bbbk} \mathcal{Q}, \mathcal{U}).$$

Prove that η is an isomorphism.

21.9. Before passing to examples, note that if the role of the graded semigroup is played by $\mathbb{Z}$ or $\mathbb{Z}_+$, then there exist only two parity homomorphisms, namely, the trivial one ρ_0, $\rho_0(n) = 0$ for any n, and the homomorphism ρ_2 *reduction modulo 2*: $\rho_2(n) = n \mod 2$ for any n.

Now let G and R be grading semigroups with parity forms $\langle \cdot, \cdot \rangle_G$ and $\langle \cdot, \cdot \rangle_R$, respectively, then on $G \oplus R$ there exist at least three parity forms:

$$\begin{aligned}
\langle (g_1, r_1), (g_2, r_2) \rangle_1 &= \langle g_1, g_2 \rangle_G, \quad g_1, g_2 \in G, \ r_1, r_2 \in R, \\
\langle (g_1, r_1), (g_2, r_2) \rangle_2 &= \langle r_1, r_2 \rangle_R, \\
\langle (g_1, r_1), (g_2, r_2) \rangle_+ &= \langle g_1, g_2 \rangle_G + \langle r_1, r_2 \rangle_R.
\end{aligned}$$

Of course, there may be other ones, and this is a good thing, as we shall see in what follows.

21.10. Auxiliary constructions. Let us introduce some constructions, which we shall need to study the algebras $\mathrm{Mat}(A, 2^n)$, where A is a commutative algebra. For an arbitrary matrix $X \in \mathrm{Mat}(A, m)$, we define the matrices $X^{i,j} \in \mathrm{Mat}(A, 2m)$, $i, j \in \mathbb{Z}_2$, by putting

$$X^{0,0} = \begin{pmatrix} X & 0 \\ 0 & X \end{pmatrix}, X^{1,1} = \begin{pmatrix} 0 & X \\ -X & 0 \end{pmatrix},$$

$$X^{0,1} = \begin{pmatrix} 0 & X \\ X & 0 \end{pmatrix}, X^{1,0} = \begin{pmatrix} X & 0 \\ 0 & -X \end{pmatrix}.$$

Let $X, Y \in \mathrm{Mat}(A, m)$ be arbitrary matrices. For the matrices $X^{k,l}$, $Y^{i,j} \in \mathrm{Mat}(A, 2m)$, let us write out the multiplication table:

$\cdot$	$Y^{0,0}$	$Y^{1,1}$	$Y^{1,0}$	$Y^{0,1}$
$X^{0,0}$	$(XY)^{0,0}$	$(XY)^{1,1}$	$(XY)^{1,0}$	$(XY)^{0,1}$
$X^{1,1}$	$(XY)^{1,1}$	$-(XY)^{0,0}$	$-(XY)^{0,1}$	$(XY)^{1,0}$
$X^{1,0}$	$(XY)^{1,0}$	$(XY)^{0,1}$	$(XY)^{0,0}$	$(XY)^{1,1}$
$X^{0,1}$	$(XY)^{0,1}$	$-(XY)^{1,0}$	$-(XY)^{1,1}$	$(XY)^{0,0}$

$$(21.8)$$

In other words,

$$X^{i,j} Y^{k,l} = (-1)^{jk} (XY)^{i+k,j+l}. \tag{21.9}$$

21.11. Second-order matrices. Above (see Section 21.2), the algebra $\mathrm{Mat}(\Bbbk, 2)$ of second-order matrices was presented as a $\mathbb{Z}_2$-graded algebra: $\mathrm{Mat}(\Bbbk, 2) = M^0 \oplus M^2$, where

$$M^0 = \left\{ \begin{pmatrix} a & b \\ -b & a \end{pmatrix} \right\}, \quad M^1 = \left\{ \begin{pmatrix} c & d \\ d & -c \end{pmatrix} \right\}.$$

It is easy to check that the matrices from M^0 do not commute and do not anticommute with each other, nor with matrices from M_1, while the matrices from M_1 do not anticommute with each other. Thus, with respect to this grading, $\mathrm{Mat}(\Bbbk, 2)$ *is not a $\mathbb{Z}_2$-graded commutative algebra*.

Now let us regard $\mathrm{Mat}(A, 2)$ as a $(\mathbb{Z}_2 \oplus \mathbb{Z}_2)$-graded algebra and choose the following generators in it:

$$E = E^{0,0} = \begin{pmatrix} 1 & 0 \\ 0 & 1 \end{pmatrix}, E^{1,1} = \begin{pmatrix} 0 & 1 \\ -1 & 0 \end{pmatrix},$$

$$E^{0,1} = \begin{pmatrix} 0 & 1 \\ 1 & 0 \end{pmatrix}, E^{1,0} = \begin{pmatrix} 1 & 0 \\ 0 & -1 \end{pmatrix},$$

The above matrices obviously form a basis of $\mathrm{Mat}(A, 2)$ that we denote by $\mathcal{E}_2$. Note that these matrices may be obtained from the construction of the previous section by putting $X = E = 1 \in A = \mathrm{Mat}(A, 2)$. Their multiplication table can be easily obtained from (21.8) by replacing the matrices X and Y by the matrix $E = 1 \in A = \mathrm{Mat}(A, 2)$

$\cdot$	$E^{0,0}$	$E^{1,1}$	$E^{1,0}$	$E^{0,1}$
$E^{0,0}$	$E^{0,0}$	$E^{1,1}$	$E^{1,0}$	$E^{0,1}$
$E^{1,1}$	$E^{1,1}$	$-E^{0,0}$	$-E^{0,1}$	$E^{1,0}$
$E^{1,0}$	$E^{1,0}$	$E^{0,1}$	$E^{0,0}$	$E^{1,1}$
$E^{0,1}$	$E^{0,1}$	$-E^{1,0}$	$-E^{1,1}$	$E^{0,0}$

$$(21.10)$$

From this we can see that all the squares of the generators are equal to $\pm E^{0,0}$, while $E^{0,0}$ commutes with the other generators, which, in turn, anticommute with each other. Comparing this table with the similar table (21.1) for quaternions, we see that they look very much alike. The main difference is that there are three units on the main diagonal of the table (21.10).

The formula 21.9 in the case under consideration acquires the form:

$$E^{i,j}E^{k,l} = (-1)^{jk}(EE)^{i+k,j+l} = (-1)^{jk}E^{i+k,j+l}. \tag{21.11}$$

Therefore,

$$E^{i,j}E^{k,l} = (-1)^{jk+il}E^{k,l}E^{i,j}. \tag{21.12}$$

On $\mathbb{Z}_2 \oplus \mathbb{Z}_2$, let us define the bilinear form $\langle \cdot, \cdot \rangle_{0,1}$ by setting

$$\langle (i,j), (k,l) \rangle_{0,1} \overset{\text{def}}{=} jk + il, \tag{21.13}$$

finally obtaining

$$E^{i,j}E^{k,l} = (-1)^{\langle (i,j),(k,l) \rangle_{0,1}} E^{k,l}E^{i,j}. \tag{21.14}$$

Thus, *the algebra*

$$\mathrm{Mat}(\Bbbk, 2) = \bigoplus_{g \in \mathbb{Z}_2 \oplus \mathbb{Z}_2} M^g$$

is a graded commutative algebra with parity form $\langle \cdot, \cdot \rangle_{0,1}$. We use the index $0, 1$ in the notation of the parity form because the latter is given by the matrix $E^{0,1}$. Indeed, let $(i,j), (k,l) \in \mathbb{Z}_2 \oplus \mathbb{Z}_2$, then

$$\langle (i,j), (k,l) \rangle_{0,1} = (i,j) \begin{pmatrix} 0 & 1 \\ 1 & 0 \end{pmatrix} \begin{pmatrix} k \\ l \end{pmatrix} = (j\ i) \begin{pmatrix} k \\ l \end{pmatrix} = jk + il.$$

Note in conclusion of this section that here elements of the form $kE^{0,0}$, $k \in A$ are even, while the other elements are neither even, nor odd.

21.12. Quaternions. The situation with the quaternion algebra is just the same as with the matrix algebra $\mathrm{Mat}(A, 2)$. Namely, *the algebra*

$$\mathbb{H} = \bigoplus_{g \in \mathbb{Z}_2 \oplus \mathbb{Z}_2} \mathbb{H}^g$$

is a graded commutative algebra, see [11], with parity form $\langle \cdot, \cdot \rangle_{0,1}$. The verification of this fact is left to the reader as an exercise.

In the case under consideration, elements of $\mathbb{H}^{0,0} = \mathbb{R}$ are even, elements of $\mathbb{H}^{1,1} = \mathbb{R} \cdot i$, $\mathbb{H}^{1,0} = \mathbb{R} \cdot j$, $\mathbb{H}^{0,1} = \mathbb{R} \cdot k$ are neither even, nor odd.

21.13. Fourth-order matrices. Let $E^{i,j}$ be an arbitrary basis matrix $E^{i,j} \in \mathcal{E}_2$ of the algebra $\mathrm{Mat}(A, 2)$. Put

$$E^{i,j,0,0} = \begin{pmatrix} E^{i,j} & 0 \\ 0 & E^{i,j} \end{pmatrix}, \quad E^{i,j,1,1} = \begin{pmatrix} 0 & E^{i,j} \\ -E^{i,j} & 0 \end{pmatrix},$$

$$E^{i,j,0,1} = \begin{pmatrix} 0 & E^{i,j} \\ E^{i,j} & 0 \end{pmatrix}, \quad E^{i,j,1,0} = \begin{pmatrix} E^{i,j} & 0 \\ 0 & -E^{i,j} \end{pmatrix}.$$

As can easily be seen, matrices of the form $E^{i,j,k,l}$ constitute a basis of the algebra $\mathrm{Mat}(\Bbbk, 4)$, which we can now regard as $\left(\bigoplus_{s=1}^{4} \mathbb{Z}_2 \right)$-graded. Denote this basis by $\mathcal{E}_4$. It follows from formula 21.9 that

$$E^{i_1,j_1,i_2,j_2} E^{k_1,l_1,k_2,l_2} = (-1)^{j_2 k_2} (E^{i_1,j_1} E^{k_1,l_1})^{i_2+k_2,j_2+l_2}$$

$$= (-1)^{j_2 k_2} ((-1)^{j_1 k_1} E^{i_1+k_1,j_1+l_1})^{i_2+k_2,j_2+l_2}$$

$$= (-1)^{j_2 k_2 + j_1 k_1} E^{i_1+k_1,j_1+l_1,i_2+k_2,j_2+l_2}.$$

Similarly,

$$E^{k_1,l_1,k_2,l_2} E^{i_1,j_1,i_2,j_2} = (-1)^{i_2 l_2 + i_1 l_1} E^{i_1+k_1,j_1+l_1,i_2+k_2,j_2+l_2}.$$

From this we finally obtain

$$E^{i_1,j_1,i_2,j_2} E^{k_1,l_1,k_2,l_2} = (-1)^{j_1 k_1 + i_1 l_1 + j_2 k_2 + i_2 l_2} E^{k_1,l_1,k_2,l_2} E^{i_1,j_1,i_2,j_2}. \tag{21.15}$$

Now let us introduce the bilinear form $\langle \cdot, \cdot \rangle_{0,1}$ on $\bigoplus_{s=1}^{4} \mathbb{Z}_2$ by setting

$$\langle (i_1, j_1, i_2, j_2), (k_1, l_1, k_2, l_2) \rangle_{0,1} \overset{\text{def}}{=} j_1 k_1 + i_1 l_1 + j_2 k_2 + i_2 l_2. \tag{21.16}$$

It is easily checked that the matrix of this bilinear form is

$$E^{0,1,0,0} = \begin{pmatrix} 0 & 1 & 0 & 0 \\ 1 & 0 & 0 & 0 \\ 0 & 0 & 0 & 1 \\ 0 & 0 & 1 & 0 \end{pmatrix}$$

Now we can say that for any two homogeneous elements U, V of the $\left(\bigoplus_{s=1}^{4} \mathbb{Z}_2 \right)$-graded algebra $\mathrm{Mat}(\Bbbk, 4)$, we have the equality

$$UV = (-1)^{\langle U, V \rangle} VU. \tag{21.17}$$

and, therefore, *the $\left(\bigoplus_{s=1}^{4} \mathbb{Z}_2 \right)$-graded algebra $\mathrm{Mat}(A, 4)$ is a graded commutative algebra* with parity form $\langle \cdot, \cdot \rangle_{0,1}$.

Exercise. Show that, for any n, the matrix algebra $\mathrm{Mat}(A, 2^n)$ is a graded commutative algebra. (In [12] it is shown that a matrix algebra $\mathrm{Mat}(A, m)$ is graded commutative only if $m = 2^n$.)

21.14. Dioles. Let P be a module over a commutative algebra A. The module P may be regarded as an A-algebra by setting $pq = 0$ for any $p, q \in P$. Now let us define the $\mathbb{Z}$-graded algebra $\mathcal{A} = A(P)$ as

$$
A^j = \begin{cases} 0, & j < 0, \\ A, & j = 0, \\ P, & j = 1, \\ 0, & j > 1. \end{cases} \tag{21.18}
$$

We take the trivial form for the parity form. Such $\mathbb{Z}$-graded commutative forms will be called *dioles*. In what follows, we shall sometimes write A instead of A^0 and P instead of A^1. Note that any $\mathbb{Z}$-graded commutative algebra for which only A^0 and A^1 are nontrivial is a diole.

The thoughtful reader has already noticed that, from the geometrical point of view, a diole appears when we consider the pair (manifold M, fibration π over M), then take the algebra of smooth functions $C^\infty(M)$ for the algebra A, and the module of sections $\Gamma(\pi)$ for the module P.

21.15. More examples.

Trivially graded algebra. Any commutative algebra A may be regarded as graded w.r.t. any Abelian semigroup G if we assume that all its components, except one, are equal to zero.

Complex numbers. The algebra of complex numbers $\mathbb{C} = \mathbb{R} \oplus \mathbb{R}i$ is $\mathbb{Z}_2$-graded commutative with trivial parity homomorphism ρ_0.

Algebras related to diffunctors. Let A be a commutative algebra and $\mathfrak{M}$ be a differentially closed category of A-modules. Then the algebra of symbols

$$
\mathcal{S}_*(A) = \bigoplus_{i=0}^{\infty} \mathcal{S}_i(A),
$$

(CM. II 10.1) and the algebra of cosymbols

$$
\mathrm{Csml}_{\mathfrak{M}}(A) = \bigoplus_{i=0}^{\infty} \mathrm{Csml}_{\mathfrak{M}}^i(A)
$$

(see Section 17.29) are commutative $\mathbb{Z}$-graded algebras with trivial parity homomorphism ρ_0. The algebra of differential forms

$$
\Lambda_{\mathfrak{M}}^* = \Lambda_{\mathfrak{M}}^*(A) = \bigoplus_{i=0}^{\infty} \Lambda_{\mathfrak{M}}^k(A),
$$

(see Section 17.29) and the de Rham cohomology algebra

$$
H_{\mathrm{dR}}^*(\mathfrak{M}) = \bigoplus_{i=0}^{\infty} H_{\mathrm{dR}}^i(\mathfrak{M})
$$

(see Section 17.29) are commutative $\mathbb{Z}$-graded algebras with parity homomorphism ρ_2 (see Section 21.9).

21.16. Grassmann algebra. Let V be an n-dimensional vector space over $\mathbb{k}$. The exterior algebra $\Lambda(V) = \bigoplus V^{\wedge k}$ has a natural $\mathbb{Z}$-graded commutative algebra structure. It is called the *Grassmann algebra* and denoted from now on by $\mathcal{F}$ (for "Fermionic"). For the category of modules over $\mathcal{F}$, we choose the category of finitely generated ones, which in this case amounts to saying that they are finite-dimensional over $\mathbb{k}$.

Fix a basis $\mathfrak{B} = \{\xi_1, \ldots, \xi_n\}$ of V; then the algebra $\mathcal{F}$ is generated by the elements $\xi_1, \ldots, \xi_n$ of degree one and is subject only to the relations $\xi_i \cdot \xi_j = -\xi_j \cdot \xi_i$. This means that $\mathcal{F}$ is free as a $\mathbb{Z}$-graded algebra. If $\sigma \subset \{1, \ldots, n\}$ is an ordered subset of length r, $\sigma = \{\sigma_1, \ldots, \sigma_r\}$, we write $\xi_\sigma = \xi_{\sigma_1} \cdots \xi_{\sigma_r}$ and set $\xi_\varnothing = 1$. Then every element of $\omega \in \mathcal{F}$ has a unique representation

$$\omega = \sum_\sigma \alpha_\sigma \xi_\sigma, \ \alpha_\sigma \in \mathbb{k}$$

21.17. Graded Lie algebras. Let $\mathcal{V} = \bigoplus_{g \in G} \mathcal{V}^g$ be a G-graded vector space over the field $\mathbb{k}$, let $\langle \cdot, \cdot \rangle$ be a parity form on G, and let $[\cdot, \cdot] \colon \mathcal{V} \bigotimes_\mathbb{k} \mathcal{V} \to \mathcal{V}$ be a bilinear operation on $\mathcal{V}$ with grading $g \in G$, i.e., an operation such that

$$[\mathcal{V}^{g_1}, \mathcal{V}^{g_2}] \subset \mathcal{V}^{g_1 + g_2 + g} \quad \forall g_1, g_2 \in G.$$

Then $\mathcal{V}$ is called a *G-graded Lie algebra* if the bracket $[\cdot, \cdot]$ is anti-symmetric

$$[X, Y] = -(-1)^{\langle X, Y \rangle + \langle g, g \rangle}[Y, X] \quad \forall X, Y \in \mathcal{V} \tag{21.19}$$

and satisfies the Jacobi identity:

$$[X, [Y, Z]] = [[X, Y], Z] + (-1)^{\langle X, Y \rangle + \langle g, Y \rangle}[Y, [X, Z]]; \tag{21.20}$$

here, as before, $\langle g, Y \rangle$ means $\langle g, \omega(Y) \rangle$. (This is the Leibnitz formula for the operator $[X, \cdot]$.)

For an even bracket $[\cdot, \cdot]$, the formulas (21.19) and (21.20) appear to be more usual. The last one can be modified by simple manipulations to the following "symmetric" form:

$$(-1)^{\langle X, Z \rangle}[X, [Y, Z]] + (-1)^{\langle Y, X \rangle}[Y, [Z, X]]$$
$$+ (-1)^{\langle Z, Y \rangle}[Z, [X, Y]] = 0. \tag{21.21}$$

It follows from the definitions that, in particular, for a G-graded commutative algebra $\mathcal{A}$, the component A^0 will be an ordinary Lie algebra over $\mathbb{k}$. Each $\mathcal{A}$-homomorphism $\phi \colon \mathcal{P} \to \mathcal{Q}$ of degree $r \in G$ may be understood as a family of A^0-homomorphisms $\phi^g \colon P^g \to Q^{g+r}$, $g \in G$ such that $\phi^{\omega(p) + \omega(a)}(ap) = (-1)^{\langle a, r \rangle} a \phi^{\omega(p)}(p)$.

21.18. Poisson dioles. Just as we transformed an arbitrary A-module P into the $\mathbb{Z}$-graded algebra $A(P)$ above (see Section 21.14), we can transform any commutative algebra into a $\mathbb{Z}$-graded Lie algebra. To do this, we consider the diole $\mathcal{A} = A(D(A))$ and define the Lie bracket $[\cdot, \cdot]$ on $\mathcal{A}$ as the extension of the commutator of vector fields to $D(A)$ by setting

$$[X, f] = [f, X] = -X(f), \quad [f, g] = 0 \quad \forall f, g \in A, \ \forall X \in D(A).$$

The bracket on $\mathcal{A}$ defined in this way has the grading -1, it is antisymmetric in the sense of (21.19) and satisfies the Jacobi identity (21.20), i.e., for any homogeneous elements x, y, z of the algebra $\mathcal{A}$, we have:

$$[x, [y, z]] = [[x, y], z] + (-1)^{(|x|+1)\cdot|y|}[y, [x, z]]. \qquad (21.22)$$

Indeed, if x, y, z are vector fields, then the sign in front of the last summand disappears and the above equality becomes the Jacobi identity for vector fields. If at least two elements of the three belong to A, then each summand of this equality vanishes. And, finally, when exactly one element belongs to A, while the two others are vector fields, the equality is a simple consequence of definitions. Indeed, let $x = f \in A$ and $y = Y$, $z = Z$ be vector fields. Let us compute the right-hand side of (21.22):

$$[[f, Y], Z] + (-1)^{(|f|+1)\cdot|Y|}[Y, [f, Z]] = [-Y(f), Z] - [Y, -Z(f)]$$
$$= Z(Y(f)) - Y(Z(f)) = -[Y, Z](f) = [f, [Y, Z]].$$

Note further that the bracket is a biderivation (actually, it is graded derivation, see Section 21.23 below) with respect to A as well:

$$[X, fg] = -X(fg) = -fX(g) - gX(f) = f[X, g] + g[X, f],$$
$$[X, fY] = X(f)Y + f[X, Y] = -[X, f]Y + f[X, Y].$$

A comment concerning the sign: it would seem that there should not be a minus sign in front of $[X, f]Y$. Actually, the minus sign is correct. The Lie bracket is odd, so is the vector field Y, hence the sign must change when we take the vector field out of the brackets.

The same construction was used for Lie algebroids, which have become a popular object of study in recent years, especially among mathematical physicists.

Let a Lie algebroid be given, i.e., a vector bundle $\pi\colon E \to M$ over a manifold M together with a Lie bracket over the module $\Gamma(\pi)$ of sections of the bundle and a homomorphism (the so-called anchor) $\rho\colon \Gamma(\pi) \to D(M)$, given by $S \mapsto S_\rho$ such that for all sections $S, T \in \Gamma(\pi)$ and $f \in C^\infty M$, we have

$$[S, fT] = S_\rho(f)T + f[S, T].$$

Just as above, define a $\mathbb{Z}$-graded algebra $\mathcal{A}$ by putting

$$A^0 = C^\infty(M), \ A^1 = \Gamma(\pi), \ A^i = 0 \text{ for } i \neq 0, 1$$

and assuming that $ST = 0$ for all $S, T \in A_1$. Define a Lie bracket $[\cdot, \cdot]$ on $\mathcal{A}$ as the extension of the given Lie bracket on $\Gamma(\pi)$ by setting

$$[S, f] = [f, S] = -S_\rho(f), \quad [f, g] = 0, \quad f, g \in A, \ X \in D(A).$$

Such a bracket on $\mathcal{A}$ has grading -1, is antisymmetric, and, for the same reasons as above, satisfies the Jacobi identity.

Definition. A diole, supplied with a Lie algebra structure, with grading -1, such that for any elements $S, T \in A^1$ and $f \in A^0$, we have

$$[S, fT] = -[S, f]T + f[S, T].$$

is called a *Poisson diole*. A Poisson diole will be called *geometrical* if $A^0 = C^\infty(M)$ for some smooth manifold M and A^1 is a projective $C^\infty(M)$-module.

Obviously, any geometrical Poisson diole can be obtained from some Lie algebroid by means of the above construction, and vice versa.

The definition of Lie algebroids seems simple enough, but this is not the case of the definitions of their morphisms. We will not define them here, but the reader will understand how complicated the definition must be, having in mind the completely different nature of the objects that must enter into it and the various ways in which these objects behave under maps of the underlying manifold. Even the fact that the composition of morphisms of Lie algebroids is again a Lie algebroid is a sufficiently cumbersome theorem.

The situation with Poisson dioles is much simpler—their morphisms are simply homomorphisms of graded algebras that preserve the Lie bracket and the fact that the composition of morphisms is a morphism is obvious.

In this situation, it is natural to restrict ourselves to the case of a fixed base manifold M, which we need not do for Poisson dioles.

21.19. Main notations. Unfortunately, at this point we will have to introduce some rather cumbersome notation. We put $I^{(n)} \overset{\text{def}}{=} (1, \ldots, n)$, by the letters I, J we will always denote ordered multiindices

$$I = (i_1, \ldots, i_k), \ 1 \leqslant i_1 < i_2 < \ldots < i_k \leqslant n,$$
$$J = (j_1, \ldots, j_l), \ 1 \leqslant j_1 < j_2 < \ldots < j_l \leqslant n.$$

Then we put $|I| = k$, $|J| = l$. If I and J satisfy $i_r \neq j_s$ for all $r \leqslant k$, $s \leqslant l$, obtaining a nonordered multiindex $(I, J) \overset{\text{def}}{=} (i_1, \ldots, i_k, j_1, \ldots, j_l)$. The sum $I + J$ will be defined as the multiindex obtained by ordering the multiindex (I, J).

Let $\mathcal{A}$ be a G-graded commutative algebra. Then we say that an l-linear map $\mathcal{L} \colon \mathcal{A} \times \cdots \times \mathcal{A} \to \mathcal{A}$ is *graded of degree g*, if for any $g_1, \ldots, g_l$, we have

$$\mathcal{L}(A^{g_1} \times \cdots \times A^{g_l}) \subseteq A^{g_1 + \cdots + g_l + g}, \quad g_1, \ldots, g_l, g \in G.$$

The degree g of a graded map $\mathcal{L}$ will be denoted by $\omega(\mathcal{L})$. Together with the degree, to any map $\mathcal{L}$ we shall assign its *total degree* defined by the formula $\overline{\omega}(\mathcal{L}) \overset{\text{def}}{=} (l - 1, \omega(\mathcal{L}))$. The vector space of all polylinear graded maps is graded with respect to the semigroup $\mathbb{Z} \times G$.

Let $a_1, \ldots, a_n \in \mathcal{A}$ be homogeneous elements and let $I = (i_1, \ldots, i_n)$ be an ordered multiindex; then the ordered set $a_{i_1}, \ldots, a_{i_n}$ will be denoted by

$a(I)$. Let $\sigma_i = (1, \ldots, i-1, i+1, i, \ldots, n)$ be a transposition. Define the parity of a permutation $a(\sigma_i) = (a_1, \ldots, a_{i-1}, a_{i+1}, a_i, \ldots, a_n)$ as $\langle a_i, a_{i+1} \rangle + 1$. This parity will be denoted by $|a(\sigma_i)|$. An l-linear graded map $\mathcal{L}$ from $\mathcal{A}$ to $\mathcal{A}$ is called *skew-symmetric* if for any transposition σ_i

$$\mathcal{L}(a(\mathrm{I}^{(n)})) = (-1)^{|a(\sigma_i)|} \mathcal{L}(a(\sigma_i)) = -(-1)^{\langle a_i, a_{i+1} \rangle} \mathcal{L}(a(\sigma_i)), \qquad (21.23)$$

and *symmetric* if

$$\mathcal{L}(a(\mathrm{I}^{(n)})) = (-1)^{\langle a_i, a_{i+1} \rangle} \mathcal{L}(a(\sigma_i)). \qquad (21.24)$$

Denote by $\mathrm{Alt}_{\Bbbk}^n \mathcal{A}$ the set of skew-symmetric n-linear maps from $\mathcal{A}$ to $\mathcal{A}$ and by $\mathrm{Sym}_{\Bbbk}^n \mathcal{A}$ the set of symmetric n-linear maps from $\mathcal{A}$ to $\mathcal{A}$. Put $\mathrm{Alt}_{\Bbbk}^* \mathcal{A} = \bigoplus_{n=1}^{\infty} \mathrm{Alt}_{\Bbbk}^n \mathcal{A}$, $\mathrm{Sym}_{\Bbbk}^* \mathcal{A} = \bigoplus_{n=1}^{\infty} \mathrm{Sym}_{\Bbbk}^n \mathcal{A}$. Obviously, the vector spaces $\mathrm{Alt}_{\Bbbk}^* \mathcal{A}$ and $\mathrm{Sym}_{\Bbbk}^* \mathcal{A}$ are graded w.r.t. the group $\mathbb{Z} \times G$.

For any element $\mathcal{L} \in \mathrm{Alt}_{\Bbbk}^n \mathcal{A}$ and an arbitrary permutation of n elements σ, we obviously have,

$$\mathcal{L}(a(\mathrm{I}^{(n)})) = (-1)^r \mathcal{L}(a(\sigma)),$$

where r is some element of $\mathbb{Z}_2$. It is easy to see that it depends only on the graded permutation $a(\sigma)$ and does not depend on $\mathcal{L}$. We will call it the *parity* of the graded permutation $a(\sigma)$ and denote it by $|a(\sigma)|$. In practice, the parity of any graded permutation can be computed by arbitrarily presenting it in the form of a composition of transpositions $a(\sigma_i)$. In particular, let I, $|I| = n$ and J, $|J| = l$ be such that the multiindex (I, J) is defined; then, as can easily be seen, we have

$$\begin{aligned} |a(I, J)| &= |a(J, I)| + nl + \langle \omega(a(I)), \omega(a(J)) \rangle \\ &= |a(J, I)| + nl + \langle a(I), a(J) \rangle, \end{aligned} \qquad (21.25)$$

where $\omega(a(I)) = \sum_{i \in I} \omega(a_i)$, $\omega(a(J)) = \sum_{j \in J} \omega(a_j)$.

Similarly, for any element $\mathcal{L} \in \mathrm{Sym}_{\Bbbk}^n \bar{A}$ and an arbitrary permutation of n elements σ, we have

$$\mathcal{L}(a(\mathrm{I}^{(n)})) = (-1)^r \mathcal{L}(a(\sigma)),$$

where r is some element of $\mathbb{Z}_2$ that depends only on the graded permutation $a(\sigma)$ and does not depend on $\mathcal{L}$. We will call $\mathcal{L}(a(\mathrm{I}^{(n)}))$ the *symmetric parity* (or *s-parity*) of the graded permutation $a(\sigma)$ and denote it by $|a(\sigma)|_s$. In practice, the symmetric parity of any graded permutation may be calculated by arbitrarily presenting the permutation as the composition of transpositions $a(\sigma_i)$. In particular, let I, $|I| = n$ and J, $|J| = l$ be such that the multiindex (I, J) is defined; then

$$|a(I, J)|_s = |a(J, I)|_s + \langle a(I), a(J) \rangle. \qquad (21.26)$$

21.20. Differential operators in graded commutative algebras. Now that we have collected enough examples of graded objects, we can pass to the main topic of this chapter—the differential calculus over graded objects.

Let $\mathcal{A}$ be a G-graded commutative algebra over $\Bbbk$, $\mathcal{P}$ and $\mathcal{Q}$ be G-graded $\mathcal{A}$-modules. To each element $a \in \mathcal{A}$, we assign the operators

$$l_a, \ r_a \ \delta_a \colon \operatorname{Hom}_{\Bbbk}(\mathcal{P}, \mathcal{Q}) \to \operatorname{Hom}_{\Bbbk}(\mathcal{P}, \mathcal{Q}),$$

defined as follows:

$$l_a(\varphi)(p) = a\,\varphi(p), \quad \varphi \in \operatorname{Hom}_{\Bbbk}(\mathcal{P}, \mathcal{Q}), \quad p \in \mathcal{P},$$
$$r_a(\varphi)(p) = (-1)^{\langle \varphi, a \rangle} \varphi(ap),$$
$$\delta_a(\varphi) = r_a(\varphi) - l_a(\varphi).$$

(Below, we will often use the old notation $a^<$ and $a^>$ instead of l_a and r_a, respectively.)

Having modified the definition of these δ_a, cf. Section 9.66, to fit the case of graded algebras, we can easily carry over to this case the whole theory of $\Bbbk$-linear differential operators. We shall now briefly repeat the outline of this theory, leaving the proofs as exercises for the reader.

Thus, the operators l_a and r_a of left and right multiplication of elements of the $\Bbbk$-module $\operatorname{Hom}_{\Bbbk}(\mathcal{P}, \mathcal{Q})$ by elements of the algebra $\mathcal{A}$ obviously commute. This allows us to define, in $\operatorname{Hom}_{\Bbbk}(\mathcal{P}, \mathcal{Q})$, the structure of a bimodule over the algebra $\mathcal{A}$.

As before, we shall write $\varphi \circ a$ instead of $r_a(\varphi)$ and $a \circ \varphi$ instead of $l_a(\varphi)$, $\varphi \in \operatorname{Hom}_{\Bbbk}(\mathcal{P}, \mathcal{Q})$. Let $a_0, \dots, a_s \in \mathcal{A}$; then we put

$$\delta_{a_0, \dots, a_s} = \delta_{a_0} \circ \dots \circ \delta_{a_s}.$$

Definition. An element $\Delta \in \operatorname{Hom}_{\Bbbk}(\mathcal{P}, \mathcal{Q})$ will be called a $\Bbbk$-*linear differential operator* (DO) of order $\leqslant s$ over $\mathcal{A}$ if, for any tuple of elements $a_0, a_1, \dots, a_s \in \mathcal{A}$,

$$\delta_{a_0, \dots, a_s}(\Delta) = 0.$$

The set of all DO's of order $\leqslant s$ from $\mathcal{P}$ to $\mathcal{Q}$ is stable with respect to the left as well as to the right multiplication by elements of the algebra $\mathcal{A}$ and is therefore supplied with two natural G-graded $\mathcal{A}$-module structures. The G-graded $\mathcal{A}$-bimodule defined by these two commuting structures will be denoted by $\operatorname{Diff}_s^{\diamondsuit}(\mathcal{P}, \mathcal{Q})$. Obviously,

$$\operatorname{Diff}_0^<(\mathcal{P}, \mathcal{Q}) = \operatorname{Diff}_0^>(\mathcal{P}, \mathcal{Q}) = \operatorname{Hom}_{\mathcal{A}}(\mathcal{P}, \mathcal{Q})$$

and for any $s \geqslant 0$, the module $\operatorname{Diff}_s^{\diamondsuit}(\mathcal{P}\mathcal{Q})$ is a submodule of the $\mathcal{A}$-bimodule $\operatorname{Hom}_{\Bbbk}(\mathcal{P}, \mathcal{Q})$. For $s < 0$, we put $\operatorname{Diff}_s^{\diamondsuit}(\mathcal{P}, \mathcal{Q}) = 0$. We also write $\operatorname{Diff}_s^{\diamondsuit}(\mathcal{A}, \mathcal{Q}) \stackrel{\text{def}}{=} \operatorname{Diff}_s^{\diamondsuit} \mathcal{Q}$. Whenever no misunderstanding can arise, we will simply write $\operatorname{Diff}_s(\mathcal{P}, \mathcal{Q})$ instead of $\operatorname{Diff}_s^<(\mathcal{P}, \mathcal{Q})$.

Exercises.   1. Verify that for the operator

$$\delta_a \colon \operatorname{Hom}_{\Bbbk}(\mathcal{P}, \mathcal{Q}) \to \operatorname{Hom}_{\Bbbk}(\mathcal{P}, \mathcal{Q}),$$

where $a \in \mathcal{A}$, $\mathcal{P}$, $\mathcal{Q}$ are $\mathcal{A}$-modules, the following version of the Leibnitz formula holds:

$$\delta_{ab} = a^< \delta_b + \delta_a \circ b^> \tag{21.27}$$

2. Prove that for all $a_i \in \mathcal{A}$, $p \in \mathcal{P}$ and $\Delta \in \operatorname{Hom}_{\Bbbk}(\mathcal{P}, \mathcal{Q})$,

$$\delta_{a^{I^{(n)}}}(\Delta)(p) = \sum_{I+J=I^{(n)}} (-1)^{|I|+\langle a(I),a(J)\rangle} a^I \Delta(a^J p). \tag{21.28}$$

The last formula shows that for $\Delta \in \operatorname{Diff}_n(\mathcal{P}, \mathcal{Q})$, we have the graded analog of formula 9.25

$$\Delta(a^{I^{(n)}} p) = - \sum_{I+J=I^{(n)}, |I|>0} (-1)^{|I|+\langle a(I),a(J)\rangle} a^I \Delta(a^J p). \tag{21.29}$$

21.21. Remark. This formula shows that if the algebra $\mathcal{A}$ is finitely generated, say by $a_1, \ldots, a_n$, and the module $\mathcal{P}$ is finitely generated by $p_1, \ldots, p_m$, then any differential operator $\Delta: \mathcal{P} \to \mathcal{Q}$ of order $\leq k$ is uniquely determined by its value on a finite number of elements of the form $a^I p_i$, where a^I is the product of at most k generators of the algebra. Moreover if the algebra and the module are free over their generators, these values can be specified arbitrarily. This gives a method for computing differential operators in practice.

Exercise. Consider the following maps $i^>$ and $i_<$, which, as maps of sets, are identities

$$i^>: \operatorname{Diff}_l^<(P,Q) \to \operatorname{Diff}_l^>(P,Q), \quad i^>(f) = f,$$
$$i_<: \operatorname{Diff}_l^>(P,Q) \to \operatorname{Diff}_l^<(P,Q), \quad i_<(f) = f.$$

Prove that they are differential operators of order $\leqslant l$.

Let $\Delta: \mathcal{P} \to \mathcal{Q}$ be a DO of order n and of grading g, and let its r-th component be $\Delta^r: P^r \to Q^{r+g}$. Obviously, Δ^r is a DO of order n in the category of A^0-modules .

If $\Delta \in \operatorname{Diff}_n^\Diamond(\mathcal{P}, \mathcal{Q})$, then, according to the definition, for any n-tuple of elements $a_1, \ldots, a_n \in \mathcal{A}$, we have

$$\delta_{a_1,\ldots,a_n}(\Delta) \in \operatorname{Hom}_{\mathcal{A}}(\mathcal{P}, \mathcal{Q}).$$

If $\mathcal{P} = \mathcal{Q} = \mathcal{A}$, then $\operatorname{Hom}_{\mathcal{A}}(\mathcal{A}, \mathcal{A}) = \mathcal{A}$ and $\delta_{a_1,\ldots,a_n}(\Delta)$ can be regarded as an element of the algebra $\mathcal{A}$. In other words, the assignment

$$(a_1, \ldots, a_n) \mapsto \delta_{a_1,\ldots,a_n}(\Delta)$$

is an n-linear map from $\mathcal{A}$ to $\mathcal{A}$. Since, as can be immediately checked, $\delta_{a,b} = (-1)^{ab}\delta_{b,a}$, it follows that this map is symmetric. Let us put

$$[a_1, \ldots, a_n]_\Delta \overset{\text{def}}{=} \delta_{a_1,\ldots,a_n}(\Delta) \tag{21.30}$$

The natural embedding of $\mathcal{A}$-modules $\mathrm{Diff}_{k-1}\,\mathcal{A} \subset \mathrm{Diff}_k\,\mathcal{A}$ allows us to define the G-graded quotient module

$$\mathcal{S}_k(\mathcal{A}) \overset{\text{def}}{=} \mathrm{Diff}_k\,\mathcal{A}/\,\mathrm{Diff}_{k-1}\,\mathcal{A},$$

which is called the module of *symbols of order k* (or *k-symbols*). The coset of the operator $\Delta \in \mathrm{Diff}_k\,\mathcal{A}$ modulo $\mathrm{Diff}_{k-1}\,\mathcal{A}$ will be denoted by $\mathrm{smbl}_k\,\Delta$. We define the *algebra of symbols* of the algebra $\mathcal{A}$ by putting

$$\mathcal{S}_*(\mathcal{A}) = \bigoplus_{n=0}^{\infty} \mathcal{S}_n(\mathcal{A}).$$

The multiplication operation in $\mathcal{S}_*(\mathcal{A})$ is induced by the composition of differential operators. More precisely, for elements $\mathrm{smbl}_l\,\Delta \in \mathcal{S}_l(\mathcal{A})$ and $\mathrm{smbl}_k\,\nabla \in \mathcal{S}_k(\mathcal{A})$, we write by definition

$$\mathrm{smbl}_l\,\Delta \cdot \mathrm{smbl}_k\,\nabla \overset{\text{def}}{=} \mathrm{smbl}_{k+l}(\Delta \circ \nabla) \in \mathcal{S}_{l+k}(\mathcal{A}).$$

This operation is well defined, since it does not depend on the choice of representatives from the classes $\mathrm{smbl}_l\,\Delta$ and $\mathrm{smbl}_k\,\nabla$. Indeed, since, say, $\mathrm{smbl}_l\,\Delta = \mathrm{smbl}_l\,\Delta'$, it follows that $\Delta - \Delta' \in \mathrm{Diff}_{l-1}\,\mathcal{A}$ and, therefore, $(\Delta - \Delta') \circ \nabla \in \mathrm{Diff}_{l+k-1}\,\mathcal{A}$.

The modules $\mathcal{S}_n(\mathcal{A})$ are graded by elements of the semigroup G, and so the algebra $\mathcal{S}_*(\mathcal{A})$ is $\mathbb{Z} \oplus G$ graded. We put $\omega(\mathrm{smbl}_l\,\Delta) = \omega(\Delta)$.

21.22 Proposition. *The algebra $\mathcal{S}_*(\mathcal{A})$ is graded commutative:*

$$\mathrm{smbl}_l\,\Delta \cdot \mathrm{smbl}_k\,\nabla = (-1)^{\langle \mathrm{smbl}_l\,\Delta, \mathrm{smbl}_l\,\nabla \rangle}\,\mathrm{smbl}_k\,\nabla \cdot \mathrm{smbl}_l\,\Delta. \quad \blacktriangleright$$

If $\Delta' \in \mathrm{Diff}_{n-1}\,\mathcal{A}$, then $[a_1, \ldots, a_n]_{\Delta'} = 0$ for any $a_1, \ldots, a_n \in \mathcal{A}$. This means that the map $[\ \ ]_\Delta$ is determined not by the operator Δ itself, but by its symbol, and, therefore, to each symbol $\varsigma \in \mathcal{S}_n$ we can assign the symmetric polylinear map

$$[\cdot, \ldots, \cdot]_\varsigma \colon \mathcal{A} \times \mathcal{A} \times \ldots \times \mathcal{A} \to \mathcal{A}.$$

21.23. Derivations. As above, we assume that P is a two-sided G-graded $\mathcal{A}$-module.

Definition. The map $\Delta \colon \mathcal{A} \to \mathcal{P}$ is called a *derivation with values in the $\mathcal{A}$-module $\mathcal{P}$* if it satisfies the "graded" Leibnitz rule

$$\Delta(ab) = \Delta(a)b + (-1)^{\langle a, \Delta \rangle} a\Delta(b) \tag{21.31}$$

The set of all derivations from $\mathcal{A}$ to $\mathcal{P}$ is denoted by $D(\mathcal{P})$. This set possesses a natural left $\mathcal{A}$-module structure. If $\Delta \in \mathrm{Diff}_1\,\mathcal{P}$, then the map

$$\bar\Delta \colon \mathcal{A} \to \mathcal{P}, \quad \bar\Delta(a) = \Delta(a) - (-1)^{\langle a, \Delta \rangle} a\Delta(1)$$

is a derivation. As can be easily seen,

$$\bar\Delta(a) = (\delta_a(\Delta))(1). \tag{21.32}$$

Suppose that $\Delta\colon \mathcal{A} \to \mathcal{Q}$ is a derivation of grading g and $\Delta^r\colon A^r \to Q^{r+g}$ is its r-th component. Obviously, in the category of A^0-modules, the map $\Delta^0\colon A^0 \to Q^g$ is a derivation, while $\Delta^r\colon A^r \to Q^{r+g}$, $r \neq 0$, are DO's of first order. If $a \in A^r$, $b \in A^g$, then

$$\Delta^{r+g}(ab) = \Delta^r(a)b + (-1)^{\langle a,\Delta\rangle} a\Delta^g(b) \tag{21.33}$$

21.24. Differential calculus over Grassmann algebras. In this section, we shall see in detail what the calculus over Grassmann algebras $\mathcal{F}$ (see Section 21.16) looks like. We start by describing the modules of differential operators, the first observation being that any $\Bbbk$-linear map between $\mathcal{F}$-modules is actually a differential operator of order $\leq 2n+1$.

Proposition. *Let P, Q be graded modules over the Grassmann algebra $\mathcal{F}$, then*

$$\mathrm{Diff}(P,Q) = \mathrm{Diff}_{2n+1}(P,Q) = \mathrm{Hom}_{\Bbbk}(P,Q)$$

◀ We need only to check that $\mathrm{Hom}_{\Bbbk}(P,Q) \subset \mathrm{Diff}_{2n+1}(P,Q)$, i.e., that

$$\delta_{v_1,\dots,v_{2n+1}}\varphi = 0$$

for all $v_1,\dots,v_{2n+1} \in \mathcal{F}$. The relation 21.27 shows that to this end it suffices to consider the case when the v_i are generators of V. In view of 21.28, we have

$$\delta_{v_1,\dots,v_{2n+1}}\varphi(u) = \delta_{v^{I^{2n+1}}}\varphi(u) = \sum_{I+J=I^{(2n+1)}} (-1)^{|I|+\langle v(I),v(J)\rangle} v^I\Delta(v^J u), \quad u \in \mathcal{F}.$$

Now we see that each term of the right-hand side of the last equation contains a product of at least $n+1$ elements v_i, which is always zero.   ▶

21.25. Remark. Since we are considering finitely generated modules, it follows that

$$\mathrm{Hom}_{\Bbbk}^{\mathbb{Z}}(P,Q) = \mathrm{Hom}_{\Bbbk}(P,Q)$$

and in particular, since $\mathrm{Hom}_{\Bbbk}^{\mathbb{Z}_2}(P,Q) = \mathrm{Hom}_{\Bbbk}(P,Q)$, we obtain the equality $\mathrm{Diff}_{\mathcal{F}}^{\mathbb{Z}} = \mathrm{Diff}_{\mathcal{F}}^{\mathbb{Z}_2}$, i.e., here the notions of $\mathbb{Z}$- or $\mathbb{Z}_2$-graded differential operator coincide.

In the case of differential operators from $\mathcal{F}$ to any A-module V, we can say more. Let $V^\circ \overset{\text{def}}{=} \mathrm{Hom}_{\Bbbk}(V,\Bbbk)$.

Proposition. *There is a canonical identification of $\Bbbk$-vector spaces*

$$\mathrm{Diff}_k\,\mathcal{Q} = \mathrm{Hom}_{\Bbbk}\Big(\bigoplus_{l\leq k}\Lambda^l V, \mathcal{Q}\Big) = \Big(\bigoplus_{l\leq k}\Lambda^l V^\circ\Big) \otimes_{\Bbbk} \mathcal{Q}$$

which is given by restricting the differential operator $\Delta\colon \mathcal{F} \to \mathcal{Q}$ of order $\leq k$ to $\bigoplus_{l\leq k}\Lambda^l V \subset \mathcal{F}$. Note that

$$\bigoplus_{l\leq k}\Lambda^l V = \mathcal{F}\Big/\bigoplus_{l>k}\Lambda^l V,$$

and therefore $\bigoplus_{l \le k} \Lambda^l V$ has a natural $\mathcal{F}$-module structure (the multiplication is just truncated). With this identification, the above isomorphism becomes an isomorphism of bimodules.

◀ This is a direct consequence of Remark 21.21.   ▶
  In particular, we obtain:

21.26 Corollary. $\operatorname{Diff} \mathcal{P} = \operatorname{Diff}_n \mathcal{P} = \operatorname{Hom}_{\Bbbk}(\mathcal{F}, \mathcal{P}) = \mathcal{F}^\circ \otimes_{\Bbbk} \mathcal{P}$.

We warn the reader that the natural inclusion $\operatorname{Diff}_k \mathcal{Q} \subset \operatorname{Diff}_{k+1} \mathcal{Q}$ does not coincide with the evident inclusion

$$\operatorname{Hom}_{\Bbbk} \Big(\bigoplus_{l \le k} \Lambda^l V, \mathcal{Q} \Big) \hookrightarrow \operatorname{Hom}_{\Bbbk} \Big(\bigoplus_{l \le k+1} \Lambda^l V, \mathcal{Q} \Big).$$

Now we are ready to describe derivations, differential forms, as well as the de Rham complex on any Grassman algebra $\mathcal{F}$.

Proposition. *If $\mathcal{P}$ is a graded $\mathcal{F}$-module, then there is a natural isomorphism of modules*

$$D_k(\mathcal{P}) = \operatorname{Hom}_{\Bbbk}(S^k V, \mathcal{P}) = S^k V^\circ \otimes_{\Bbbk} \mathcal{P}$$

obtained by restricting the arguments of the multi-derivation to elements of $V \subset \mathcal{F}$.

Corollary. *The representative object of the functors D_k is $\Lambda^k = S^k V \otimes_{\Bbbk} \mathcal{F}$, so that $\Lambda(\mathcal{F}) = SV \otimes_{\Bbbk} \mathcal{F}$.*

Note that there are differential forms of arbitrary high order, which indicates that an attempt to construct integration theory on the basis of differential forms fails in the graded situation.

Recall that in general if $\mathcal{A}$ is G-graded commutative, then $\Lambda(\mathcal{A})$ is $G \oplus \mathbb{Z}$-graded commutative, with the parity form

$$\langle \cdot, \cdot \rangle_{G \oplus \mathbb{Z}} = \langle \cdot, \cdot \rangle_G + \langle \cdot, \cdot \rangle_{\rho_{\mathbb{Z}}}.$$

Moreover, the exterior differential d is of degree $(0, 1)$ and the algebra $\Lambda(\mathcal{A})$ is generated by the elements $a \in \mathcal{A} = \Lambda^0$ and $da \in \Lambda^1$. Using this and the previous corollary, we obtain:

Proposition. *The module $\Lambda(\mathcal{F})$ is freely generated as a $\mathbb{Z} \oplus \mathbb{Z}$-algebra by the elements ξ_i of degree $(1, 0)$, where the ξ_i are the generators of V, and by the elements $d\xi_i$ of degree $(1, 1)$. Explicitly, this means that we have the relations:*

$$\xi_i \xi_j = -\xi_j \xi_i, \qquad d\xi_i d\xi_j = d\xi_j d\xi_i, \qquad \xi_i d\xi_j = -d\xi_j \xi_i. \quad ▶$$

Consider the polynomial algebra

$$B \stackrel{\text{def}}{=} SV = \bigoplus_{i \ge 0} S^i V,$$

where $S^i V$ is i-th symmetric power of V, which physicists call "bosonic". Now note that if we start with this algebra, we also obtain the algebra $\Lambda(B) = SV \otimes_{\Bbbk} \Lambda V$. So in the case of $\Bbbk$-vector spaces, there is a canonical identification $\phi \colon \Lambda(\mathcal{F}) \to \Lambda(B)$. If we denote our basis of V in the bosonic picture by $x_1, \dots, x_n$, then this isomorphism is given by

$$x_i \leftrightarrow d\xi_i, \qquad dx_i \leftrightarrow \xi_i.$$

But this is not an isomorphism of graded algebras, since in $\Lambda(B)$ we have $x_i dx_j = dx_j \cdot x_i$, while in $\Lambda(\mathcal{F})$ a sign appears when we exchange these two elements.

Nevertheless, it is useful to interpret the operations from one picture in terms of operators in the other one, and we shall use this to compute the de Rham cohomology of $\Lambda(\mathcal{F})$. In this connection, recall that the Liouville field on a vector space V is $l = \sum_i x_i \frac{\partial}{\partial x_i}$.

Proposition. *Let $d_{\mathcal{F}}$ be the differential over $\Lambda(\mathcal{F})$, then*

$$d_{\mathcal{F}} \colon \Lambda(B) \to \Lambda(B)$$

is an inclusion of the Liouville vector field in $\Lambda(B)$, i.e., $d_{\mathcal{F}}(\beta)$ is an inclusion of l into β, $\beta \in \Lambda(B)$.

◄ We shall denote the exterior differential in $\Lambda(B)$ by d. Using the isomorphism ϕ, we see that

$$d_{\mathcal{F}}(x_i) = d_{\mathcal{F}}(d_{\mathcal{F}}\xi_i) = 0 = i_l(x_i)$$
$$d_{\mathcal{F}}(dx_i) = d_{\mathcal{F}}\xi_i = x_i = i_l(dx_i)$$

It remains to show that $d_{\mathcal{F}}$ is a derivation of degree -1 on the algebra $\Lambda(B)$. We leave this verification to the reader as an exercise.   ►

21.27 Corollary. *The de Rham cohomology of $\mathcal{F}$ is*

$$H^i(\mathcal{F}) = \begin{cases} \Bbbk, & i = 0, \\ 0, & i \neq 0. \end{cases} \quad ►$$

21.28. The Lie algebra of derivations.

Definition. The *commutator* of two derivations $\Delta, \nabla \in D(\mathcal{A})$ is the operator

$$[\Delta, \nabla] = \Delta \circ \nabla - (-1)^{\langle \Delta, \nabla \rangle} \nabla \circ \Delta.$$

Exercises. Prove that:

1. $[\Delta, \nabla] \in D(\mathcal{A})$ and $\omega([\Delta, \nabla]) = \omega(\Delta) + \omega(\nabla)$, ω denotes the value of the grading.

2. With respect to the commutator, $D(\mathcal{A})$ is a G-graded Lie algebra.

21.29. The Batalin–Vilkovisky bracket. Already from the algebraic definition of a differential operator, it is possible to extract notions and constructions which are extremely useful for physical applications. I. Batalin and G. Vilkovisky discovered that to any operator Δ of second order and odd grading such that $\Delta \circ \Delta = 0$ it is possible to assign a bracket $\mathcal{A} \times \mathcal{A} \to \mathcal{A}$ satisfying a kind of Jacobi identity. In the framework of the algebraic approach, we will extend the class of operators for which such a bracket can be applied, and also construct its n-ary version (see [27]).

Definition. A differential operator of odd grading $\Delta \in \mathrm{Diff}_n \, \mathcal{A}$ subject to the condition

$$\Delta \circ \Delta \in \mathrm{Diff}_{2n-2} \, \mathcal{A}. \qquad (21.34)$$

will be called a *Batalin–Vilkovisky operator*. Further, $\mathrm{smbl}_n \, \Delta$ is called the *Batalin–Vilkovisky symbol*. The bracket

$$[\cdot, \ldots, \cdot]_\varsigma : A \otimes \ldots \otimes A \to A,$$

where $\varsigma = \mathrm{smbl}_n \, \Delta$ (see Exercise 3 from Section 10.2), will be called the *Batalin–Vilkovisky bracket*.

21.30 Theorem. *The Batalin–Vilkovisky bracket satisfies the following n-ary version of the Jacobi identity:*

$$\sum_{\substack{I+J=I^{(2n-1)} \\ |I|=n}} (-1)^{|a(I,J)|_s + |a(J)|} [[a(I)]_\varsigma, a(J)]_\varsigma = 0. \qquad (21.35)$$

◄ For the proof of the theorem we need the next lemma, which can be established by induction, using the fact that the operator δ_a acts on the composition of DO's as a derivation:

$$\delta_a(\Delta \circ \nabla) = \delta_a(\Delta) \circ \nabla + (-1)^{\langle a, \Delta \rangle} \Delta \circ \delta_a(\nabla). \qquad (21.36)$$

21.31 Lemma.

$$\delta_{a(I^{(l)})}(\Delta \circ \nabla) = \sum_{\substack{I+J=I^{(l)} \\ 0 \leqslant |I| \leqslant l}} (-1)^{|a(I,J)|_s + \langle (a(J), \Delta \rangle} \delta_{a(I)}(\Delta) \circ \delta_{a(J)}(\nabla) \qquad (21.37)$$

Now assume that $\nabla = \Delta \in \mathrm{Diff}_n \, \mathcal{A}$ and $l = 2n-1$; then in the decomposition (21.37) only the terms with $|I| = n, |J| = n-1$, and $|I| = n-1, |J| = n$ will be nonzero, and so in this case the equality (21.37) can be given the following form:

$$\delta_{a(I^{(2n-1)})}(\Delta \circ \Delta)$$
$$= \sum_{\substack{I+J=I^{(2n-1)} \\ |I|=n}} (-1)^{|a(I,J)|_s + \langle a(J), \Delta \rangle + \langle \Delta, \Delta \rangle} \Big((-1)^{\langle \Delta, \Delta \rangle} \delta_{a(I)}(\Delta) \circ \delta_{a(J)}(\Delta)$$
$$+ (-1)^{\langle \delta_{a(J)}(\Delta), \delta_{a(I)}(\Delta) \rangle} \delta_{a(J)}(\Delta) \circ \delta_{a(I)}(\Delta) \Big) \qquad (21.38)$$
$$= \sum_{\substack{I+J=I^{(2n-1)} \\ |I|=n}} (-1)^{|a(I,J)|_s + \langle a(J), \Delta \rangle + \langle \Delta, \Delta \rangle} \Big((-1)^{\langle \Delta, \Delta \rangle} [a(I)]_\Delta \circ \delta_{a(J)}(\Delta)$$
$$+ (-1)^{\langle \delta_{a(J)}(\Delta), [a(I)]_\Delta \rangle} \delta_{a(J)}(\Delta) \circ [a(I)]_\Delta \Big)$$

If Δ is an operator of odd grading, then the last expression in the big brackets is nothing other than

$$\delta_{[a(I)]_\Delta}\big(\delta_{a(J)}(\Delta)\big) = \delta_{[a(I)]_\Delta, a(J)}(\Delta) = [[a(I)]_\Delta, a(J)]_\Delta,$$

and equality (21.38) acquires the form

$$\delta_{a(I^{(2n-1)})}(\Delta \circ \Delta) = -\sum_{\substack{I+J=I^{(2n-1)} \\ |I|=n}} (-1)^{|a(I,J)|_s + |a(J)|}[[a(I)]_\Delta, a(J)]_\Delta$$

The remark asserting that for the Batalin–Vilkovisky bracket of the Batalin–Vilkovisky operator the left-hand side of the equality vanishes concludes the proof of the theorem. ▶

The last equality also shows that the condition $\Delta \circ \Delta \in \mathrm{Diff}_{2n-2}\,\mathcal{A}$ is a necessary and sufficient condition for the validity of the Jacobi identity 21.35.

21.32. Proposition. If Δ is a Batalin–Vilkovisky operator of order n and ∇ is an even DO of order k such that $\nabla \circ \nabla \in \mathrm{Diff}_{2k-2}\,\mathcal{A}$, then the composition $\Delta \circ \nabla$ is a Batalin–Vilkovisky operator of order $n+k$.

◀ Since $\mathrm{smbl}(\Delta \circ \nabla) = \mathrm{smbl}(\Delta \circ \nabla - \frac{1}{2}[\Delta, \nabla])$, it sufffices to prove that $\Delta \circ \nabla - \frac{1}{2}[\Delta, \nabla]$ is a Batalin–Vilkovisky operator.

$$\left(\Delta \circ \nabla - \frac{1}{2}[\Delta, \nabla]\right) \circ \left(\Delta \circ \nabla - \frac{1}{2}[\Delta, \nabla]\right)$$

$$= \Delta \circ \nabla \circ \Delta \circ \nabla - \frac{1}{2}[\Delta, \nabla] \circ \Delta \circ \nabla - \frac{1}{2}\Delta \circ \nabla \circ [\Delta, \nabla] + \frac{1}{4}[\Delta, \nabla] \circ [\Delta, \nabla]$$

$$= \Delta \circ \nabla \circ \Delta \circ \nabla - \frac{1}{2}\Delta \circ \nabla \circ \Delta \circ \nabla + \frac{1}{2}\nabla \circ \Delta \circ \Delta \circ \nabla - \frac{1}{2}\Delta \circ \nabla \circ \Delta \circ \nabla$$

$$+ \frac{1}{2}\Delta \circ \nabla \circ \nabla \circ \Delta + \frac{1}{4}[\Delta, \nabla] \circ [\Delta, \nabla]$$

$$= \frac{1}{2}\nabla \circ \Delta \circ \Delta \circ \nabla + \frac{1}{2}\Delta \circ \nabla \circ \nabla \circ \Delta + \frac{1}{4}[\Delta, \nabla] \circ [\Delta, \nabla]$$

Each of the summands in the last expression is a DO of order $n+k-2$. ▶

21.33. The Batalin–Vilkovisky bracket. Case $n = 2$. In that case

$$|(a_1, a_2, a_3)|_s = 0,$$
$$|(a_1, a_3, a_2)|_s = |a_2| \cdot |a_3|,$$
$$|(a_2, a_3, a_1)|_s = |a_1| \cdot |a_2| + |a_1| \cdot |a_3|$$

and Jacobi identity 21.35 acquires the form

$$\begin{aligned} &(-1)^{|a_3|}[[a_1, a_2]_\varsigma, a_3]_\varsigma + (-1)^{\langle a_2, a_3 \rangle + |a_2|}[[a_1, a_3]_\varsigma, a_2]_\varsigma \\ &+ (-1)^{\langle a_1, a_2 \rangle + \langle a_1, a_3 \rangle + |a_1|}[[a_2, a_3]_\varsigma, a_1]_\varsigma = 0 \end{aligned} \qquad (21.39)$$

To each element $a \in \mathcal{A}$, we can assign the Hamiltonian vector field (=Hamiltonian derivation) $X_a \colon \mathcal{A} \to \mathcal{A}$ by putting

$$X_a = (-1)^{|a|+1}[\,\cdot\,, a]_\varsigma, \quad X_a(b) = (-1)^{|a|+1}[b, a]_\varsigma \tag{21.40}$$

21.34. Exercise. Verify that X_a is a derivation.

Proposition.

$$[X_{a_2}, X_{a_1}] = -(-1)^{|a_2|} X_{[a_1,a_2]_\varsigma} \tag{21.41}$$

◄ In each summand of (21.39), let us place a_3 in the first position and cancel out the expression $(-1)^{\langle a_3, a_1 \rangle + \langle a_3, a_2 \rangle}$, obtaining

$$[a_3, [a_1, a_2]_\varsigma]_\varsigma + (-1)^{|a_2|}[[a_3, a_1]_\varsigma, a_2]_\varsigma$$
$$+ (-1)^{\langle a_2, a_1 \rangle + |a_1|}[[a_3, a_2]_\varsigma, a_1]_\varsigma = 0 \tag{21.42}$$

According to definition (21.40), we have

$$[a_3, a_2]_\varsigma = (-1)^{|a_2|+1} X_{a_2}(a_3)$$
$$[[a_3, a_2]_\varsigma, a_1]_\varsigma = (-1)^{|a_1|+|a_2|} X_{a_1}(X_{a_2}(a_3))$$
$$[[a_3, a_1]_\varsigma, a_2]_\varsigma = (-1)^{|a_2|+|a_1|} X_{a_2}(X_{a_1}(a_3))$$
$$[a_3, [a_1, a_2]_\varsigma]_\varsigma = (-1)^{|[a_1,a_2]_\varsigma|+1} X_{[a_1,a_2]_\varsigma}(a_3)$$
$$= (-1)^{|a_2|+|a_1|} X_{[a_1,a_2]_\varsigma}(a_3)$$

Substituting this expression into (21.42) and multiplying by $(-1)^{|a_1|}$, we conclude that

$$(-1)^{|a_2|} X_{[a_1,a_2]_\varsigma}(a_3) + X_{a_2}(X_{a_1}(a_3))$$
$$- (-1)^{\langle a_2, a_1 \rangle + |a_2| + |a_1| + 1} X_{a_1}(X_{a_2}(a_3))$$
$$= (-1)^{|a_2|} X_{[a_1,a_2]_\varsigma}(a_3) + [X_{a_2}, X_{a_1}](a_3) = 0,$$

as claimed. ▶

21.35. Derivation of dioles. Let $\mathcal{A} = A(P)$ be a diole. Let us try to understand the meaning of the $\mathcal{A}$-module $D(\mathcal{A})$. We have $A^i = 0$, and if $i \neq 0, 1$, then any derivation Δ of grading r is a pair (Δ^0, Δ^1), where $\Delta^0 \in D(A^r)$ and $\Delta^1 \in \mathrm{Diff}_A(P, A^{r+1})$, and

$$\Delta^0(ab) = \Delta^0(a)b + a\Delta^0(b), \quad \forall a, b \in A, \tag{21.43}$$
$$\Delta^1(ap) = \Delta^0(a)p + a\Delta^1(p), \quad \forall a \in A, \; p \in P. \tag{21.44}$$

Obviously, $D(\mathcal{A})^r$ is nontrivial only if r equals -1, 0 and 1. Consider these three cases.

(1) $D(\mathcal{A})^1$. In that case, $\Delta^1 \in \mathrm{Diff}_A(P, A^2) = 0$ and condition (21.44) degenerates. Therefore, $D(\mathcal{A})^1 = D(P)$.

(2) $D(\mathcal{A})^0$. In that case,

$$\Delta^0 \in D(A^0) = D(A) \text{ and } \Delta^1 \in \mathrm{Diff}_A(P, A^1) = \mathrm{Diff}_A(P, P)$$

and conditions (21.43), (21.44) are nondegenerate. The A-module consisting of pairs (Δ_0, Δ_1), where $\Delta^1 : P \to P$ is a DO of first order, while the map $\Delta^0 : A \to A$ is a derivation satisfying condition (21.44), is customarily denoted by Der P. Thus, $D(\mathcal{A})^0 = \text{Der } P$. Note that if $\Delta^0 = 0$, then Δ^1 is an A-homomorphism from P to P, and so $\text{Hom}_A(P, P) \subset \text{Der } P$.

(3) $D(\mathcal{A})^{-1}$. We have

$$\Delta^0 \in D(A^{-1}) = 0 \text{ and } \Delta^1 \in \text{Diff}_A(P, A^0) = \text{Diff}_A(P, A).$$

The conditions (21.43), (21.44) degenerate into the following single condition

$$\Delta^1(ap) = a\Delta^1(p), \ \forall a \in A, \ p \in P,$$

so that $D(\mathcal{A})^{-1} = \text{Hom}_A(P, A) = P^*$.

Thus, we have proved the following.

21.36 Proposition.

$$D(\mathcal{A})^j = \begin{cases} 0, & j > 1, \\ D(P), & j = 1, \\ \text{Der } P, & j = 0, \\ \text{Hom}_A(P, A) = P^*, & j = -1, \\ 0, & j < -1. \end{cases} \quad (21.45) \quad \blacktriangleright$$

Exercises.   1. Describe the $\mathcal{A}$-module structure in $D(\mathcal{A})$.

2. Describe the $\mathcal{A}$-modules $D(\mathcal{Q})$ to in the general case.

21.37. The module Der P from the geometrical viewpoint. To any diole $\mathcal{A} = (A, P)$ we can assign its value at the point $h \in |A|$ (where $|A|$, as before, denotes the $\Bbbk$-spectrum of the algebra A) by writing $\mathcal{A}_h \overset{\text{def}}{=} (A_h = \Bbbk, P_h)$. Then the derivations with values in A_h are pairs of the form (X_h^0, X_h^1), where X_h^0 is the tangent vector at a point $h \in |A|$, while X_h^1 is a map from the module P to the vector space P_h.

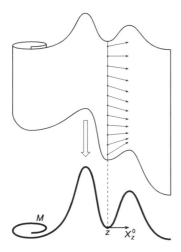

Figure 21.1. The fiber π_z of the bundle π over z.

In the case when A is the algebra of smooth functions on a smooth manifold M and P is the module of sections of a vector bundle π over M, we can identify M and $|A|$. Here P_z, $z \in M$, can be interpreted as the fiber π_z of π over the point z. Further, X_z^0 is the tangent vector at the point z, and X_z^1 is "the set of tangent vectors to points of the fiber π_z that are projected on X_z^0."

21.38. Representing objects. As in the nongraded case, the functors D and $\mathrm{Diff}_n(\mathcal{P}, \cdot)$, where $\mathcal{P}$ is a graded module over $\mathcal{A}$, are representable. Indeed, we have the following theorem.

Theorem. (1) *There exists a unique (up to isomorphism) $\mathcal{A}$-module $\Lambda(\mathcal{A})$ and a derivation of zero grading $\delta \colon \mathcal{A} \to \Lambda(\mathcal{A})$ such that for any $\mathcal{A}$-module $\mathcal{Q}$ the following isomorphism exists:*

$$\mathrm{Hom}(\Lambda(\mathcal{A}), \mathcal{Q}) = D(\mathcal{Q}), \quad \mathrm{Hom}(\Lambda(\mathcal{A}) \ni h \mapsto h \circ \delta \in D(\mathcal{Q}).$$

(2) *For any $\mathcal{A}$-module $\mathcal{P}$, there exists a unique (up to an isomorphism) $\mathcal{A}$-module $\mathcal{J}^n(\mathcal{P})$ and a DO of zero grading $\mathrm{j}_n \colon \mathcal{P} \to \mathcal{J}^n(\mathcal{P})$ such that for any $\mathcal{A}$-module $\mathcal{Q}$ the following isomorphism exists:*

$$\mathrm{Hom}(\mathcal{J}^n(\mathcal{P}), \mathcal{Q}) = \mathrm{Diff}_n(\mathcal{P}, \mathcal{Q}),$$
$$\mathrm{Hom}(\mathcal{J}^n(\mathcal{P}) \ni h \mapsto h \circ \mathrm{j}_n \in \mathrm{Diff}_n(\mathcal{P}, \mathcal{Q}).$$

◀ The proof of this theorem is left to the reader as an exercise.   ▶

The module $\Lambda(\mathcal{A})$ is called the *module of differential forms of first degree* of the algebra $\mathcal{A}$, the module $\mathcal{J}^n(\mathcal{P})$, the *module of jets of order n* of the module $\mathcal{P}$, while δ and j_n are the *universal derivations* and *universal DO of order n*, respectively. Elements of the module $\Lambda(\mathcal{A})$ are called *differential forms of first degree*, while elements $\mathcal{J}^n(\mathcal{P})$ are *jets of order n* of the $\mathcal{A}$-module $\mathcal{P}$.

If $a, b \in \mathcal{A}$, $p \in \mathcal{P}$ are homogeneous elements, then $a\delta b \in \Lambda(\mathcal{A})$ and $a\mathrm{j}_n(p) \in \mathcal{J}^n(\mathcal{P})$ will also be homogeneous elements, and for the grading ω we will have

$$\omega(a\delta b) = \omega(a) + \omega(b), \ \omega(a\mathrm{j}_n(p)) = \omega(a) + \omega(p).$$

21.39. Representative objects and the category of modules over dioles. Let $\mathcal{A} = (A, P)$ be a diole and $\mathcal{Q} = (Q^0, Q^1)$ be an $\mathcal{A}$-module. This means that

(1) Q^0 and Q^1 are A-modules;
(2) there exist an A-homomorphism $\circledast_{\mathcal{Q}} \colon P \bigotimes_A Q^0 \to Q^1$; the element $\circledast_{\mathcal{Q}}(p \bigotimes_A q)$, $q \in Q^0$ is said to be the *product* of p by q and is denoted by $p \cdot q$.

In other words, in order to transform the pair of A-modules (Q^0, Q^1) into an $\mathcal{A}$-module, it suffices to define the A-homomorphism $\circledast \colon P \otimes_A Q^0 \to Q^1$.

For simplicity of notation, let us put $\Lambda = \Lambda(A)$ and $\mathcal{J}^1 = \mathcal{J}^1(A)$, consider the pair $(\Lambda, \mathcal{J}^1)$, and define the map $\circledast_\Lambda \colon P \otimes_A \Lambda \to \mathcal{J}^1$ by setting

$$\circledast_\Lambda(p \otimes_A da) = p \cdot da \overset{\mathrm{def}}{=} \mathrm{j}_1(ap) - a\mathrm{j}_1(p).$$

Proposition. 1. $bp \cdot da = b(p \cdot da)$.

2. $p \cdot d(ab) = p \cdot adb + p \cdot bda$.

◀ By definition,

$$bp \cdot da = j_1(abp) - aj_1(bp) = aj_1(bp) + bj_1(ap) - abj_1(p) - aj_1(bp)$$
$$= bj_1(ap) - abj_1(p) = b(p \cdot da).$$

Further,

$$j_1(abp) = aj_1(bp) + bj_1(ap) - abj_1(p).$$

Therefore,

$$p \cdot d(ab) = j_1(abp) - abj_1(p) = aj_1(bp) + bj_1(ap) - 2abj_1(p)$$
$$= a(bj_1(p) - j_1(bp)) + b(aj_1(p) - j_1(ap))$$
$$= p \cdot adb + p \cdot bda. \qquad \blacktriangleright$$

This proposition means that the above-constructed map is an $\mathcal{A}$-homomorphism and so transforms $\Omega = (\Lambda(A), \mathcal{J}^1(P))$ into an $\mathcal{A}$-module. By definition,

$$j_1(ap) = p \cdot da + aj_1(p),$$

i.e., $\widetilde{d} = (d, j_1) \in D(\Omega)$.

Now let $\Delta = (\Delta^0, \Delta^1) \in D(\mathcal{Q})^l$. By the universality properties of the operators d and j_1, there exist A-homomorphisms $h_0 \colon D(A) \to Q^l$ and $h_1 \colon \mathcal{J}^1(P) \to Q^{l+1}$ such that $\Delta^0 = h_0 \circ d$ and $\Delta_1 = h_1 \circ j_1$. In order that the pair $h = (h_0, h_1)$ be an $\mathcal{A}$-homomorphism from Ω to $\mathcal{Q}$, it is necessary and sufficient that $h_1(da \cdot p) = h_0(da) \cdot p$. Let us verify this:

$$p \cdot h_0(da) = p \cdot \Delta_0(a) = \Delta_1(ap) - a\Delta_1(p)$$
$$= h_1(j_1(ap)) - h_1(aj_1(p)) = h_1\big(j_1(ap)) - aj_1(p)\big) = h_1(p \cdot da).$$

Note that if $Q^l = 0$, then $\Delta_1 \in \operatorname{Hom}_A(P, Q^{l+1}) \subset \operatorname{Diff}_1(P, Q^{l+1})$ and $h_1 = \Delta_1 \circ \pi_{1,0}$, where $\pi_{1,0} \colon \mathcal{J}^1(P) \to \mathcal{J}^0(P) = P$ is the natural projection.

Actually, we have proved that Ω is the representing object of the functor D, while $\widetilde{d} = (d, j_1)$ is the corresponding derivation. In view of the uniqueness of the representing object, we obtain the following.

Theorem. $\Lambda(\mathcal{A}) = (\Lambda(A), \mathcal{J}^1(P))$, $\delta = \widetilde{d} = (d, j_1)$. $\qquad \blacktriangleright$

Exercises.   1. Denote by $\Lambda^1(P)$ the submodule of the module $\mathcal{J}^1(P)$ generated by all elements of the form $j_1(ap) - aj_1(p)$. Prove the equality $\Lambda^1(P) = \Lambda^1 \otimes P$.

2. Prove the exactness of the short sequence

$$0 \to \Lambda^1(P) \to \mathcal{J}^1(P) \to P \to 0. \qquad (21.46)$$

To any $\mathcal{A}$-module $\mathcal{Q} = \bigoplus_{g \in \mathbb{Z}} Q^g$, we can assign the family of $\mathcal{A}$-modules

$$\mathfrak{J}^n(\mathcal{Q}) \stackrel{\text{def}}{=} \bigoplus_{g \in \mathbb{Z}} \mathcal{J}^n(Q^g), \; n \in \mathbb{Z}_+,$$

where $p \cdot j_n(q) \stackrel{\text{def}}{=} j_n(p \cdot q)$, $p \in P$, $q \in Q^g$.

Theorem–Exercise. Prove that
1. *The $\mathcal{A}$-modules $\mathfrak{J}^n(\mathcal{Q})$ are well defined.*
2. $\mathfrak{J}^n(\mathcal{Q}) = \mathcal{J}^n(\mathcal{Q})$.

21.40. Adjoint operators and integral forms. All the constructions of sections 20.34–20.38 can be easily carried over to the case of graded commutative algebras. In the definition of the map $\widetilde{\Delta}$ (see (20.20)), an appropriate plus or minus sign appears:

$$\widetilde{\Delta}(\nabla) \stackrel{\text{def}}{=} (-1)^{\langle \Delta, \nabla \rangle} \nabla \circ \Delta \in \text{Diff}^{>}(P, \Lambda^k), \ \nabla \in \text{Diff}^{+}(Q, \Lambda^k), \quad (21.47)$$

and formulas (20.21)–(20.22) take the following form:

$$(\nabla \circ \Delta)_n^* = (-1)^{\langle \Delta, \nabla \rangle} \Delta_n^* \circ \nabla_n^*, \quad (21.48)$$

$$\int \langle \Delta(p), \widehat{q} \rangle = (-1)^{\langle \Delta, p \rangle} \int \langle p, \Delta^*(\widehat{q}) \rangle. \quad (21.49)$$

Exercise. Construct the cochain complex of integral forms for the Grassman algebra $\mathcal{F}$.

21.41. The Berezinian for supermanifolds. In conclusion of this chapter, let us show how, for supermanifolds, the Berezinian (defined earlier for nongraded commutative algebras in 20.36) naturally arises under the conceptually correct approach, and not as *deus ex machina* or a rabbit out of the magician's hat in the classical approach.

We will not give a formal definition of the notion of supermanifold (this would lead us out of the framework of this, book limiting ourselves to some "hand-waving." As the simplest example of such a manifold, consider the superspace—the pair $(\mathbb{R}^{n+k}, \mathcal{A}^{n|k})$, where $\mathcal{A}^{n|k} = C^{\infty}(\mathbb{R}^n) [\xi_1, \ldots, \xi_k]$ and the elements ξ_i are odd: $\xi_i \xi_j = -\xi_j \xi_i$. In other words, $\mathcal{A}^{n|k}$ consists of polynomials of odd variables $\xi_1, \ldots, \xi_k$ with coefficients from the algebra of smooth functions on $\mathbb{R}^n$. This superspace will be denoted by $\mathbb{R}^{n|k}$, and $n|k$ will be its superdimension.

In the general case, the given supermanifold $(\mathcal{M}, \mathcal{A})$ of superdimension $n|k$ can be represented as a smooth manifold of dimension $n + k$ with a ring $\mathcal{A}$ of functions on it such that for any coordinate domain $\mathcal{U}$ the pair $(\mathcal{U}, \mathcal{A}|_{\mathcal{U}})$ is isomoprphic to the superspace $\mathbb{R}^{n|k}$. (We omit the details.) A typical example, in fact the one from which the study of supermanifolds began, is the pair $(T^*(M), \Lambda^*(M))$, where M is a smooth manifold, while $\Lambda^*(M)$ is the de Rham algebra of differential forms.

We have the following.

Theorem. *Let $(\mathcal{M}, \mathcal{A})$ be a supermanifold of superdimension $n|k$. Then*
(1) $\mathcal{B}_i = \widehat{\mathcal{A}}_i = 0$, *if $i \neq n$.*
(2) $\mathcal{B}_n = \widehat{\mathcal{A}}_n$ *is a module of rank 1.*

(3) *For any coordinate domain $\mathcal{U}$, the module is $\mathcal{B}_n|_{\mathcal{U}}$ free. To each coordinate system $\{x, \xi\} = \{x_1, \ldots, x_n, \xi_1, \ldots, \xi_k\}$ on $\mathcal{U}$ a generator $e_{x,\xi}$ of the module $\mathcal{B}_n|_{\mathcal{U}}$ can be naturally associated so that under the change of coordinates $\{x, \xi\} \to \{v, \psi\}$ the following equality holds:*

$$e_{x,\xi} = \operatorname{Ber} J\left(\frac{x, \xi}{v, \psi}\right) e_{v,\psi};$$

here $J\left(\frac{x, \xi}{v, \psi}\right)$ is the Jacobi matrix of the coordinate change, while Ber is the Berezin determinant, which in the case of even matrices can be calculated by the following rule:

$$\operatorname{Ber}\begin{pmatrix} B C \\ D E \end{pmatrix} = \det(B - CE^{-1}D)(\det E)^{-1},$$

where B and E are square matrices of orders n and k, respectively.

We will not prove this theorem. We have stated it here in order to demonstrate how the notion of Berezin determinant arises and to show what role it plays in "supermathematics."

Note that under the traditional approach, the appearance of the Berezin determinant is veiled in mystery: first the Berezin determinant of an even matrix $\begin{pmatrix} B & C \\ D & E \end{pmatrix}$ is defined as the expression $\det(B - CE^{-1}D)(\det E)^{-1}$, then $\operatorname{Ber} \mathcal{M}$ is defined as a one-dimensional fibration for which the local section over the domain $\mathcal{U}$ is of the form $f(x)\,\mathrm{D}(x)$, where $f \in C^\infty(\mathcal{U})$ and D is the basis local section, which, under the change of coordinates D is multiplied by the Berezin determinant of the corresponding Jacobi matrix.

Afterword

If we continue on the path traced out by this book and analyze to what extent contemporary mathematics corresponds to the observability principle, we see that many things in our science are simply conceptually unfounded. This unavoidably leads to serious difficulties, which are usually ignored from force of habit even when they contradict our experience. If, for example, measure theory is the correct theory of integration, then why is it that all attempts to construct the continual integral on its basis have failed, although the existence of such integrals is experimentally verified?

As the result of this, physicists are forced to use "unobservable" mathematics in their theories; as a result, considerable difficulties arise, say, in quantum field theory, which some experts even regard as an inherent aspect of the theory. It is generally believed that the mathematical basis of quantum mechanics is the theory of self-adjoint operators in Hilbert space. But then why does Dirac writes that "physically significant interactions in quantum field theory are so strong that they throw any Schrödinger state vector out of Hilbert space in the shortest possible time interval"?

Having noted this, one must either avoid writing the Schrödinger equation in the context of quantum fields theory or refuse to consider Hilbert spaces as the foundation of quantum mechanics. Dirac reluctantly chose the first alternative, and this refusal was forced, since the mathematics of that time allowed him to talk about solutions of differential equations only in a very limited language (see the quotation at the beginning of the Introduction). On the other hand, since the Hilbert space formalism contains no procedure for distinguishing one vector from another, the observability

© Springer Nature Switzerland AG 2020
J. Nestruev, *Smooth Manifolds and Observables*, Graduate Texts
in Mathematics 220, https://doi.org/10.1007/978-3-030-45650-4

principle is not followed here. Thus the second alternative seems moreappropriate, but it requires specifying many other points, e.g., finding out how one can observe solutions of partial differential equations; this question, however, is outside the sphere of interests of the PDE experts: To them, even setting the question seems strange, to say the least.

Thus the systematic mathematical formalization of the observability principle requires rethinking many branches of mathematics that seemed established once and for all. The main difficult step that must be taken in this direction is to find solutions in the framework of the differential calculus, avoiding the appeal of functional analysis, measure theory, and other purely set-theoretical constructions. In particular, we must refuse measure theory as integration theory in favor of the purely cohomological approach. One page suffices to write out the main rules of measure theory. The number of pages needed to explain de Rham cohomology is much larger. The conceptual distance between the two approaches shows what serious difficulties must be overcome on this road.

At present it is clear that this road leads to the secondary differential calculus (already mentioned in the Introduction) and its main applications, e.g., cohomological physics. The reader may obtain an idea of what has already been done, and what remains to be done, in this direction by consulting the references appearing below.

Appendix

A. M. Vinogradov
Observability Principle, Set Theory and the "Foundations of Mathematics"

The following general remarks are meant to place the questions discussed in this book from the perspective of observable mathematics.

Propositional and Boolean algebras. While the physicist describes nature by means of measuring devices with $\mathbb{R}$-valued scales, the layman does so by means of statements. Using the elementary operations of conjunction, disjunction, and negation, new statements may be constructed from given ones. A system of statements (propositions) closed with respect to these operations is said to be a *propositional algebra*. Thus, the means of observation of an individual not possessing any measuring devices is formalized by the notion of propositional algebra. Let us explain this in more detail.

Let us note, first of all, that the individual observing the world without measuring devices was considered above only as an example of the main, initial mechanism of information processing, which in the sequel we shall call *primitive*. Thus, we identify propositional algebras with primitive means of observation.

Further, let us recall that any propositional algebra A may be transformed into an unital commutative algebra over the field $\mathbb{Z}_2$ of residues modulo 2 by introducing the operations of multiplication and addition as follows:

$$pq \overset{\text{def}}{=} p \wedge q,$$

$$p + q \overset{\text{def}}{=} (p \wedge \bar{q}) \vee (\bar{p} \wedge q),$$

where $\vee$ and $\wedge$ are the propositional connectives conjunction and disjunction, respectively, while the bar over a letter denotes negation. All elements

© Springer Nature Switzerland AG 2020
J. Nestruev, *Smooth Manifolds and Observables*, Graduate Texts in Mathematics 220, https://doi.org/10.1007/978-3-030-45650-4

of the algebra thus obtained are idempotent, i.e., $a^2 = a$. Let us call any unital commutative $\mathbb{Z}_2$-algebra *Boolean* if all its elements are idempotent. Conversely, any Boolean algebra may be regarded as a propositional algebra with respect to the operations

$$p \wedge q \overset{\text{def}}{=} pq,$$

$$p \vee q \overset{\text{def}}{=} p + q + pq,$$

$$\bar{p} \overset{\text{def}}{=} 1 + p.$$

This shows that there is no essential difference between propositional and Boolean algebras, and the use of one or the other only specifies what operations are involved in the given context. Thus we can restate the previous remarks about means of observation as follows: *Boolean algebras are primitive means of observation.*

Boolean spectra. The advantage of the previous formulation is that it immediately allows us to discern the remarkable analogy with the observation mechanism in classical physics as interpreted in this book. Namely, in this mechanism one must merely replace the $\mathbb{R}$-valued measurement scales by $\mathbb{Z}_2$-valued ones (i.e., those that say either "yes" or "no") and add the idempotence condition. This analogy shows that *what we can observe by means of a Boolean algebra A is its $\mathbb{Z}_2$-spectrum, i.e., the set of all its homomorphisms as an unital $\mathbb{Z}_2$-algebra to the unital $\mathbb{Z}_2$-algebra $\mathbb{Z}_2$.*

Let us denote this spectrum by $\mathrm{Spec}_{\mathbb{Z}_2}$ and endow it with the natural topology, namely the Zariski one. Then we can say, more precisely, that Boolean algebras allow us to observe topological spaces of the form $\mathrm{Spec}_{\mathbb{Z}_2}$, which we shall call, for this reason, *Boolean spaces.*

In connection with the above, one may naturally ask whether the spectra of Boolean algebras possess any structure besides the topological one, say, a smooth structure, as was the case for spectra of $\mathbb{R}$-algebras. The reader who managed to do Exercise 4 from Section 9.45 already knows that the differential calculus over Boolean algebras is trivial in the sense that any differential operator on such an algebra is of order zero, i.e., is a homomorphism of modules over this algebra. This means, in particular, that *the phenomenon of motion cannot be adequately described and studied in mathematical terms by using only logical notions* or, to put it simply, by using everyday language (recall the classical logical paradoxes on this topic).

The Stone theorem stated below, which plays a central role in the theory of Boolean algebras, shows that the spectra of Boolean algebras possess only one independent structure: the topological one. In the statement of the theorem it is assumed that the field $\mathbb{Z}_2$ is supplied with the discrete topology.

Stone's theorem. *Any Boolean space is an absolutely disconnected compact Hausdorff space and, conversely, any Boolean algebra coincides with* *the algebra of open-and-closed sets of its spectrum with respect to the set-theoretic operations of symmetric difference and intersection.*

Recall that the absolute disconnectedness of a topological space means that the open-and-closed sets form a base of its topology. The appearance of these simultaneously open-and-closed sets in Boolean spaces is explained by the fact that any propositional algebra possesses a natural duality. Namely, the negation operation maps it onto itself and interchanges conjunction and disjunction. Note also that Stone's theorem is an identical twin of the Spectrum theorem (see Sections 7.2 and 7.7). Their proofs are based on the same idea, and differ only in technical details reflecting the specifics of the different classes of algebras under consideration. The reader may try to prove this theorem as an exercise, having in mind that the elements of the given Boolean algebra can be naturally identified with the open-and-closed subsets of its spectrum, while the operations of conjunction, disjunction, and negation then become the set-theoretical operations of intersection, union, and complement, respectively. It is easy to see that the spectrum of a finite Boolean algebra is a finite set supplied with the discrete topology. Thus any finite Boolean algebra turns out to be isomorphic to the algebra of all subsets of a certain set.

"Eyes" and "ears." After all these preliminaries, the role of "eyes" and "ears" in the process of observation may be described as follows. First of all, the "crude" data absorbed by our senses are written down by the brain and sent to the corresponding part of our memory. One may think that in the process of writing down, the crude data are split up into elementary blocks, "macros," and so on, which are marked by appropriate expressions of everyday language. These marks are needed for further processing of the stored data. The system of statements constituting some description generates an ideal of the controlling Boolean algebra, thus distinguishing the corresponding closed subset in its spectrum. Supposing that to each point of the spectrum an elementary block is assigned, and this block is marked by the associated maximal ideal, we come to the conclusion that to each closed subset of the spectrum one can associate a certain image, just as a criminalist creates an identikit from individual details described by witnesses. Thus, if we forget about the "material" content of the elementary blocks (they may be "photographs" of an atomic fragment of a visual or an audio image, etc.) that corresponds (according to the above scheme) to points of the spectrum of the controlling Boolean algebra, we may assume that everything that can be observed on the primitive level is tautologically expressed by the points of this spectrum.

Boolean algebras corresponding to the primitive level. It is clear that any rigorous mathematical notion of observability must come from some notion of observer, understood as a kind of mechanism for gathering and processing information. In other words, the notion of observability must be formalized approximately in the same way as Turing machines formalize the notion of algorithm. So as not to turn out to be an a priori formalized metaphysical scheme, such a formalization must take into account "experimental data." The latter may be found in the construction and evolution of computer hardware and in the underlying theoretical ideas. Therefore it is useful to regard the individual mathematician, or better still, the mathematical community, in the spirit of the "noosphere" of Vernadskii, as a kind of computer. Then, having in mind that the operational system of any modern computer is a program written in the language of binary codes, we can say that there is no alternative to Boolean algebras as the mechanism describing information on the primitive level. For practical reasons, as well as for considerations of theoretical simplicity, it would be inconvenient to limit the size of this algebra by some concrete number, say the number of elementary particles in the universe. Hence it is natural to choose the free algebra in a countable number of generators. The notion of level of observability is apparently important for the mathematical analysis of the notion of observability itself, and we shall return to it below.

How set theory appeared in the foundations of mathematics. As we saw above, any propositional algebra is canonically isomorphic to the algebra of all subsets of the spectrum of the associated Boolean algebra. If this spectrum is finite, then its topology is discrete. So we can forget about the topology without losing anything. Moreover, any concrete individual, especially if he/she is not familiar with Boolean algebras, feels sure that what she/he is observing are just subsets or, more precisely, the identikits which she/he defined. Therefore, such an immediate "material" feeling leads us to the idea that the initial building blocks of precise abstract thinking are "points" ("elements") grouped together in "families," i.e., sets. Having accepted or rather having experienced this feeling of primitivity of the notion of set under the pressure of our immediate feelings, we are forced to place set theory at the foundation of exact knowledge, i.e., of mathematics. On the primitive level of finite sets, this choice, in view of what was explained above, does not contradict the observability principle, since any finite set can be naturally and uniquely interpreted as the spectrum of some Boolean algebra.

However, if we go beyond the class of finite sets, the situation changes radically: The notion of observable set, i.e., of Boolean space, ceases to coincide with the general notion of a set without any additional structure. Therefore, our respect for the observability principle leads us to abandon the notion of a set as the formal-logical foundation of mathematics and leave the paradise so favored by Hilbert. One of the advantages of such a step, among others, is that it allows us to avoid many of the paradoxes inherent to set theory. For example, the analog of the "set of all sets" in observable mathematics is the "Boolean space of all Boolean spaces." But this last construction is clearly meaningless, because it defines no topology in the "Boolean space of all Boolean spaces." Or the "observable" version of the "set (not) containing itself as an element," i.e., the "Boolean space (not) containing itself as an element" is so striking that no comment is needed. In this connection we should additionally note that in order to observe Boolean spaces (on the primitive level!) as individual objects, a separate Boolean space that distinguishes them is required.

Observable mathematical structures (Boole groups). Now it is the time to ask what observable mathematical structures are. If we are talking about groups observable in the "Boolean" sense, then we mean topological groups whose set of elements constitutes a Boolean space. Such a group should be called Boolean. In other words, a *Boole group* is a group structure on the spectrum of some Boolean algebra. If we replace in this definition the notion of Boolean observability by that of classical observability, we come to the notion of Lie group, i.e., of a group structure on the spectrum of the classical algebra of observables.

Observing observables: different levels of observability. Just as the operating system in a computer manipulates programs of the next level, one can imagine a Boolean algebra of the primitive level (see above) with the points of its spectrum marking other Boolean algebras. In other words, this is a Boolean algebra observing other Boolean algebras. Iterating this procedure, we come to "observed objects," which, if one forgets the multistep observation scheme, can naively be understood as sets of cardinality higher than finite or countable. For instance, starting from the primitive level, we can introduce into observable mathematics things that in "nonobservable" mathematics are related to sets of continual cardinality. In this direction, one may hope that there is a constructive formalization of the observability of smooth $\mathbb{R}$-algebras, which, in turn, formalize the observation procedure in classical physics.

Down with set theory? The numerous failed attempts to construct mathematics on the formal-logical foundations of set theory, together with the considerations related to observability developed above, lead us to refuse this idea altogether. We can note that it also contradicts the physiological basis of human thought, which ideally consists in the harmonious interaction of the left and right hemispheres of the brain. It is known that the left hemisphere is responsible for rational reasoning, computations, logical analysis, and pragmatic decision-making. Dually, the right hemisphere answers for "irrational" thought, i.e., intuition, premonitions, emotions, imagination, and geometry. If the problem under consideration is too hard for direct logical analysis, we ask our intuition what to do. We also know that in order to obtain a satisfactory result, the intuitive solution must be controlled by logical analysis and, possibly, corrected on its basis. Thus, in the process of decision-making, in the search for the solution of a problem, etc., the switching of control from one hemisphere to the other takes place, and such iterations can be numerous.

All this, of course, is entirely relevant to the solution of mathematical problems. The left hemisphere, i.e., the algebro-analytical part of our brain, is incapable of finding the solution to a problem whose complexity is higher than, say, the possibilities of human memory. Indeed, from any assumption one can deduce numerous logically correct consequences. Therefore, in the purely logical approach, the number of chains of inference grows at least exponentially with their length, while those that lead to a correct solution constitute a vanishingly small part of that number. Thus if the correct consequence is chosen haphazardly at each step, and the left hemisphere knows no better, then the propagation of this "logical wave" in all directions will overfill our memory before it reaches the desired haven.

The only way out of this situation is to direct this wave along an appropriate path, i.e., to choose at each step the consequences that can lead in a more or less straight line to the solution. But what do we mean by a "straight line"? This means that an overall picture of the problem must be sketched, a picture on which possible ways of solution could be drawn. The construction of such an overall picture, in other words, of the geometrical image of the problem, takes place in the right hemisphere, which was created by nature precisely for such constructions. The basic building blocks for them, at least when we are dealing with mathematics, are sets. These are sets in the naive sense, since they live in the right hemisphere. Hence any attempt to formalize them, moving them from the right hemisphere to the left one, is just an outrage against nature. So let us leave set theory in the right hemisphere in its naive form, thanks to which it is has been so useful.

Infinitesimal observability. Above we considered Boolean algebras as analog of smooth algebras. But we can interchange our priorities and do things the other way around. From this point of view, the operations or, better, the functors of the differential calculus, will appear as the analog of logical operations, and the calculus itself as a mechanism for manipulating infinitesimal descriptions. In this way we would like to stress the infinitesimal aspect related to observability.

Some of the "primary" functors were described in this book. Their complete list should be understood as the logic algebra of the differential calculus. The work related to the complete formalization of this idea is still to be completed.

In conclusion let us note, expressing ourselves informally, that in our imaginary computer, working with stored knowledge, the program called "differential calculus" is not part of its operating system, and so is located at a higher level than the primitive one (see above). This means that the geometrical images built on its basis cannot be interpreted in a material way. They should retain their naive status in the sense explained above. The constructive differential calculus, developed in the framework of "constructive mathematical logic," illustrates what can happen if this warning is ignored.

References

[1] V.N. Chetverikov, A.B. Bocharov, S.V. Duzhin, N.G. Khor'kova, I.S. Krasil'shchik, A.V. Samokhin, Y.N. Torkhov, A.M. Verbovetsky, A.M. Vinogradov, in *Symmetries and Conservation Laws for Differential Equations of Mathematical Physics*, vol. 182, Translations of Mathematical Monographs, ed. by J. Krasil'shchik, A. Vinogradov (American Mathematical Society, Providence, 1999)

[2] H. Goldschmidt, D. Spencer, On the nonlinear cohomology of Lie equations IV. J. Differ. Geom. **13**(4), 455–526 (1978)

[3] H. Goldschmidt, Sur la structure des equations de Lie, I, II, III. J. Differ. Geom. **6**(3), 357–373 (1972); **7**(1-2), 67–95 (1972); **11**(2) 167–223 (1976)

[4] M. Henneaux, I.S. Krasil'shchik, A.M. Vinogradov (eds.), *Secondary Calculus and Cohomological Physics, Contemporary Mathematics*, vol. 219 (American Mathematical Society, Providence, 1998)

[5] I. Krasil'shchik, Algebraic theories of brackets and related (co)homologies. Acta Appl. Math. **109**, 137–150 (2010). https://doi.org/10.1007/s10440-009-9445-1, (Preprint. arXiv:0812.4676)

[6] I.S. Krasil'shchik, V.V. Lychagin, A.M. Vinogradov, *Geometry of Jet Spaces and Nonlinear Differential Equations*, vol. 1, Advanced Studies in Contemporary Mathematics (Gordon and Breach, London, 1986)

© Springer Nature Switzerland AG 2020
J. Nestruev, *Smooth Manifolds and Observables*, Graduate Texts
in Mathematics 220, https://doi.org/10.1007/978-3-030-45650-4

[7] I.S. Krasil'shchik, A.M. Verbovetsky, *Homological Methods in Equations of Mathematical Physics* (Open Education, Opava, 1998)

[8] I.S. Krasil'shchik, A.M. Vinogradov (eds.), Algebraic aspects of differential calculus. A dedicated issue of Acta. Appl. Math. **49**(3) (1997)

[9] I.S. Krasil'shchik, A.M. Vinogradov, Nonlocal trends in the geometry of differential equations: symmetries, conservation laws, and Bäcklund transformations. Acta Appl. Math. **15**, 161–209 (1989)

[10] I.S. Krasil'shchik, A.M. Vinogradov, What is the Hamiltonian formalism? Uspekhi Mat. Nauk **30**(1) (1975) (in Russian, English translation in: London Math. Soc. Lecture Notes Ser., **60** (1981))

[11] V.V. Lychagin, Colour calculus and colour quantizations. Acta Appl. Math. **41**, 193–226 (1995)

[12] S. Morier-Genoud, V. Ovsienko, Simple graded commutative algebras. J. Alg. **323**, 1649–1664 (2010)

[13] J.A. Schouten, Uber Differentialkomitanten zweier kontravarianter Grossen. Proc. Ned. Akad. Wet. Amst. **43**, 449–452 (1940)

[14] A. Verbovetsky, Differential Operators Over Quantum Spaces. Acta Appl. Math. **49**, 339–361 (1997)

[15] A. Verbovetsky, Lagrangian formalism over graded algebras. J. Geom. Phys. **18**, 195–214 (1996)

[16] A.M. Vinogradov, A common generalization of the Schouten and Nijenhuis brackets, cohomology, and superdifferential operators (Russian). Mat. Zametki **47**(6), 138–140 (1990). English translation in Math. Notes, **47** (1990)

[17] A.M. Vinogradov, *Cohomological Analysis of Partial Differential Equations and Secondary Calculus*, vol. 204, Translations of Mathematical Monographs (American Mathematical Society, Providence, 2001)

[18] A.M. Vinogradov, From symmetries of partial differential equations towards secondary ("quantized") calculus. J. Geom. Phys. **14**, 146–194 (1994)

[19] A.M. Vinogradov, Geometry of nonlinear differential equations. J. Sov. Math. **17**, 1624–1649 (1981)

[20] A.M. Vinogradov, Local symmetries and conservation laws. Acta Appl. Math. **2**(1), 21–78 (1984)

[21] A.M. Vinogradov, Logic of differential calculus and the zoo of geometric structures. Geom. Jets Fields **110**, 257–285 (Banach Center Publications, 2016), arXiv:1511.06861v1 (2015) [math.DG]

[22] A.M. Vinogradov, A. Blanco, Green formula and legendre transformation. Acta Appl. Math. **83**, 149–166 (2004)

[23] A.M. Vinogradov, A. De Paris, *Fat Manifolds and Linear Connections* (World Scientific, 20081), xii+297 pp

[24] A.M. Vinogradov, G. Vezzosi, On higher order analogues of de Rham cohomology. Diff. Geom. Appl. **19**, 29–59 (2003)

[25] A.M. Vinogradov, L. Vitagliano, Iterated Differential Forms I–V, Dokl. Akad. Nauk, **407**(1), 16–18 (2006); **407**(2), 151–153 (2006); **413**(1), 7–10 (2007); **414**(1), 447–450 (2007); **416**(2), 161–165 (2007). English translation: Doklady Mathematics, **73**(2), 169–171 (2006); **73**(2), 182–184 (2006); **75**(2), 177–180 (2007); **75**(3), 403–406 (2007); **76**(2), 673–677 (2007)

[26] M.M. Vinogradov, Residues without duality. Acta Appl. Math. **49**(3), 281–291 (1997)

[27] M.M. Vinogradov, *n-ary Batalin–Vilkovisky brackets* [seminar talk], Geometry of Differential Equations Seminar (Independent University of Moscow, Moscow, Russia, 2010). https://gdeq.org/Seminar_talk,_3_March_2010

Index

© Springer Nature Switzerland AG 2020
J. Nestruev, *Smooth Manifolds and Observables*, Graduate Texts
in Mathematics 220, https://doi.org/10.1007/978-3-030-45650-4

theorem, 105
tangent vector field along a map, 128
tangent vector field along a
        submanifold, 127
tautological bundle, 160, 162, 186
Taylor expansion, 17
thermodynamics of an ideal gas, 8
topology
    in $|P|$, 178
    in the dual space $M = |\mathcal{F}|$, 25
    Zariski, 94
total space (of a fibration), 152
triviality criterion for bundles, 167

U
universal
    1-form, 227
    biderivation, 304
    polyderivation, 317
universal derivation
    algebraic case, 228
    descriptive definition, 225
    geometrical case, 227
universal differential operator, 137,
        235
universal vector field, 128
    of an algebra, 290

V
vector bundle, 172

adapted coordinates, 172
direct sum, 184
module of sections, 174
morphisms, 173
tensor product, 192
$\mathbb{I}_M$, 172
Whitney sum, 184
$\mathbb{O}_M$, 172
vector field, 122
    along maps, 127
    as a smooth section, 123
    on submanifolds, 126
    transformation, 123
    universal, 128
volume form, 343

W
Whitney sum, 163
    of vector bundles, 184

Z
Zariski topology, 85, 94
zero section, 174